Assignment Problems

Assignment Problems

Rainer Burkard
Graz University of Technology
Graz, Austria

Mauro Dell'Amico
University of Modena and Reggio Emilia
Reggio Emilia, Italy

Silvano Martello
University of Bologna
Bologna, Italy

siam. Society for Industrial and Applied Mathematics • Philadelphia

10 9 8 7 6 5 4 3 2 1

The accompanying Web page contains directly downloadable software, as well as links to software that can be downloaded from the Web. Use of such software is generally allowed for scientific and teaching purposes but not for professional use, for which the authors should be contacted.

Cover art of "The Arnolfini Portrait" by Jan van Eyck used with permission of the National Gallery. Besides its immediate relationship to matching (marriage theorem), this painting is notable for its "dual" content: The convex mirror hanging on the back wall reflects not only the Arnolfinis' backs but also the two people facing them, one of which is probably the portrait's artist.

Library of Congress Cataloging-in-Publication Data

Burkard, Rainer E.
Assignment problems / Rainer Burkard, Mauro Dell'Amico, Silvano Martello.
p. cm.
Includes bibliographical references and index.
ISBN 978-0-898716-63-4
1. Assignment problems (Programming) 2. Computer science—Mathematics.
I. Dell'Amico, Mauro. II. Martello, Silvano. III. Title.
QA402.6.B87 2008
519.7'2—dc22

2008032587

Rainer Burkard dedicates his work to Heidi.

Mauro Dell'Amico dedicates his work to Francesca, Lorenzo, Laura, and Lucia.

Silvano Martello dedicates his work to Italiana.

Contents

List of Figures

List of Tables

List of Algorithms

Preface

Half a century ago Harold W. Kuhn published two famous articles presenting the Hungarian algorithm, the first polynomial-time method for the assignment problem. This historic result allowed for the first time an easy solution of real-world instances that no computer on earth could then handle. The Hungarian algorithm and other fundamental results on integer and linear programming, obtained in the same years, gave birth to an exciting new research area, today known as combinatorial optimization. Over the next fifty years the assignment problem, its linear variations, and its quadratic counterpart have attracted hundreds of researchers, accompanying and sometimes anticipating the development of combinatorial optimization.

This volume presents a comprehensive view of this huge area, starting from the conceptual foundations laid down since the 1920s by the studies on matching problems, and examining in detail theoretical, algorithmic, and practical developments of the various assignment problems. Although the covered area is wide, each of the ten chapters is essentially self contained, and the readers can easily follow a single chapter in which they are interested by encountering a few pointers to the essential background given in previous parts.

This book has been developed with the ambition of providing useful instruments to a variety of users: researchers, practitioners, teachers, and students.

Researchers will find an up-to-date detailed exposition of the theoretical and algorithmic state of the art, not only of the basic linear sum assignment problem but also of its many variations, for which there is plenty of room for improvements: bottleneck, algebraic, balanced, quadratic, and multi-index assignment problems are promising areas for new investigations. In particular, the quadratic assignment problem still lacks effective exact solution methods: after decades of investigations, instances of size 30 require CPU *years* to be solved to optimality. Although this problem is $\mathcal{NP}$-hard, such results could indicate that its combinatorial structure has not yet been fully understood.

Practitioners need clear expositions of successful applications, information on the practical performance of exact and heuristic algorithms, and pointers to high quality software. Almost all of the chapters have one or more sections devoted to the description of real-world situations managed through the described methodologies. The experimental behavior of various algorithms is examined on the basis of computational experiments. The associated home page, http://www.siam.org/books/ot106/assignmentproblems.html, makes available a number of efficient computer codes, either through direct downloads or through links.

Teachers and students will have a potential textbook for advanced courses in discrete mathematics, integer programming, combinatorial optimization, and algorithmic computer science. The theoretical background of each argument is presented in a rigorous way. The algorithms are introduced through an intuitive explanation of their essential features. In most cases, the presentation is completed by a detailed pseudo-code implementation. The main algorithmic techniques are illustrated through a number of exercises presented in the form of fully-developed numerical examples. A further didactic tool is provided by the applets available on the associated home page, which allow step-by-step execution of some basic algorithms.

We are indebted to several friends who have helped us in the preparation of this volume. Alberto Caprara, Bettina Klinz, Ulrich Pferschy, Franz Rendl, and Paolo Toth read preliminary parts of this book, providing valuable suggestions and correcting errors. The authors obviously retain the sole responsibility for any remaining errors. Constructive comments were also made by Peter Hahn, Alexander Korbut, Catherine Roucairol, and Alexander Schrijver. Silvano Martello thanks Letizia Cheng Cheng Sun and Andrea Bergamini, students at the University of Bologna, for the implementation of the home page and its original applets.

Graz, Reggio Emilia, Bologna
May 2008

RAINER BURKARD
MAURO DELL'AMICO
SILVANO MARTELLO

Chapter 1
Introduction

1.1 Assignments

Assignment problems deal with the question of how to assign n items (jobs, students) to n other items (machines, tasks). There are different ways in mathematics to describe an assignment: we can view an assignment as a bijective mapping φ between two finite sets U and V of n elements. By identifying the sets U and V we get the representation of an assignment by a *permutation*. We can write a permutation φ as

$$\begin{pmatrix} 1 & 2 & \dots & n \\ \varphi(1) & \varphi(2) & \dots & \varphi(n) \end{pmatrix},$$

which means that 1 is mapped to $\varphi(1)$, 2 is mapped to $\varphi(2), \dots, n$ is mapped to $\varphi(n)$. In the following, we shall describe a permutation φ just by $\varphi(1), \varphi(2), \dots, \varphi(n)$. For example, the permutation

$$\begin{pmatrix} 1 & 2 & 3 & 4 & 5 & 6 & 7 \\ 2 & 1 & 3 & 7 & 6 & 4 & 5 \end{pmatrix}$$

is denoted as $\varphi = (2, 1, 3, 7, 6, 4, 5)$.

Permutations can also be written in *cyclic form*, i.e., as a family of one or more cycles

$$\langle 1, \varphi(1), \varphi(\varphi(1)), \dots \rangle \quad \dots \quad \langle k, \varphi(k), \varphi(\varphi(k)), \dots \rangle$$

such that the first and last elements, say, k and l, of each cycle satisfy $\varphi(l) = k$. For example, the permutation above has the cyclic representation $(1, 2)(3)(4, 7, 5, 6)$. A permutation represented by a single cycle is called *cyclic permutation*.

Every permutation φ of the set $\{1, 2, \dots, n\}$ corresponds in a unique way to an $n \times n$ *permutation matrix* $X_\varphi = (x_{ij})$ with

$$x_{ij} = \begin{cases} 1 & \text{if } j = \varphi(i), \\ 0 & \text{otherwise.} \end{cases}$$

$$\varphi = (\ 2,\quad 4,\quad 3,\quad 1\ )$$

$$X_\varphi = \begin{pmatrix} 0 & 1 & 0 & 0 \\ 0 & 0 & 0 & 1 \\ 0 & 0 & 1 & 0 \\ 1 & 0 & 0 & 0 \end{pmatrix}$$

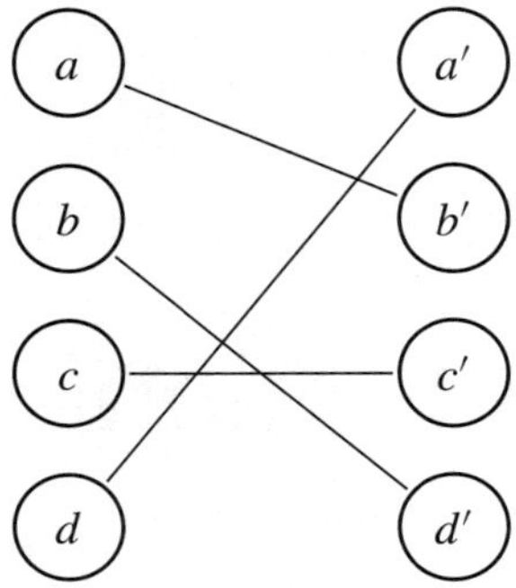

Figure 1.1. *Different representations of assignments.*

Permutation matrices are characterized by the system of linear equations

$$\sum_{j=1}^{n} x_{ij} = 1 \qquad (i = 1, 2, \ldots, n), \tag{1.1}$$

$$\sum_{i=1}^{n} x_{ij} = 1 \qquad (j = 1, 2, \ldots, n), \tag{1.2}$$

and the condition

$$x_{ij} \in \{0, 1\} \qquad \text{for all } i, j = 1, 2, \ldots, n. \tag{1.3}$$

The first set of equations (1.1) says that every row of a permutation matrix sums to 1. The second set of equations (1.2) says that every column of a permutation matrix has a sum of 1. Finally, (1.3) says that a permutation matrix has only the entries 0 and 1. In particular, a permutation matrix has exactly n 1-entries, one in every row and in every column. The relations (1.1)–(1.3) are called *assignment constraints*.

Let $\mathcal{S}_n$ denote the set of all assignments (permutations) of n items. This set has $n!$ elements. In Section 2.2 we shall investigate the structure of $\mathcal{S}_n$ in more detail.

Bipartite graphs offer another possibility for describing assignments. A graph $G = (U, V; E)$ with disjoint vertex sets U and V and edge set E is called *bipartite* if every edge connects a vertex of U with a vertex of V and there are no edges which have both endpoints in U and in V. A *matching* M in G is a subset of the edges such that every vertex of G meets at most one edge of the matching. Let us suppose that the number of vertices in U and V equals n, i.e., $|U| = |V| = n$. If in this case every vertex of G coincides with an edge of the matching M, the matching M is called a *perfect matching*. Obviously, every assignment can be represented as a perfect matching, as shown in Figure 1.1.

A first important question is the following. Suppose that a bipartite graph $G = (U, V; E)$ with edge set E and vertex sets U and V is given. We ask whether there exists a perfect matching $M \subseteq E$. In other words, does there exist a set of n edges in G such that every vertex coincides with exactly one edge? This problem is known as *bipartite perfect matching problem* and plays a fundamental role in the theory of assignment problems.

In 1935 Philip Hall (1904–1982) gave a necessary and sufficient condition for the existence of a perfect matching. In 1949 the famous mathematician Hermann Weyl (1885–

1955) gave a nice interpretation of the perfect matching problem. Let us view the vertices in U as young ladies and the vertices in V as young men. An edge $[i, j]$ indicates that lady i is a friend of j. A perfect matching corresponds to a marriage of all young ladies and young men where a couple can only be married if the partners are friends. The 1935 result by Hall is known as the *marriage theorem*. This theorem, however, does not directly provide an efficient method for finding a perfect matching. In the early years of mathematical programming, labeling techniques were used to construct a perfect matching in $O(n^3)$ time. These methods were refined later by several authors. Well known is the approach of Hopcroft and Karp [376], who showed that a perfect matching can be found in $O(n^{5/2})$ time. We shall discuss their algorithm and improvements thereof in detail in Chapter 3.

Flows in networks offer another opportunity for describing assignments. Let G be the bipartite graph introduced above. We embed G in a network $\mathcal{N} = (N, A, q)$ with *node set* N, *arc set* A, and *arc capacities* q. The node set N consists of a *source* s, a *sink* t, and the vertices of $U \cup V$. The source is connected to every node in U by an arc of capacity 1, and every node in V is connected to the sink by an arc of capacity 1. Moreover, every edge in E is directed from U to V and supplied with capacity 1. A *flow* in network $\mathcal{N}$ is a function $f : A \to \mathbb{R}$ with

$$\sum_{(i,j)\in A} f(i, j) = \sum_{(j,k)\in A} f(j, k) \quad \text{for all } j \in U \cup V, \tag{1.4}$$

$$0 \le f(i, j) \le q(i, j) \quad \text{for all } (i, j) \in A. \tag{1.5}$$

The constraints (1.4) and (1.5) are called *flow conservation constraints* and *capacity constraints*, respectively. The equalities (1.4) say that the total incoming flow in node j, $j \neq s, t$, equals the total flow which leaves that node. The capacity constraints say that the flow is nonnegative and must obey the arc capacities. The *value* $z(f)$ of flow f is defined as $z(f) = \sum_{(s,i)\in A} f(s, i)$. The *maximum network flow problem* asks for a flow with maximum value $z(f)$. Obviously, an integral flow in the special network constructed above corresponds to a matching in G. If the flow has the value n, then it corresponds to a perfect matching in G. Figure 1.2 shows a maximum (perfect) matching in a bipartite graph and the corresponding maximum flow in a network.

Another possibility for describing assignments is offered by matroid theory. Let E be a finite nonempty set, called *ground set*, and let $\mathcal{F}$ be a collection of subsets $F \subseteq E$, called *independent sets*. The system $(E, \mathcal{F})$ is called a *matroid* if it fulfills the three following matroid axioms:

(M1) $\emptyset \in \mathcal{F}$.

(M2) $F' \subseteq F$ and $F \in \mathcal{F}$ imply $F' \in \mathcal{F}$.

(M3) For any two sets F and F' in $\mathcal{F}$ with $|F| < |F'|$ there exists an $e \in F' \setminus F$ such that $F \cup \{e\} \in \mathcal{F}$.

To give an example, let $G = (U, V; E)$ be a bipartite graph. We define $\mathcal{F}_1$ as the collection of all subsets F of E which have the property that every vertex of U coincides with at most one edge of F. It is obvious that $(E, \mathcal{F}_1)$ fulfills the matroid axioms (M1)–(M3).

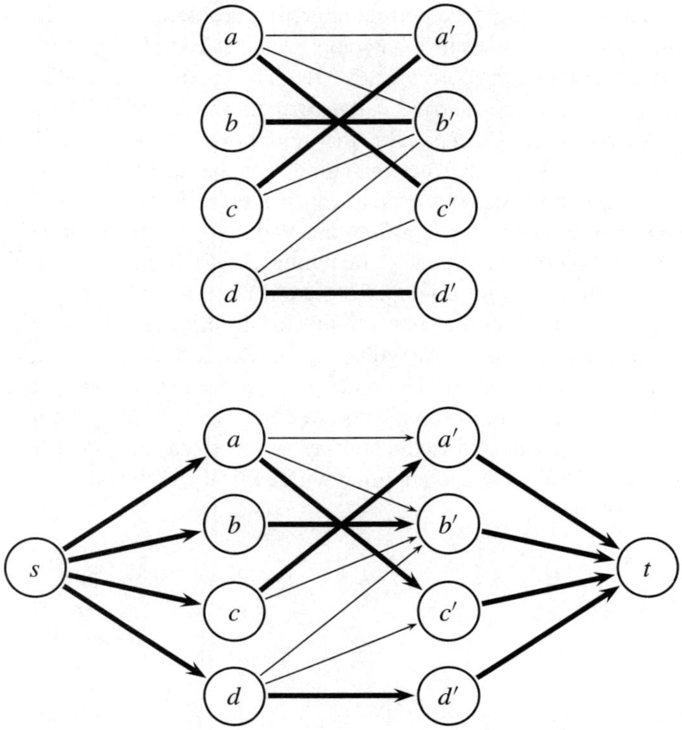

Figure 1.2. *Perfect matching and corresponding network flow model.*

Now, let $(E, \mathcal{F}_1)$ and $(E, \mathcal{F}_2)$ be two matroids defined on the same ground set E. A set F lies in the *intersection of the two matroids* $(E, \mathcal{F}_1)$ and $(E, \mathcal{F}_2)$ if $F \in \mathcal{F}_1$ and $F \in \mathcal{F}_2$.

To continue our example, let $\mathcal{F}_2$ be the collection of all subsets F of E which have the property that every vertex of V coincides with at most one edge of F. Thus any matching corresponds to a set $F \in \mathcal{F}_1 \cap \mathcal{F}_2$, and, vice versa, every set $F \in \mathcal{F}_1 \cap \mathcal{F}_2$ corresponds to a matching. If we assume $|U| = |V| = n$, then every perfect matching (assignment) corresponds to a set $F \in \mathcal{F}_1 \cap \mathcal{F}_2$ with $|F| = n$.

More about matroids can be found for example in the book by Welsh [662].

1.2 Linear assignment problems

So far we have discussed different ways to describe assignments. Moreover, the perfect matching problem mentioned above settles the question of the existence of an assignment. Now we are going to discuss optimization problems with respect to assignments. Since there are in general many assignments possible, we are interested in the best suited assignment for the problem under investigation. Therefore, we must state our goal by specifying an objective function. Given an $n \times n$ cost matrix $C = (c_{ij})$, where c_{ij} measures the cost of assigning i to j, we ask for an assignment with minimum total cost, i.e., the objective function

$$\sum_{i=1}^{n} c_{i\varphi(i)}$$

is to be minimized. The *linear sum assignment problem (LSAP)* can then be stated as

$$\min_{\varphi \in S_n} \sum_{i=1}^{n} c_{i\varphi(i)}. \tag{1.6}$$

A typical case in which such a linear sum assignment problem occurs is the following situation.

Example 1.1. Suppose that n jobs are to be assigned to n machines (or workers) in the best possible way. Let us assume that machine j needs c_{ij} time units in order to process job i. We want to minimize the total completion time. If we assume that the machines work in series, we have to minimize the linear sum objective function $\sum_{i=1}^{n} c_{i\varphi(i)}$. If we assume that the machines work in parallel, we have to minimize the bottleneck objective function $\max_{1\le i\le n} c_{i\varphi(i)}$. ∎

This example shows that there are different objective functions of interest. When a cost is to be minimized, usually a sum objective is used. If a time is to be minimized, a so-called *bottleneck objective function* of the form $\max_{1\le i\le n} c_{i\varphi(i)}$ is often used. Although this function is not written in linear form, the optimization problem with such an objective function is called "linear" in contrast to the quadratic problems introduced in Section 1.3. The *linear bottleneck assignment problem (LBAP)* can be written as

$$\min_{\varphi \in S_n} \max_{1\le i\le n} c_{i\varphi(i)}. \tag{1.7}$$

If we describe permutations by the corresponding permutation matrices $X = (x_{ij})$, an LSAP can be modeled as

$$\begin{aligned}
\min \quad & \sum_{i=1}^{n}\sum_{j=1}^{n} c_{ij}x_{ij} \\
\text{s.t.} \quad & \sum_{j=1}^{n} x_{ij} = 1 \qquad (i = 1, \ldots, n), \\
& \sum_{i=1}^{n} x_{ij} = 1 \qquad (j = 1, \ldots, n), \\
& x_{ij} \in \{0, 1\} \qquad (i, j = 1, \ldots, n).
\end{aligned}$$

Similarly, an LBAP can be written as

$$\begin{aligned}
\min \quad & \max_{1\le i,j\le n} c_{ij}x_{ij} \\
\text{s.t.} \quad & \sum_{j=1}^{n} x_{ij} = 1 \qquad (i = 1,\ldots,n), \\
& \sum_{i=1}^{n} x_{ij} = 1 \qquad (j = 1,\ldots,n), \\
& x_{ij} \in \{0,1\} \qquad (i,j = 1,\ldots,n).
\end{aligned}$$

In Section 6.3 on algebraic assignment problems, it will be pointed out that both the LSAP and the LBAP can be viewed as special cases of a more general model, the so-called *algebraic assignment problem*. In that section we shall state a general method for solving algebraic assignment problems.

Sometimes linear assignment problems are stated as *weighted perfect matching problems*. Consider, for example, a bipartite graph $G = (U, V; E)$ where every edge $e \in E$ has a nonnegative weight $c(e)$. We may ask for a perfect matching in G where the sum of all weights is a minimum. This weighted perfect matching problem is nothing else than a linear assignment problem where the cost matrix may have some forbidden entries. The forbidden entries correspond to edges which are not present in the edge set E of G. On the other hand, any LSAP can be formulated as a weighted perfect matching problem: these two problems are equivalent.

The LSAP can also be seen as a special case of the famous Hitchcock-Koopmans transportation problem (see Hitchcock [369]). Given the bipartite graph above, let the vertices $i \in U$ represent *sources* capable of supplying positive integer amounts a_i and the vertices $j \in V$ represent *sinks* having positive integer demands b_j with $\sum_i a_i = \sum_j b_j$. The *transportation problem* is to find the least cost transportation pattern from the sources to the sinks. The special case where all the supply amounts and demands are equal to one is thus LSAP.

Linear assignment problems can also be viewed as min-cost network flow problems in special graphs. Let a network $\mathcal{N} = (N, A, q)$ with *node set* N, *arc set* A, and *arc capacities* q be given. We assume in addition that a unit flow along arc $(i, j) \in A$ incurs cost c_{ij}. The *cost of a flow* f is defined as

$$c(f) = \sum_{(i,j)\in A} c_{ij} f(i, j). \tag{1.8}$$

Let us use the flow representation of assignments and let us assume that a unit flow along any arc (i, j) with $i \neq s$, $j \neq t$ in the corresponding network incurs cost c_{ij}, whereas it incurs cost 0 along any arc (s, i) leaving the source and any arc (j, t) entering the sink. Then the assignment problem is equivalent to finding a maximum flow of value n with minimum cost. Such problems are called *min-cost network flow problems*.

Linear assignment problems can also be viewed as special matroid intersection problems. Let two matroids $(E, \mathcal{F}_1)$ and $(E, \mathcal{F}_2)$ be defined on the same finite ground set E. Moreover, every $e \in E$ has a cost $c(e)$. The *matroid intersection problem* asks for a set $F \in \mathcal{F}_1 \cap \mathcal{F}_2$ with maximum cost:

$$\max_{F \in \mathcal{F}_1 \cap \mathcal{F}_2} \sum_{e \in F} c(e). \tag{1.9}$$

In order to relate this problem to a linear assignment problem, we consider a complete bipartite graph $G = (U, V; E)$ with $|U| = |V| = n$. If $(E, \mathcal{F}_1)$ and $(E, \mathcal{F}_2)$ are defined as special matroids as in Section 1.1, then the linear assignment problem can be written as a matroid intersection problem. More about matroid intersection problems can be found for example in the book by Lawler [448].

Matching problems, network flow problems, and matroid problems are three important areas of combinatorial optimization which lead in different directions, such as matchings in non-bipartite graphs, flows in general networks, and general matroid theory. Assignment problems lie in the intersection of these fields. Therefore, methods from all of these fields can be used for and applied to assignment problems.

Efficient methods for solving linear sum and bottleneck assignment problems will be described in Chapter 4 and Chapter 6, respectively.

1.3 Quadratic assignment problems

Linear objective functions are not the only important objectives for assignment problems. In practice, assignment problems with a quadratic objective function play an important role.

Example 1.2. A set of n facilities has to be allocated to a set of n locations. We are given three $n \times n$ input matrices: $A = (a_{ik})$, $B = (b_{jl})$, and $C = (c_{ij})$, where a_{ik} is the flow between facility i and facility k, b_{jl} is the distance between location j and location l, and c_{ij} is the cost of placing facility i at location j. We assume that the total cost depends on the flow between facilities multiplied by their distance and on the cost for placing a facility at a certain site. Each product $a_{ik} b_{\varphi(i)\varphi(k)}$ represents the flow between facilities i and k multiplied by their distance when facility i is assigned to location $\varphi(i)$ and facility k is assigned to location $\varphi(k)$. The objective is to assign each facility to a location such that the total cost is minimized. ■

The objective function of Example 1.2 can be written as

$$\min_{\varphi \in \mathcal{S}_n} \sum_{i=1}^{n} \sum_{k=1}^{n} a_{ik} b_{\varphi(i)\varphi(k)} + \sum_{i=1}^{n} c_{i\varphi(i)}. \tag{1.10}$$

When we express the permutations by permutation matrices $X = (x_{ij})$, the *quadratic assignment problem* (QAP) can be modeled as a quadratic integer program of the form

$$\begin{aligned}
\min \sum_{i=1}^{n}\sum_{j=1}^{n}\sum_{k=1}^{n}\sum_{l=1}^{n} a_{ik}b_{jl}x_{ij}x_{kl} &+ \sum_{i=1}^{n}\sum_{j=1}^{n} c_{ij}x_{ij} \\
\text{s.t.} \sum_{j=1}^{n} x_{ij} = 1 & \qquad (i = 1, 2, \ldots, n), \\
\sum_{i=1}^{n} x_{ij} = 1 & \qquad (j = 1, 2, \ldots, n), \\
x_{ij} \in \{0, 1\} & \qquad (i, j = 1, 2, \ldots, n).
\end{aligned}$$

In Chapters 7 and 8 we deal with quadratic assignment problems in detail.

There is an important difference between linear and quadratic assignment problems. For linear assignment problems, there exist efficient, polynomial-time solution methods. These allow us to solve problems with large values of n. On the other hand, quadratic assignment problems are so-called $\mathcal{NP}$-hard problems (see, e.g., Garey and Johnson [300]). This means that an optimal solution can only be found by (implicit) enumeration of all possibilities unless $\mathcal{P} = \mathcal{NP}$. Quadratic assignment problems belong even to the core of $\mathcal{NP}$-hard problems in the sense that it is even impossible to find an approximate solution within some constant factor from the optimum value in polynomial time unless $\mathcal{P} = \mathcal{NP}$. The largest benchmark instances of quadratic assignment problems for which an optimal solution has been determined have a size around $n \approx 30$. For this reason, many heuristics have been proposed for quadratic assignment problems, which yield a more or less good suboptimal solution. Several of these heuristics are discussed in Chapter 8. Computational experiments show that many of the heuristics yield good suboptimal solutions. This effect can be explained by the asymptotic behavior of quadratic assignment problems, which is discussed in Section 9.2.

Sometimes, cubic and bi-quadratic (quartic) objective functions are minimized over the set of assignments. This leads to cubic and bi-quadratic assignment problems, which are discussed in Section 9.3.

1.4 Multi-index assignment problems

In a *timetabling problem* the assignment of n courses to n time slots and to n rooms is required. Let c_{ijk} denote the cost for assigning course i to time slot j in room k. We want to find an assignment φ of the courses to time slots and an assignment ψ of the courses to the rooms such that the total cost is at a minimum. This leads to the so-called *axial 3-index assignment problem*

$$\min_{\varphi,\psi \in \mathcal{S}_n} \sum_{i=1}^{n} c_{i\,\varphi(i)\,\psi(i)}. \tag{1.11}$$

Figure 1.3. *Axial* (a) *and planar* (b) *3-index assignment problems.*

We can write this 3-index assignment problem as an integer linear program in the following way:

$$
\begin{aligned}
\min \quad & \sum_{i=1}^{n}\sum_{j=1}^{n}\sum_{k=1}^{n} c_{ijk}x_{ijk} \\
\text{s.t.} \quad & \sum_{j=1}^{n}\sum_{k=1}^{n} x_{ijk} = 1 \qquad (i = 1, 2, \ldots, n), \\
& \sum_{i=1}^{n}\sum_{k=1}^{n} x_{ijk} = 1 \qquad (j = 1, 2, \ldots, n), \\
& \sum_{i=1}^{n}\sum_{j=1}^{n} x_{ijk} = 1 \qquad (k = 1, 2, \ldots, n), \\
& x_{ijk} \in \{0, 1\} \qquad (i, j, k = 1, 2, \ldots, n).
\end{aligned}
$$

Figure 1.3(a) gives a three-dimensional intuition of the constraints: a "1" on a face of the matrix means that exactly one 1 must be in that face.

Karp [407] showed that the axial 3-index assignment problem is $\mathcal{NP}$-hard. We outline axial 3-index assignment problems in Section 10.2.

A similar problem arises if we are looking for *Latin squares*, i.e., square arrays of size n where every position is filled by one of the numbers $1, 2, \ldots, n$ such that every row and column of the square contains all numbers. For example, a Latin square of size 3 may have the form

2	1	3
1	3	2
3	2	1

.

Latin squares are feasible solutions of so-called *planar 3-index assignment problems*, which can be formulated in the following way. We say that n permutations $\varphi_1, \varphi_2, \ldots, \varphi_n$ are *mutually distinct* if $\varphi_r(i) \neq \varphi_s(i)$ for any $i = 1, 2, \ldots, n$ and $r \neq s$. The problem is to find n mutually distinct permutations such that

$$\sum_{i=1}^{n} \sum_{k=1}^{n} c_{i\varphi_k(i)k} \tag{1.12}$$

is a minimum. The corresponding integer linear program is

$$\begin{aligned} \min \quad & \sum_{i=1}^{n} \sum_{j=1}^{n} \sum_{k=1}^{n} c_{ijk} x_{ijk} \\ \text{s.t.} \quad & \sum_{k=1}^{n} x_{ijk} = 1 \qquad (i, j = 1, 2, \ldots, n), \\ & \sum_{i=1}^{n} x_{ijk} = 1 \qquad (j, k = 1, 2, \ldots, n), \\ & \sum_{j=1}^{n} x_{ijk} = 1 \qquad (i, k = 1, 2, \ldots, n), \\ & x_{ijk} \in \{0, 1\} \qquad (i, j, k = 1, 2, \ldots, n). \end{aligned}$$

Let $L = (l_{ij})$ be the Latin square of size n. Then, for $i, j = 1, \ldots, n$, l_{ij} is the (unique) index value k such that $x_{ijk} = 1$ in a feasible solution.

This problem is even more difficult to solve than the axial 3-index assignment problem. The name "planar assignment problem" stems from the following fact: the variables x_{ijk} can be viewed as cells of a cube. A solution x_{ijk} is feasible for the planar assignment problem if its 1-entries form, in every horizontal or vertical plane of this cube, an assignment. For example, if we fix index $i = 3$, then x_{3jk} is a permutation matrix. The same holds for x_{i2k} if we fix, for instance, index $j = 2$.

Figure 1.3(b) gives a three-dimensional intuition of the constraints: a "1" on a line of the matrix means that exactly one 1 must be in that line. For example, the 3×3 Latin square above provides the following solution to the corresponding planar 3-index assignment problem: $x_{12k} = x_{21k} = x_{33k} = 1$ for $k = 1$, $x_{11k} = x_{23k} = x_{32k} = 1$ for $k = 2$, $x_{13k} = x_{22k} = x_{31k} = 1$ for $k = 3$, and $x_{ijk} = 0$ elsewhere.

We discuss planar 3-index assignment problems in Section 10.3. Recently, k-index assignments with $k > 3$ have been considered. They arise, for example, when flying objects are tracked. We discuss them shortly in Section 10.4.

1.5 Research lines for assignment problems

In the previous sections we discussed optimization problems in connection with assignments. This field is still active and comprises several lines of research. Perhaps the most important, and also the main task of this book, is to provide efficient and fast solution methods for assignment problems. As was pointed out, linear assignment problems can be solved by fast, polynomial-time algorithms, whereas quadratic assignment problems and multi-index assignment problems are $\mathcal{NP}$-hard. Therefore, only implicit enumeration algorithms are known for solving them exactly. Since enumeration algorithms are not very efficient, there is a particular need for good heuristics, i.e., heuristics which yield a near optimal solution in a reasonable amount of time. Such heuristics for QAPs are discussed in Section 8.2. There is also an interesting second approach. One can ask whether assignment problems are solved more quickly if their cost matrices have special properties. Several such matrix properties are known, both for linear and quadratic assignment problems. There are also many open questions in this field. We discuss special cases of linear assignment problems in Section 5.2 and special cases of the QAP in Section 8.4.

Another line of research concerns the asymptotic behavior of assignment problems. Assume that the cost coefficients of an assignment problem are independent and identically distributed random variables in [0, 1]. What can be said about the expected optimal solution value, provided the size of the problems tends to infinity? In Section 5.1 we deal with the expected optimum value of the linear sum assignment problem, and in Section 9.2 we address the asymptotic behavior of quadratic assignment problems. There is a large qualitative difference between linear and quadratic assignment problems: whereas for linear assignment problems the difference between best and worst solution value tends to infinity as the problem size goes to infinity, quadratic assignment problems show the strange behavior that the ratio between best and worst objective function value tends to 1 as the size of the problem approaches infinity. We identify a condition which is responsible for that phenomenon. This strange asymptotic behavior, which can also be explained by tools of statistical mechanics, is discussed in Section 9.2.

Chapter 2

Theoretical foundations

2.1 The marriage theorem and the existence of perfect matchings

Let $G = (U, V; E)$ be a bipartite graph with vertex sets $U = \{1, 2, \dots, n\}$ and $V = \{1, 2, \dots, s\}$ and edge set E. Every edge $[i, j]$ has one vertex in U and the other vertex in V. A subset M of E is called a *matching* if every vertex of G coincides with at most one edge from M. An edge $e = [i, j] \in M$ matches vertex i with vertex j. In this case, the vertices i and j are called *matched*. The cardinality $|M|$ of M is called *cardinality of the matching*. A matching is called *maximal* if it cannot be enlarged by any edge of the graph. A matching is called *maximum* if it has as many edges as possible, i.e., it has maximum cardinality. Note that a maximal matching may not be maximum (see Figure 2.1), but every maximum matching is obviously maximal. A matching M is called a *perfect matching* if *every* vertex of G coincides with exactly one edge from M, i.e., every vertex of G is matched. An obvious condition for the existence of a perfect matching in the bipartite graph $G = (U, V; E)$ is that the vertex sets U and V have the same number of elements, i.e., $|U| = |V|$. Clearly, every perfect matching is a maximum matching and a maximal matching.

Matchings, maximum matchings, and perfect matchings can be defined in a straightforward way for any finite, undirected graph. Due to a famous result by Edmonds [247], a maximum cardinality matching in an arbitrary graph can be found in polynomial time. Since this part of matching theory is not directly connected with assignments, we refer the interested reader to Schrijver [599] for a thorough treatment of this subject.

The *adjacency matrix* of a bipartite graph $G = (U, V; E)$ is a $|U| \times |V|$ matrix $A = (a_{ij})$ defined by

$$a_{ij} = \begin{cases} 1 & \text{if } [i, j] \in E, \\ 0 & \text{otherwise.} \end{cases} \tag{2.1}$$

If we are interested in a perfect matching, we must require that $|U| = |V| = n$ holds. A perfect matching corresponds in this case to a selection of n 1-entries in A, one in every row and column.

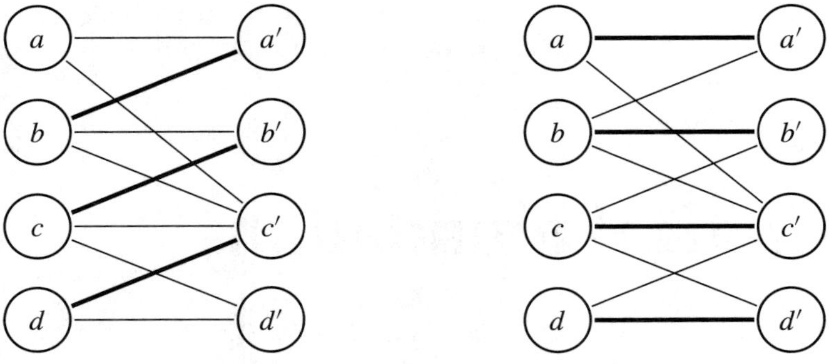

Figure 2.1. *The left figure shows a maximal matching which cannot be extended by any edge. The right figure shows a maximum matching of larger size.*

The *permanent* of matrix A, $\mathrm{per}(A)$, is defined by

$$\mathrm{per}(A) = \sum_{\varphi \in \mathcal{S}_n} a_{1\varphi(1)} a_{2\varphi(2)} \dots a_{n\varphi(n)}.$$

Since the product $a_{1\varphi(1)} a_{2\varphi(2)} \dots a_{n\varphi(n)}$ equals 1 if and only if φ is an assignment (perfect matching), the permanent $\mathrm{per}(A)$ counts the different perfect matchings in G. The numerical evaluation of a permanent (and therefore finding the number of different perfect matchings in a graph G) is $\#\mathcal{P}$-complete (see Valiant [645]). This implies that finding the number of different perfect matchings in G is at least as hard as any $\mathcal{NP}$-complete problem.

Hall's *marriage theorem* states a necessary and sufficient condition for the existence of a perfect matching in a bipartite graph. We state it first in a slightly more general form. For a vertex $i \in U$, let $N(i)$ denote the set of its neighbors, i.e., the set of all vertices $j \in V$ which are connected with i by an edge in E. When we view the vertices in U as young ladies and the vertices in V as young men, the set $N(i)$ contains the friends of i. Moreover, for any subset U' of U let $N(U') = \bigcup_{i \in U'} N(i)$.

Theorem 2.1. (Hall [360], 1935.) *Let $G = (U, V; E)$ be a bipartite graph. It is possible to match every vertex of U with a vertex of V if and only if for all subsets U' of U*

$$|U'| \le |N(U')| \qquad \textit{(Hall's condition)}. \tag{2.2}$$

Remark: The theorem says that, in our interpretation, each lady can marry one of her friends. If we assume in addition that $|U| = |V|$, we can express Theorem 2.1 as follows.

Theorem 2.2. (Marriage theorem.) *Let $G = (U, V; E)$ be a bipartite graph with $|U| = |V|$. There exists a perfect matching (marriage) in G if and only if G fulfills Hall's condition (2.2).*

Remark: This means that, under the additional assumption $|U| = |V|$, all ladies and men can marry.

Proof. We prove Theorem 2.1. Hall's condition is obviously necessary for the existence of a matching which matches all vertices in U. Therefore, we have only to show that this condition is also sufficient. We prove the sufficiency by induction on the number of elements in $|U|$. If $|U| = 1$, the only lady can surely marry one of her friends. So, let us suppose that the theorem holds for all bipartite graphs with $|U| = k$ and let $\overline{G} = (\overline{U}, \overline{V}; \overline{E})$ be a bipartite graph with $|\overline{U}| = k + 1$. We assume that $\overline{G}$ fulfills Hall's condition (2.2). We consider two cases.

Case 1: $|U'| < |N(U')|$ holds for all nonempty proper subsets U' of $\overline{U}$. In this case we match an arbitrary vertex i with one of its neighbors, say, j. By deleting i and j and all incident edges, we get a new graph $G = (U, V; E)$ with $|U| = k$. Graph G still fulfills Hall's condition, since every vertex i lost at most one neighbor. Therefore, G contains a matching of size $|\overline{U}| - 1$. Adding the edge $[i, j]$ we get a matching of size $|\overline{U}|$ in $\overline{G}$.

Case 2: There is a nonempty proper subset U' for which Hall's condition holds with equality: $|U'| = |N(U')|$. Due to our induction, we can match each $i \in U'$ with an appropriate $j \in N(U')$. Now we delete the sets U' in $\overline{U}$, $N(U')$ in $\overline{V}$ and the edges of $\overline{E}$ incident with these vertices. We get a new graph $\widetilde{G} = (\widetilde{U}, \widetilde{V}; \widetilde{E})$ and have to show that Hall's condition still holds in the new graph. Suppose it does not. Then there exists a subset W of $\widetilde{U}$ whose neighborhood $N_{\widetilde{G}}(W)$ in $\widetilde{G}$ fulfills $|W| > |N_{\widetilde{G}}(W)|$. Thus we get, using $N_{\widetilde{G}}(W) = N(W) \setminus N(U')$,

$$|N(U' \cup W)| = |N(U')| + |N(W) \setminus N(U')| < |U'| + |W| = |U' \cup W|.$$

But this means that Hall's condition does not hold in $\overline{G}$, which is a contradiction. □

Remark: When every man and every lady ranks the members of the opposite sex with a total ordering, we can define the *stable marriage (matching) problem* as follows: Assign each man to a lady so that there is no pair of a man and a lady who both prefer the other to their actual partner. In other words, for any pair of a man m and a woman w not married together, either m prefers his wife to w or w prefers her husband to m. A polynomial-time algorithm for this problem is due to Gale and Shapley [298]. We refer the interested reader to the book by Gusfield and Irving [350] for a thorough treatment of this subject.

Theorem 2.2 immediately yields the following observation. A graph is called *k-regular* if every vertex i has degree $d(i) = k$.

Corollary 2.3. *Every k-regular bipartite graph* ($k \geq 1$) *has a perfect matching.*

Proof. Let $G = (U, V; E)$ be a k-regular bipartite graph. As

$$k|U| = \sum_{j \in U} d(j) = \sum_{j \in V} d(j) = k|V|$$

we have $|U| = |V|$. Now let $\emptyset \neq U' \subset U$. Let $E_{U'}$ be the set of edges incident with U' and let $E_{N(U')}$ be the set of edges incident with the neighbors of set U'. Clearly $E_U \subseteq E_{N(U')}$. Therefore, $k|U'| = |E_{U'}| \leq |E_{N(U')}| = k|N(U')|$, which implies $|U'| \leq |N(U')|$. Thus Hall's condition (2.2) is fulfilled. □

We can easily express Hall's theorem in the language of $0-1$ matrices by using the adjacency matrix of graph G. Let us assume that $|U| = |V| = n$. If the adjacency matrix A contains n 1-entries, one in every row and column (which correspond to a perfect matching), we say that matrix A *contains* a permutation matrix. Hall's condition (2.2) says that for $k = 0, 1, \ldots, n-1$ matrix A does not contain a $(k+1) \times (n-k)$ submatrix of 0 elements, since otherwise the $k+1$ vertices corresponding to the rows of this submatrix would have less than $k+1$ neighbors. So we get the following theorem, equivalent to Theorem 2.1.

Theorem 2.4. (Frobenius [290], 1917.) *Let A be an arbitrary $n \times n$ matrix with entries 0 and 1. Matrix A contains a permutation matrix if and only if, for $k = 0, 1, \ldots, n-1$, matrix A does not contain a $(k+1) \times (n-k)$ submatrix of 0 elements.*

In other words, we can express Theorem 2.4 in the following way.

Corollary 2.5. *Every $(k+1) \times (n-k)$ submatrix of any $n \times n$ permutation matrix P contains at least one 1-entry.*

In 1916 König proved a theorem on matchings which turns out to be one of the cornerstones of algorithms for assignment problems. Before we state König's matching theorem and prove that it is equivalent to Hall's marriage theorem, we need the following definition (see Figure 2.2).

Definition 2.6. *Given a bipartite graph G, a* vertex cover *(or* transversal set*) C in G is a subset of the vertices of G such that every edge of G coincides with at least one vertex in this set C.*

Theorem 2.7. (König's matching theorem [425], 1916.) *In a bipartite graph the minimum number of vertices in a vertex cover equals the maximum cardinality of a matching:*

$$\min_{C \text{ vertex cover}} |C| = \max_{M \text{ matching}} |M|.$$

Historical note. Hall's marriage theorem developed from a question in algebra. In 1910 Miller [492] showed that the left and right residual classes of an arbitrary subgroup of a finite group have a common system of representatives. On the other hand, in 1917 Frobenius [290] showed Theorem 2.4 in connection with determinants. He already noticed the relationship with König's matching theorem of 1916, but considered it of *little value*. In 1924 van der Waerden [655] noticed that Miller's theorem is based on a combinatorial argument and is closely related to König's matching theorem [425]. Based on this observation, Hall proved, in 1935, the above Theorem 2.1 which we call today, according to an interpretation of Weyl [663], the *marriage theorem*. For more details see Schrijver [600].

We now show that König's matching theorem is equivalent to Hall's theorem.

Theorem 2.8. *Theorems 2.1 and 2.7 are equivalent.*

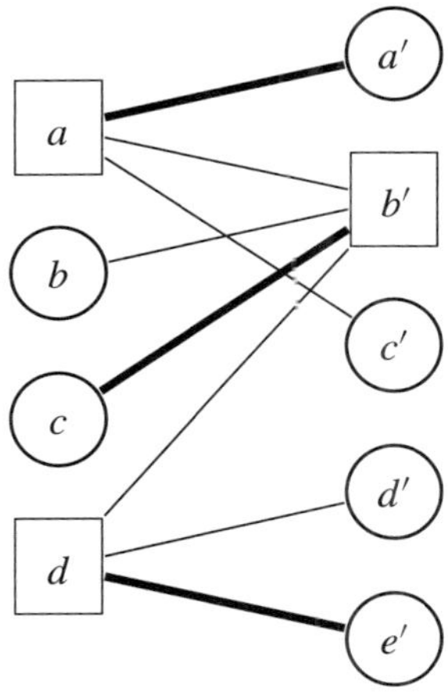

Figure 2.2. *Maximum matching (bold edges) and minimum vertex cover (square vertices).*

Proof.
1. Hall's theorem implies König's matching theorem. Trivially, any matching M and any vertex cover C fulfill $|M| \le |C|$. Thus we have to prove that there is a matching M whose size is equal to the size of a minimum vertex cover, i.e., a vertex cover of minimum cardinality.

Let C be a minimum vertex cover in the bipartite graph G. We define a new bipartite graph $G' = (U', V'; E')$ by $U' = U \cap C$, $V' = V \setminus C$, and $E' = \{[i, j] : [i, j] \in E \cap (U' \times V')\}$. Assume that this graph G' does not fulfill Hall's condition (2.2). In this case, according to Hall's theorem, there is a subset $W \subseteq U'$ with $|W| > |N(W)|$. Now, $C' = (U' \setminus W) \cup N(W) \cup (V \cap C)$ is a vertex cover in G. Namely, every edge with one vertex in $U \cap C$ has either one vertex in $U' \setminus W$ or one vertex in $N(W)$. All other edges have one vertex in $V \cap C$. Thus we get

$$|C'| = |U' \setminus W| + |N(W)| + |V \cap C| < |U' \setminus W| + |W| + |V \cap C| = |U \cap C| + |V \cap C| = |C|.$$

But this is in contradiction to C being a minimum vertex cover.

Therefore, every vertex of $U \cap C$ can be matched with a vertex of $V \setminus C$. An analogous argument shows that every vertex of $V \cap C$ can be matched with a vertex of $U \setminus C$. This proves that there exists a matching M with $|M| = |C|$.

2. König's matching theorem implies Hall's theorem. Let $G = (U, V; E)$ be a bipartite graph which fulfills Hall's condition (2.2). This means in particular, that every vertex $i \in U$ coincides with an edge $e \in E$. We know that in G the cardinality of a minimum vertex cover equals the cardinality of a maximum matching. We show that U is a minimum vertex cover. Let C be any vertex cover in G. If $U' = U \setminus C \neq \emptyset$, then no edge with one endpoint in U' can have its other endpoint in $V \setminus C$, since C is a vertex cover. Thus we get $N(U') \subseteq V \cap C$ and

$$|C| = |U \cap C| + |V \cap C| \ge |U \cap C| + |N(U')| \ge |U \cap C| + |U'| = |U|.$$

According to König's matching theorem there exists a matching with $|U|$ edges, i.e., every vertex of U will be matched. □

In algorithms for the linear assignment problem we will make use of the following equivalent formulation of König's matching theorem. Given a bipartite graph $G = (U, V; E)$ with $|U| = |V| = n$, we define the *complementary adjacency matrix* B of G as an $n \times n$ matrix $B = (b_{ij})$ where

$$b_{ij} = \begin{cases} 0 & \text{if } [i, j] \in E, \\ 1 & \text{otherwise.} \end{cases} \tag{2.3}$$

This means that if the adjacency matrix A has an entry $a_{ij} = 0$, then $b_{ij} = 1$ and vice versa. A *zero cover* is a subset of the rows and columns of matrix B which contains all the 0 elements of B. A row (resp. column) which is an element of a zero cover is called a *covered row* (resp. *covered column*). A *minimum zero cover* is a zero cover with a minimum number of covered rows and columns. A minimum zero cover is directly connected to a vertex cover in the bipartite graph $G = (U, V; E)$. If the vertex cover contains a vertex $i \in U$, then the corresponding row of matrix B is covered. If the vertex cover contains a vertex $j \in V$, then the corresponding column of B is covered. Thus every vertex cover immediately leads to a zero cover in B and vice versa. Now we get the following.

Proposition 2.9. *There exists an assignment φ with $b_{i\varphi(i)} = 0$ for all $i = 1, \ldots, n$ if and only if the minimum zero cover has n elements.*

König's theorem is a special case of Ford and Fulkerson's *max-flow min-cut theorem* [213]. Let us consider a network $\mathcal{N} = (N, A, q)$ with node set N and arc set A. Every arc (i, j) has a capacity $q(i, j) \geq 0$. (Infinite capacities $q(i, j) = \infty$ are admitted.) We distinguish two special nodes, the source s and the sink t. Without loss of generality we may assume that there is no arc entering the source and there is no arc leaving the sink. An *(s,t)-flow* in the network $\mathcal{N}$ is a function $f : A \to \mathbb{R}$ with

$$\sum_{(i,j)\in A} f(i, j) = \sum_{(j,k)\in A} f(j, k) \quad \text{for all } j \in N \setminus \{s, t\}, \tag{2.4}$$

$$0 \leq f(i, j) \leq q(i, j) \quad \text{for all } (i, j) \in A. \tag{2.5}$$

In the following, we call an (s, t)-flow a flow. Constraints (2.4) (*flow conservation constraints*) impose that the total flow entering any node j, $j \neq s, t$, be equal to the total flow leaving that node. Constraints (2.5) (*capacity constraints*) impose that the flow in any arc be nonnegative and does not exceed the arc capacity. The *maximum network flow problem* asks for a flow with maximum value $z(f)$ where

$$z(f) = \sum_{(s,i)\in A} f(s, i) \left(= \sum_{(i,t)\in A} f(i, t) \right). \tag{2.6}$$

An *(s,t)-cut* C in network $\mathcal{N}$ is induced by a partition $(X, \overline{X})$ of node set N such that $s \in X$ and $t \in \overline{X}$. We will denote by $\delta^+(X)$ and $\delta^-(X)$ the sets of arcs crossing the cut in the two directions:

$$\delta^+(X) = \{(i, j) \in A : i \in X, j \in \overline{X}\},$$
$$\delta^-(X) = \{(i, j) \in A : i \in \overline{X}, j \in X\}.$$

The *value* $v(C)$ of the cut is then defined as

$$v(C) = \sum_{(i,j)\in\delta^+(X)} q(i, j). \tag{2.7}$$

Note that every directed path from the source to the sink contains at least one arc of $\delta^+(X)$ and that the flow conservation constraints imply

$$z(f) = \sum_{(i,j)\in\delta^+(X)} f(i, j) - \sum_{(i,j)\in\delta^-(X)} f(i, j). \tag{2.8}$$

Lemma 2.10. *The value $z(f)$ of an arbitrary (s,t)-flow is always bounded by the value $v(C)$ of an arbitrary (s,t)-cut.*

Proof. All paths going from s to t use at least one of the arcs of $\delta^+(X)$. The thesis follows from (2.7) and (2.8). □

Now we state the following.

Theorem 2.11. (Ford-Fulkerson's max-flow min-cut theorem [276], 1956.) *The value of a maximum (s,t)-flow equals the value of a minimum (s,t)-cut.*

Proof. The theorem is trivial for $z(f) = \infty$. Therefore, we assume that $z(f)$ is finite. Due to Lemma 2.10 we have only to show that there are a flow f and a cut C induced by $(X, \overline{X})$, with $z(f) = v(C)$. Let f be a maximum flow. We define a set X by the following three conditions:

1. $s \in X$;
2. if $i \in X$ and $f(i, j) < q(i, j)$, then $j \in X$ (arc (i, j) is called a *forward arc*);
3. if $i \in X$ and $f(j, i) > 0$, then $j \in X$ (arc (j, i) is called a *backward arc*).

In other words, X is the subset of nodes of N that can be reached from s by traversing unsaturated arcs (forward arcs) in their original direction and arcs with positive flow (backward arcs) in the reverse direction.

First, we prove that $(X, \overline{X})$ defines a cut, i.e., that $t \in \overline{X}$. Assume to the contrary that $t \in X$. In this case there is a sequence of nodes in X with $s = n_0, n_1, n_2, \ldots, n_\ell = t$. Either (n_k, n_{k+1}) is a forward arc in A or (n_{k+1}, n_k) is a backward arc in A. Let F denote the set of all forward arcs and let B denote the set of all backward arcs in this sequence. We define

$$\varepsilon_1 = \min_{(i,j)\in F} q(i, j) - f(i, j)$$

and

$$\varepsilon_2 = \min_{(i,j)\in B} f(i,j).$$

Let $\varepsilon = \min(\varepsilon_1, \varepsilon_2)$. Due to the construction we get $\varepsilon > 0$. Now we modify the flow f by increasing the flow values on all forward arcs in F by ε and by decreasing the flow on all backward arcs in B by ε. It is easy to check that we get another feasible flow which obeys the flow conservation constraints and the capacity constraints, but now has a value $z(f) + \varepsilon$. This is a contradiction to the assumption that flow f has been a maximum flow.

According to the definition of set X we have:

- an arc (i, j) with $i \in X,\ j \in \overline{X}$ fulfills $f(i, j) = q(i, j)$;
- an arc (i, j) with $i \in \overline{X},\ j \in X$ fulfills $f(i, j) = 0$.

Thus $\sum_{(i,j)\in\delta^+(X)} f(i, j) = v(C)$ and $\sum_{(i,j)\in\delta^-(X)} f(i, j) = 0$, and therefore (2.8) gives $z(f) = v(C)$. □

The cut C induced by $(X, \overline{X})$ in the network $\mathcal{N} = (N, A, q)$ also offers a way to construct a vertex cover in the underlying bipartite graph $G = (U, V; E)$, as well as a zero cover of the complementary adjacency matrix. Recall that the node set of $\mathcal{N}$ is $N = U \cup V \cup \{s, t\}$, while its arc set A is produced by all edges of E when directed from U to V (with infinite capacity), plus arcs (s, i) $(i \in U)$ and (j, t) $(j \in V)$ (all with unit capacity).

Proposition 2.12. *Let G be a bipartite graph and let C be the cut induced by $(X, \overline{X})$ in the corresponding network $\mathcal{N}$.*

1. *A vertex cover of G is given by all vertices $i \in U \cap \overline{X}$ and all vertices $j \in V \cap X$.*
2. *A minimum zero cover of the complementary adjacency matrix of G is given by all rows i with $i \in \overline{X}$ and all columns j with $j \in X$.*

Example 2.13. Consider the bipartite graph given in Figure 2.3. The corresponding complementary adjacency matrix has the form

$$B = \begin{pmatrix} 0 & 0 & 0 & 0 \\ 1 & 0 & 1 & 1 \\ 1 & 0 & 1 & 1 \\ 1 & 0 & 1 & 1 \end{pmatrix}. \tag{2.9}$$

Obviously, all zero elements can be covered by the first row (corresponding to vertex $a \in U$) and the second column (corresponding to vertex $b' \in V$). Now, a minimum cut of the corresponding network is shown in Figure 2.4: the minimum cut is given by $X = \{s, b, c, d, b'\}$ and $\overline{X} = \{a, a', c', d', t\}$. Set $\overline{X}$ contains only one index of U, namely, a; set X contains only one index of V, namely, b'. Thus the vertex cover is given by vertex a and vertex b'. ∎

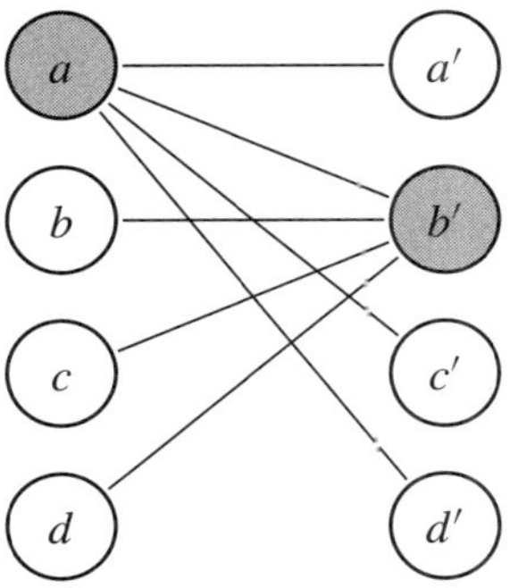

Figure 2.3. *Vertex cover in Example 2.13.*

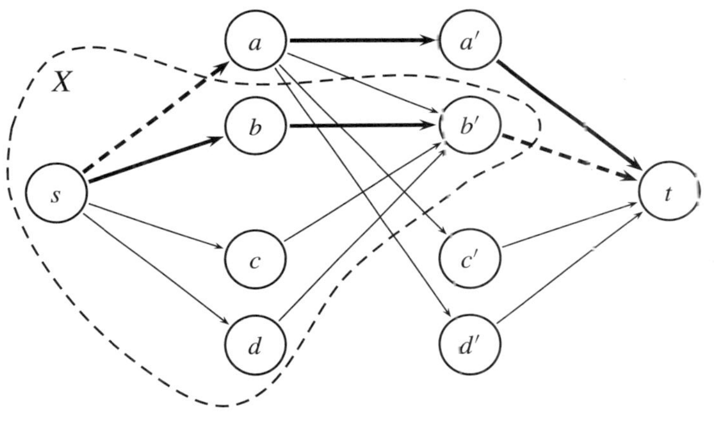

Figure 2.4. *A maximum flow and a minimum cut in the network of Example 2.13. The forward arcs crossing the cut are dotted.*

Now we are going to show that Ford-Fulkerson's max-flow min-cut theorem implies König's theorem. Let $G = (U, V; E)$ be a bipartite graph. We embed G in the network $\mathcal{N} = (N, A, c)$. The node set N consists of a *source* s, a *sink* t, and the vertices of $U \cup V$. The source is connected to every node in U by an arc of capacity 1, every node in V is connected to the sink by an arc of capacity 1, and every edge in E is directed from U to V and supplied with infinite capacity. A maximum flow in this network is integral valued and corresponds to a matching in G with maximum cardinality $z(f)$. Due to Ford-Fulkerson's theorem there is a cut C in the network $\mathcal{N}$ with a finite value which equals $z(f)$. Since all arcs (i, j) with $i \in U$ and $j \in V$ have an infinite capacity, the minimum cut can only contain arcs from the source to a node in U or from a node in V to the sink. This cut corresponds to a vertex cover of the bipartite graph G: If there were an edge $[i, j] \in G$ where neither vertex i nor vertex j are covered, then there would be a path (s, i, j, t) in the network $\mathcal{N}$ which has no arc in the cutset C in contradiction to C being a cut in $\mathcal{N}$.

We are closing this section by showing that any matching can be enlarged in a certain way, which turns out to be crucial for maximum matching algorithms. Namely, let any young lady $i \in U' \subseteq U$ marry one of her friends. If this matching M is not already a maximum cardinality matching in $G = (U, V; E)$, then there always exists a matching M' of larger cardinality where all ladies $i \in U'$ remain married, but now possibly to other friends. More formally, we have the following.

Theorem 2.14. *Given any matching M in $G = (U, V; E)$, there always exists a maximum matching in G such that all vertices that are matched in M remain matched.*

This theorem, whose proof will be given later, shows that though a maximal matching cannot be extended by just adding another edge, one can find a maximum matching such that all previously matched vertices remain matched.

Theorem 2.14 can be seen as a consequence of the augmentation lemma, Lemma 3.3, discussed in Section 3.2, as well as a special case of the following theorem by Mendelsohn and Dulmage. Recall that the *symmetric difference* of two sets A and B is $(A \setminus B) \cup (B \setminus A)$.

Theorem 2.15. (Mendelsohn-Dulmage [487], 1958.) *Let M_1 and M_2 be two matchings in the bipartite graph $G = (U, V; E)$. Then there exists a matching $M \subseteq M_1 \cup M_2$ such that M matches all vertices of U which are matched by M_1 and all vertices of V which are matched by M_2.*

Proof. Let G' be that subgraph of G whose edges are given by the symmetric difference of M_1 and M_2. The components of G' are of five types:

1. (even) cycles whose edges alternate between M_1 and M_2;
2. paths of odd length, starting from a vertex in U which is matched by M_1 to a vertex in V which is not matched by M_2;
3. paths of odd length, starting from a vertex in V which is matched by M_2 to a vertex in U which is not matched by M_1;
4. paths of even length, starting from a vertex in U which is matched by M_1 to a vertex in U which is not matched by M_1;
5. paths of even length, starting from a vertex in V which is matched by M_2 to a vertex in V which is not matched by M_2.

In all five cases the edges alternate between the two given matchings. We construct the matching M which fulfills the requirements of the theorem in the following way. First, we set $M = M_1 \cap M_2$. For each component in G' which is a cycle C, we add the matching edges $C \cap M_1$ to M. Further, for each path in G' we add the first, third, fifth, ... edge of the path to M. It is easy to see that all vertices in U which were originally matched by M_1 remain matched. Moreover, all vertices in V which were originally matched by M_2 remain also matched by M. □

The following example illustrates the proof of the Mendelsohn-Dulmage theorem.

Example 2.16. Figure 2.5(a) contains a graph $G = (U, V; E)$ with two matchings: matching M_1 is drawn with solid lines, matching M_2 is drawn with dashed lines. The shaded

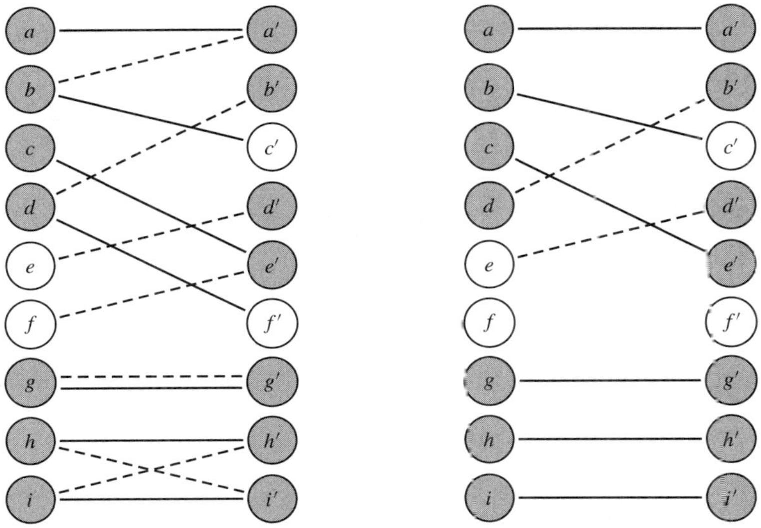

(a) (b)

Figure 2.5. *Example for the Mendelsohn-Dulmage theorem:* (a) *original matchings M_1 (solid lines) and M_2 (dashed lines);* (b) *final matching.*

vertices on the left side are incident with edges of matching M_1, the shaded vertices on the right side are incident with edges of matching M_2. We want to find a matching in this graph such that all shaded vertices are matched.

The bipartite graph with edges of the symmetric difference of M_1 and M_2 is obtained by deleting the edge from vertex g to vertex g' in the graph shown above. The symmetric difference contains

- the cycle (h, h', i, i');
- the odd-length path (a, a', b, c') starting in the M_1-matched vertex $a \in U$ and leading to an unmatched vertex in V;
- the odd-length path (d', e) starting from an M_2-matched vertex in V and leading to an unmatched vertex of U;
- the even-length path (c, e', f) starting from an M_1-matched vertex of U and leading to an unmatched vertex of U;
- the even-length path (b', d, f') starting from an M_2-matched vertex of V and leading to an unmatched vertex of V.

We initialize the new matching with edge $[g, g']$ $(= M_1 \cap M_2)$, then we add $[h, h']$ and $[i, i']$ from the cycle, and the first, third, ... edges of the paths, thus obtaining a matching with all shaded vertices matched, as shown in Figure 2.5(b). ∎

We are now going to prove Theorem 2.14.

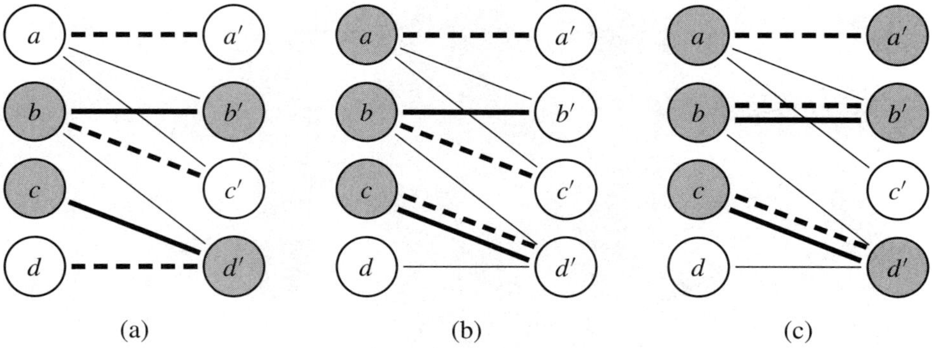

Figure 2.6. *Construction of a maximum matching which contains all matched vertices of an arbitrary given matching.*

Proof of Theorem 2.14. First, we apply Theorem 2.15 to the given matching $\overline{M}$ and an arbitrary maximum matching $\widehat{M}$. Thus we get a maximum matching $\widetilde{M}$ which matches all vertices in U which were previously matched by $\overline{M}$. Next, we apply once more Theorem 2.14 to $\widetilde{M}$ and the original matching $\overline{M}$. Now we get (possibly another) maximum matching contained in $\widetilde{M} \cup \overline{M}$ which keeps matched all vertices of U and those vertices of V which were already matched by the given $\overline{M}$. Thus a maximum matching has been found which matches all vertices which were previously matched by $\overline{M}$. □

The following example illustrates the procedure that produces a maximum matching which leaves matched the originally matched vertices.

Example 2.17. Consider the bipartite graph shown in Figure 2.6(a). The bold lines show the given matching M. The matched vertices are shaded. The dashed lines in this figure show an arbitrary maximum matching. A first application of Theorem 2.14 leads to the dashed maximum matching of Figure 2.6(b): all originally matched vertices on the left side are now also matched in the maximum matching. Now we apply Theorem 2.14 a second time to the maximum matching of Figure 2.6(b). The result is shown by the dashed lines and shaded vertices in Figure 2.6(c). ∎

2.2 The assignment polytope

A *doubly stochastic* matrix X is an $n \times n$ matrix whose entries fulfill

$$\sum_{j=1}^{n} x_{ij} = 1 \qquad (i = 1, 2, \ldots, n), \tag{2.10}$$

$$\sum_{i=1}^{n} x_{ij} = 1 \qquad (j = 1, 2, \ldots, n), \tag{2.11}$$

$$x_{ij} \geq 0 \qquad (i, j = 1, 2, \ldots, n). \tag{2.12}$$

Every permutation matrix is a doubly stochastic matrix. Every assignment φ of n items can thus be described by a doubly stochastic matrix with $x_{ij} \in \{0, 1\}$ for $i, j = 1, 2, \ldots, n$. The set of all doubly stochastic matrices forms the so-called *assignment polytope* P_A. Birkhoff [100] showed that the assignments uniquely correspond to the vertices of P_A. Thus every doubly stochastic matrix can be written as a convex combination of permutation matrices.

Theorem 2.18. (Birkhoff [100], 1946.) *The vertices of the assignment polytope uniquely correspond to permutation matrices.*

Historical note. The Birkhoff theorem is implicitly contained in a 1931 theorem due to Egerváry [253]. See Dell'Amico and Martello [217].

Before we prove Birkhoff's theorem we show the following lemma (see also Corollary 2.5).

Lemma 2.19. *For any $k \in \{0, 1, \ldots, n-1\}$, if a square matrix R with nonnegative entries and equal row and column sums, say, equal to α, contains a $(k+1) \times (n-k)$ submatrix of 0 elements, then $R = 0$.*

Proof. By permuting the rows and columns of R we can arrange the entries of R in the following form

$$\begin{array}{c} \\ n-k-1 \\ k+1 \end{array} \begin{array}{c} \begin{array}{cc} n-k & \quad k \end{array} \\ \left(\begin{array}{c|c} R_1 & R_2 \\ \hline 0 & R_3 \end{array} \right) \end{array} .$$

The sum of all coefficients in the first $n-k$ columns is $(n-k)\alpha$; the sum of all coefficients in the last $k+1$ rows is $(k+1)\alpha$. Further, the sum of all elements in R is $n\alpha$. We get

$$n\alpha = \sum_{i=1}^{n} \sum_{j=1}^{n} r_{ij} \geq \sum_{i=1}^{n} \sum_{j=1}^{n-k} r_{ij} + \sum_{i=n-k}^{n} \sum_{j=1}^{n} r_{ij} = (n-k)\alpha + (k+1)\alpha = (n+1)\alpha.$$

This implies $\alpha = 0$ and therefore $R = 0$. □

Proof of Birkhoff's theorem. Let X be an arbitrary doubly stochastic matrix. Matrix X does not contain any $(k+1) \times (n-k)$ submatrix of 0 elements due to Lemma 2.19, since otherwise $X = 0$ contradicting that matrix X is doubly stochastic. Then, according to the Frobenius theorem, Theorem 2.4, there exists a permutation matrix $P_1 = (p_{ij})$ with the following property: if $p_{ij} = 1$, then $x_{ij} > 0$. Let λ_1 be the smallest positive entry of X for which $p_{ij} = 1$ holds. Then $X - \lambda_1 P_1$ is a matrix with nonnegative entries and equal row and column sums. We repeat this process as long as there is no $(k+1) \times (n-k)$ submatrix of 0 elements. Thus we get

$$X = \lambda_1 P_1 + \lambda_2 P_2 + \cdots + \lambda_r P_r + R \tag{2.13}$$

with positive values $\lambda_1, \lambda_2, \ldots, \lambda_r$ and permutation matrices $P_1, P_2, \ldots, P_r$. The matrix R has nonnegative entries, equal row and column sums $\alpha = 1 - \lambda_1 - \lambda_2 - \cdots - \lambda_r (\geq 0)$, and contains a $(k+1) \times (n-k)$ submatrix of 0 elements. Thus, due to Lemma 2.19, $R = 0$. Therefore, $\lambda_1 + \lambda_2 + \cdots + \lambda_r = 1$ and matrix X is a convex combination of the permutation matrices $P_1, P_2, \ldots, P_r$. □

The following example shows the decomposition of a doubly stochastic matrix as explained in the proof to Birkhoff's theorem.

Example 2.20. Consider the doubly stochastic matrix

$$X = \begin{pmatrix} 1/2 & 1/3 & 1/6 \\ 1/3 & 1/3 & 1/3 \\ 1/6 & 1/3 & 1/2 \end{pmatrix}.$$

We choose as first permutation matrix

$$P_1 = \begin{pmatrix} 1 & 0 & 0 \\ 0 & 1 & 0 \\ 0 & 0 & 1 \end{pmatrix}.$$

The corresponding value λ_1 becomes $\lambda_1 = \min(1/2, 1/3, 1/2) = 1/3$. Thus we get

$$X - \lambda_1 P_1 = \begin{pmatrix} 1/6 & 1/3 & 1/6 \\ 1/3 & 0 & 1/3 \\ 1/6 & 1/3 & 1/6 \end{pmatrix}.$$

Next we choose

$$P_2 = \begin{pmatrix} 1 & 0 & 0 \\ 0 & 0 & 1 \\ 0 & 1 & 0 \end{pmatrix}$$

and get for the corresponding value $\lambda_2 = \min(1/6, 1/3, 1/3) = 1/6$. Now we get

$$X - \lambda_1 P_1 - \lambda_2 P_2 = \begin{pmatrix} 0 & 1/3 & 1/6 \\ 1/3 & 0 & 1/6 \\ 1/6 & 1/6 & 1/6 \end{pmatrix}$$

which can be decomposed into

$$1/6 \cdot \begin{pmatrix} 0 & 1 & 0 \\ 1 & 0 & 0 \\ 0 & 0 & 1 \end{pmatrix} + 1/6 \cdot \begin{pmatrix} 0 & 1 & 0 \\ 0 & 0 & 1 \\ 1 & 0 & 0 \end{pmatrix} + 1/6 \cdot \begin{pmatrix} 0 & 0 & 1 \\ 1 & 0 & 0 \\ 0 & 1 & 0 \end{pmatrix},$$

so the doubly stochastic matrix X can be decomposed in a convex combination of five permutation matrices. ■

The *dimension* $\dim P$ *of a polytope* $P = \{x \in \mathbb{R}^d : Ax = b, x \geq 0\}$ is the dimension of the smallest affine subspace of $\mathbb{R}^d$ containing P. Geometrically, any single equation of the system $Ax = b$ corresponds to a hyperplane in $\mathbb{R}^d$ of dimension $d-1$. Therefore, the intersection of k linearly independent hyperplanes in $\mathbb{R}^d$ is an affine subspace of dimension $d-k$ and we get

$$\dim P = d - \text{rank}(A).$$

The assignment polytope

$$P_A = \left\{ x \in \mathbb{R}^{n^2} : \sum_{j=1}^{n} x_{ij} = 1 \text{ for } 1 \le i \le n;\ \sum_{i=1}^{n} x_{ij} = 1 \text{ for } 1 \le j \le n;\ x \ge 0 \right\}$$

has the coefficient matrix

$$A = \left(\begin{array}{ccc|ccc|c|ccc}
1 & 1 \cdots & 1 & & & & & & & \\
 & & & 1 & 1 \cdots & 1 & & & & \\
 & & & & & & \cdots & & & \\
 & & & & & & & 1 & 1 \cdots & 1 \\
\hline
1 & & & 1 & & & & 1 & & \\
 & 1 & & & 1 & & & & 1 & \\
 & & \cdots & & & \cdots & \cdots & & & \cdots \\
 & & 1 & & & 1 & & & & 1
\end{array} \right). \tag{2.14}$$

Matrix A can be viewed as the vertex-edge incidence matrix of the complete bipartite graph $K_{n,n} = (U, V; E)$ with $|U| = |V| = n$ and $E = [i, j]$ for all $i \in U,\ j \in V$. The first n rows correspond to the vertices of U; the last n rows correspond to the vertices in V. Every column corresponds to an edge. The sum of the first n rows of A equals the sum of the last n rows. So, $\text{rank}(A) < 2n$. On the other hand, it is easy to see that the first $2n - 1$ rows are linearly independent: (i) delete the last row of A; (ii) the first n columns, together with columns $kn + 1$ $(k = 1, 2, \ldots, n - 1)$, form a system of $2n - 1$ linearly independent column vectors.

Therefore, $\text{rank}(A) = 2n - 1$ and $\dim P = n^2 - (2n - 1) = (n - 1)^2$. Due to a famous result of Carathéodory, any $x \in P$ can be written as a convex combination of at most $\dim P + 1$ vertices of P. Thus we get the following.

Proposition 2.21. *The assignment polytope has the dimension $(n - 1)^2$. Any doubly stochastic matrix can be written as a convex combination of at most $n^2 - 2n + 2$ permutation matrices.*

The following proposition relies on the fact that the matrix matroid generated by A and the graphic matroid corresponding to $K_{n,n}$ are isomorphic. (The interested reader is referred to Welsh [662].)

Proposition 2.22. *Every subset of linearly independent columns of matrix A corresponds to a subgraph of $K_{n,n}$ which does not contain any cycle.*

Proof. We show that the columns are linearly dependent if the corresponding subgraph G contains a cycle and vice versa.

1. Assume that G contains a cycle. Since G is bipartite, the cycle has an even number of edges which we can color red and blue in an alternating way. Let $a_{[ij]}$ denote the column

of matrix A which corresponds to edge $[i, j]$. Now we get

$$\sum_{[i,j]\text{ blue}} a_{[ij]} = \sum_{[i,j]\text{ red}} a_{[ij]} \tag{2.15}$$

since every vertex of the cycle coincides with a red and a blue edge. This shows that the corresponding column vectors are linearly dependent.

2. Let $\{a_{[ij]} : [i, j] \in D\}$ be a set of linearly dependent columns of matrix A. Then there are coefficients $\alpha_{[ij]}$, not all equal to 0, such that

$$\sum_{[i,j]\in D} \alpha_{[ij]} a_{[ij]} = 0.$$

Let $D' = \{[i, j] : \alpha_{[ij]} \neq 0\}$. Then every vertex which coincides with an edge in D' must also coincide with another edge in D'. Thus starting from an arbitrary edge in D' one can form an infinite sequence of edges where the endpoint of one edge is the starting point of the next edge. Since $|D'| < \infty$, this sequence of edges must contain a cycle. □

A subgraph T of a graph G is called a *spanning tree* if it is *connected* (i.e., there is a path between any two vertices) and has no cycle. Since a spanning tree of a graph with r vertices has $r - 1$ edges, a spanning tree T of $K_{n,n}$ has $2n - 1$ edges. According to Proposition 2.22, there is a one-to-one correspondence between the spanning trees of $K_{n,n}$ and the submatrices of A with $2n - 1$ linearly independent columns. This one-to-one correspondence plays a fundamental role when we solve linear assignment problems by linear programming techniques, as can be seen in detail in Section 4.5.2.

Given an $m \times n$ matrix A and the polytope $P = \{x \in \mathbb{R}^n : Ax = b,\ x \geq 0\}$, let us consider a linear program of the form

$$\min\{c'x : Ax = b,\ x \geq 0\}. \tag{2.16}$$

We assume that matrix A has full rank m, i.e., the rows of matrix A are linearly independent. A subset B of m indices corresponding to m linearly independent columns of A is called a *basis*. The columns of B can be joined to form the *basic matrix* A_B. Similarly, the components x_j of the vector x such that column j is in B form the vector x_B. Obviously, a basic matrix is a regular $m \times m$ matrix. Therefore, the equation system $A_B x_B = b$ has a unique solution x_B. The basis B is called *feasible* if $x_B \geq 0$. A solution x_B and $x_j = 0$ for $j \notin B$ is called a *basic solution*. If the basis B is feasible, the corresponding *basic solution* is also called *feasible* and corresponds to a vertex of the polytope P. Different bases may correspond to the same vertex of P. In this case we call the basic solution *degenerated.*

The main theorem of linear programming says that if a linear program attains a finite optimum solution, then there exists a vertex of the underlying polyhedral set $\{x : Ax = b,\ x \geq 0\}$ in which the optimum is attained. Thus Birkhoff's theorem has an important consequence for linear sum assignment problems. It enables us to relax the constraints

$$x_{ij} \in \{0, 1\} \text{ for } i, j = 1, 2, \ldots, n \tag{2.17}$$

to

$$0 \leq x_{ij} \leq 1 \text{ for } i, j = 1, 2, \ldots, n$$

since every feasible basic solution of the linear program

$$\min \quad \sum_{i=1}^{n}\sum_{j=1}^{n} c_{ij}x_{ij} \tag{2.18}$$

$$\text{s.t.} \quad \sum_{j=1}^{n} x_{ij} = 1 \qquad (i = 1, 2, \dots, n), \tag{2.19}$$

$$\sum_{i=1}^{n} x_{ij} = 1 \qquad (j = 1, 2, \dots, n), \tag{2.20}$$

$$0 \le x_{ij} \le 1 \qquad (i, j = 1, 2, \dots, n) \tag{2.21}$$

corresponds to a vertex of the assignment polytope, i.e., it is a permutation matrix which fullfills the integrality constraints (2.17). Due to this property, linear sum assignment problems can be solved via linear programming techniques. The next proposition provides a graph theoretic illustration of basic solutions of linear assignment problems.

Proposition 2.23. *There is a one-to-one correspondence between the basic solutions of a linear assignment problem* (2.18)–(2.21) *and the spanning trees of the complete bipartite graph $K_{n,n}$. Every feasible basis corresponds to a spanning tree which contains a perfect matching of $K_{n,n}$.*

Proof. The first part is an immediate consequence of Proposition 2.22 and the remarks following its proof. Thus every basis B corresponds uniquely to a spanning tree T_B and vice versa. In a basic solution only variables of B can have a positive value. Since a feasible basic solution corresponds to a vertex of the assignment polytope, we have $x_{[ij]} = 1$ if and only if the edges $[i, j]$ form a perfect matching in $K_{n,n}$. Therefore, the perfect matching defined by $x_{[ij]} = 1$ is a subset of the edges of T_B. □

The last two propositions are closely related to another proof that every basic solution of a linear assignment problem is integer valued. We say that a regular $n \times n$ matrix A is *unimodular* if $\det(A) = \pm 1$. An $m \times n$ matrix A is called *totally unimodular* if every regular $k \times k$ submatrix has a determinant 1 or -1. If the right-hand side vector b of an equation system $A_B x = b$ is integer valued and the regular matrix A_B is unimodular, then Cramer's rule implies immediately that the solution x of this equation system is also integer valued. We are going to show that the coefficient matrix (2.14) of a linear assignment problem is totally unimodular. Therefore, every basic matrix is unimodular and every basic solution of the assignment problem is integer valued. This also yields another proof of Birkhoff's theorem.

Proposition 2.24. *The coefficient matrix (2.14) of an assignment problem is totally unimodular.*

Proof. Certainly, any regular 1×1 submatrix of A has determinant 1. We prove this proposition by induction and assume that it holds for all matrices of size $k \times k$ or smaller. Let D be any $(k+1) \times (k+1)$ submatrix of matrix A. If matrix D contains a 0-column, then D is singular, i.e., $\det D = 0$. Now assume that every column contains two different 1 entries, implying that the first of these 1 entries comes from the upper part of A (first n rows) and the second comes from the last n rows. Then the sum of the rows coming from the upper part equals the sum of the remaining rows. This shows that the row vectors are linearly dependent and again $\det D = 0$. So, let us assume that one column of C contains only one 1 entry, say, a_{ij}. Let D_{ij} be the minor formed by deleting row i and column j from matrix D. Then $\det D = \pm a_{ij} \det(D_{ij}) = \pm \det(D_{ij})$. But $\det(D_{ij})$ is either 0 or ± 1 according to the induction. Therefore $\det D \in \{0, \pm 1\}$. □

An alternative proof of Proposition 2.24 is given in Section 4.1.1.

The next propositions, due to Balinski and Russakoff [65], describe the structure of the assignment polytope in greater detail. We shall answer the question how many basic solutions correspond to one vertex of the assignment polytope, and we will describe the neighborhood structure of the assignment polytope. First, we need a classical result by Cayley [175] on the number of different spanning trees in a complete graph K_n.

Theorem 2.25. (Cayley [175], 1889.) *The complete graph K_n has n^{n-2} different spanning trees.*

Proof. (This elegant proof was given by Pitman [550] 110 years after the first proof by Cayley.) Instead of trees we consider *labeled rooted trees* on n vertices: a labeled rooted tree is a tree with one distinguished vertex as the root. The arcs are oriented such that all paths in the tree lead to the root. Every arc has a label from $\{1, 2, \ldots, n-1\}$ and no two arcs have the same label. There are $(n-1)!\, n$ possibilities to transform a tree in a labeled rooted tree as there are n choices for the root and $(n-1)!$ choices to distribute the labels on the arcs. If we denote by T_n the total number of different spanning trees in K_n, we get as the total number of different labeled spanning rooted trees

$$(n-1)!\; n\; T_n. \tag{2.22}$$

There is another way to count all labeled rooted trees: we start with n isolated vertices and have $n(n-1)$ choices to draw an arc with label 1. After having inserted k arcs, we have the following situation (see Figure 2.7): We have $r = n-k$ connected components where every component has just one root. When we insert the next arc with label $k+1$, this arc must start in a vertex with outdegree 0 of one of the components (i.e., a root) and can end in any vertex of a different component. Let $n_1, n_2, \ldots, n_r$ be the number of vertices in the current connected components. Therefore, we have

$$(n-n_1) + (n-n_2) + \cdots + (n-n_r) = rn - (n_1 + n_2 + \cdots + n_r) = (n-k)n - n$$

possibilities to draw an arc with label $k+1$. This yields

$$\prod_{k=0}^{n-2} n(n-k-1) = n^{n-1}(n-1)! \tag{2.23}$$

labeled rooted trees in total. Comparing (2.22) with (2.23) we get

$$T_n = n^{n-2}. \quad \square$$

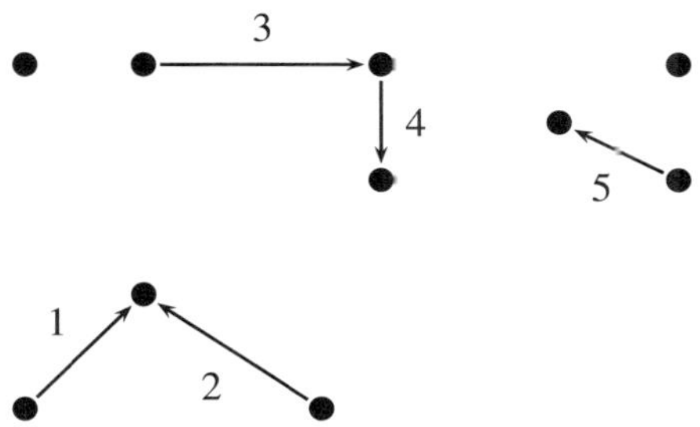

Figure 2.7. *Construction of a labeled rooted tree after the insertion of 5 arcs. As $n = 10$, we have $10 - 5 = 5$ connected components.*

With the help of Cayley's theorem on the number of spanning trees in a complete graph, Balinski and Russakoff [65] computed the number of different bases which correspond to the same vertex of the assignment polytope as follows.

Proposition 2.26. *Each vertex of the assignment polytope corresponds to $2^{n-1}n^{n-2}$ different feasible bases.*

Proof. Let X be a vertex of the assignment polytope P_A, and let M_X be the corresponding perfect matching in the complete bipartite graph $K_{n,n}$. Every feasible basis associated with vertex X corresponds to a spanning tree in $K_{n,n}$ which contains all matched edges of M_X. Now we contract every matched edge to a vertex and obtain by this operation the complete graph K_n. A basis corresponds now to a spanning tree in K_n. Due to Cayley's theorem, the complete graph K_n contains n^{n-2} spanning trees. Every edge e of such a spanning tree connects two vertices of K_n, i.e., two matching edges of M_X. This can be done in two different ways. Namely, if the edge e connects the matching edges $[i, j]$ and $[k, l]$, then either $e = [i, l]$ or $e = [k, j]$. Thus there are $2^{n-1}n^{n-2}$ different spanning trees containing the perfect matching M_X in the complete bipartite graph $K_{n,n}$. □

Proposition 2.26 shows that the basic solutions of the linear assignment problem are highly degenerated. This leads to some difficulties when applying a primal simplex method to linear assignment problems, as can be seen in Section 4.5.2.

In [65], Balinski and Russakoff investigated the polyhedral structure of the assignment polytope. In particular they identified the adjacency of vertices on the polytope and the diameter of P_A. In the following we give an account on their main results.

Two distinct vertices X and Y of the assignment polytope are said to be *adjacent* (or to be *neighbors*) if there are bases B_X and B_Y (called *adjacent bases*) corresponding to X and Y, respectively, which differ in exactly one column. An *edge* of the polytope connects two adjacent vertices. Let M_X and M_Y denote the perfect matchings corresponding to X and Y, respectively. Then we get the following.

Proposition 2.27. *Two distinct vertices X and Y of the assignment polytope are adjacent if and only if the edges of $M_X \cup M_Y$ contain only one cycle.*

Proof. For every basis B_X of vertex X, let E_X be the edge set of the corresponding spanning tree. Obviously, $M_X \subset E_X$.

1. Let X and Y be two adjacent vertices on the assignment polytope. Since $X \neq Y$, the subgraph of $K_{n,n}$ consisting of all edges of M_X and M_Y contains at least one cycle. Let us assume that it contains more than one cycle. Then the subgraph formed by the edges of E_X and E_Y for any bases B_X of vertex X and B_Y of vertex Y also contains more than one cycle. But this contradicts the definition of adjacent vertices.

2. Conversely, let the subgraph G of $K_{n,n}$ consisting of all edges of M_X and M_Y contain just one cycle C and possibly isolated edges $[i, j] \in I$. Choose a subset F of further edges from $K_{n,n}$ which connect the isolated edges and the cycle C so that the graph consisting of the edges $C \cup I \cup F$ is connected and still has the only cycle C. Let $[r, s] \in C \cap M_X$. Then there exists an adjacent edge $[s, t]$ in the cycle which belongs to M_Y. The set $E_X = C \cup I \cup F \setminus \{[s, t]\}$ corresponds to a feasible basis of vertex X. On the other hand, the set $E_Y = C \cup I \cup F \setminus \{[r, s]\}$ corresponds to a feasible basis of vertex Y. The bases B_X and B_Y differ in only one element. Therefore, the vertices X and Y are adjacent. □

Next we enumerate the neighbors of a vertex.

Proposition 2.28. *Every vertex of the assignment polytope P_A has*

$$\sum_{k=0}^{n-2} \binom{n}{k} (n - k - 1)! \tag{2.24}$$

neighbors.

Proof. The assignment polytope is vertex symmetric, i.e., every vertex has the same number of neighbors, since by renaming the right-hand vertices of the bipartite graph $K_{n,n}$ any perfect matching can be mapped to the perfect matching $M = \{[i, i'] : i = 1, 2, \ldots, n\}$. Due to Proposition 2.27 we have to count the number of perfect matchings which have $0, 1, 2, \ldots, n - 2$ edges in common with M and form just one cycle with the edges of M. The number of perfect matchings which have no edge in common with M equals $(n - 1)!$ (the number of different cycles with n vertices). Similarly, there are $\binom{n}{1}(n - 2)!$ perfect matchings which have one edge in common with M and form just one cycle with the edges of M. In general, there are $\binom{n}{k}(n-k-1)!$ perfect matchings which have k edges in common with M and form just one cycle with M. Adding up these numbers, we get (2.24). □

Adjacent vertices on a polytope are joined by an edge of the polytope. Let $d(X, Y)$ be the smallest number of edges on the polytope needed to reach vertex Y from vertex X. The *diameter* diam(P) of a polytope P is defined as

$$\text{diam}(P) = \max_{X,Y} d(X, Y). \tag{2.25}$$

Obviously, P_A has a diameter of 1 for $n = 2$ and $n = 3$ since in these cases any two different vertices of P_A are neighbors.

Proposition 2.29. *For $n \geq 4$, the diameter of the assignment polytope is 2.*

Proof. Let us consider two different vertices X and Y of an assignment polytope with $n \geq 4$ which are not adjacent. We have to show that there exists a vertex Z which is a neighbor of X and a neighbor of Y. Recall that the perfect matchings corresponding to X and Y are denoted by M_X and M_Y, respectively. We distinguish two cases.

Case 1. $M_X \cap M_Y = \emptyset$, i.e., all components of the bipartite graph with edge set $M_X \cup M_Y$ are cycles C_k, $k = 1, 2, \dots, p$. Since X and Y are not adjacent, we get at least two different cycles according to Proposition 2.27. So, $p \geq 2$. We remove an arbitrary edge $e_k \in M_Y$ from each cycle C_k. Thus we get a subgraph of $K_{n,n}$ which consists of single paths P_k, $k = 1, 2, \dots, p$. By adding p appropriate edges f_k from $K_{n,n}$ these single paths can be joined together to form one large cycle C which contains all $2n$ vertices. Note that the edges e_k and f_k also form one cycle C_1 in which edges e_k and edges f_k alternate. Now, by deleting an arbitrary edge f_k in cycle C, we get a spanning tree which contains all edges of M_X. Thus this spanning tree defines a basis for vertex X. If we remove an arbitrary edge $e \in M_X$ from cycle C, we get another spanning tree which contains a perfect matching. This perfect matching consists of the edges f_k ($k = 1, 2, \dots, p$) and the edges of M_Y which still belong to cycle C. We denote the corresponding vertex of the assignment polytope by Z. Obviously, Z is adjacent to X since the corresponding two spanning trees differ in only one edge. But Z is also adjacent to Y since $M_Z \cup M_Y$ contains one single cycle, namely, C_1.

Case 2. $M_X \cap M_Y \neq \emptyset$, i.e., the bipartite graph with edge set $M_X \cup M_Y$ contains at least two cycles and single edges. In this case we fix the assignments of the single edges and apply the procedure of Case 1 to the remaining vertices. This again yields a vertex Z which is adjacent to both X and Y. □

Example 2.30. Let vertex X of the assignment polytope correspond to the permutation $\varphi_1 = (2, 1, 4, 3)$, and let vertex Y correspond to the permutation $\varphi_2 = (1, 2, 3, 4)$. The two vertices X and Y are not adjacent on the assignment polytope since the set $M_X \cup M_Y$ decomposes in two cycles. According to the construction in the proof of Proposition 2.29, we delete the matching edges $[2, 2']$ and $[3, 3']$ of M_Y and introduce two new edges $[2, 3']$ and $[3, 2']$. This leads to the cycle C defined in the proof of Proposition 2.29. By removing edge $[1, 2'] \in M_X$ we obtain the new vertex Z which corresponds to the permutation $\varphi_3 = (1, 3, 2, 4)$. It can be seen that on the assignment polytope vertex Z is adjacent both to vertex X and to vertex Y. ■

Balinski and Russakoff [65] showed, moreover, that any pair of feasible bases of the linear assignment problem is connected by a path of at most $2n - 1$ neighboring feasible bases. This proves that the Hirsch conjecture holds for the linear assignment problem.

Hirsch conjecture. *Given any two bases B_1 and B_2 of a linear program, there exists a sequence of* rank(B_1) *adjacent feasible bases leading from B_1 to B_2.*

Naddef [508] proved that the Hirsch conjecture is true for any 0-1 polytope, while Klee and Walkup [420] showed that it is false for unbounded polyhedra. For general bounded polyhedra this conjecture is still open. For a survey on the Hirsch conjecture see Klee and Kleinschmidt [419].

A polytope is called *Hamiltonian* if there exists a path along the edges of the polytope which visits all vertices exactly once and returns to the original starting point. Balinski and Russakoff [65] show by an explicit construction of such a Hamiltonian cycle that the assignment polytope is Hamiltonian. In a series of papers, Brualdi and Gibson [116, 117, 118, 119, 120] derived further results on the assignment polytope.

Chapter 3

Bipartite matching algorithms

3.1 Bipartite matching

In the introductory chapters it turned out that finding a maximum cardinality matching in a bipartite graph plays a crucial role for assignment problems. Therefore, we discuss in this chapter various efficient methods for finding maximum matchings in bipartite graphs. Let $G = (U, V; E)$ be a bipartite graph with vertex sets $U = \{1, 2, \ldots, n\}$ and $V = \{1, 2, \ldots, s\}$ and assume $n \le s$. The classical method uses simple labeling strategies for finding augmenting paths which finally lead to a maximum matching. This basic method has time complexity $O(|U||E|)$. In the next section we discuss the labeling method and give some hints on how this algorithm can be sped up in practice by some simple heuristics. Hopcroft and Karp [376] improved on the time complexity by constructing augmenting paths in parallel. They proved that a maximum cardinality matching in G can be found with $O(|E|\sqrt{|U|})$ operations. We discuss the approach by Hopcroft and Karp in Section 3.3. In 1991, Alt, Blum, Mehlhorn, and Paul [27] further improved on the complexity of maximum cardinality algorithms by using bit-operations. We discuss their approach in Section 3.4.

The maximum cardinality bipartite matching problem becomes particularly simple if the underlying bipartite graph is convex. A graph G is called *convex* if the existence of edges $[i, j]$ and $[k, j]$ with $i < k$ implies that G also contains all edges $[h, j]$ with $i < h < k$. In this case Glover [316] stated a simple algorithm which finds a maximum cardinality matching in $O(n + s)$ time. Since many problems in practice lead to convex bipartite graphs, this case is of special interest, see Section 3.5.

In 1947 Tutte [643] pointed out a close relationship between perfect matchings and determinants of certain skew symmetric matrices. In Section 3.6 we investigate the relation between maximum matchings and the rank of certain matrices. This leads to fast (probabilistic) algorithms for determining the cardinality of a maximum matching and for finding a minimum vertex cover. We close this chapter by dealing with applications of maximum bipartite matchings, namely, the fleet assignment problem in the airline industry (Section 3.8.1) and the time-slot assignment problem arising in telecommunication via satellites (Section 3.8.2).

Unless otherwise specified, throughout this chapter we will assume, without loss of generality, that $n = |U| \leq |V|$. We will denote $|E|$ by m.

3.2 A labeling method for finding a maximum cardinality matching

Let a bipartite graph $G = (U, V; E)$ and a matching M in this graph be given (M might even be empty). Remember that an edge of E is called a *matching edge* if it belongs to M, and a *non-matching edge* otherwise. A path whose edges are alternately matching and non-matching is called an *alternating path.*

Definition 3.1. (Augmenting path.) *An alternating path $i_1, j_1, i_2, j_2, \dots, i_k, j_k$ with $i_1, i_2, \dots, i_k \in U$ and $j_1, j_2, \dots, j_k \in V$ is called an* augmenting path *P for the matching M if the vertices i_1 and j_k are unmatched, i.e., no matching edge meets these vertices.*

Since the first and the last vertex in an augmenting path are unmatched, an augmenting path starts and ends with an edge not in M. Every augmenting path $i_1, j_1, i_2, j_2, \dots, i_k, j_k$ has an odd length, namely, it contains k non-matching edges and $k - 1$ matching edges. A single non-matching edge whose endpoints are not matched is an augmenting path of length 1. The name *augmenting path* stems from the augmentation operation described in Definition 3.2 below.

In the following we frequently use operations between matchings and augmenting paths. For the sake of simplicity, we will use the symbol P to denote both the ordered sequence of the vertices met by the augmenting path (as in $P = (i_1, j_1, i_2, j_2, \dots, i_k, j_k)$; see Definition 3.1) and the corresponding edge set, i.e., $P = \{[i_1, j_1], [j_1, i_2], [i_2, j_2], \dots, [i_k, j_k]\}$.

Let us also recall that the *symmetric difference* of two sets A and B is defined as

$$A \ominus B = (A \setminus B) \cup (B \setminus A).$$

Definition 3.2. (Augmentation.) *Let P be an augmenting path with respect to the matching M. Then the* matching augmented by P *is obtained by the following two rules:*

1. *the non-matching and matching edges in P change their role: all previously non-matching edges of $M \cap P$ now become matching and all previously matching edges of $M \cap P$ become non-matching;*
2. *all matching edges of M which do not lie on the path P remain matching edges.*

By observing that augmenting a matching M by a path P consists of nothing more than taking the symmetric difference of the two involved edge sets M and P, we denote this operation as $M \ominus P$.

Figure 3.1(a) shows a matching M. Figure 3.1(b) shows the augmented matching $M \ominus P$ produced by path $P = (b, b', c, c', d, d')$.

Lemma 3.3. (Augmentation lemma, Berge [85], 1957.) *If M is not a maximum matching in G, then there exists an augmenting path P with respect to M, and $M' = M \ominus P$ is a matching in G with $|M'| = |M| + 1$.*

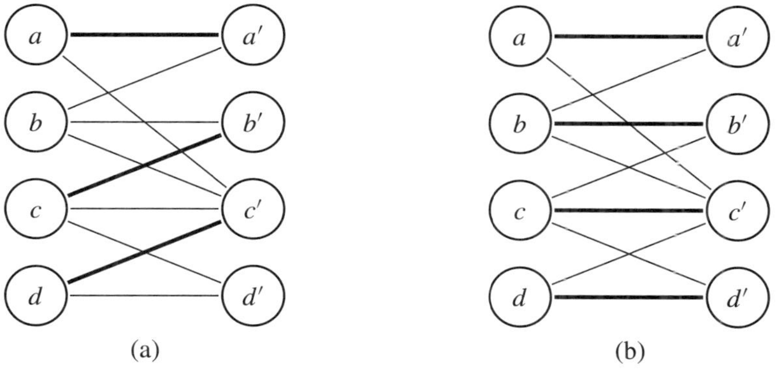

Figure 3.1. (a) *Matching M;* (b) *augmented matching $M \ominus P$, for path $P = (b, b', c, c', d, d')$.*

Proof. Let M be the given matching and let $\overline{M}$ be any maximum cardinality matching. If M is not maximum, then the symmetric difference of M and $\overline{M}$ is not empty and cannot consist only of cycles or paths of even length, since $\overline{M}$ has more edges than M. Therefore, there must exist a path P in the symmetric difference with odd length which starts and ends with an edge of $\overline{M}$. This is an augmenting path with respect to the given matching M. Due to the definition of an augmenting path, it is straightforward that $M' = M \ominus P$ is again a matching. Since $k - 1$ matching edges of M are exchanged against k edges not in M, the new matching M' has size $|M| + 1$. □

Remark: This lemma provides an alternative proof of Theorem 2.14.

Corollary 3.4. *A matching M has a maximum cardinality if and only if there is no augmenting path with respect to M.*

Due to Lemma 3.3 and Corollary 3.4, we can start with an arbitrary matching M and augment this matching step-by-step by means of augmenting paths until a maximum cardinality matching is reached. This can be done by labeling the vertices of the graph G in a proper way. Let L (L stands for **L**eft side) contain the unmatched vertices $i \in U$. Labeled vertices on the **R**ight side are collected in the set R. Initially, $R := \emptyset$. We start from an $i \in L$ on the left side, label all the unlabeled vertices j on the right side such that $[i, j] \in E$ by i, and add them to R. If a labeled vertex j on the right-hand side is unmatched, we have found an augmenting path and we get a new matching $M \ominus P$ with one more edge. Otherwise we remove a vertex j from R: j is matched by an edge $[\bar{\imath}, j]$. Next, vertex $\bar{\imath}$ is labeled by j and $\bar{\imath}$ is added to L. Now we try to continue the augmenting path from the newly labeled vertex $\bar{\imath}$. In this way either we find an augmenting path or we conclude that no such path exists. In the latter case the matching already has maximum cardinality.

Augmenting paths are used extensively in the algorithms for the linear sum assignment problem discussed in Chapter 4. In particular, Algorithm 3.1 is closely related to Procedure Alternate(k) (Algorithm 4.2).

Algorithm 3.1. Cardinality_matching.
$O(nm)$ implementation of a labeling algorithm for finding a maximum cardinality matching.

```
let M be a matching in graph G = (U, V; E) (possibly M = ∅);
let L contain all unmatched vertices of U;
R := ∅;
while L ∪ R ≠ ∅ do
    choose a vertex x from L ∪ R;
    if x ∈ L then Scan_leftvertex(x) else Scan_rightvertex(x)
endwhile

Procedure Scan_leftvertex(x)
L := L \ {x};
while there exists an edge [x, j] with j unlabeled do
    label j as l(j) := x;
    R := R ∪ {j}
endwhile

Procedure Scan_rightvertex(x)
R := R \ {x};
if there is a matching edge [i, x] then
    label i as r(i) := x;
    L := L ∪ {i}
else [comment: augmentation of the matching]
    starting from x, find the alternating path P by backtracking the labels:
    P := (..., r(l(x)), l(x), x);
    M := M ⊖ P;
    let L contain all unmatched vertices of U;
    R := ∅;
    cancel all labels
endif
```

At every augmentation, one additional vertex of U is matched. Therefore, there are at most n augmentations. Moreover, every vertex is labeled at most once per augmentation. Therefore, finding an augmenting path requires at most $O(m)$ steps, and this algorithm finds a maximum cardinality matching in $O(nm)$ time.

Example 3.5. Consider the bipartite graph given in Figure 3.2. We start with the empty matching $M = \emptyset$. $L = \{a, b, c, d, e\}$ consists of all vertices of U. We choose $a \in L$; its neighbor a' is unlabeled. Therefore, $l(a') = a$. Now we choose $a' \in R$ and we get an augmenting path $P = (a, a')$. Thus the edge $[a, a']$ becomes matched and we continue in the same way by matching the vertices b, c, and d with b', c', and d', respectively. Now, $L = \{e\}$ and we label $l(b') = e$. Thus $R = \{b'\}$ and b is labeled by b': $r(b) = b'$, $L = \{b\}$. So we continue with vertex b and label $l(a') = b$, $l(d') = b$. We now get $R = \{a', d'\}$. If we choose a' and then d', we get $r(a) = a'$, $r(d) = d'$ and $L = \{a, d\}$. We continue with $a \in L$, but cannot find a non-matching edge starting in a. So next, d is chosen and we label $l(c') = d$, $l(e') = d$. Thus $R = \{c', e'\}$. If we now select

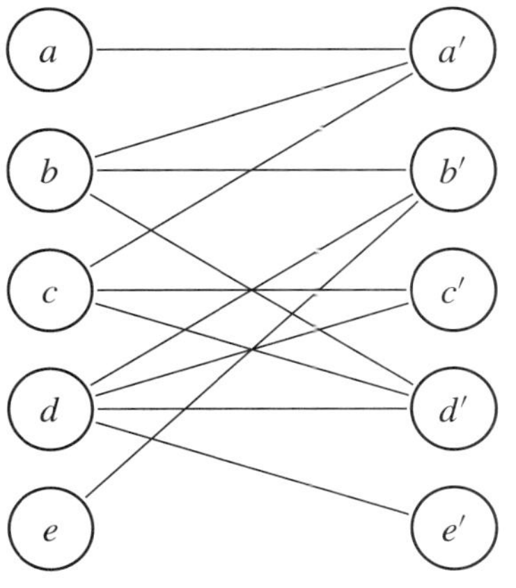

Figure 3.2. *Bipartite graph for Examples 3.5, 3.13, and 3.14.*

e', we have found an augmenting path P which can be retrieved by following the labels starting with e'. We get $P = (e, b', b, d', d, e')$. The augmentation leads to the matching $M = \{[a, a'], [b, d'], [c, c'], [d, e'], [e, b']\}$. Since now $L = \emptyset$, we are done. M is a maximum matching. ∎

As was pointed out above, one can start the algorithm with the empty matching M. In practice, however, it is often quite simple to find a matching of large cardinality. This can accelerate the performance of the algorithm considerably, since in most cases only a few augmentation steps will be necessary to find an optimal solution. A straightforward way to create the first matching M could be to scan the vertices of U one by one and match an edge $[i, j]$ if $j \in V$ is not already matched. A better method is the following *greedy procedure*. Scan the vertices in increasing order of their degrees. Suppose that $i \in U$ is the next unmatched vertex. Choose as matching edge starting from i an edge $[i, j]$ where vertex j is unmatched and has minimum degree. Similarly, starting from a vertex $j \in V$, choose a matching edge $[i, j]$ such that i is an unmatched vertex with minimum degree. After having constructed a first matching in the way described above, the labeling algorithm finds a maximum cardinality matching after growing only a few augmenting paths.

If we use this greedy approach in the example above, we immediately get a maximum matching by scanning, in sequence, the vertices a, e, e' (with degree 1), c' (with degree 2), and b (with degree 3). After having constructed a first matching in the way described above, the labeling algorithm finds a maximum cardinality matching after growing only a few augmenting paths.

Note that the labeling algorithm described above also provides a minimum vertex cover (see Definition 2.6) in graph G. The vertex cover is given by the unlabeled vertices of U and the labeled vertices of V. This is trivial for Example 3.5, for which the minimum vertex cover consists of all vertices of U. Consider instead the bipartite graph of Figure 2.3. At the last iteration, the only unlabeled vertex of U would be a, and the only labeled vertex of V would be b': $\{a, b'\}$ is a minimum vertex cover.

The labeling algorithm discussed above can also be interpreted in terms of network flows. We embed G in a network $\mathcal{N} = (N, A, q)$ with *node set* N, *arc set* A, and *arc capacities* q. The node set N consists of a *source* s, a *sink* t, and the vertices of $U \cup V$. The source is connected to every node in U by an arc of capacity 1, every node in V is connected to the sink by an arc of capacity 1, and every edge in E is directed from U to V and supplied with infinite capacity (see, e.g., the graph of Figure 2.4). As we know, a maximum cardinality matching in G corresponds to a maximum flow in network $\mathcal{N}$. The labeling also leads to a minimum cut $(X, \overline{X})$ in the network corresponding to the graph $G = (U, V; E)$. Those vertices which are labeled in the last round of the algorithm form, together with the source s, the set X. The set $\overline{X}$ contains the unlabeled vertices and the sink (see Proposition 2.12). By considering again the graph of Figure 2.4, we have: $X = \{b, c, d, b', s\}$ and $\overline{X} = \{a, a', c', d', t\}$.

Let f be an arbitrary flow in $\mathcal{N}$. We define the incremental network $\mathcal{N}_f$ with respect to flow f as follows.

Definition 3.6. (Incremental network with respect to flow f.) *The incremental network with respect to flow f in the network $\mathcal{N}$ is the network $\mathcal{N}_f = (N, A_f, q_f)$. It has two kinds of arcs:*

- forward arcs (i, j) *if* $f(i, j) < q(i, j)$. *Their capacity is defined as* $q_f(i, j) = q(i, j) - f(i, j)$;
- backward arcs (j, i) *if* $f(i, j) > 0$. *The capacity of a backward arc is defined as* $q_f(j, i) = f(i, j)$.

Figure 3.3 shows a feasible flow (bold arcs), the corresponding incremental network, and those arcs that are relevant for finding a maximum flow.

A directed path in $\mathcal{N}_f$ from source s to sink t is again called an *augmenting path*. Ford-Fulkerson's max-flow min-cut Theorem 2.11 implies that if f is a maximum flow, then every path from the source to the sink contains an arc (i, j) with $f(i, j) = q(i, j)$. This means that if f is a maximum flow in $\mathcal{N}$, then the incremental network $\mathcal{N}_f$ does not contain a directed path from s to t, i.e., there is no augmenting path in $\mathcal{N}_f$.

A flow Δf in the incremental network is called an *incremental flow*. Let f be a flow in $\mathcal{N}$ and let Δf be a flow in the incremental network $\mathcal{N}_f$. Then the augmented flow $f \ominus \Delta f$ is defined by

$$f \ominus \Delta f(i, j) = \begin{cases} f(i, j) + \Delta f(i, j) & \text{if } (i, j) \text{ is a forward arc,} \\ f(i, j) - \Delta f(j, i) & \text{if } (j, i) \text{ is a backward arc.} \end{cases} \tag{3.1}$$

Now we get the following.

Lemma 3.7. *Let f be a flow in network N.*

1. *If Δf is a flow in $\mathcal{N}_f$, then $f \ominus \Delta f$ is a flow in $\mathcal{N}$.*
2. *For any two flows f and g in $\mathcal{N}$ whose flow values $z(f)$ and $z(g)$ fulfill $z(g) > z(f)$, there exists an incremental flow Δf in $\mathcal{N}_f$ such that $g = f \ominus \Delta f$.*

(a)

(b)

(c)

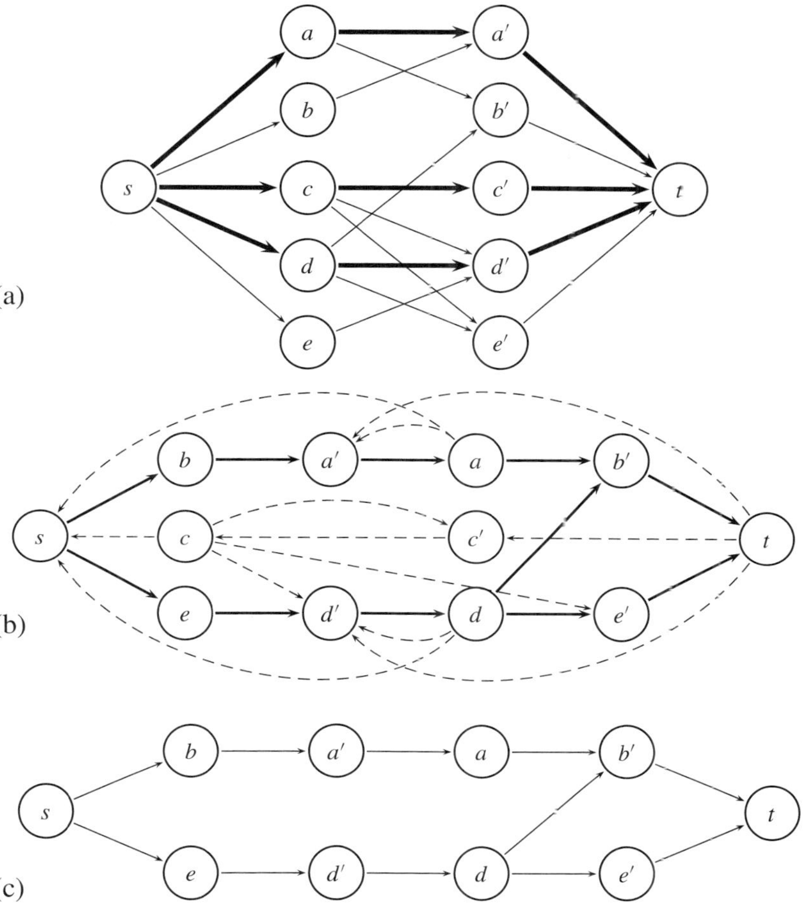

Figure 3.3. *The upper figure* (a) *shows a flow in a network stemming from a maximum matching problem. Figure* (b) *shows the complete incremental network including dashed arcs that never occur in an augmenting path. Figure* (c) *shows only those arcs that are relevant for finding a maximum flow.*

Proof.

1. Due to the definition of capacities of arcs in the incremental network and to (3.1), the flow $f \ominus \Delta f$ fulfills the capacity constraints and the flow nonnegativity in network $\mathcal{N}$. Both the flow f and the incremental flow Δf fulfill the flow conservation constraints in all nodes of $U \cup V$. Therefore, an easy calculation shows that $f \ominus \Delta f$ also fulfills the flow conservation constraints in network $\mathcal{N}$.

2. Define the incremental flow Δf by

$$\Delta f(i,j) = \begin{cases} g(i,j) - f(i,j) & \text{if } g(i,j) \geq f(i,j), \\ f(j,i) - g(j,i) & \text{if } g(j,i) < f(j,i). \end{cases} \tag{3.2}$$

It is easy to show that Δf is an incremental flow and fulfills $f \ominus \Delta f = g$. □

The elaborations above show that there is a one-to-one correspondence between augmenting paths with respect to a matching M and augmenting paths in the corresponding incremental network $\mathcal{N}_f$. There is only one difference between the augmentation steps $M \ominus P$ and $f \ominus \Delta f$: in the latter the flow f (the matching M) is augmented along several augmenting paths in one step. This idea is used in the next section to show that the labeling method can be sped up by creating several augmenting paths in parallel.

3.3 The Hopcroft–Karp algorithm

The idea of Hopcroft and Karp [376] was to augment a matching not only by one augmenting path, but by a maximal system of vertex-disjoint augmenting paths of the same minimum length. We call an augmenting path P with respect to a matching M a *shortest augmenting path* if the number of edges in P is a minimum. Since the symmetric difference is an associative and commutative operation, we have

$$(M \ominus P_1) \ominus P_2 = (M \ominus P_2) \ominus P_1.$$

If two augmenting paths P_1 and P_2 with respect to M are vertex disjoint, the set $M' = (M \ominus P_1) \ominus P_2$ is again a matching. So we can augment M simultaneously by P_1 and P_2 and get a larger new matching M'. This can easily be generalized to k vertex disjoint paths.

Definition 3.8. *Let $\Delta M = (P_1, P_2, \dots, P_k)$ be a system of k pairwise vertex disjoint augmenting paths with respect to a matching M. Then*

$$M \ominus \Delta M = (\dots((M \ominus P_1) \ominus P_2)\dots) \ominus P_k.$$

Since $\ominus$ is associative and commutative, the definition does not depend on the numbering of the paths. An augmentation is a special case of taking the symmetric difference where the two involved sets have to fulfill additional properties. It will be clear from the context whether we speak of an augmentation or just of taking a symmetric difference. In particular, since $P_1, P_2, \dots, P_k$ are vertex disjoint, we get

$$M \ominus \Delta M = M \ominus (P_1 \cup P_2 \cup \dots \cup P_k).$$

Lemma 3.9. *Let M be a matching in the bipartite graph G and let $\Delta M = (P_1, P_2, \dots, P_k)$ be a system of k vertex disjoint shortest augmenting paths with respect to M. Then any augmenting path P' with respect to $M' = M \ominus \Delta M$ has a length*

$$|P'| \geq |P_1| + 2|\Delta M \cap P'|. \tag{3.3}$$

Proof. Since $P_1, P_2, \dots, P_k$ are shortest augmenting paths, they all have the same length. Let $\overline{M} = M' \ominus P'$. According to the Augmentation Lemma 3.3 we get $|\overline{M}| = |M| + k + 1$. Therefore, the symmetric difference $M \ominus \overline{M}$ contains $k+1$ vertex disjoint augmenting paths

$Q_1, Q_2, \ldots, Q_{k+1}$ with respect to M. Due to the associativity of the symmetric difference we get

$$M \ominus \overline{M} = M \ominus ((M \ominus \Delta M) \ominus P') = (M \ominus M) \ominus (\Delta M \ominus P') = \Delta M \ominus P'.$$

Thus we get

$$|M \ominus \overline{M}| = |\Delta M \ominus P'| \geq |Q_1| + |Q_2| + \cdots + |Q_{k+1}| \geq (k+1)|P_1|,$$

as all paths in ΔM have the same length $|P_1|$. Since $|\Delta M \ominus P'| = |\Delta M| + |P'| - 2|\Delta M \cap P'|$ and $|\Delta M| = k|P_1|$, we get $|P'| \geq |P_1| + 2|\Delta M \cap P'|$. □

Now we consider a *maximal* system of vertex disjoint shortest augmenting paths with respect to the matching M. This means that there does not exist any further vertex disjoint augmenting path P with respect to M of the same minimum length. We do not assume that this system is maximum, i.e., that it consists of a maximum number of vertex disjoint shortest augmenting paths. Then we get the following lemma as consequence of Lemma 3.9.

Lemma 3.10. *Let $\Delta M = (P_1, P_2, \ldots, P_k)$ be a maximal system of vertex disjoint shortest augmenting paths with respect to a matching M and let $M' = M \ominus \Delta M$. If P' is a shortest augmenting path with respect to M', then $|P'| \geq |P_1| + 2$.*

Proof. Let P' be a shortest path with respect to M'. Let us assume that $|P'| = |P_1|$. Lemma 3.9 implies in this case that $\Delta M \cap P' = \emptyset$ and, therefore, ΔM and P' are edge disjoint. Since ΔM is a maximal system of vertex disjoint shortest augmenting paths, the path P' has a vertex v in common with one of the paths of ΔM, say, with path P_1, since otherwise P' would be another shortest augmenting path with respect to M, which contradicts the maximality of ΔM. This common vertex cannot be the first or the last vertex of the path because P' is an augmenting path with respect to $M \ominus P_1$. Due to the structure of the augmenting paths the common vertex is a matched vertex of set V. But then P_1 and P' would also share an edge incident with v, namely, the corresponding edge of $M \ominus P_1$. This contradicts the fact that the two paths are edge disjoint. Therefore, we have $|P'| > |P_1|$. Since every augmenting path has an odd length, we get $|P'| \geq |P_1| + 2$. □

For constructing a maximal system of shortest augmenting paths we introduce the notion of a layered graph. Let $G = (U, V; E)$ be a bipartite graph with a matching M. The *layered graph* LM with respect to the matching M is defined as follows.

The first layer L_0 contains all unmatched vertices of the set U. The second layer L_1 contains all vertices $j \in V$ which are endpoints of a non-matching edge $[i, j]$ in G with $i \in L_0$. The corresponding edges are collected in the edge set E_1. If L_1 contains an unmatched vertex of V, we stop and clean up the layered graph (see below for the meaning of "clean up"). Otherwise every vertex j of L_1 is matched by an edge $[i, j] \in M$. These edges form the edge set E_2. The corresponding vertices i are now the vertices of layer L_2. Thus $|L_2| = |L_1|$. Now, we scan all non-matching edges $[i, j]$ with $i \in L_2$ and $j \notin L_1$ and add them to edge set E_3. The corresponding vertices v form the vertices of layer L_3. We stop this process with the first layer L_h, h odd, which either contains an unmatched vertex $j \in V$ or is empty. In the latter case there does not exist an augmenting path with respect to the given matching M, implying that matching M is maximum. Otherwise we start to clean up the layered graph: in the last layer all matched vertices are deleted together with all incident edges. Thus in layer L_{h-1} there may now be vertices from which no edge leads to a vertex in layer L_h. Again, all such vertices and their incoming edges are deleted. This process is continued until layer L_0. Obviously, the cleaned layered graph can be constructed in $O(m)$ steps since every edge is considered at most twice. Figure 3.4, where we consider the same example as in Figure 3.3, shows a matching and the corresponding layered graph. In this case no cleaning occurs.

Now we orient all edges from layer L_k to layer L_{k+1} ($k = 0, 1, \ldots, h-1$). Note that every path from the first layer L_0 to a vertex of the last layer L_h uses non-matching and matching edges alternately and has a length equal to h. Since h is odd and every $j \in L_h$ is unmatched, every such path is an augmenting path with respect to the given matching M. The length h is called the *depth of the layered graph*.

Next, we determine a maximal system of pairwise vertex disjoint augmenting paths in the layered graph. This can be done by scanning the vertices of the layered graph in a depth first search. We start the first path with an arbitrary edge between the first two layers and continue it from layer to layer, until a vertex j in the last layer L_h is reached. This path is the first augmenting path P_1. For every vertex v_k ($k = 0, 1, \ldots, h$) on path P_1, we delete all edges not belonging to P_1 which are incident with vertex v_k. So, some vertices of the layers $L_1, \ldots, L_h$ may now have no incoming edge and some vertices of the layers $L_0 \ldots, L_{h-1}$ may have no outgoing edge. We delete these vertices and their incident edges. This step is repeated as many times as possible. Then we start again, if possible, from a new vertex in the first layer to obtain another augmenting path. By scanning all edges of the layered graph G just once, a maximal system ΔM of pairwise vertex disjoint augmenting paths can be found.

From the layered graph of Figure 3.4 we first obtain path $P_1 = (b, a', a, b')$. Then we delete edge $[d, b']$ and obtain the second path $P_2 = (e, d', d, e')$. Augmenting the matching by P_1 and P_2 produces the maximum matching.

All steps which lead from M to $M' = M \ominus \Delta M$ are summarized as one iteration. Thus one iteration starts with the construction of the layered network and the determination of ΔM and takes $O(m)$ operations. Whereas Algorithm 3.1 uses $O(n)$ augmenting paths in the worst case, the following theorem shows that we have only $O(\sqrt{n})$ augmentations if we use maximal systems of vertex disjoint shortest augmenting paths.

Theorem 3.11. (Hopcroft–Karp [376], 1973.) *Let $G = (U, V; E)$ be a bipartite graph with m edges. Then a matching of maximum cardinality in G can be found within $O(m\sqrt{n})$ operations.*

Proof. We are going to prove that the number of iterations is bounded by $\sqrt{n}$. Let M^* be a maximum cardinality matching in G. At each iteration the size of the matching increases by at least 1. Thus if $|M^*| \le \sqrt{n}$, there is nothing to show. So let us assume that $|M^*| > \sqrt{n}$. Then there is an iteration k^* with $|M_{k^*}| < |M^*| - \sqrt{n}$ in which the matching M_{k^*} reaches for the first time a size of at least $|M^*| - \sqrt{n}$. After this iteration, there are at most $\sqrt{n}$ additional iterations. If we can show that M_{k^*} has been obtained in $\sqrt{n}$ iterations, then the theorem is proven. We will prove this fact by showing that the layered graph at iteration k^* has a depth less than $2\sqrt{n} + 1$.

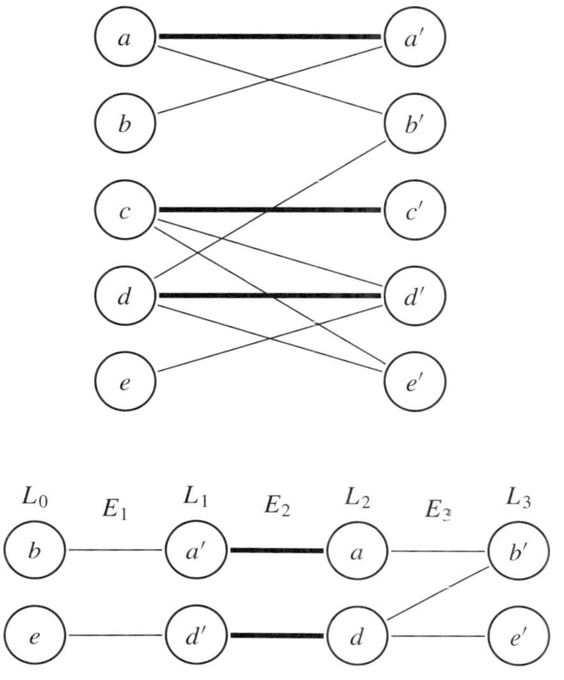

Figure 3.4. *Layered graph.*

According to Theorem 2.14 there exists a maximum cardinality matching M^* which leaves matched all vertices of U which are matched in M_{k^*}. Therefore, the symmetric difference of M_{k^*} and M^* contains more than $\sqrt{n}$ vertex disjoint augmenting paths which start and end in an unmatched vertex with respect to M_{k^*}. Not all of these paths can have a length greater than $2\sqrt{n}+1$ since every path with such length contains more than $\sqrt{n}$ vertices of U. Since there are more than $\sqrt{n}$ paths, this would mean that the augmenting paths contain more than n different vertices of U, which is impossible. Therefore, there exists at least one augmenting path with a length less than $2\sqrt{n}+1$. Since the layered graph with respect to a matching contains only the shortest augmenting paths, the depth of the layered graph with respect to the matching M_{k^*} is less than $2\sqrt{n}+1$, which proves the theorem. □

The same proof technique can be used to show the following proposition (see Gabow and Tarjan [293]). If we replace $\sqrt{n}$ by n/k in the proof of Theorem 3.11, we obtain the following.

Proposition 3.12. *Let $G = (U, V; E)$ be a bipartite graph with m edges. For any integer k, $k \geq 1$, a matching whose size differs from the size of a maximum matching by at most n/k can be computed in $O(km)$ time.*

We can summarize Hopcroft–Karp's method in the following algorithm.

ALGORITHM 3.2. Hopcroft_Karp.
Hopcroft–Karp $O(m\sqrt{n})$ algorithm for finding a maximum cardinality matching.

let $G = (U, V; E)$ be a bipartite graph with initial (possibly empty) matching M;
let U_0 contain all unmatched vertices of U;
let V_0 contain all unmatched vertices of V;
while $U_0 \neq \emptyset$ **do**
    Construct_layered_ graph;
    Find_set_ΔM;
    $M := M \ominus \Delta M$;
    update sets U_0, V_0 of unmatched vertices
endwhile

Procedure Construct_layered_ graph
$L_0 := U_0, k^* := k := 0$;
while $L_k \neq \emptyset$ **do**
    for each $i \in L_k$ **do** $N(i) := \{j : [i, j] \in E \setminus M, j \notin L_1 \cup L_3 \cup \cdots \cup L_{k-1}\}$;
    $L_{k+1} := \bigcup_{i \in L_k} N(i)$;
    if $L_{k+1} = \emptyset$ **then** stop; [**comment:** M is a maximum matching]
    if $L_{k+1} \cap V_0 \neq \emptyset$ **then**
        $k^* := k + 1$;
        $L_{k+2} := \emptyset$
    else
        $L_{k+2} := \{\bar{\imath} : [\bar{\imath}, j] \in M, \; j \in L_{k+1}\}$
    endif;
    $k := k + 2$
endwhile

Procedure Find_set_ΔM
comment: find a maximal set ΔM of disjoint augmenting paths;
delete all vertices in $L_{k^*} \setminus V_0$;
mark all remaining vertices as unscanned;
$k := 1$; [**comment:** path counter]
while $L_0 \neq \emptyset$ **do**
    choose $x_0 := i \in L_0$ and delete it in L_0;
    $\ell := 0$;
    while $\ell \geq 0$ **do**
        while x_ℓ has an unscanned neighbor in $L_{\ell+1}$ **do**
            choose unscanned neighbor $x_{\ell+1}$;
            mark $x_{\ell+1}$ as scanned;
            $\ell := \ell + 1$;
            if $\ell = k^*$ **then**
                $P_k := (x_0, x_1, \ldots, x_{k^*})$;
                $k := k + 1$
            endif

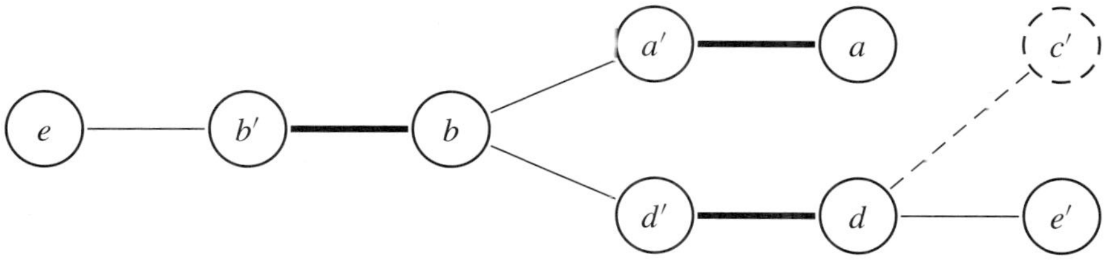

Figure 3.5. *Incremental graph in the second iteration of Example 3.13.*

endwhile;
if $\ell < k^*$ **then** $\ell := \ell - 1$ **else** $\ell = -1$
endwhile
endwhile;
return $\Delta M := (P_1, P_2, \ldots, P_{k-1})$

Example 3.13. We illustrate this algorithm on the graph of Figure 3.2. We start with the empty matching $M = \emptyset$. The first layered graph coincides with the given graph $G = (U, V; E)$ with length $k^* = 1$.

Procedure Find_set_ΔM finds as first path $P_1 = (a, a')$. Vertex a is then deleted in L_0, and vertex a' is scanned. Next we find path $P_2 = (b, b')$, vertex b is deleted, vertex b' is scanned. Similarly we find the paths $P_3 = (c, c')$ and $P_4 = (d, d')$. Finally, vertex e has no unscanned neighbor. Thus ΔM becomes the union of paths P_1, P_2, P_3, P_4, the augmentation step yields, as first nonempty matching, $M = \{[a, a'], [b, b'], [c, c'], [d, d']\}$, and sets U_0 and V_0 become $U_0 = \{e\}$ and $V_0 = \{e'\}$.

The next layered graph starts with $L_0 = \{e\}$ and continues with $L_1 = \{b'\}$, $L_2 = \{b\}$, $L_3 = \{a', d'\}$. The next neighbor of a' is a; the next neighbor of d' is d. Thus $L_4 = \{a, d\}$. Vertex a has no neighbor and vertex d has two neighbors, c' and e' where $e' \in V_0$. Thus $L_5 = \{c', e'\}$ and $k^* = 5$.

Procedure Find_set_ΔM first deletes the matched vertex c' in L_5 (see Figure 3.5) and starts with $x_0 = e$, $x_1 = b'$, $x_2 = b$, $x_3 = a'$, $x_4 = a$ growing a first augmenting path. Since x_4 has no neighbor, we investigate x_3, which also has no neighbor. Thus we come to x_2, whose only unscanned neighbor is vertex d'. We get $x_3 = d'$, $x_4 = d$ and the only remaining neighbor of x_4, namely, vertex $x_5 = e'$. Thus we obtain the single augmenting path $\Delta M = P_1 = (e, b', b, d', d, e')$ of length 5. The new augmentation step leads to the matching $M = \{[a, a'], [b, d'], [c, c'], [d, e'], [e, b']\}$. Since we now have $U_0 = \emptyset$, we are done: M is a maximum matching. ∎

3.4 Improvements by Alt, Blum, Mehlhorn, and Paul

Alt, Blum, Mehlhorn, and Paul [27] improved Hopcroft and Karp's method for the case of "dense" graphs by using a fast adjacency matrix scanning technique due to Cheriyan, Hagerup, and Mehlhorn [183]. Thus they obtained an algorithm which finds a maximum bipartite matching in $O(n^{1.5}\sqrt{m/\log n})$ time, where $n = |U|+|V|$. Throughout this section we number the vertices in U by $\{1, 2, \dots, n_1\}$ and the vertices in V by $\{n_1+1, n_1+2, \dots, n\}$.

The heart of this algorithm is a function $c(j)$, which either returns an arc (i, j) such that vertex i lies in the layer immediately preceding the layer of vertex j in the layered graph or states that no such vertex exists. Since the concept of a layered graph relies on oriented graphs, we orient the non-matching edges $[i, j]$ from left to right, i.e., (i, j), and the matching edges from right to left, i.e., (j, i). In order to get $c(j)$ fast, we store two binary matrices, namely, a $(\lceil\sqrt{n}\,\rceil + 2) \times n$ matrix $L = (l_{kj})$, the *level matrix*, defined by

$$l_{kj} = \begin{cases} 1 & \text{if vertex } j \text{ lies in layer } k, \\ 0 & \text{otherwise} \end{cases} \tag{3.4}$$

and the adjacency matrix A of the directed graph defined above. We have $c(j) = i$ if and only if $a_{ij}l_{(k-1),i} = 1$, where k is the level of vertex j. The trick is to store the two binary matrices L and A in blocks of length $\log n$, as RAM words. In this way function $c(j)$ can be evaluated in $O(n/\log n)$ time by using elementary binary operations.

Alt, Blum, Mehlhorn, and Paul start from the empty matching in $G = (U, V; E)$ with n vertices and m edges. All edges are oriented from U to V. They determine augmenting paths until the layered graph has a length of $\sqrt{m \log n/n}$. With the matching found so far they continue with Hopcroft and Karp's method. Their method is summarized in Algorithm 3.3.

Algorithm 3.3. Revised_cardinality_matching.
Alt, Blum, Mehlhorn, and Paul $O(n^{1.5}\sqrt{m/\log n})$ algorithm for finding a maximum cardinality matching.

```
let a bipartite graph G = (U, V; E) be given;
n := |U| + |V|, m := |E|;
M := ∅; [comment: all vertices are unmatched]
K := 1; [comment: K is the index of the highest level]
for each pair (k, j) with k ∈ {0, 1, 2, 3} and j ∈ U ∪ V do l_kj := 0;
comment: L = (l_kj) is the level matrix;
for each i ∈ U do layer(i) := 0, l_0i := 1;
for each j ∈ V do layer(j) := 1, l_1j := 1;
for each i ∈ U do for each j ∈ V do
    if [i, j] ∈ E then a_ij := 1 else a_ij := 0;
comment: the n × n matrix A is the adjacency matrix of the graph;
while K ≤ √(m log n/n) do
    while ∃ unmatched r ∈ V with layer(r) = K do
        P := (r); [comment: P is the current path]
        while P ≠ ∅ do
            r := last vertex of P;
            if r is unmatched and layer(r) = 0 then [comment: P is augmenting]
                M := M ⊖ P;
                comment: reverse the direction of the arcs;
```

```
                for each arc (i, r) ∈ P do a_ri := 1, a_ir := 0;
                P := ∅
            else
                if c(r) ≠ nil then add i = c(r) as last vertex to path P
                else
                    remove r from P;
                    l_{layer(r),r} := 0;
                    layer(r) := layer(r) + 2;
                    l_{layer(r),r} := 1
                endif
            endif
        endwhile
    endwhile;
    if both U and V contain unmatched vertices then
        K := K + 2;
        comment: provide two new rows for the level matrix;
        for each x ∈ U ∪ V do l_{K+1,x} := l_{K+2,x} := 0;
    else
        stop; [comment: M is a maximum matching]
    endif
endwhile;
comment: find the remaining augmenting paths;
execute Algorithm Hopcroft_Karp

Procedure c(r)
if a_ir · l_{layer(r)-1,i} = 0 for all i ∈ U ∪ V then
    c(r) := nil;
else
    c(r) := i for a vertex i with a_ir · l_{(layer(r)-1),i} = 1;
endif
```

Example 3.14. We illustrate Algorithm 3.3 as before on the graph depicted in Figure 3.2. We start with the empty matching $M = \emptyset$. At the beginning, the level matrix L and the layer array have the form

$$L = \begin{array}{c} \begin{array}{cccccccccc} a & b & c & d & e & a' & b' & c' & d' & e' \end{array} \\ \left(\begin{array}{ccccc|ccccc} 1 & 1 & 1 & 1 & 1 & 0 & 0 & 0 & 0 & 0 \\ 0 & 0 & 0 & 0 & 0 & 1 & 1 & 1 & 1 & 1 \\ 0 & 0 & 0 & 0 & 0 & 0 & 0 & 0 & 0 & 0 \\ 0 & 0 & 0 & 0 & 0 & 0 & 0 & 0 & 0 & 0 \end{array}\right) \end{array},$$

$$layer = (\ 0\ \ 0\ \ 0\ \ 0\ \ 0\ \ 1\ \ 1\ \ 1\ \ 1\ \ 1\ ).$$

The first adjacency matrix is

$$A = \left(\begin{array}{ccccc|ccccc}
0 & 0 & 0 & 0 & 0 & 1 & 0 & 0 & 0 & 0 \\
0 & 0 & 0 & 0 & 0 & 1 & 1 & 0 & 1 & 0 \\
0 & 0 & 0 & 0 & 0 & 1 & 0 & 1 & 1 & 0 \\
0 & 0 & 0 & 0 & 0 & 0 & 1 & 1 & 1 & 1 \\
0 & 0 & 0 & 0 & 0 & 0 & 1 & 0 & 0 & 0 \\
\hline
0 & 0 & 0 & 0 & 0 & 0 & 0 & 0 & 0 & 0 \\
0 & 0 & 0 & 0 & 0 & 0 & 0 & 0 & 0 & 0 \\
0 & 0 & 0 & 0 & 0 & 0 & 0 & 0 & 0 & 0 \\
0 & 0 & 0 & 0 & 0 & 0 & 0 & 0 & 0 & 0 \\
0 & 0 & 0 & 0 & 0 & 0 & 0 & 0 & 0 & 0
\end{array}\right).$$

Moreover, the bound on the length of the layered graph to be considered is equal to $\sqrt{m \log n/n} \approx 1.66$ (by computing log on the natural basis e). This means that this algorithm turns to the Hopcroft–Karp procedure as soon as the layered graph has two or more layers. In order to illustrate the way the algorithm constructs layer by layer, we do not use this bound in this example, but continue until we have reached a maximum matching.

We start with the first unmatched vertex on level 1, namely, with vertex a', and we initially get the path $P = (a')$. Procedure $c(a')$ returns $c(a') = a$. Thus P becomes $P = (a', a)$. Hence, we obtain a first matching edge $[a, a']$ which we orient from a' to a. The adjacency matrix is updated accordingly. In the next steps we get the matching edges $[b, b']$, $[c, c']$, and $[d, d']$. All corresponding arcs are directed from right to left.

The following iteration starts with $r := e'$ and we obtain $P = (e', d)$. Since $c(d) = nil$, the vertex d is deleted from P and its level is increased to 2. Thus we obtain the new level matrix

$$L = \left(\begin{array}{ccccc|ccccc}
1 & 1 & 1 & 0 & 1 & 0 & 0 & 0 & 0 & 0 \\
0 & 0 & 0 & 0 & 0 & 1 & 1 & 1 & 1 & 1 \\
0 & 0 & 0 & 1 & 0 & 0 & 0 & 0 & 0 & 0 \\
0 & 0 & 0 & 0 & 0 & 0 & 0 & 0 & 0 & 0
\end{array}\right).$$

Next vertex e' is investigated. Since $c(e') = nil$, vertex e' is deleted in P and its level is increased to 3.

The new level matrix is

$$L = \left(\begin{array}{ccccc|ccccc}
1 & 1 & 1 & 0 & 1 & 0 & 0 & 0 & 0 & 0 \\
0 & 0 & 0 & 0 & 0 & 1 & 1 & 1 & 1 & 0 \\
0 & 0 & 0 & 1 & 0 & 0 & 0 & 0 & 0 & 0 \\
0 & 0 & 0 & 0 & 0 & 0 & 0 & 0 & 0 & 1
\end{array}\right)$$

and the current adjacency matrix is

$$A = \left(\begin{array}{ccccc|ccccc} 0 & 0 & 0 & 0 & 0 & 0 & 0 & 0 & 0 & 0 \\ 0 & 0 & 0 & 0 & 0 & 1 & 0 & 0 & 1 & 0 \\ 0 & 0 & 0 & 0 & 0 & 1 & 0 & 0 & 1 & 0 \\ 0 & 0 & 0 & 0 & 0 & 0 & 1 & 1 & 0 & 1 \\ 0 & 0 & 0 & 0 & 0 & 0 & 1 & 0 & 0 & 0 \\ \hline 1 & 0 & 0 & 0 & 0 & 0 & 0 & 0 & 0 & 0 \\ 0 & 1 & 0 & 0 & 0 & 0 & 0 & 0 & 0 & 0 \\ 0 & 0 & 1 & 0 & 0 & 0 & 0 & 0 & 0 & 0 \\ 0 & 0 & 0 & 1 & 0 & 0 & 0 & 0 & 0 & 0 \\ 0 & 0 & 0 & 0 & 0 & 0 & 0 & 0 & 0 & 0 \end{array}\right).$$

Since there is no unmatched vertex in layer 1, we increase K to 3 and continue again with e'. We get $P = (e', d, d', b)$. Now vertex b is evaluated, and we get $c(b) = nil$. Therefore, the layer of b is increased to 2 and b is deleted in P. Since $c(d') = c$, we obtain $P = (e', d, d', c)$. The evaluation of vertex c yields again nil. Therefore, c is deleted in P and its level is increased to 2. Now d', d, and e' are consecutively deleted in P, the level of d' is increased to 3, the level of d is increased to 4, and the level of e' is increased to 5.

Thus we obtain the new level matrix

$$L = \left(\begin{array}{ccccc|ccccc} 1 & 0 & 0 & 0 & 1 & 0 & 0 & 0 & 0 & 0 \\ 0 & 0 & 0 & 0 & 0 & 1 & 1 & 1 & 0 & 0 \\ 0 & 1 & 1 & 0 & 0 & 0 & 0 & 0 & 0 & 0 \\ 0 & 0 & 0 & 0 & 0 & 0 & 0 & 0 & 1 & 0 \\ 0 & 0 & 0 & 1 & 0 & 0 & 0 & 0 & 0 & 0 \\ 0 & 0 & 0 & 0 & 0 & 0 & 0 & 0 & 0 & 1 \end{array}\right).$$

We continue with $K = 5$ and $P = (e')$. Successively we get $P = (e', d, d', b, b', e)$. Thus another augmenting path is reached, which after augmentation yields the maximum matching. ∎

The validity of Algorithm 3.3 relies on the following facts.

Lemma 3.15. *During execution of Algorithm 3.3 the following statements remain true.*

1. *$i \in U$ if and only if $layer(i)$ is even.*
2. *Any $P = (v_0, v_1, \dots, v_q)$ is a path in the current layered graph which starts in an unmatched vertex $v_0 \in V$. Moreover, $layer(v_\ell) = K - \ell$ for $\ell = 0, 1, \dots, q$.*
3. *All unmatched vertices of V are in layers K or $K + 2$.*
4. *Every arc (i, j) of the incremental graph fulfills $layer(i) \le layer(j) + 1$.*
5. *The algorithm increases the matching M along a shortest augmenting path.*

Proof. Obviously, all points of Lemma 3.15 hold at the beginning of the algorithm. Since the layers are always increased by 2, the first statement is true. The second statement holds due to the definition of Procedure $c(r)$. When a vertex is relabeled, that is, its layer is changed, it must be on the path P. Thus no unmatched vertex in layer $K+2$ can be relabeled. When K is increased by 2, there is no unmatched vertex in layer K. This settles the third assertion. It is also easy to see that the relabeling strategy keeps the property that every arc (i, j) of the incremental graph fulfills $layer(i) \le layer(j) + 1$. Namely, before the relabeling we have either $layer(i) + 1 = layer(j)$ or $layer(j) + 1 = layer(i)$. In the first case vertex i and in the second case vertex j might be relabeled. In both cases, after the relabeling of the corresponding vertex we get $layer(i) \le layer(j) + 1$. Therefore, the fourth statement is still fulfilled after the relabeling.

When the last vertex of path P reaches an unmatched vertex in layer 0, we get an augmenting path. This is a shortest augmenting path since due to the relabeling procedure any shorter augmenting path would have been found in an earlier iteration. (A vertex is only relabeled if it has no predecessor in the previous layer.) Thus the fifth statement is also true. □

Let us turn to the complexity of Algorithm 3.3. When the while-loop of the algorithm ends with matching M, then all remaining augmenting paths, if any, have a length of at least $\sqrt{m \log n/n} + 1$. The incremental graph with respect to M must contain at least $|M^*| - |M|$ vertex disjoint paths, where M^* denotes a maximum cardinality matching. Therefore, we get $(|M^*| - |M|)\sqrt{m \log n/n} < n$ and $|M| > |M^*| - n/\sqrt{m \log n/n}$. Thus the matching M can be completed by Hopcroft–Karp's algorithm with at most $n^{1.5}/\sqrt{m \log n}$ phases, each of which takes $O(m)$ time. For fixed layer K we call at most $O(n)$ times Procedure $c(r)$ due to the fourth statement of Lemma 3.15. Since Procedure $c(r)$ takes $O(n/\log n)$ time, we get as total complexity

$$O(n^2/\log n \cdot \sqrt{m \log n/n} + n^{1.5} m/\sqrt{m \log n}) = O(n\sqrt{mn/\log n}).$$

Thus we have shown the following.

Theorem 3.16. (Alt, Blum, Mehlhorn, and Paul [27], 1991.) *Let $G = (U, V; E)$ be a bipartite graph with n vertices. Then a matching of maximum cardinality in G can be found within $O(n^{1.5}\sqrt{m/\log n})$ operations.*

A different algorithm, based on the compression of graph G via its partition into bipartite cliques, was proposed by Feder and Motwani [268]. Its time complexity is $O(\sqrt{n}m \log(n^2/m)/\log n)$.

3.5 Matchings in convex bipartite graphs

In the case of convex bipartite graphs the maximum cardinality matching problem can be solved by simple linear time algorithms.

Throughout this section we assume that $G = (U, V; E)$ is a convex bipartite graph with $|U| = n$, $|V| = s$ and $|E| = m$.

Definition 3.17. (Convex bipartite graph.) *The bipartite graph $G = (U, V; E)$ is called* convex *if the vertices of set V can be re-numbered such that the neighbors of every vertex $i \in U$ form an interval, i.e., the neighborhood $N(i)$ of any vertex $i \in U$ has the form $N(i) = \{j, j+1, j+2, \dots, h\} \subseteq V$.*

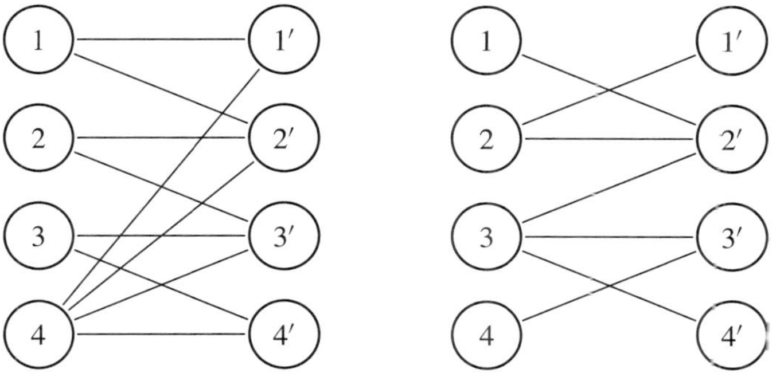

Figure 3.6. *Convex bipartite graph at left and doubly convex bipartite graph at right.*

Graph G is a doubly convex bipartite graph *if G is a convex bipartite graph, and also the vertices of set U can be renumbered such that the neighbors of every vertex $j \in V$ form an interval, i.e., $N(j) = \{k, k+1, k+2, \ldots, l\}$.*

Figure 3.6 shows a convex and a doubly convex bipartite graph. Obviously, the property of being convex depends on a suited numbering of the vertices in V (and also of U in the doubly convex case). Such a numbering can be found in $O(m+n+s)$ time with the help of PQ-trees introduced by Booth and Lueker [108]. For the left graph, there is no way to rearrange the vertices of U such that the graph becomes doubly convex. Thus doubly convex graphs form a strict subset of convex graphs.

3.5.1 Algorithms

Glover [316] developed a simple algorithm for finding a maximum matching in a convex bipartite graph provided the graph is already given such that all sets of neighbors of vertices in U form intervals.

ALGORITHM 3.4. Convex.
Glover's algorithm for finding a maximum cardinality matching in a convex bipartite graph.

let $G = (U, V; E)$ be a bipartite graph where the vertices of $V = \{1, 2, \ldots, s\}$ are such that the neighborhoods of vertices of U form intervals;
$M := \emptyset$;
for each $i \in U$ **do** $\alpha(i) := \max\{j \in V : [i, j] \in E\}$;
for $j := 1$ **to** s **do** [**comment:** $j \in V$]
    if j has an unmatched neighbor **then**
        find an unmatched neighbor $i(j) \in U$ with minimum value $\alpha(i)$;
        add edge $[i(j), j]$ to M
    endif
endfor

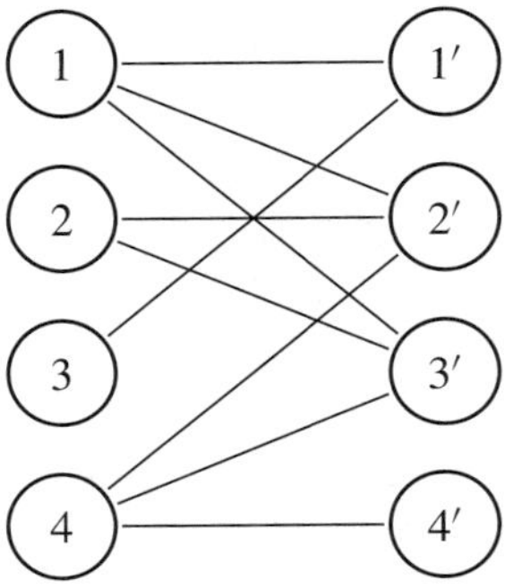

Figure 3.7. *Convex bipartite graph used in Example 3.18.*

The following example illustrates Algorithm 3.4 on the convex graph of Figure 3.7.

Example 3.18. For the vertices in U we get $\alpha(1) = 3,\ \alpha(2) = 3,\ \alpha(3) = 1$, and $\alpha(4) = 4$. Thus the maximum matching becomes $M = ([3, 1'],\ [1, 2'],\ [2, 3'],\ [4, 4'])$. ∎

Proposition 3.19. *Glover's algorithm finds a maximum cardinality matching in a convex graph.*

Proof. We will show that there exists a maximum matching M that contains the first edge, say, $[i, \overline{j}]$ with $\overline{j} = 1$, found by Glover's algorithm. If we delete in G the vertices i and $\overline{j}$ and their incident edges, we get again a convex bipartite graph $\overline{G}$ which contains all edges of $\overline{M} = M \setminus \{[i, \overline{j}]\}$. Thus $\overline{M}$ is a maximum matching in $\overline{G}$. Moreover, the relative order of the values $\alpha(i)$ is kept for the remaining vertices. Thus we can apply the same argument as in the previous step, which shows that the matching found by Glover's algorithm is a maximum cardinality matching.

Let $[i, \overline{j}]$ be the first edge determined by Algorithm 3.4 and assume that the maximum matching M' does not contain this edge. If vertex i is unmatched in M', then there is a matching edge $[k, \overline{j}]$ in M'. By replacing this edge in M' by the edge $[i, \overline{j}]$, we get another maximum matching. A similar transformation can be performed if vertex $\overline{j}$ is unmatched in M'. Now assume that both vertices i and $\overline{j}$ are matched in M', i.e., M' contains the two edges $[i, l]$ and $[k, \overline{j}]$. Since G is convex, and due to the rule that i is chosen as neighbor of vertex $\overline{j}$ with minimum value $\alpha(i)$, we get

$$\overline{j} < l \leq \alpha(i) \leq \alpha(k).$$

Due to the convexity of G, there exists the (non-matching) edge $[k, l]$ in G. Now we exchange in M' the edges $[k, \overline{j}]$ and $[i, l]$ against the edges $[i, \overline{j}]$ and $[k, l]$ and get in this way a maximum cardinality matching which contains the edge $[i, \overline{j}]$. □

Lipski and Preparata [462] describe a detailed implementation for Glover's algorithm and prove that their implementation solves the problem in $O(nA(n)+s)$ time, where $A(n)$ is the extremely slow growing inverse Ackermann function. By using a special data structure, Gabow and Tarjan [296] improved on this complexity and showed that the convex maximum cardinality matching problem can be solved in linear time $O(n+s)$, provided for every $i \in U$ the first and last entries in $N(i)$ are given.

In the case of doubly convex graphs, Glover's algorithm can still be simplified (see Glover [316]).

Proposition 3.20. *At any iteration of Algorithm 3.4, for the current vertex $j \in V$ either the first or the last unmatched neighbor has the minimum $\alpha(i)$ value among the matched neighbors.*

Proof. If a vertex of V has only one or two unmatched neighbors in U, there is nothing to prove. Otherwise, let $j \in V$ be a vertex whose first and last unmatched neighbors in U are i_{first} and i_{last}, respectively, and for which we have

$$i_{\text{first}} < i < i_{\text{last}} \text{ and } \alpha(i) < \min(\alpha(i_{\text{first}}), \alpha(i_{\text{last}})).$$

Let a' be that vertex of V which corresponds to $\min(\alpha(i_{\text{first}}), \alpha(i_{\text{last}}))$. Vertex a' is connected with i_{first} and i_{last} and, therefore, due to the convexity of the neighbors of vertices in V, also with vertex i. This is a contradiction to $\alpha(i) < \min(\alpha(i_{\text{first}}), \alpha(i_{\text{last}}))$. □

Using this property, Lipski and Preparata [462] described a comparatively simple algorithm which solves the maximum matching problem in doubly convex bipartite graphs in linear time $O(n+s)$.

3.5.2 Applications

Convex bipartite graphs arise in several applications. One of them is the following *terminal assignment problem* in the design of layouts for integrated circuits. In the terminal assignment problem, any of n entry terminals positioned in the upper row of a channel are to be assigned to one of s exit terminals positioned in the lower row in such a way that the maximum number of connections (nets) that cross any vertical line is minimum; see Figure 3.8. This maximum number of connections is called the *density* of the assignment. The density is a measure of the width of the channel that is required to route the nets, see, e.g., Preparata and Lipski [559].

A terminal assignment problem is defined by the sequence of entry positions $p_1 < p_2 < \cdots < p_n$ on a horizontal line and the sequence of exit positions $q_1 < q_2 < \cdots < q_s$ with $n \le s$ on another horizontal line. Entry and exit positions can be viewed as vertex sets U and V of a bipartite graph G with edge set $\{[i, j] : i \in U, j \in V\}$. A terminal assignment φ corresponds to a maximum matching in this graph G: every entry position p_i is matched to an exit position $q_j = \varphi(p_i)$. Since G is complete and $n \le s$, there is always a maximum matching of cardinality n. The *local density* d_x of a terminal assignment at position x is defined as the number of nets (matching edges) which cross position x (see Figure 3.8):

$$d_x = |\{i : p_i < x < \varphi(p_i) \text{ or } p_i > x > \varphi(p_i)\}|.$$

The density d of an assignment φ is the maximum local density over all positions x:

$$d = \max_x d_x.$$

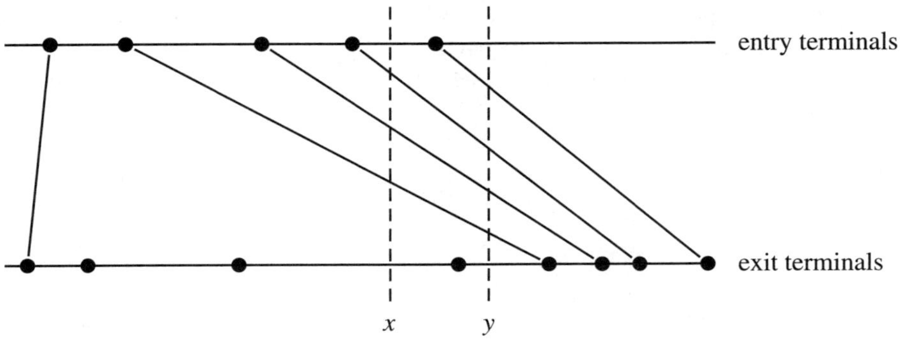

Figure 3.8. *The local density at location x is 3; the local density at location y is 4. The density d of this assignment is $d = 4$.*

It is easy to see that one can assume that two matching edges do not cross in an optimal terminal assignment. Uncrossing does not increase the density. Therefore, we can assume that in an optimal terminal assignment $i < k$ implies $\varphi(p_i) < \varphi(p_k)$. This remark can be used to show the following lemma by Rendl and Rote [590].

Lemma 3.21. *An uncrossed terminal assignment has density not greater than d if and only if for all i the number of entries between $x = p_i$ and $y = \varphi(p_i)$ is less than d:*

$$|\{k : p_i < p_k < \varphi(p_i)\}| < d$$

and

$$|\{k : \varphi(p_i) < p_k < p_i\}| < d.$$

The proof of this simple lemma is left to the reader. Suppose now that we are interested in checking whether a terminal assignment of density d does exist. Lemma 3.21 enables us to say beforehand which exit terminals can possibly be assigned to each entry terminal. The possible exit terminals form an interval (see Figure 3.9). The entry terminals which are connected to a specific exit terminal also form an interval. Thus we get a doubly convex bipartite graph and we have to check whether it is possible to find in this graph a matching of cardinality n. As mentioned in Section 3.5, this can be performed by Lipski and Preparata's algorithm [462] in linear time $O(n+s)$. This result is used by Rendl and Rote [590] to show that the multilayer terminal assignment problem, where the entries may lie in an arbitrary number of different layers, can be solved in linear time.

Motivated by the terminal assignment problem, Atallah and Hambrusch [48] investigated bipartite matchings of minimum density. Let again a sequence of entry positions $p_1 < p_2 < \cdots < p_n$ and a sequence of exit positions $q_1 < q_2 < \cdots < q_s$ be given on two (different) horizontal lines. In addition, let E be a given set of edges which connect entries with exits. We ask for a maximum matching with minimum density which only uses edges of the given edge set E. Atallah and Hambrusch showed by reduction from the monotone 3-SAT problem that this problem is $\mathcal{NP}$-hard even if the degree of every entry vertex is

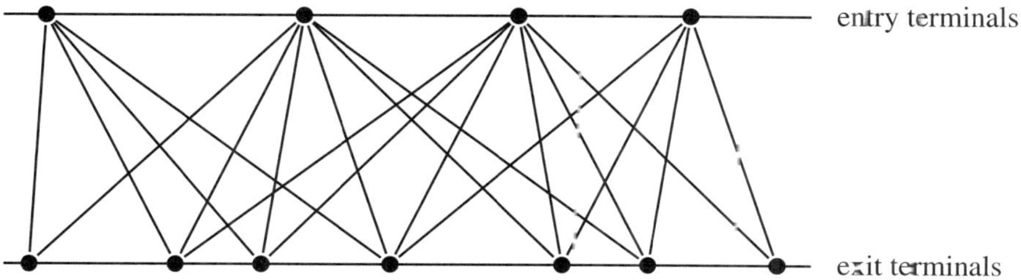

Figure 3.9. *This graph shows entry vertices on the upper line and exit vertices on the lower line. The edges are chosen according to Lemma 3.21 for testing whether a terminal assignment with density $d = 2$ does exist.*

2, i.e., every entry vertex is connected to two exit vertices. In the case that the underlying bipartite graph is complete, Atallah and Hambrusch designed an efficient algorithm which finds the minimum density in linear time $O(n + s)$.

Malucelli [477] considered a problem of *scheduling synchronization in transit networks*. Given n transportation lines, we want to define the departure time of each line i, within a given time window $[s_i, e_i]$, so that the average waiting time of the passengers at the transit points is minimized. The problem, which has a quadratic objective function (hence is discussed in Section 9.4.2), can be modeled on a convex bipartite graph $G = (U, V; E)$ where U contains one vertex for each line, V contains one vertex for each line departure time, and E has an edge $[i, j]$ for each line i and each feasible starting time $j \in [s_i, e_i]$.

3.6 Maximum matchings and matrix algorithms

There are close connections between maximum matchings and the rank of matrices. Tutte [643] pointed out that the existence of a perfect matching in a graph G can be checked by evaluating the determinant of a certain skew symmetric matrix, the so-called Tutte matrix of G. We exploit here an idea of Edmonds [248] which leads not only to fast probabilistic algorithms for finding maximum matchings in bipartite graphs, but also enables us to find a minimum vertex cover in a fast way.

Let $G = (U, V; E)$ be a bipartite graph with $|U| = n$, $|V| = s$ and $|E| = m$. Following an idea by Edmonds [248], we associate with G an $(n \times s)$ matrix $A(G) = (a_{ij})$ with indeterminate entries x_{ij} if $[i, j] \in E$ and $a_{ij} = 0$ if $[i, j] \notin E$. Obviously, matrix $A(G)$ is a generalized adjacency matrix and there exists a matching of cardinality c if we can rearrange c rows and columns in A such that each diagonal element of the corresponding $c \times c$ submatrix is an indeterminate entry x_{ij}. In other words, the maximum rank of matrix A over the field $F[x_{i_1 j_1}, \dots, x_{i_m j_m}]$ equals the cardinality of a maximum matching (Edmonds [248]).

Ibarra, Moran, and Hui [385] showed that simultaneously finding a maximal independent set of rows and a maximal independent set of columns of a real $(n \times s)$ matrix A (with $n \le s$) can be performed by $O(n^{\beta-1}s)$ arithmetic operations on the matrix elements. Here, $O(n^\beta)$ is the complexity for multiplying two $n \times n$ matrices. Coppersmith and Winograd [197] showed that matrix multiplication is possible with $\beta = 2.376$.

Based on these observations, Ibarra and Moran [384] showed the following theorem.

Theorem 3.22. *In each bipartite graph $G = (U, V; E)$ with $|U| = n$ and $|V| = s$, the cardinality of a maximum matching and the matched vertices can be found by $O(n^{\beta-1}s)$ arithmetic operations, where $O(n^\beta)$ is the time complexity for multiplying two $n \times n$ matrices.*

Several comments are due. First, since the algorithm of Ibarra, Moran, and Hui [385] cannot handle indeterminates as matrix entries, Ibarra and Moran [384] proposed replacing the indeterminates by large integers (see below). Thus the arithmetic operations involve large integers. Therefore, the bitwise complexity for determining the size of a maximum matching using exact arithmetic becomes $\tilde{O}(n^{\beta-1}s)$, where $\tilde{O}(f(\cdot))$ is defined as $O(f(\cdot)\log^k(f(\cdot)))$ for some constant k. It follows that $O(n^{\beta-1}s\log^k(n^{\beta-1}s))$ is asymptotically the best complexity currently known for finding the size of a maximum matching in a bipartite graph. Second, the algorithm of Ibarra and Moran does not provide a maximum matching, as it only finds a regular submatrix of A of maximum size. Thus we can determine the size of a maximum matching and find the matched vertices.

As mentioned above, the results of [385] cannot be applied directly to the matrix $A(G)$, whose entries are 0's and indeterminates. Therefore, Ibarra and Moran proposed replacing the indeterminates $x_{i_1j_1}, \ldots, x_{i_mj_m}$ by numbers $c_1, c_2, \ldots, c_m$ with $c_m \geq 2$ and $c_i \geq c_{i+1}^2$ for $i = 1, 2, \ldots, m-1$.

Example 3.23. The matrix

$$A = \begin{pmatrix} x_1 & x_2 & 0 \\ 0 & 0 & x_3 \\ x_4 & 0 & x_5 \end{pmatrix}$$

can be replaced by the matrix

$$A = \begin{pmatrix} 422500 & 625 & 0 \\ 0 & 0 & 25 \\ 4 & 0 & 2 \end{pmatrix} \quad \blacksquare$$

The validity of the method is based on a lemma by Ibarra and Moran [384].

Lemma 3.24. *If the $n \times n$ matrix $A(G)$ has rank n, then the matrix A, where all indeterminates $x_{i_kj_k}$ of $A(G)$ are replaced by numbers c_k fulfilling $c_m \geq 2$ and $c_l \geq c_{l+1}^2$ for $l = 1, 2, \ldots, m-1$, also has rank n.*

Based on Theorem 3.22, Ibarra and Moran [384] developed a probabilistic algorithm for finding the cardinality of a maximum matching in $G = (U, V; E)$ with $|U| = |V| = n$. When the algorithm performs K iterations, it takes $O(Kn^{2.376}\log^3 n)$ time and finds the cardinality of a maximum matching with probability at least $(1 - 1/2^K)$.

A simpler approach can be used for testing whether a given bipartite graph has a perfect matching. The following probabilistic algorithm by Lovász [466] tests whether a given bipartite graph $G = (U, V; E)$, with $|U| = |V| = n$ and $|E| = m$, contains a perfect matching and provides the correct answer with probability at least $(1 - 1/m^K)$.

ALGORITHM 3.5. Test_perfect_matching.
Probabilistic algorithm for testing whether a bipartite graph has a perfect matching.

let $G = (U, V; E)$ be a bipartite graph with edge set $E = \{e_1, e_2, \dots, e_m\}$;
$K :=$ maximum number of iterations;
for $k := 1$ **to** K **do**
  randomly generate values of x_e, $e \in E$, from the set $\{1, 2, \dots, m^2\}$;
  for each $e = [i, j] \in E$ **do** $a_{ij} := x_e$;
  for each $[i, j] \notin E$ **do** $a_{ij} := 0$;
  if det $A(G) \neq 0$ **then return** `yes` [**comment:** G contains a perfect matching]
endfor;
comment: G contains with probability $(1 - 1/m^K)$ no perfect matching;
return `no`

Due to Coppersmith and Winograd [197] the determinant of an $n \times n$ matrix can be computed in $O(n^{2.376})$ time. The algorithm errs if it yields the value 0 for the determinant but det $A(G)$ is not identically equal to 0. This happens if the generated random numbers hit by chance a root of the polynomial det $A(G)$. According to a result by Schwartz [604] this happens with probability equal to $1/|E|$. Thus the algorithm errs after K repetitions with probability $(1/|E|)^K$.

Note that this probabilistic algorithm does not provide a perfect matching either. It only asserts with high probability the existence of a perfect matching in the given bipartite graph G. A similar parallel algorithm, which provides the matching, too, is due to Mulmuley, Vazirani, and Vazirani [501]. Their algorithm requires $O(n^{3.5} m)$ processors to find a perfect matching in $O(\log^2 n)$ time.

3.7 Perfect matchings in bipartite random graphs

The probability that a sparse bipartite graph $G = (U, V; E)$ with $|U| = |V| = n$ contains a perfect matching is very high provided it contains no isolated vertices, i.e., vertices which are not incident with any edge. In this section we specify two models for bipartite random graphs and prove results on the existence of a perfect matching. These results will be used later for the construction of fast algorithms for linear assignment problems which rely on thinning out the edge set of G.

For bounding probabilities we make use of *Stirling's formula*

$$\sqrt{2\pi n}\left(\frac{n}{e}\right)^n \leq n! \leq e^{\frac{1}{12n}}\sqrt{2\pi n}\left(\frac{n}{e}\right)^n, \tag{3.5}$$

which implies

$$n! \geq 2\sqrt{n}\left(\frac{n}{e}\right)^n \geq \left(\frac{n}{e}\right)^n. \tag{3.6}$$

Therefore, we get for any integers k and n with $1 \leq k \leq n$

$$\binom{n}{k} \leq \frac{n(n-1)\cdots(n-k+1)}{k!} \leq \frac{n^k}{k!} \leq \left(\frac{en}{k}\right)^k. \tag{3.7}$$

In the first model we consider vertex sets U and V with $|U| = |V| = n$. Any of the possible n^2 edges exists with probability p. For example, if $p = 1/2$, we may toss a coin for each of the n^2 possible edges and insert the edge in G whenever we see "head." Instead of specifying the probability for an edge, we may alternately insert N edges randomly. Then we get a bipartite random graph with edge probability $p = N/n^2$. Now we can prove the following result, which goes back to Bollobás and Thomason [105]. It is stated in a form which allows an immediate application to random bottleneck assignment problems (Pferschy [543]).

Theorem 3.25. *Let $G = (U, V; E)$ with $|U| = |V| = n$ be a bipartite random graph without isolated vertices and at least $n \log n$ edges. Then G contains a perfect matching with probability greater than $1 - O(1/\sqrt{n \log n})$.*

Proof. Let G be a bipartite random graph without isolated vertices (which can easily be attained by inserting at least one edge for every vertex) and let $p = \log n/n$ be the probability for an edge. We assume that G has no perfect matching. According to Hall's theorem, Theorem 2.1, there exists a subset U' of U with $|U'| > |N(U')|$ and, by symmetry, there also exists a subset V' of V with $|V'| > |N(V')|$. Let A be a subset of U or V of smallest cardinality which violates Hall's condition (2.2). It is easy to see that the subgraph of G induced by the vertices of A and $N(A)$ is connected since otherwise A could be replaced by a proper subset. Moreover, $|N(A)| = |A| - 1$ due to the minimum cardinality of set A (as otherwise we could delete vertices of A). As a connected set contains a spanning tree, the induced bipartite graph $(A, N(A); E_A)$ has at least $2|A| - 2$ edges. Now we show

$$|A| \le \lfloor (n+1)/2 \rfloor. \tag{3.8}$$

Namely, if $A \subset U$, then $B = V \setminus N(A)$ violates Hall's condition (2.2) as $N(B) \subset U \setminus A$. Hence,

$$|A| \le |B| = |V| - |N(A)| = n - |A| + 1,$$

which shows (3.8).

Now let F_k denote the event that there is a subset A of U or V in G such that $k = |A| \le (n+1)/2$, $|A| = |N(A)| + 1$ and the subgraph induced by A and $N(A)$ is connected. We have to show that

$$\mathbb{P}\left(\bigcup_{k=2}^{n_1} F_k\right) \le O(1/\sqrt{n \log n}), \tag{3.9}$$

where $\mathbb{P}(e)$ denotes the probability of event e and $n_1 = \lfloor (n+1)/2 \rfloor$. The probability that there exist subsets A_1 of U and A_2 of V in G with $k = |A_1| = |A_2| + 1$ such that the induced subgraph $(A_1, A_2; E_{A_1})$ has at least $2k - 2$ edges and no vertex of A_1 is joined to a vertex of $V \setminus A_2$ is at most

$$\binom{k(k-1)}{2k-2} p^{2k-2} (1-p)^{k(n-k+1)}. \tag{3.10}$$

Since we have $2\binom{n}{k}$ choices for set A_1 with $|A_1| = k$ and $\binom{n}{k-1}$ more choices for A_2, we get

$$\sum_{k=2}^{n_1} \mathbb{P}(F_k) \le 2 \sum_{k=2}^{n_1} \binom{n}{k} \binom{n}{k-1} \binom{k(k-1)}{2k-2} p^{2k-2} (1-p)^{k(n-k+1)}. \tag{3.11}$$

Using (3.7) and the fact that $p = \log n/n$ we get

$$\sum_{k=2}^{n_1} \mathbb{P}(F_k) \le 2\sum_{k=2}^{n_1} \left(\frac{en}{k}\right)^k \left(\frac{en}{k-1}\right)^{k-1} \left(\frac{ek}{2}\right)^{2k-2} \left(\frac{\log n}{n}\right)^{2k-2} \left(1-\frac{\log n}{n}\right)^{k(n-k+1)}. \tag{3.12}$$

Next we bound the last factor as follows:

$$k(n-k+1) = n\left(k - \frac{k^2}{n} + \frac{k}{n}\right)$$

and, as

$$\left(1+\frac{a}{n}\right)^n \le e^a \text{ for every } a, \tag{3.13}$$

we have

$$\left(1-\frac{\log n}{n}\right)^n \le e^{-\log n} = \frac{1}{n}.$$

Thus

$$\left(1-\frac{\log n}{n}\right)^{k(n-k+1)} \le \left(\frac{1}{n}\right)^{k-\frac{k^2}{n}+\frac{k}{n}} \le n^{\frac{k^2}{n}-k}. \tag{3.14}$$

Summarizing, we get

$$\sum_{k=2}^{n_1} \mathbb{P}(F_k) \le \sum_{k=2}^{n_1} \frac{e^{4k-3}k^{k-2}(\log n)^{2k-2}n^{1-k+\frac{k^2}{n}}}{(k-1)^{k-1}2^{2k-3}}. \tag{3.15}$$

Since, for $k \ge 2$,

$$\frac{8}{e^2}\left(\frac{e}{4}\right)^k \le 1$$

and

$$\frac{k^{k-2}}{(k-1)^{k-1}} = \frac{1}{k}\left(1+\frac{1}{k-1}\right)^{k-1} < \frac{e}{k} < e,$$

we get

$$\sum_{k=2}^{n_1} \mathbb{P}(F_k) \le \sum_{k=2}^{n_1} (e\log n)^{3k} n^{1-k+\frac{k^2}{n}}. \tag{3.16}$$

As

$$\lim_{n\to\infty} n^{1/n} = 1,$$

by straightforward manipulations of

$$a_k = (e\log n)^{3k} n^{1-k+\frac{k^2}{n}}$$

it can be shown that there exist constants C_1 and C_2 such that

$$a_2 < \frac{C_1}{\sqrt{n}\log n}$$

and

$$a_k < \frac{C_2}{n\sqrt{n \log n}} \quad \text{for } k = 3, \ldots, n_1,$$

which concludes the proof. □

Another possibility to generate bipartite random graphs is the following. For every vertex of the sets U and V, we randomly insert d ($d \geq 1$) *directed* arcs. In this way the random graph will have no isolated vertex. Since $|U| = |V| = n$, a *perfect matching* in this directed bipartite graph is a subset of n arcs such that every vertex has either indegree 1 and outdegree 0 or outdegree 1 and indegree 0. Note that it does not make sense to randomly generate an undirected regular bipartite graph (i.e., an undirected graph in which every vertex has the same degree d) since every such graph contains a perfect matching (see Corollary 2.3). Walkup [657] proved the following.

Theorem 3.26. *Let $G = (U, V; E)$ with $|U| = |V| = n$ be a bipartite random graph generated as above with outdegree d at each node. For $d = 1$ the probability $P(n, 1)$ that G contains a perfect matching goes to 0 as n tends to infinity, whereas for $d \geq 2$ the probability $P(n, d)$ that G contains a perfect matching goes to 1 as n approaches ∞. In particular, we have the following.*

1. $P(n, 1) \leq 3\sqrt{n}\left(\frac{2}{e}\right)^n$.

2. $1 - P(n, 2) \leq \frac{1}{5n}$.

3. $1 - P(n, d) \leq \frac{1}{122}\left(\frac{d}{n}\right)^{(d+1)(d-2)}$ *for* $d \geq 3$.

Proof. We will fully prove statements 1 and 3, while for statement 2 we will only prove a weaker bound.

Case $d = 1$. There are $n!$ perfect matchings and, for each of them, there are 2^n ways of assigning directions to the edges. Since $d = 1$, the probability of a single arc is $1/n$. Thus the probability $P(n, 1)$ of the existence of a perfect matching is bounded by

$$2^n n! \left(\frac{1}{n}\right)^n.$$

Using (3.5) we get

$$P(n, 1) \leq e^{\frac{1}{12n}} \sqrt{2\pi n} \left(\frac{2}{e}\right)^n \leq 3\sqrt{n}\left(\frac{2}{e}\right)^n.$$

This shows that $\lim_{n\to\infty} P(n, 1) = 0$.

Case $d \geq 2$. Let $\beta_n^d(k)$ be the probability that, in the undirected bipartite graph $G = (U, V; E)$ where every vertex has the outdegree d, a set A of k vertices in U has at most

$k-1$ neighbors in V. According to Hall's theorem the probability that G has no perfect matching is bounded by

$$\sum_{k=1}^{n} \beta_n^d(k).$$

There are $\binom{n}{k}$ possibilities for subsets A of U with k vertices and $\binom{n}{k-1}$ possibilities for subsets $N(A)$ of V with $k-1$ vertices. For any vertex of a fixed set A,

$$\frac{\binom{k-1}{d}}{\binom{n}{d}}$$

is the probability that all arcs leaving this vertex terminate in a given set B with $k-1$ vertices. Similarly, for every vertex not in B,

$$\frac{\binom{n-k}{d}}{\binom{n}{d}}$$

is the probability that all arcs leaving this vertex do not point to the set A. Thus we get for $\beta_n^d(k)$

$$\beta_n^d(k) = \binom{n}{k}\binom{n}{k-1}\left(\frac{\binom{k-1}{d}}{\binom{n}{d}}\right)^k \left(\frac{\binom{n-k}{d}}{\binom{n}{d}}\right)^{n-k+1}. \tag{3.17}$$

Note that, due to the construction of the random graphs, $\beta_n^d(k) = 0$ for $k \le d$ and $k \ge n-d+1$. Moreover,

$$\beta_n^d(k) = \beta_n^d(n-k+1).$$

For $d = 2$, in particular, (3.17) reads

$$\binom{n}{k}^2 \frac{k}{n-k+1}\left(\frac{\binom{k-1}{2}^k \binom{n-k}{2}^{n-k+1}}{\binom{n}{2}^{n+1}}\right). \tag{3.18}$$

Since for $d = 2$ and $1 \le n \le 4$ we have $\beta_n^2(k) = 0$, we may assume that $n \ge 5$. Since for $0 < x < 1$ we have $e^x \le 1/(1-x)$, we get

$$e^{\frac{1}{12n}} \le e^{\frac{1}{60}} \le \frac{60}{59}.$$

Thus Stirling's formula (3.5) applied to the numerator and the denominator of $\binom{n}{k}^2 = \left(\frac{n!}{k!(n-k)!}\right)^2$ yields

$$\beta_n^2(k) \le \frac{1}{2\pi}\left(\frac{60}{59}\right)^2 \left(\frac{k-1}{k}\right)^k \left(\frac{k-2}{k}\right)^k \left(\frac{n-k-1}{n-k}\right)^{n-k} \left(\frac{n}{n-1}\right)^{n-1}$$
$$\times \left(\frac{(n-k-1)n}{(n-k+1)(n-1)^2}\right).$$

Four applications of (3.13) yield for $d = 2$

$$\beta_n^2(k) \leq \frac{1}{122n}. \tag{3.19}$$

We have thus proved that, for $d = 2$, we have $1 - P(n, 2) \leq \frac{1}{122}$. For the sharper bound the reader is referred to Walkup [657].

Now let $d \geq 3$. In the following computations we assume $2 \leq s < d < r \leq n$. From

$$n(r - s) \leq r(n - s)$$

we get immediately

$$n^{d-2}(r-2)\cdots(r-d+1) \leq r^{d-2}(n-2)\cdots(n-d+1).$$

This yields

$$\frac{\binom{r}{d}}{\binom{n}{d}} \leq \left(\frac{r}{n}\right)^{d-2} \frac{\binom{r}{2}}{\binom{n}{2}}. \tag{3.20}$$

Thus we get

$$\beta_n^d(k) \leq \left(\frac{k-1}{n}\right)^{k(d-2)} \left(\frac{n-k}{n}\right)^{(n-k+1)(d-2)} \beta_n^2(k). \tag{3.21}$$

Let

$$b(k) = (k-1)^k (n-k)^{n-k+1}.$$

We are going to show that $b(k)$ is nonincreasing for $d + 1 \leq k \leq (n+1)/2$ by showing that $\log b(k) = k \log(k-1) + (n-k+1)\log(n-k)$ is nonincreasing. To this end it is enough to evaluate the first derivative of function $\log b(k)$:

$$\frac{\mathrm{d}}{\mathrm{d}k} \log b(k) = -\log\left(\frac{n-k}{k-1}\right) + \frac{1}{k-1}\left(1 - \frac{k-1}{n-k}\right). \tag{3.22}$$

Since $\log x \geq 1 - 1/x$ for $x \geq 1$, we have

$$\frac{\mathrm{d}}{\mathrm{d}k} \log b(k) \leq 0,$$

and, therefore, $b(k)$ is nonincreasing, which implies that the same holds for $\beta_n^d(k)$. This means that

$$\sum_{k=1}^{n} \beta_n^d(k) \leq n\beta_n^d(d+1) \leq n\left(\frac{d}{n}\right)^{(d+1)(d-2)} \left(\frac{n-d-1}{n}\right)^{(n-d-1)(d-2)} \frac{1}{122n}$$

$$\leq \frac{1}{122}\left(\frac{d}{n}\right)^{(d+1)(d-2)}. \quad \square$$

3.8 Applications of maximum matching problems

Besides being a fundamental theoretical problem, maximum matching arises in some interesting real-world applications.

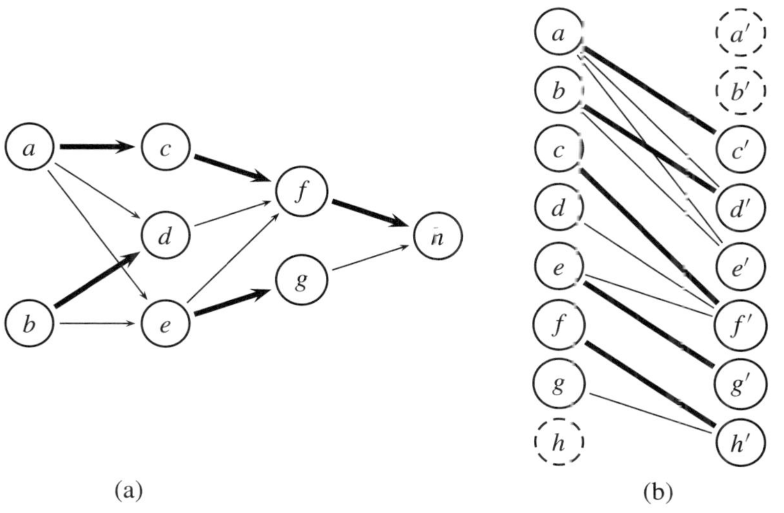

Figure 3.10. *Figure* (a) *shows a network with (a minimum number) of node disjoint paths (bold) which cover all nodes; Figure* (b) *shows the corresponding maximum matching.*

3.8.1 Vehicle scheduling problems

In transportation science the following problem plays an important role. In an airline network (or a railway network or a bus network) a certain set of trips must be served. What is the minimum number of vehicles (aircrafts in this case) needed to serve all trips?

This problem can be modeled as a disjoint path problem in a network $\mathcal{N} = (N, A)$ with *node set* N and *arc set* A. Usually, one has a daily or weekly schedule. Every trip (e.g., Monday morning trip from A to B) is modeled as a node in this network. Two nodes i and j are joined by an arc (i, j) if it is possible to serve trip j immediately after trip i by the same vehicle. The trips which are served by one vehicle form, therefore, a directed path in the graph. In order to serve all trips with as few vehicles as possible, we have to find a minimum number of node-disjoint paths in $\mathcal{N}$ which cover all nodes. An example is shown in Figure 3.10(a).

We now transform this problem to a maximum matching problem in a bipartite graph $G = (U, V; E)$: For every node $i \in N$ in which some arc starts we generate a vertex $i \in U$, and for every node $j \in N$ in which some arc ends we generate a vertex $j \in V$. Every arc $(i, j) \in A$ leads to an edge $[i, j] \in E$. A matching in G leads to arcs in $\mathcal{N}$ which form node disjoint paths, since the matched vertices correspond to nodes of $\mathcal{N}$ whose indegree and outdegree is at most 1. Thus we get the following.

Proposition 3.27. *Every system of node disjoint paths in $\mathcal{N}$ corresponds to a matching in G and vice versa.*

Let us assume that k node disjoint paths in $\mathcal{N}$ cover all nodes. Every path in $\mathcal{N}$ leaves two vertices of G unmatched. Thus minimizing the number of paths amounts to maximizing the cardinality of the corresponding matching in G. Thus we have shown that a maximum matching in G corresponds to a minimum number of node disjoint paths in the network $\mathcal{N}$. Figure 3.10(b) illustrates the connection between the node disjoint paths of Figure 3.10(a) and matchings.

It is also possible and meaningful to weight the connections (i, j). For example, it makes a difference if trip j starts 30 minutes after the arrival of the vehicle in the terminal point of trip i or if it starts 3 hours later. To deal with this issue, we can introduce weights c_{ij} where a small weight describes a good connection. This leads to a maximum matching problem with minimum total weight, i.e., to the linear assignment problem treated in Chapter 4.

The reader is referred to Fischetti, Lodi, Martello, and Toth [271] for a general treatment of vehicle scheduling problems and to Cordeau, Toth, and Vigo [198] for a survey on applications to train scheduling. A case study in connection with German intercity trains was described by Neng [511]. Recently, Grönkvist [341] gave a fleet management model, along with a number of possible solution approaches, for an airline company.

3.8.2 Time slot assignment problem

In telecommunication systems using satellites the data to be remitted are first buffered in the ground station, then remitted in very short data bursts to the satellite where they are amplified and sent back to Earth. Onboard the satellite there are n transponders which connect the sending stations with the receiving stations. In the so-called *time division multiple access* (TDMA) technique these $n \times n$ connections change in very short time intervals of variable length λ. The *switch mode* describes, for each $n \times n$ connection, which sending stations are momentarily connected with which receiving stations. We can model the switch mode by a permutation matrix with exactly one 1-entry in every row and column. The row and columns of the permutation matrix $P = (p_{ij})$ correspond to the (currently) participating Earth stations. Since we have n transponders onboard the satellite, only n Earth stations can send and remit data at the same time. Thus we can assume that P has n rows and columns. The entry $p_{ij} = 1$ means that currently the ith Earth station remits data to the jth participating Earth station. The $n \times n$ traffic matrix $T = (t_{ij})$, $T \geq 0$ (elementwise), contains the information about the times needed to transfer the required data from station i to station j. After applying switch mode P_k for a time interval of length λ_k, the traffic matrix is changed to $T - \lambda_k P_k$. The *time slot assignment problem* asks for switch modes and lengths of the corresponding time intervals such that all data are remitted in the shortest possible time:

$$\min \quad \sum_k \lambda_k \tag{3.23}$$

$$\text{s.t.} \quad \sum_k \lambda_k P_k \geq T \qquad \text{elementwise,} \tag{3.24}$$

$$\lambda_k \geq 0. \tag{3.25}$$

Historical note. The algorithm we are going to introduce was first discovered in 1931 by Egerváry [253] (see Dell'Amico and Martello [217]) and includes an idea to prove the famous Birkhoff theorem, Theorem 2.18. As also shown in [217], the same algorithm was later rediscovered many times in different contexts, like open shop scheduling and telecommunications.

First, we show how to modify the traffic matrix T so that it has constant row and column sums. Let t^* be the maximum value of the row and column sums and observe that no two elements of the same row or column can be remitted at the same time, so t^* is a lower bound for min $\sum_k \lambda_k$. We can fill up the matrix in the following straightforward way by adding dummy elements s_{ij} such that in matrix $T + S$ all row and column sums are equal to t^*. We will prove that the solution we obtain for the problem has value t^*; hence, it is not affected by the transformation of matrix T.

Algorithm 3.6. Equalize_sums.
Obtain a matrix with equal row and column sums.

for $i := 1$ **to** n **do** $R_i := \sum_{j=1}^n t_{ij}$;
for $j := 1$ **to** n **do** $C_j := \sum_{i=1}^n t_{ij}$;
$t^* := \max\ (\max_i R_i,\ \max_j C_j)$;
for $i := 1$ **to** n **do** $a_i := t^* - R_i$;
for $j := 1$ **to** n **do** $b_j := t^* - C_j$;
for $i := 1$ **to** n **do**
  for $j := 1$ **to** n **do**
    $s_{ij} := \min(a_i, b_j)$;
    $a_i := a_i - s_{ij}$;
    $b_j := b_j - s_{ij}$
  endfor
endfor
$T := T + S$

Therefore, the matrix $(1/t^*) \cdot T$ produced by the algorithm is doubly stochastic. Due to Birkhoff's theorem, Theorem 2.18, a doubly stochastic matrix is the convex combination of permutation matrices. Thus we can write the traffic matrix T as a nonnegative linear combination (weighted sum) of switch modes. The sum of weights equals t^*, i.e., the optimal solution value of (3.23)–(3.25). Since every point of a convex set in $\mathbb{R}^d$ can be written as a linear combination of $d + 1$ extreme points of this convex set and the dimension of the assignment polytope equals $d = (n - 1)^2$, matrix T is decomposed in at most $n^2 - 2n + 2$ different switch modes (which are as permutation matrices extreme points of the assignment polytope).

The actual decomposition of the traffic matrix produced by Algorithm 3.6 can be found as follows.

Algorithm 3.7. Decompose.
Decomposition of traffic matrices.

$k := 1$;

while $T \neq 0$ **do**
  construct a bipartite graph G with $|U| = |V| = n$ and an edge $[i, j]$ iff $t_{ij} > 0$;
  find a perfect matching φ_k in G, corresponding to a switch mode P_k;
  $\lambda_k := \min\{t_{i\varphi_k(i)} : i = 1, 2, \ldots, n\}$;
  $T := T - \lambda_k P_k$;
  $k := k + 1$
endwhile

Note that at every iteration we have a matrix T with constant row and column sums. Therefore, there exists a perfect matching and it can be found by any maximum matching algorithm. The first maximum matching can be found in $O(n^{5/2})$ steps. For every edge which is deleted an augmenting path has to be determined which takes $O(n^2)$ operations. Since we have $O(n^2)$ steps we get a total complexity of $O(n^4)$ operations. The following example illustrates this decomposition method.

Example 3.28. Let the traffic matrix be

$$T = \begin{pmatrix} 4 & 7 & 1 \\ 3 & 1 & 0 \\ 2 & 1 & 1 \end{pmatrix}.$$

The maximum row and column sum is $t^* = 12$. Using Algorithm 3.6 we define $a = (0, 8, 8)$ and $b = (3, 3, 10)$ and we add the following matrix S to T

$$S = \begin{pmatrix} 0 & 0 & 0 \\ 3 & 3 & 2 \\ 0 & 0 & 8 \end{pmatrix}.$$

We get a new traffic matrix with equal row and column sums:

$$T = \begin{pmatrix} 4 & 7 & 1 \\ 6 & 4 & 2 \\ 2 & 1 & 9 \end{pmatrix}.$$

We choose as the first perfect matching $\varphi_1 = (1, 2, 3)$, i.e., the switch mode

$$P_1 = \begin{pmatrix} 1 & 0 & 0 \\ 0 & 1 & 0 \\ 0 & 0 & 1 \end{pmatrix}.$$

Thus $\lambda_1 := \min\{4, 4, 9\} = 4$. We now set $T = T - 4 \cdot P_1$ and get

$$T = \begin{pmatrix} 0 & 7 & 1 \\ 6 & 0 & 2 \\ 2 & 1 & 5 \end{pmatrix}.$$

As the next perfect matching we use $\varphi_2 = (2, 3, 1)$, i.e., the switch mode

$$P_2 = \begin{pmatrix} 0 & 1 & 0 \\ 0 & 0 & 1 \\ 1 & 0 & 0 \end{pmatrix}.$$

Thus $\lambda_2 := \min\{7, 2, 2\} = 2$. The reduction of T leads to

$$T = \begin{pmatrix} 0 & 5 & 1 \\ 6 & 0 & 0 \\ 0 & 1 & 5 \end{pmatrix}.$$

The next two steps yield $\varphi_3 = (2, 1, 3)$ with $\lambda_3 = 5$ and $\varphi_4 = (3, 1, 2)$ with $\lambda_4 = 1$. Thus we get an optimal decomposition

$$T = \begin{pmatrix} 4 & 7 & 1 \\ 6 & 4 & 2 \\ 2 & 1 & 9 \end{pmatrix}$$

$$= 4 \cdot \begin{pmatrix} 1 & 0 & 0 \\ 0 & 1 & 0 \\ 0 & 0 & 1 \end{pmatrix} + 2 \cdot \begin{pmatrix} 0 & 1 & 0 \\ 0 & 0 & 1 \\ 1 & 0 & 0 \end{pmatrix} + 5 \cdot \begin{pmatrix} 0 & 1 & 0 \\ 1 & 0 & 0 \\ 0 & 0 & 1 \end{pmatrix} + 1 \cdot \begin{pmatrix} 0 & 0 & 1 \\ 1 & 0 & 0 \\ 0 & 1 & 0 \end{pmatrix}. \quad \blacksquare$$

Unfortunately this mathematically nice decomposition does not yield a technically feasible solution, since it uses too many switch modes which cannot be realized. Therefore, one has to restrict the number of switch modes. But with the additional restriction that the number of switch modes is $O(n)$ the time slot assignment problem becomes $\mathcal{NP}$-hard, as shown by Rendl [571] (see Section 10.3.1). Thus one has to solve the constraint time slot assignment problem by heuristics (see, e.g., the comments on balanced linear assignment problems at the end of Section 6.5).

Another possibility consists in considering a different model. According to a proposal of Lewandowski, Liu, and Liu [454], a system of $2n$ switch modes $P_1, \ldots, P_n$ and $Q_1, \ldots, Q_n$ could be fixed onboard the satellite, where any pair of matrices P_k and Q_l has just one 1-element in common and

$$\sum_{k=1}^{n} P_k = \sum_{l=1}^{n} Q_l = \mathbf{1},$$

where $\mathbf{1}$ is the matrix with 1-entries only. Let $p_{ij}^{(k)}$ and $q_{ij}^{(l)}$ denote the elements of P_k and Q_l, respectively. Both sets of matrices $P_k (k = 1, 2, \ldots, n)$ and $Q_l (l = 1, 2, \ldots, n)$ describe feasible solutions to the planar 3-index assignment problem (see Section 1.4) given by $x_{ijk} = p_{ij}^{(k)}$ (or $x_{ijk} = q_{ij}^{(k)}$) for $i, j, k = 1, 2, \ldots, n$, hence, they correspond to two Latin squares L_P and L_Q. Two Latin squares A and B are *orthogonal* if the pairs (a_{ij}, b_{ij}) are all distinct. Since any pair of matrices P_k and Q_l has just one 1-element in common, L_P and L_Q are *orthogonal Latin squares*.

Historical note. Euler tried to find a pair of orthogonal Latin squares for $n = 6$, but failed. He conjectured that no such pair exists for $n = 4k + 2$ ($k \geq 1$ integer). In 1901 Tarry [634] showed that the conjecture holds for $n = 6$. However, in 1960, Bose, Shrikhande, and Parker [110] proved that Euler's conjecture is false by showing that there exists a pair of orthogonal Latin squares for all $n \geq 3$, $n \neq 6$.

The problem of decomposing a traffic matrix T into a weighted sum of permutation matrices corresponding to two orthogonal Latin squares can also be formulated as

$$\min \quad \sum_{k=1}^{n} (\lambda_k + \mu_k) \tag{3.26}$$

$$\text{s.t.} \quad \sum_{k=1}^{n} (\lambda_k p_{ij}^{(k)} + \mu_k q_{ij}^{(k)}) \geq t_{ij} \qquad (i, j = 1, 2, \ldots, n), \tag{3.27}$$

$$\lambda_k, \mu_k \geq 0 \qquad (k = 1, 2, \ldots, n). \tag{3.28}$$

Following Burkard [129] we can dualize this linear program by introducing n^2 dual variables s_{ij}. Thus we get

$$\begin{aligned} \max \quad & \sum_{i=1}^{n} \sum_{j=1}^{n} t_{ij} s_{ij} & \\ \text{s.t.} \quad & \sum_{i=1}^{n} \sum_{j=1}^{n} p_{ij}^{(k)} s_{ij} \leq 1 & (k = 1, 2, \ldots, n), \\ & \sum_{i=1}^{n} \sum_{j=1}^{n} q_{ij}^{(l)} s_{ij} \leq 1 & (l = 1, 2, \ldots, n), \\ & s_{ij} \geq 0 & (i, j = 1, 2, \ldots, n). \end{aligned}$$

According to Theorem 4.2 of Heller and Tompkins [368], the coefficient matrix of this problem is totally unimodular. Therefore, we can replace $s_{ij} \geq 0$ with $s_{ij} \in \{0, 1\}$. This and the orthogonality of the Latin squares enables us to introduce new variables v_{kl} defined by

$$v_{kl} = \begin{cases} 1 & \text{if } \exists\, (i, j) : s_{ij} = 1 \text{ and } p_{ij}^k = q_{ij}^l = 1, \\ 0 & \text{otherwise.} \end{cases}$$

By introducing new cost coefficients

$$u_{kl} = t_{ij} \text{ for } p_{ij}^k = q_{ij}^l,$$

we get the following linear assignment problem, which is equivalent to (3.26)–(3.28):

$$\begin{aligned} \max \quad & \sum_{k=1}^{n} \sum_{l=1}^{n} u_{kl} v_{kl} & \\ \text{s.t.} \quad & \sum_{k=1}^{n} v_{kl} \leq 1 & (l = 1, \ldots, n), \\ & \sum_{l=1}^{n} v_{kl} \leq 1 & (k = 1, \ldots, n), \\ & v_{kl} \in \{0, 1\} & (k, l = 1, \ldots, n). \end{aligned}$$

Thus we have proven the following.

Proposition 3.29. *The optimal decomposition of a traffic matrix into a system of* $2n$ *permutation matrices corresponding to two orthogonal Latin squares can be found by means of a linear sum assignment problem in* $O(n^3)$ *time.*

Chapter 4

Linear sum assignment problem

4.1 Introduction

The *linear sum assignment problem* (LSAP) is one of the most famous problems in linear programming and in combinatorial optimization. Informally speaking, we are given an $n \times n$ *cost matrix* $C = (c_{ij})$ and we want to match each row to a different column in such a way that the sum of the corresponding entries is minimized. In other words, we want to select n elements of C so that there is exactly one element in each row and one in each column and the sum of the corresponding costs is a minimum.

Alternatively, one can define LSAP through a graph theory model. Define a bipartite graph $G = (U, V; E)$ having a vertex of U for each row, a vertex of V for each column, and cost c_{ij} associated with edge $[i, j]$ $(i, j = 1, 2, \ldots, n)$: The problem is then to determine a minimum cost perfect matching in G (*weighted bipartite matching problem*: find a subset of edges such that each vertex belongs to exactly one edge and the sum of the costs of these edges is a minimum).

Without loss of generality, we assume that the costs c_{ij} are nonnegative. Cases with negative costs can be handled by adding to each element of C the value $\chi = -\min_{i,j}\{c_{ij}\}$: Since we need to select one element per row, any solution of value z for the original cost matrix corresponds to a solution of value $z + n\chi$ for the transformed cost matrix. In this way we can manage the maximization version of the problem by solving LSAP on a transformed instance having costs $\widetilde{c}_{ij} = -c_{ij}$. Let us also observe that most preprocessing algorithms (see, e.g., Algorithm 4.1), which are preliminary executed on LSAP instances in order to accelerate the solution algorithms, produce a nonnegative cost matrix.

We also assume in general that the values in C are finite, with some c_{ij} possibly having a very large value ($< \infty$) when assigning i to j is forbidden. In the case of sparse matrices, where a very large number of (i, j) assignments is forbidden, we denote by m the number of admitted (i, j) assignments and, in graph $G = (U, V; E)$ above, we only include in E the edges corresponding to them, so $|E| = m$.

Most of the algorithms we present work for general cost matrices, although special approaches, such as those based on cost-scaling (see Section 4.2.3), require an integer cost matrix. For such cases we denote by C the maximum c_{ij} value.

4.1.1 Mathematical model

By introducing a binary matrix $X = (x_{ij})$ such that

$$x_{ij} = \begin{cases} 1 & \text{if row } i \text{ is assigned to column } j, \\ 0 & \text{otherwise,} \end{cases}$$

LSAP can be modeled as

$$\min \quad \sum_{i=1}^{n} \sum_{j=1}^{n} c_{ij} x_{ij} \tag{4.1}$$

$$\text{s.t.} \quad \sum_{j=1}^{n} x_{ij} = 1 \qquad (i = 1, 2, \ldots, n), \tag{4.2}$$

$$\sum_{i=1}^{n} x_{ij} = 1 \qquad (j = 1, 2, \ldots, n), \tag{4.3}$$

$$x_{ij} \in \{0, 1\} \qquad (i, j = 1, 2, \ldots, n). \tag{4.4}$$

Note that the resulting matrix X is a permutation matrix (see Section 1.1).

The $2n \times n^2$ matrix defined by the left-hand sides of the assignment constraints (4.2) and (4.3) is the incidence matrix of the bipartite graph G introduced above. It has a row i for each vertex $i \in U$, a row $n + j$ for each vertex $j \in V$, and a column for each edge $[i, j] \in E$: The entries in this column are 1 in rows i and $n+j$ and 0 elsewhere. This matrix, shown in Figure 4.1, has the nice combinatorial property of being totally unimodular.

$$\begin{array}{r} \\ 1 \\ \\ \\ n \\ n+1 \\ \\ \\ 2n \end{array}
\begin{array}{c} \begin{array}{cccc|cccc|c|cccc} x_{11} & & \cdots & x_{1n} & x_{21} & & \cdots & x_{2n} & \cdots & x_{n1} & & \cdots & x_{nn} \end{array} \\
\left(\begin{array}{cccc|cccc|c|cccc}
1 & 1 & \cdots & 1 & & & & & & & & & \\
 & & & & 1 & 1 & \cdots & 1 & & & & & \\
 & & & & & & & & \cdots & & & & \\
 & & & & & & & & & 1 & 1 & \cdots & 1 \\
\hline
1 & & & & 1 & & & & & 1 & & & \\
 & 1 & & & & 1 & & & & & 1 & & \\
 & & \cdots & & & & \cdots & & \cdots & & & \cdots & \\
 & & & 1 & & & & 1 & & & & & 1
\end{array}\right) \end{array}$$

Figure 4.1. *Node-edge incidence matrix of G.*

Definition 4.1. *An integer matrix is* totally unimodular *if the determinant of every square submatrix has value 0, $+1$, or -1.*

We will use the following sufficient condition due to Heller and Tompkins [368].

Theorem 4.2. *An integer matrix A with $a_{ij} = 0, \pm 1$ is totally unimodular if no more than two non-zero entries appear in any column and if its rows can be partitioned into two sets, I_1 and I_2, such that*

1. *if a column has two entries of the same sign their rows are in different sets;*
2. *if a column has two entries of different signs, their rows are in the same set.*

It is then easy to see that the matrix defined by (4.2) and (4.3) satisfies the condition above with $I_1 = \{1, 2, \ldots, n\}$ and $I_2 = \{n+1, n+2, \ldots, 2n\}$. Since it is known that a linear program with integer right-hand sides and totally unimodular constraint matrix always has an integer optimal solution, this shows that LSAP is equivalent to its continuous relaxation, given by (4.1)–(4.3) and

$$x_{ij} \geq 0 \qquad (i, j = 1, 2, \ldots, n). \tag{4.5}$$

The same result was proved, in a different way, by Birkhoff [100] (see Theorem 2.18 and Proposition 2.24).

4.1.2 Complementary slackness

By associating dual variables u_i and v_j with assignment constraints (4.2) and (4.3), respectively, the dual problem is

$$\max \quad \sum_{i=1}^{n} u_i + \sum_{j=1}^{n} v_j \tag{4.6}$$

$$\text{s.t.} \quad u_i + v_j \leq c_{ij} \qquad (i, j = 1, 2, \ldots, n). \tag{4.7}$$

By duality theory, a pair of solutions respectively feasible for the primal and the dual is optimal if and only if (*complementary slackness*)

$$x_{ij}(c_{ij} - u_i - v_j) = 0 \qquad (i, j = 1, 2, \ldots, n). \tag{4.8}$$

The values

$$\overline{c}_{ij} = c_{ij} - u_i - v_j \qquad (i, j = 1, 2, \ldots, n) \tag{4.9}$$

are the linear programming *reduced costs*. This transformation from C to $\overline{C}$ is a special case of what is known as "admissible transformation," which is formally stated in Chapter 6, Definition 6.19. Indeed, for any feasible primal solution X, the transformed objective function is

$$\begin{aligned}
\sum_{i=1}^{n}\sum_{j=1}^{n}(c_{ij} - u_i - v_j)x_{ij} &= \sum_{i=1}^{n}\sum_{j=1}^{n} c_{ij}x_{ij} - \sum_{i=1}^{n} u_i \sum_{j=1}^{n} x_{ij} - \sum_{j=1}^{n} v_j \sum_{i=1}^{n} x_{ij} \\
&= \sum_{i=1}^{n}\sum_{j=1}^{n} c_{ij}x_{ij} - \sum_{i=1}^{n} u_i - \sum_{j=1}^{n} v_j,
\end{aligned} \tag{4.10}$$

i.e., for any feasible solution X the objective function values induced by C and $\overline{C}$ differ by the same constant $\sum_{i=1}^{n} u_i + \sum_{j=1}^{n} v_j$.

The algorithms for LSAP are based on different approaches: a first class of methods directly solves the primal problem, a second one solves the dual, and a third one uses an intermediate approach (primal-dual). Most of these methods adopt a *preprocessing phase* to determine a feasible dual solution and a partial primal solution (where less than n rows are assigned) satisfying the complementary slackness conditions. A basic $O(n^2)$ time implementation of this phase is given in Algorithm 4.1, which stores the partial assignment both in X and in

$$row(j) = \begin{cases} i & \text{if column } j \text{ is assigned to row } i, \\ 0 & \text{if column } j \text{ is not assigned,} \end{cases} \qquad (j = 1, 2, \dots, n). \tag{4.11}$$

Note that the reduced costs given by the resulting dual variables are nonnegative.

ALGORITHM 4.1. Procedure Basic_preprocessing.
Feasible dual solution and partial primal solution.

```
for i := 1 to n do for j := 1 to n do x_ij := 0;
comment: row and column reduction;
for i := 1 to n do u_i := min{c_ij : j = 1, 2, ..., n};
for j := 1 to n do v_j := min{c_ij − u_i : i = 1, 2, ..., n};
comment: find a partial feasible solution;
for j := 1 to n do row(j) := 0;
for i := 1 to n do
    for j := 1 to n do
        if (row(j) = 0 and c_ij − u_i − v_j = 0) then
            x_ij := 1, row(j) := i;
            break
        endif
    endfor
endfor
```

In the following we will frequently store a (partial) assignment in φ that implements the inverse of *row* (see Section 1.1), i.e.,

$$\varphi(i) = \begin{cases} j & \text{if row } i \text{ is assigned to column } j, \\ 0 & \text{if row } i \text{ is not assigned,} \end{cases} \qquad (i = 1, 2, \dots, n). \tag{4.12}$$

The u_i and v_j values determined by the first two statements satisfy the dual constraints (4.7). The x_{ij} values subsequently obtained ensure satisfaction of complementary slackness conditions (4.8), while for the primal constraints (4.1) and (4.2) the $\leq$ sign holds instead of $=$. Note that an alternative algorithm could first perform the column reduction and then the row reduction, generally obtaining different reduced costs and assignments. We illustrate Basic_preprocessing through a numerical example, to be resumed several times in the following.

Example 4.3. Given the input matrix C below, we obtain the dual variables u and v (shown on the left and on the top) and the corresponding reduced cost matrix $\overline{C}$:

$$
\begin{array}{c}
\\
7\\
2\\
1\\
2
\end{array}
\begin{array}{c}
\begin{array}{cccc} 0 & 2 & 0 & 0 \end{array}\\
\begin{pmatrix}
7 & 9 & 8 & 9\\
2 & 8 & 5 & 7\\
1 & 6 & 6 & 9\\
3 & 6 & 2 & 2
\end{pmatrix}
\end{array}
\qquad
\begin{pmatrix}
\underline{0} & 0 & 1 & 2\\
0 & 4 & 3 & 5\\
0 & 3 & 5 & 8\\
1 & 2 & \underline{0} & 0
\end{pmatrix}.
$$

$$C \qquad\qquad\qquad\qquad \overline{C}$$

We then obtain $row = (1, 0, 4, 0)$ (thus $\varphi = (1, 0, 0, 3)$) and the partial assignment shown by the underlined zeroes in $\overline{C}$. ■

4.1.3 Historical notes, books, and surveys

The first algorithm for LSAP was presented in 1946 by Easterfield [245]: it is a non-polynomial $O(2^n n^2)$ time approach, based on iterated application of a particular class of admissible transformations (see Section 6.3). Easterfield did not give a name to the problem (the title of the paper is "A combinatorial algorithm"), nor did it Thorndike [638] who, in a 1950 paper ("The problem of classification of personnel"), proposed three heuristic algorithms for LSAP. The current name appeared for the first time in 1952, in the paper "The personnel assignment problem" by Votaw and Orden [653]. Further historical details on the early years of LSAP and the Hungarian algorithm can be found in Schrijver [599, Chapter 17], Frank [277], and Schrijver [600].

Primal-dual algorithms have been the first polynomial-time methods for LSAP. The famous Hungarian algorithm, presented in the mid-1950s by Kuhn [438, 440], in its original formulation solves the problem in $O(n^4)$ time. A variation of this algorithm was presented by Munkres [502]: he showed that his method requires at most $(11n^3 + 12n^2 + 31n)/6$ operations, where, however, some operations are "scan a line," thus leaving an $O(n^4)$ overall time complexity. Other $O(n^4)$ time algorithms were later proposed by Iri [386] and Desler and Hakimi [232]. In 1960 Silver [609, 610] gave the first computer code for LSAP, a modified version of the Munkres algorithm, written in Algol. The best time complexity for a Hungarian algorithm is $O(n^3)$ (see the implementation proposed by Lawler [448] in 1976). The first $O(n^3)$ algorithm for LSAP had, however, appeared in a 1969 paper by Dinic and Kronrod [235] that was ignored for many years. In the early 1970s Tomizawa [640] and Edmonds and Karp [250] showed that shortest path computations on the reduced costs produce an $O(n^3)$ time algorithm for LSAP. Computer implementations of the Hungarian algorithm adopting shortest path techniques (e.g., the codes proposed in the 1980s by Burkard and Derigs [145], Jonker and Volgenant [392], and Carpaneto, Martello, and Toth [165]) have for many years been the most successful tools for the practical solution of LSAP instances. In the mid-1980s Gabow [295] used the *cost scaling* technique, developed in the early 1970s by Edmonds and Karp [250], for obtaining a cost-scaling Hungarian algorithm for LSAP. The weakly polynomial time complexity of the resulting algorithm, $O(n^{3/4} m \log C)$, was later improved by Gabow and Tarjan [297] to $O(\sqrt{n}\, m \log(nC))$. Detailed descriptions of primal-dual methods are given in Sections 4.2 and 4.4.

The first *primal simplex algorithms*, proposed in the mid-1970s by Cunningham [205] and Barr, Glover, and Klingman [68], required exponential time due to the high degeneracy of the linear program associated with LSAP. A few years later Roohy-Laleh [589] modified the Cunningham algorithm to obtain an $O(n^5)$ time complexity, while in 1993 Akgül [19] proposed an $O(n^3)$ time primal simplex algorithm. Primal methods are described in Section 4.5. The first *primal* (non-simplex) *algorithm* was proposed in 1964 by Balinski and Gomory [64]. It iteratively improves, through alternating paths, a feasible assignment and a dual (infeasible) solution satisfying complementary slackness and solves the problem in $O(n^4)$ time. Other primal algorithms were given in the following years by Srinivasan and Thompson [619, 620] and Klein [421]. The latter paper introduced the "cycle canceling" technique, which was very important for the solution of min-cost flow problems. (Such a technique had however been studied for the first time by Robinson [587] in 1949.) An $O(n^3)$ primal algorithm was obtained in 1978 by Cunningham and Marsh [208] by generalizing the Klein idea.

The first *dual* (non-simplex) *algorithm* for LSAP appeared in the already mentioned 1969 paper by Dinic and Kronrod [235], discussed in Section 4.3. This method is also the basis of the dual algorithm presented in 1980 by Hung and Rom [382], in which a series of relaxed problems (where constraints (4.3) are disregarded) is solved by updating the current solution through shortest paths until the solution becomes feasible for LSAP. This algorithm too has time complexity $O(n^3)$. In 1981 Bertsekas [86] proposed a dual algorithm having pseudo-polynomial time complexity but high average efficiency in practice (the *auction algorithm*). Polynomial-time auction algorithms were later obtained by Bertsekas and Eckstein [92] and by Orlin and Ahuja [515], who combined auction and a particular scaling technique (known as *ε-relaxation*). The time complexity of the latter algorithm is $O(\sqrt{n}\ m \log(nC))$, equal to that of the primal-dual scaling algorithm by Gabow and Tarjan [297]. The same time complexity characterizes the computationally very effective algorithms developed in the mid-1990s by Goldberg and Kennedy [328], who adopted a scaling technique (*pseudoflow*) originally developed for min-cost flow problems. The most famous *dual simplex algorithms* for LSAP are the so-called *signature* methods, proposed in the mid-1980s by Balinski [62] and Goldfarb [334]. These algorithms have $O(n^3)$ time complexity, and it can be shown that they are substantially equivalent to the dual (non-simplex) approach by Hung and Rom [382]. Dual methods are discussed in Section 4.6.

The latest relevant theoretical result for LSAP was obtained in the new millennium by Kao, Lam, Sung, and Ting [403] who closed a long standing gap between the time complexity of LSAP and that of the maximum cardinality matching problem. Their result is discussed in Section 4.7.

The 1980s saw the diffusion of parallel computers. In the following years many sequential methods for LSAP (especially auction, shortest path, and primal simplex algorithms) have been parallelized and computationally tested on parallel machines. We describe these topics in Section 4.11.

Starting in the late 1970s, many books and surveys on LSAP have been proposed in the literature. The first survey was presented by Burkard [128], who included a summary of results on the structure of the associated polytope. The book by Burkard and Derigs [145] considers various assignment-type problems and includes, among others, a Fortran program implementing a variant of the shortest augmenting path algorithm proposed by Tomizawa [640] (see Sections 4.4.1 and 4.9). Derigs [226] presented an extensive survey on the shortest augmenting path technique (see Section 4.4), discussing and relating to it all classical algorithms and examining the results of an extensive computational experience, performed over 14 Fortran codes. Martello and Toth [481] reviewed LSAP and other linear assignment-type problems and analyzed the performance of different algorithms through computational experiments. The book by Bertsekas [88] on relaxation and auction techniques (see Section 4.6.3) includes several implementations of algorithms for LSAP and the corresponding Fortran listings (also downloadable from the internet). The survey by Akgül [18] analyzes the main solution approaches and discusses their relationships. The volume edited by Johnson and McGeoch [390] includes several papers on implementations of algorithms for LSAP proposed at the first *DIMACS Implementation Challenge*. A specialized survey on the probabilistic analysis of simple online and offline heuristic algorithms for LSAP can be found in Faigle [264].

In 1997 Dell'Amico and Martello [219] presented an annotated bibliography, with special attention to results obtained in the 1980s and the 1990s. The extensive survey by Burkard and Çela [138] gives the state-of-the-art on LSAP and other linear assignment problems with other objective functions like the algebraic, bottleneck, balanced, axial, and planar assignment problems (see Chapters 6 and 10). Dell'Amico and Toth [220] presented extensive computational experiments with the eight most popular computer codes for dense instances of LSAP. Burkard [132] surveyed recent developments in the fields of bipartite matchings (see Chapter 3), LSAP, and quadratic assignment problems (see Chapter 7). Linear and non-linear assignment problems are discussed in a recent survey by Pentico [541].

Chapters on LSAP can also be found, e.g., in Murty [507, Chapter 3], Ahuja, Magnanti, and Orlin [11, Chapter 12], Jungnickel [399, Chapter 13], Korte and Vygen [428, Chapter 11], and Schrijver [599, Chapters 17 and 18].

4.2 Primal-dual algorithms

The genesis of these methods was reported by the author, H. W. Kuhn [441]. In 1953, while reading the classical graph theory book by König [426], *Theorie der Endlichen und Unendlichen Graphen*, he encountered an efficient algorithm for determining the maximum cardinality matching of a bipartite graph. This problem is equivalent to an LSAP with 0-1 costs, and a footnote in König's book pointed to a paper by Egerváry [253] (in Hungarian) for a generalization of the result to general cost matrices. Kuhn then spent two weeks translating Egerváry's paper, with the aid of a Hungarian dictionary, finding a method to reduce a general LSAP to a finite number of LSAPs with 0-1 costs. (The translated paper can be found in Kuhn [439].) The combination of Egerváry's reduction and König's maximum cardinality matching algorithm produced an efficient algorithm for solving LSAP [438, 440] that Kuhn called the "Hungarian method" in honor of these two Hungarian mathematicians. These two famous papers were recently reprinted in the 50th anniversary of the original publication, together with an editorial note from H. W. Kuhn [442, 443].

(We mention here that Egerváry's paper [253] also contains a second nice result that directly provides an efficient algorithm for solving the preemptive open shop scheduling problem. This algorithm was independently rediscovered many times in the following years. See Section 3.8.2 and Dell'Amico and Martello [217] for more details.)

4.2.1 The Hungarian algorithm

The Hungarian algorithm is recognized as a predecessor of the *primal-dual method* for linear programming, designed one year later by Dantzig, Ford, and Fulkerson [212]. It starts with a feasible dual solution u, v satisfying (4.7) and a partial primal solution (in which less than n rows are assigned) satisfying the complementary slackness conditions (4.8) with respect to u, v. Each iteration solves a restricted primal problem independent of the costs, trying to increase the cardinality of the current assignment by operating on the partial graph of $G = (U, V; E)$ that only contains the edges of E having zero reduced costs. If the attempt is successful, a new primal solution in which one more row is assigned is obtained. Otherwise, the current dual solution is updated so that new edges having zero reduced costs are obtained.

In order to describe the algorithm we need some basic definitions. The current partial assignment defines on G a set of *assigned edges* and the corresponding set of *assigned vertices*. An *alternating path* (see Section 3.2) is an elementary path whose edges are alternately not assigned and assigned. An *alternating tree* rooted in a vertex k is a tree in which all paths emanating from k are alternating. An *augmenting path* is an alternating path whose initial and terminal edges (and, hence, vertices) are not assigned. The restricted problem of the Hungarian algorithm consists of searching for an augmenting path in the bipartite partial graph $G^0 = (U, V; E^0)$ that only contains edges $[i, j]$ such that $\overline{c}_{ij} = 0$. If such a path P is found, the improved assignment is obtained by interchanging unassigned and assigned edges along P, i.e., by setting $x_{ij} = 1$ for the $(\lfloor |P|/2 \rfloor + 1)$ unassigned edges $[i, j]$ of P (odd edges) and $x_{ij} = 0$ for the $\lfloor |P|/2 \rfloor$ assigned edges $[i, j]$ of P (even edges).

The "Dijkstra-like" procedure given in Algorithm 4.2 (see Dijkstra [234]) looks for an alternating and possibly augmenting path starting at a given unassigned vertex $k \in U$ by progressively growing an alternating tree rooted in k. It is also closely related to Algorithm 3.1 of Section 3.2 for bipartite matching. At any iteration, a vertex is *labeled* (if it belongs to a path emanating from k) or *unlabeled*. A labeled vertex of V can be *scanned* (if it has been used for extending the tree) or *unscanned*. In this implementation, labeling and scanning coincide for a vertex of U. Set LV stores the currently labeled vertices of V, while sets SU and SV store the currently scanned vertices of U and V, respectively.

ALGORITHM 4.2. Procedure Alternate(k).
Find an alternating tree rooted at an unassigned vertex $k \in U$.

$SU := LV := SV := \emptyset$;
fail := `false`, *sink* $:= 0$, $i := k$;
while (*fail* = `false` **and** *sink* $= 0$) **do**
    $SU := SU \cup \{i\}$;
    for each $j \in V \setminus LV \ : \ c_{ij} - u_i - v_j = 0$ **do** $pred_j := i$, $LV := LV \cup \{j\}$;
    if $LV \setminus SV = \emptyset$ **then** *fail* := `true`
    else
        let j be any vertex in $LV \setminus SV$;
        $SV := SV \cup \{j\}$;
        if $row(j) = 0$ **then** $sink := j$ **else** $i := row(j)$
    endif
endwhile
return *sink*

Each iteration consists of two phases. The unique candidate vertex $i \in U$ is first labeled and scanned ($i = k$ at the first iteration). The scanning consists in extending the tree

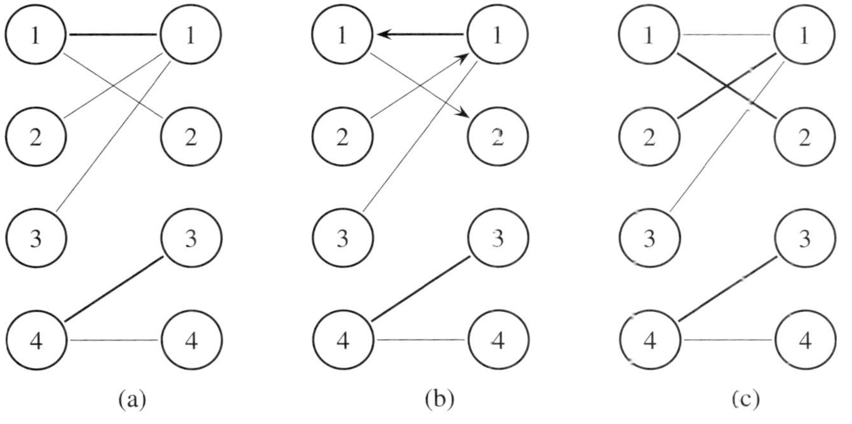

Figure 4.2. (a) *Graph G^0; *(b) *alternating tree;* (c) *new graph G^0.*

by assigning i as predecessor to all unlabeled vertices $j \in V$ for which $[i, j] \in E^0$ and in labeling these vertices. In the second phase, a labeled unscanned vertex $j \in V$ is selected and scanned by adding the unique edge $[row(j), j] \in E^0$ to the tree. Vertex $row(j) \in U$ then becomes the candidate for the next iteration. The execution can terminate, in this second phase, in two possible situations: (i) if the selected vertex $j \in V$ is not assigned, we have obtained an augmenting path from k to j; (ii) if V contains no labeled unscanned vertex the current tree cannot grow any longer.

Each iteration of the main loop requires $O(n)$ time. At each iteration, a different vertex $j \in LV \setminus SV$ is selected and added to SV so a different vertex $i = row(j)$ is considered. It follows that the time complexity of Alternate(k) is $O(n^2)$.

Example 4.4. We continue from Example 4.3. Let us start from the reduced cost matrix $\overline{C}$ and arrays u, v, and *row* we have obtained and assume that $k = 2$. Figure 4.2(a) shows the bipartite partial graph $G^0 = (U, V; E^0)$, with thick lines denoting the current partial assignment. Procedure Alternate(k) produces:

$SU = LV = SV = \emptyset, fail =$ `false`, $sink = 0$;

$i = 2$: $SU = \{2\}, pred_1 = 2, LV = \{1\}$;
  $j = 1$: $SV = \{1\}$;

$i = 1$: $SU = \{2, 1\}, pred_2 = 1, LV = \{1, 2\}$;
  $j = 2$: $SV = \{1, 2\}, sink = 2$.

We have thus obtained an augmenting tree (in this case a path) which is shown by the arrows in Figure 4.2(b). A new solution (see Figure 4.2(c)), obtained by interchanging unassigned and assigned edges along the augmenting path is: $x_{12} = x_{21} = x_{43} = 1$ (and $x_{ij} = 0$ elsewhere). ∎

The Hungarian method is given in Algorithm 4.3. Let (u, v) be the current feasible dual solution. The current assignment is stored in arrays *row* and φ (see (4.11),(4.12)). On exit, the solution matrix X is

$$x_{ij} = \begin{cases} 1 & \text{if } \varphi(i) = j, \\ 0 & \text{otherwise.} \end{cases}$$

Arrays u, v, *row*, and φ can either be initialized to zero or through a preprocessing phase such as, e.g., Procedure Basic_preprocessing of Section 4.1.2. We denote with $\overline{U} \subseteq U$ the set of *assigned vertices* of U.

ALGORITHM 4.3. Hungarian.
Basic $O(n^4)$ Hungarian algorithm.

```
initialize u, v, row, φ and Ū;
while |Ū| < n do
    let k be any vertex in U \ Ū;
    while k ∉ Ū do
        sink := Alternate(k);
        if sink > 0 then [comment: increase the primal solution]
            Ū := Ū ∪ {k}, j := sink;
            repeat
                i := pred_j, row(j) := i, h := φ(i), φ(i) := j, j := h
            until i = k
        else [comment: update the dual solution]
            δ := min{c_ij − u_i − v_j : i ∈ SU, j ∈ V \ LV};
            for each i ∈ SU do u_i := u_i + δ;
            for each j ∈ LV do v_j := v_j − δ
        endif
    endwhile
endwhile
```

At each iteration of the outer loop, an unassigned vertex of U is selected and assigned through the inner loop, i.e., through a series of calls to Procedure Alternate(k) followed by an updating of the dual variables, until an augmenting path is obtained. Whenever Alternate(k) fails in producing an augmenting path, the dual variables corresponding to the labeled vertices are updated by determining the minimum reduced cost δ ($\delta > 0$) of an edge connecting a labeled vertex of U to an unlabeled vertex of V. We show that this updating is such that the next execution of Alternate(k) will produce an enlarged tree.

Proposition 4.5. *The dual updating of Procedure Hungarian is such that*

1. *the edges of E^0 that connect pairs of labeled vertices or pairs of unlabeled vertices maintain zero reduced cost;*
2. *at least one new edge, connecting a labeled vertex of U to an unlabeled vertex of V, enters E^0;*
3. *no reduced cost becomes negative.*

Proof. Recall that labeling and scanning coincide for a vertex of U. We consider the four sets of edges $[i, j]$ whose costs are updated in a different way:

$- i \notin SU, j \notin LV$: no updating occurs;

$- i \in SU, j \in LV$: the updating produces $\overline{c}_{ij} := \overline{c}_{ij} - \delta + \delta$, hence 1. follows;

$- i \in SU, j \notin LV$: the updating produces $\overline{c}_{ij} := \overline{c}_{ij} - \delta$. By definition of δ, the resulting reduced costs are nonnegative, and at least one of them has value zero, and hence 2. follows;

$- i \notin SU, j \in LV$: the updating produces $\overline{c}_{ij} := \overline{c}_{ij} + \delta$, and 3. is proved as well. □

It follows that, at the next execution, Alternate(k) will label at least one more vertex of V, so an augmenting path will result after at most n calls to Alternate(k). The correctness of algorithm Hungarian follows.

The initialization step of Hungarian requires $O(n^2)$ time if performed through Basic_preprocessing or a similar method. The outer loop of Hungarian is executed $O(n)$ times. At each iteration, Procedure Alternate(k) and the dual updating are executed $O(n)$ times in the inner loop. We have already observed that each execution of Alternate(k) finds an alternating tree in $O(n^2)$ time. The value of δ is also computed in $O(n^2)$ time. The overall time complexity of algorithm Hungarian is thus $O(n^4)$. The Fortran listing of this algorithm can be found in Carpaneto and Toth [166], while a QuickBasic implementation was later presented by Lotfi [465].

If the data are integer or rational numbers, one can show that any primal-dual algorithm, independently of the labeling technique used, will terminate in a finite number of iterations. Aráoz and Edmonds [42] considered an LSAP instance including irrational numbers. They showed that a primal-dual algorithm using a "non-Hungarian" labeling may run forever on such instance without finding the optimal solution.

Example 4.6. We make use of the instance introduced in Example 4.3 and assume that the initialization is performed through Procedure Basic_preprocessing, as already shown. We have obtained $u = (7, 2, 1, 2)$, $v = (0, 2, 0, 0)$, and $row = (1, 0, 4, 0)$; hence, we have $\varphi = (1, 0, 0, 3)$ and $\overline{U} = \{1, 4\}$.

As $|\overline{U}| = n - 2$, the outer loop will be executed twice. The first call to Procedure Alternate(k) with $k = 2$ (already illustrated in the present section) returns $sink = 2$ and $pred = (2, 1, -, -)$; hence, Hungarian increases the primal solution by setting $\overline{U} = \{1, 4, 2\}$, $row = (2, 1, 4, 0)$, $\varphi = (2, 1, 0, 3)$.

Alternate(k) is then executed for $k = 3$ as follows:

$SU = LV = SV = \emptyset$, $fail$ = `false`, $sink = 0$;

$i = 3$: $SU = \{3\}$, $pred_1 = 3$, $LV = \{1\}$;
  $j = 1$: $SV = \{1\}$;

$i = 2$: $SU = \{3, 2\}$, $fail$ = `true`.

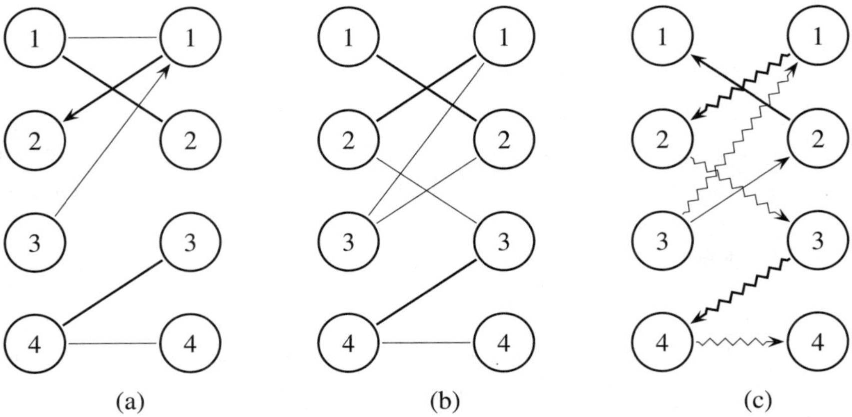

Figure 4.3. (a) *Alternating tree;* (b) *new graph* G^0*;* (c) *augmenting tree.*

In Figure 4.3(a) the arrows indicate the resulting alternating tree (in this case, too, a path). On return, the dual solution is updated as $\delta = 3$, $u = (7, 5, 4, 2)$, $v = (-3, 2, 0, 0)$, and hence,

$$\overline{C} = \begin{pmatrix} 3 & 0 & 1 & 2 \\ 0 & 1 & 0 & 2 \\ 0 & 0 & 2 & 5 \\ 4 & 2 & 0 & 0 \end{pmatrix}.$$

Figure 4.3(b) shows the new bipartite partial graph $G^0 = (U, V; E^0)$, with thick lines denoting the current partial assignment.

Alternate(k) is then executed again for $k = 3$, producing the augmenting tree shown by the arrows in Figure 4.3(c), where the zig-zag lines show the augmenting path. On return, the primal solution is increased producing $\overline{U} = \{1, 4, 2, 3\}$, $row = (3, 1, 2, 4)$, $\varphi = (2, 3, 1, 4)$.

We have thus obtained the optimal solution, of value 17, defined by $x_{12} = x_{23} = x_{31} = x_{44} = 1$ (and $x_{ij} = 0$ elsewhere). ∎

Note that the way the tree is grown by Alternate(k) depends on the way the statement "let j be any vertex in $LV \setminus SV$" is implemented: Selecting j according to a First-In First-Out (FIFO) rule produces a breadth-first like growth of the tree, while use of a Last-In First-Out (LIFO) rule grows it in a depth-first way.

Finally, we mention Mack's so-called Bradford method [471], an approach developed in the 1960s that is in a sense equivalent to the Hungarian algorithm, but is more comprehensible and suitable for undergraduate teaching. A discussion of Mack's algorithm, that also includes possible ways to improve it, can be found in Jonker and Volgenant [393].

4.2.2 An O(n^3) implementation of the Hungarian algorithm

We have shown in Proposition 4.5 (point 1) that, after a dual updating, the previous alternating tree rooted in k still belongs to the new graph $G^0 = (U, V; E^0)$. This observation leads to an implementation (developed by Lawler [448] in the 1970s) that grows the current tree without resetting to empty the sets of scanned and labeled vertices (as done at each call to Alternate(k)). The procedure given in Algorithm 4.4 iterates tree extensions and dual updates until an augmenting path emanating from the input vertex k is obtained. The meaning of all variables is the same as in Section 4.2.1. In addition we make use of π_j ($j \in V$) to denote the minimum reduced cost of an edge connecting a labeled vertex $i \in U$ to j. Note that

(i) at each iteration the current vertex $i \in SU$ is assigned as predecessor to vertex $j \in V \setminus LV$ whenever the reduced cost of $[i, j]$ (even if it is greater than zero) improves on the current π_j value. However, j is added to LV only if such reduced cost is zero;

(ii) after each dual update the procedure adds to set LV all the unlabeled vertices of V for which a new incident edge having reduced cost equal to zero has been obtained.

Algorithm 4.4. Procedure Augment(k).
Find an augmenting tree rooted at an unassigned vertex $k \in U$

for each $j \in V$ **do** $\pi_j := +\infty$;
$SU := LV := SV := \emptyset$;
$sink := 0, i := k$;
while $sink = 0$ **do**
  $SU := SU \cup \{i\}$;
  for each $j \in V \setminus LV \; : \; c_{ij} - u_i - v_j < \pi_j$ **do**
    $pred_j := i, \pi_j := c_{ij} - u_i - v_j$;
    if $\pi_j = 0$ **then** $LV := LV \cup \{j\}$
  endfor;
  if $LV \setminus SV = \emptyset$ **then** [**comment:** dual update]
    $\delta := \min\{\pi_j \; : \; j \in V \setminus LV\}$;
    for each $i \in SU$ **do** $u_i := u_i + \delta$;
    for each $j \in LV$ **do** $v_j := v_j - \delta$;
    for each $j \in V \setminus LV$ **do**
      $\pi_j := \pi_j - \delta$;
      if $\pi_j = 0$ **then** $LV := LV \cup \{j\}$
    endfor
  endif;
  let j be any vertex in $LV \setminus SV$;
  $SV := SV \cup \{j\}$;
  if $row(j) = 0$ **then** $sink := j$ **else** $i := row(j)$
endwhile;
return $sink$

The Hungarian algorithm then reduces to iterated executions of Augment(k), each one followed by an increase of the primal solution, as shown in Algorithm 4.5

ALGORITHM 4.5. Hungarian_3.
$O(n^3)$ implementation of the Hungarian algorithm.

initialize u, v, row, φ and $\overline{U}$;
while $|\overline{U}| < n$ **do**
    let k be any vertex in $U \setminus \overline{U}$;
    $sink$:= Augment(k);
    $\overline{U} := \overline{U} \cup \{k\}$, $j := sink$;
    repeat
        $i := pred_j$, $row(j) := i$, $h := \varphi(i)$, $\varphi(i) := j$, $j := h$
    until $i = k$
endwhile

Example 4.7. We resume from the beginning the numerical instance developed in Examples 4.3, 4.4, and 4.6. The initialization produced by Basic_preprocessing is obviously the same, and the first call to Augment(k) (with $k = 2$) basically performs the same operations as Alternate(2), thus producing solution $x_{12} = x_{21} = x_{43} = 1$ (and $x_{ij} = 0$ elsewhere), shown in Figure 4.2(c), and $\overline{U} = \{1, 2, 4\}$, $row = (2, 1, 4, 0)$, $\varphi = (2, 1, 0, 3)$.

Augment(k) is then executed for $k = 3$. The two first iterations are very similar to the iterations performed by Alternate(3) when it fails (see Section 4.2.1) and produce the same alternating tree:

$\pi = (\infty, \infty, \infty, \infty)$, $SU = LV = SV = \emptyset$, $sink = 0$;

$i = 3$: $SU = \{3\}$, $pred = (3, 3, 3, 3)$, $\pi = (0, 3, 5, 8)$, $LV = \{1\}$;
    $j = 1$: $SV = \{1\}$;

$i = 2$: $SU = \{3, 2\}$, $pred = (3, 3, 2, 2)$, $\pi = (0, 3, 3, 5)$.

The first dual update follows: $\delta = 3$, $u = (7, 5, 4, 2)$, $v = (-3, 2, 0, 0)$, and $\pi = (0, 0, 0, 2)$, so $LV = \{1, 2, 3\}$. The execution of Augment(k) continues with

    $j = 2$: $SV = \{1, 2\}$;

$i = 1$: $SU = \{3, 2, 1\}$;
    $j = 3$: $SV = \{1, 2, 3\}$;

$i = 4$: $SU = \{3, 2, 1, 4\}$, $pred = (3, 3, 2, 4)$, $\pi = (0, 0, 0, 0)$, $LV = \{1, 2, 3, 4\}$;
    $j = 4$: $SV = \{1, 2, 3, 4\}$, $sink := 4$.

On return, Hungarian_3 produces the same optimal solution as before. ∎

The main loop of Augment(k) is executed $O(n)$ times, since (similarly to what happens for Alternate(k)) at each iteration a different vertex $j \in LV \setminus SV$ is selected and added to

SV. Each iteration is thus performed for a different i and requires $O(n)$ time, since the π_j values allow computation of δ in $O(n)$ time. The time complexity of Augment(k) is thus $O(n^2)$. Since the main loop of Hungarian_3 is executed $O(n)$ times, the algorithm has an overall $O(n^3)$ time complexity.

With respect to Hungarian, algorithm Hungarian_3 can be seen as a special primal-dual method in which, for each root k, a series of restricted primal problems independent of the costs is solved: each of these problems is not solved from scratch, but starting from the solution of the previous restricted primal. In addition, values π_j depending on the costs are computed in order to accelerate the subsequent dual update.

An improved $O(n^3)$ Hungarian algorithm was developed by Jonker and Volgenant [391]. Fortran implementations of Hungarian algorithms were proposed by McGinnis [485], Carpaneto and Toth [166], and Carpaneto, Martello, and Toth [165]. The Carpaneto and Toth [166] paper, which includes the Fortran listing of their code, provides computational comparisons with the primal simplex algorithm by Barr, Glover, and Klingman [68] (see Section 4.5). See Section 4.9 for links to available software in the public domain.

4.2.3 Cost-scaling algorithms

The *scaling* technique was developed in 1972 by Edmonds and Karp [250] for solving a minimum cost flow problem by treating a sequence of derived problems with scaled down capacities. The method was relatively ignored during the next decade, until in 1985 Gabow [295] adopted it for obtaining cost-scaling algorithms for a series of network problems, including a cost-scaling approach to LSAP based on the Hungarian method. This algorithm, as well as its improved version developed a few years later later by Gabow and Tarjan [297], had considerable theoretical interest, but did not produce computer implementations characterized by high efficiency in the practical solution of LSAP instances. Effective algorithms were however obtained in the 1990s by combining cost-scaling techniques and push-relabel methods, as can be seen in Section 4.6.

In order to introduce the cost-scaling approach, we need to assume that the costs c_{ij} are nonnegative integers with $\mathcal{C} = \max_{i,j}\{c_{ij}\}$. We already observed in Section 4.1 that the case of negative costs can be handled by subtracting the minimum (negative) value from all entries of C. If the costs are rational, they must be scaled up to integers

The basic idea of the cost-scaling approach is easily caught through a recursive description. Given a problem instance, halve the costs (i.e., a number a is replaced by $\lfloor a/2 \rfloor$) and solve the resulting instance recursively (directly solving the problem if all costs are zero). The solution obtained is near-optimal for the original instance: Transform it to an optimal one.

In iterative terms the approach can be described as follows. Let each cost be represented by $r = \lfloor \log \mathcal{C} \rfloor + 1$ bits. Start by optimally solving the problem with zero costs, and assume, for the sake of description simplicity, that they are represented by no bit at all. Then execute r iterations as follows. At iteration p, add the pth most significant bit of the original costs to the right of the bits representing the current costs: The current solution (previously optimal) is now near-optimal, so derive from it an optimal solution to the new problem. For example, the second row of the cost matrix used in the examples of the previous sections, (2, 8, 5, 7), would evolve as shown in Table 4.1.

Table 4.1. *Cost evolution in a scaling algorithm.*

i	Decimal				Binary			
0	0	0	0	0	–	–	–	–
1	0	1	0	0	0	1	0	0
2	0	2	1	1	00	10	01	01
3	1	4	2	3	001	100	010	011
4	2	8	5	7	0010	1000	0101	0111

The Gabow algorithm can be sketched as follows.

0. initialize the $n \times n$ current cost matrix $\widehat{C}$ to zero;
   choose any complete matching on the induced bipartite graph $G = (U, V; E)$, and set $u_i = 0$ for all $i \in U$ and $v_j = 0$ for all $j \in V$;

1. **for** $p := 1$ **to** r **do**

   1.1 scale up the current costs $\hat{c}_{ij}$;
   start with an empty matching on G;
   set $u_i = 2u_i$ for all $i \in U$ and $v_j = 2v_j$ for all $j \in V$;

   1.2 **repeat**

   find an augmenting path (see Section 4.2.2) for costs $\hat{c}_{ij}$, and let u_i,v_j be the updated dual variables;

   let $G^0 = (U, V; E^0)$ be the bipartite partial graph of G that only contains edges $[i, j]$ such that $\hat{c}_{ij} - u_i - v_j = 0$;

   transform the current matching to a maximum cardinality matching on G^0 through the algorithm by Hopcroft and Karp [376]

   until the current matching is complete.

endfor

Observe that at each iteration the dual variables (u_i, v_j) are doubled, and the costs $\hat{c}_{ij}$ are at least doubled, so the current optimal dual solution (satisfying $\hat{c}_{ij} - u_i - v_j \geq 0$) is transformed into a feasible solution for the scaled costs. The algorithm runs in $O(n^{3/4} m \log C)$ (weakly polynomial) time. The overall correctness of the algorithm and its time complexity are formally proved in [295].

The time complexity of the Gabow [295] algorithm was later improved by Gabow and Tarjan [297]. Instead of computing an optimal solution at each of $\log C$ iterations, the new algorithm computes an approximate solution at each of $\log(nC)$ iterations, but the additional $\log n$ iterations ensure that the last approximate solution is optimal. The time complexity is $O(\sqrt{n}\, m \log(nC))$. This is the best time bound currently known for a cost-scaling algorithm for LSAP. The same time bound was later obtained by Orlin and Ahuja [515] through a hybrid scaling algorithm (see Section 4.6.3).

4.3 The Dinic–Kronrod algorithm

Well before the $O(n^3)$ implementation of the Hungarian algorithm, Dinic and Kronrod [235] had proposed a different approach, totally independent of linear programming duality theory. Although ignored for a long time (the algorithm is presented in a very sketchy way, with several inaccuracies in the English translation), this is the first $O(n^3)$ time algorithm for LSAP. The key observation in [235] is the following.

Theorem 4.8. (Dinic and Kronrod [235], 1969.) *Given n values Δ_j ($j = 1, 2, \ldots, n$), let an element c_{ij} be called Δ-minimal if $c_{ij} - \Delta_j \le c_{ik} - \Delta_k$ for all k. Then a set of n Δ-minimal elements $c_{i\varphi(i)}$ ($i = 1, 2, \ldots, n$) such that $\varphi(i) \neq \varphi(k)$ if $i \neq k$ is an optimal solution to LSAP.*

Proof. The value of the considered solution is

$$\sum_{i=1}^{n} c_{i\varphi(i)} = \sum_{j=1}^{n} \Delta_j + \sum_{i=1}^{n} (c_{i\varphi(i)} - \Delta_{\varphi(i)}). \tag{4.13}$$

The first sum in the right-hand side is a constant. The second sum is minimal by definition. □

Observe that, under the hypothesis of Theorem 4.8, by setting $v_j = \Delta_j$ ($j = 1, 2, \ldots, n$) and $u_i = c_{i\varphi(i)} - \Delta_{\varphi(i)}$ ($i = 1, 2, \ldots, n$) we obtain a feasible dual solution satisfying the complementary slackness conditions (4.8), thus proving in a different way the optimality of solution φ.

The Dinic–Kronrod algorithm can be described as follows. Given a (possibly zero) vector Δ, a Δ-minimal element is initially selected in each row. If all the selected elements are in different columns, then the solution is optimal by Theorem 4.8. If this is not the case, let us define the *deficiency* of the solution as the number of columns with no assigned element. One such column is then selected, one or more Δ_j values are increased, and a new solution with deficiency reduced by one is obtained. Let s be the selected column (and set $SV = \{s\}$). Value Δ_s is increased by a quantity δ such that (i) all currently selected elements remain Δ-minimal; (ii) there exists a row i for which $c_{i\varphi(i)} - \Delta_{\varphi(i)} = c_{is} - (\Delta_s + \delta)$, i.e., we have an alternative element selection for row i. If column $\varphi(i)$ has more than one row assigned, then the deficiency is reduced by selecting, in row i, element c_{is} instead of $c_{i\varphi(i)}$. Otherwise, column $\varphi(i)$ is added to set SV and Δ_s and $\Delta_{\varphi(i)}$ are increased by a new value δ such that condition (i) above holds and there is one new row with alternative element selection involving a column of SV. The process is iterated until a column $\varphi(i)$ with more than one row assigned is encountered, and the solution is updated by appropriately shifting the alternative assignments.

Dinic and Kronrod proposed different ways to compute quantity δ, used for the updating of the Δ_j values. We start with a possible implementation of the simplest one in order to catch the basic idea. In Algorithm 4.6, s denotes the selected unassigned column, while sets LU and SV store the currently labeled rows and scanned columns, respectively.

Algorithm 4.6. Procedure DK_basic(s).
Basic method for reducing the deficiency.

$SV := LU := \emptyset$;
$sink := 0,\ j := s$;
while $sink = 0$ **do**
  $SV := SV \cup \{j\}$;
  $\delta := \min\{(c_{ij} - \Delta_j) - (c_{i\varphi(i)} - \Delta_{\varphi(i)})\ :\ i \in U \setminus LU,\ j \in SV\}$;
  let i^* be the row i that determines δ, and set $LU := LU \cup \{i^*\}$;
  for each $j \in SV$ **do** $\Delta_j := \Delta_j + \delta$;
  $j := \varphi(i^*)$;
  if column j has two or more rows assigned **then** $sink := i^*$
endwhile;
return $sink$

One can immediately see that the computation of δ ensures that, after the updating of Δ, all elements currently selected in rows $i \notin LU$ remain Δ-minimal and that an alternative assignment results for row i^* in a column of SV. The currently selected elements in rows $i \in LU$ remain Δ-minimal as well, since

(i) all the corresponding columns have been added to SV, and

(ii) only the Δ_j values of such columns have been updated, and all have been increased by the same quantity δ.

The time complexity of DK_basic(s) depends on the way δ is computed. A straightforward implementation, requiring $O(n^2)$ time, would produce an $O(n^3)$ time procedure, hence an $O(n^4)$ time algorithm for LSAP. The best way given in [235] can be implemented as in the $O(n^2)$ procedure introduced in Algorithm 4.7, where δ is computed in $O(n)$ time. The statements used for the computation of δ are those involving vector q, where each q_i value ($i \notin LU$) stores the minimum $c_{ij} - \Delta_j$ value for $j \in SV$. The remaining notations are similar to those adopted in the previous sections, with a few differences due to the fact that the Dinic–Kronrod algorithm operates on columns instead of rows. The column $j \in SV$ from which row $i \in U$ has been labeled is stored in $pred_i$.

Algorithm 4.7. Procedure DK_reduce(s).
$O(n^2)$ method for reducing the deficiency.

for each $i \in U$ **do** $q_i := +\infty$;
$SV := LU := \emptyset$;
$sink := 0,\ \delta := 0,\ j := s$;
while $sink = 0$ **do**
  $SV := SV \cup \{j\}$;
  for each $i \in U \setminus LU$ **do**
    if $c_{ij} - \Delta_j < q_i$ **then** $pred_i := j,\ q_i := c_{ij} - \Delta_j$
  endfor;
  $\delta := \min\{q_i - (c_{i\varphi(i)} - \Delta_{\varphi(i)})\ :\ i \in U \setminus LU\}$;

```
    let i* be the row i that determines δ, and set LU := LU ∪ {i*};
    for each j ∈ SV do Δ_j := Δ_j + δ;
    for each i ∈ U \ LU do q_i := q_i − δ;
    j := φ(i*);
    if column j has two or more rows assigned then sink := i*
endwhile;
return sink
```

The complete $O(n^3)$ method, shown in Algorithm 4.8, consists of a main loop that applies the above procedure to each column with no assigned row. Array *row* has non-zero values only for the columns that have exactly one row assigned.

ALGORITHM 4.8. Dinic_Kronrod.
Dinic–Kronrod algorithm.

```
initialize Δ; [comment: possibly Δ_j = 0 for j = 1, 2, ..., n]
for i := 1 to n do φ(i) := arg min{c_ij − Δ_j : j = 1, 2, ..., n};
for j := 1 to n do row(j) := 0;
for i := 1 to n do
    if column φ(i) has exactly one row assigned then row(φ(i)) := i;
V̄ := {j : column j has at least one row assigned};
while |V̄| < n do
    let k be any column in V \ V̄;
    sink := DK_reduce(k);
    V̄ := V̄ ∪ {k}, i := sink, ĵ := φ(i);
    repeat
        j := pred_i, φ(i) := j, h := row(j), row(j) := i, i := h
    until j = k
    if column ĵ has exactly one row, say i, assigned then row(ĵ) := i
endwhile
```

Example 4.9. We start from the very beginning with the nonreduced input matrix C introduced in Example 4.3 and reproduced here in Figure 4.4(a). The Δ values are shown in row zero. The algorithm starts by setting $\Delta = (0, 0, 0, 0)$ and computing $\varphi = (1, 1, 1, 3)$, $row = (0, 0, 4, 0)$, $\overline{V} = \{1, 3\}$. The assignment is given by the underlined entries: The deficiency of the solution is 2 (columns 2 and 4 are not matched). The first iteration is performed by calling DK_reduce(2):

$q = (\infty, \infty, \infty, \infty)$, $SV = LU = \emptyset$, $sink = 0$, $\delta = 0$;
$j = 2$: $SV = \{2\}$, $pred = (2, 2, 2, 2)$, $q = (9, 8, 6, 6)$;
    $\delta = 2$, $i^* = 1$: $LU = \{1\}$, $\Delta = (0, 2, 0, 0)$, $q = (9, 6, 4, 4)$;
$j = 1$: $sink = 1$.

Figure 4.4(b) shows the current $c_{ij} - \Delta_j$ values, with c_{11} and c_{12} now both Δ-minimal. The new solution is then determined by $\Delta = (0, 2, 0, 0)$, $\varphi = (2, 1, 1, 3)$, $row = (0, 1, 4, 0)$,

$$
\begin{array}{cccc}
\begin{array}{cccc} 0 & 0 & 0 & 0 \end{array} & \begin{array}{cccc} 0 & 2 & 0 & 0 \end{array} & \begin{array}{cccc} 0 & 2 & 1 & 1 \end{array} & \begin{array}{cccc} 0 & 4 & 3 & 3 \end{array} \\
\begin{pmatrix} \underline{7} & 9 & 8 & 9 \\ \underline{2} & 8 & 5 & 7 \\ \underline{1} & 6 & 6 & 9 \\ 3 & 6 & \underline{2} & 2 \end{pmatrix} &
\begin{pmatrix} 7 & \underline{7} & 8 & 9 \\ \underline{2} & 6 & 5 & 7 \\ \underline{1} & 4 & 6 & 9 \\ 3 & 4 & \underline{2} & 2 \end{pmatrix} &
\begin{pmatrix} 7 & \underline{7} & 7 & 8 \\ \underline{2} & 6 & 4 & 6 \\ \underline{1} & 4 & 5 & 8 \\ 3 & 4 & \underline{1} & 1 \end{pmatrix} &
\begin{pmatrix} 7 & \underline{5} & 5 & 6 \\ 2 & 4 & \underline{2} & 4 \\ \underline{1} & 2 & 3 & 6 \\ 3 & 2 & -1 & \underline{-1} \end{pmatrix} \\
\text{(a)} & \text{(b)} & \text{(c)} & \text{(d)}
\end{array}
$$

Figure 4.4. *Values $c_{ij} - \Delta_j$ during execution of the Dinic–Kronrod algorithm.*

$\overline{V} = \{1, 3, 2\}$. A new iteration is thus performed by calling DK_reduce(4):

$q = (\infty, \infty, \infty, \infty)$, $SV = LU = \emptyset$, $sink = 0$, $\delta = 0$;
$j = 4$: $SV = \{4\}$, $pred = (4, 4, 4, 4)$, $q = (9, 7, 9, 2)$;
  $\delta = 0$, $i^* = 4$: $LU = \{4\}$, $\Delta = (0, 2, 0, 0)$;
$j = 3$ $SV = \{4, 3\}$, $pred = (3, 3, 3, 4)$, $q = (8, 5, 6, 2)$;
  $\delta = 1$, $i^* = 1$: $LU = \{4, 1\}$, $\Delta = (0, 2, 1, 1)$, $q = (8, 4, 5, 2)$ (Figure 4.4(c));
$j = 2$ $SV = \{4, 3, 2\}$, $pred = (3, 3, 2, 4)$, $q = (8, 4, 4, 2)$;
  $\delta = 2$, $i^* = 2$: $LU = \{4, 1, 2\}$, $\Delta = (0, 4, 3, 3)$, $q = (8, 4, 2, 2)$;
$j = 1$: $sink = 2$.

Figure 4.4(d) shows the new $c_{ij} - \Delta_j$ values. The complete solution is then produced by $\Delta = (0, 4, 3, 3)$: $\varphi = (2, 3, 1, 4)$, $row = (3, 1, 2, 4)$. ■

The Hungarian algorithm operates on a partial feasible assignment: At each iteration, Procedure Augment(k) produces a new partial feasible solution with one additional assignment. The Dinic–Kronrod algorithm operates on a complete but infeasible row assignment: At each iteration, Procedure DK_reduce(s) produces a new solution where one more column is assigned and a multiply-assigned column has one assignment less.

In order to further analyze differences and similarities between the Dinic–Kronrod and the $O(n^3)$ implementation of the Hungarian algorithm, let us consider the usual underlying bipartite graph $G = (U, V; E)$. Procedures Augment(k) and DK_reduce(s) are apparently very different. Augment(k) determines an alternating path going from an unassigned vertex of one shore of G to an unassigned vertex of the other shore (U and V, respectively, in our implementation, but the opposite choice would be equivalent). The path produced by DK_reduce(s) goes instead from an unassigned vertex of one shore (V) to a multiply-assigned vertex of the same shore. We already observed that Augment(k) obtains the required path by growing an alternating tree rooted in the starting vertex, and it is not difficult to recognize that the same holds for DK_reduce(s). Indeed, at each iteration it selects the best edge going from a scanned vertex of V (initially, $SV = \{s\}$) to a vertex $i^* \in U$ not belonging to the tree: Both this edge and the assigned edge $[i^*, \varphi(i^*)]$ are added to the tree, while vertex $\varphi(i^*)$ is added to SV.

A closer look at the two procedures shows that the choice of the next edge to be added to the tree is equivalent. Consider indeed the alternative implementation of Augment(k) that starts from a vertex of V, thus using sets LU, SU, and SV: the value of π_i, minimum reduced cost of an edge connecting a labeled vertex $j \in V$ to $i \in U$, would be $\pi_i = c_{ij} - u_i - v_j$ for each edge $[i, j]$ evaluated for addition to the current tree, and the choice would be produced by $\delta = \min\{\pi_i : i \in U \setminus LU\}$. Consider now DK_reduce(s) and recall that a feasible dual solution is given by $v_j = \Delta_j$ $(j = 1, 2, \dots, n)$ and $u_i = c_{i\varphi(i)} - \Delta_{\varphi(i)}$ $(i = 1, 2, \dots, n)$. Hence, when $pred_i = j$, we have $q_i = c_{ij} - \Delta_j = \pi_i + u_i$ and we can write

$$\delta = \min\{q_i - u_i - (c_{i\varphi(i)} - u_i - \Delta_{\varphi(i)}) : i \in U \setminus LU\}. \tag{4.14}$$

Since $c_{i\varphi(i)} - u_i - \Delta_{\varphi(i)} = 0$, the same next edge is selected for growing the current tree.

Finally, we observe that the idea of evolving relaxed problems by updating a dual solution through shortest paths, until the solution becomes feasible, is also the basis of the dual algorithm by Hung and Rom [382], treated in Section 4.6.1.

4.4 Shortest path implementation of the Hungarian algorithm

Most efficient algorithms for LSAP are based on shortest augmenting path techniques. In the early 1960s, Hoffman and Markowitz [373] observed that an LSAP can be solved through a sequence of n shortest paths on cost matrices of increasing size from 1×1 to $n \times n$. Such matrices, however, could include negative costs, hence, each shortest path search would require $O(n^3)$ time. In the early 1970s Tomizawa [640] and Edmonds and Karp [250], studying shortest path algorithms for the min-cost flow problem, observed that, by using the reduced costs, one can apply the Dijkstra algorithm to obtain an $O(n^3)$ time algorithm for LSAP.

4.4.1 Transformation to a minimum cost flow problem

Relations between assignment problems and min-cost flow problems were already discussed in Sections 1.1, 1.2, 2.1, and 3.2. Given a digraph $D = (N; A)$ with integer *capacity* $q(i, j)$ and *unit cost* c_{ij} associated with each arc $(i, j) \in A$, the *minimum cost flow problem* (MCFP) consists in transmitting a prefixed integer quantity ζ of flow from a given *source* vertex $s \in N$ to a given *sink* vertex $t \in N$ so that the flow in each arc does not exceed the corresponding capacity and the total cost is a minimum. Let $f(i, j)$ denote the flow along arc (i, j). The problem can be formally modeled as

$$\min \sum_{(i,j)\in A} c_{ij} f(i, j) \tag{4.15}$$

$$\text{s.t.} \quad \sum_{(h,j)\in A} f(h, j) - \sum_{(i,h)\in A} f(i, h) = \begin{cases} \zeta & \text{if } h = s, \\ -\zeta & \text{if } h = t, \\ 0 & \text{otherwise} \end{cases} \qquad (h = 1, 2, \dots, |N|), \tag{4.16}$$

$$0 \le f(i, j) \le q(i, j) \qquad ((i, j) \in A). \tag{4.17}$$

It easily follows from Theorem 4.2 that the constraint matrix of MCFP is totally unimodular, since we can set I_1 to the set of all rows and I_2 to the empty set. Hence, a feasible flow will be integral valued, since ζ and the capacities are integers.

An LSAP instance defined on a graph $G = (U, V; E)$ can be transformed into an equivalent instance of MCFP as follows (see also Section 1.1). Let

$$N = \{s\} \cup U \cup V \cup \{t\}, \tag{4.18}$$
$$A = \{(s, i) : i \in U\} \cup \{(i, j) : [i, j] \in E, i \in U, j \in V\} \cup \{(j, t) : j \in V\}. \tag{4.19}$$

Now assign zero cost to the arcs emanating from s and to those entering t, associate the matching costs c_{ij} with the remaining arcs (i, j) of A ($i \in U$, $j \in V$), and set all capacities to one. The solution to the LSAP instance is then obtained by transmitting, at minimum cost, a quantity n of flow from s to t.

Shortest path algorithms for MCFP operate on the so-called incremental graph. Given a feasible flow $\mathcal{X} = (f(i, j))$, the *incremental digraph* $D^r = (N; A^r)$ is obtained from D as follows: $A^r = A^f \cup A^b$, with $A^f = \{(i, j) \in A : f(i, j) < q(i, j)\}$ (*forward arcs*) and $A^b = \{(j, i) : (i, j) \in A \text{ and } f(i, j) > 0\}$ (*backward arcs*). A forward arc $(i, j) \in A^f$ has cost $c^r_{ij} = c_{ij}$ and *residual capacity* $q^r(i, j) = q(i, j) - f(i, j)$, while a backward arc $(j, i) \in A^b$ has cost $c^r_{ji} = -c_{ij}$ and residual capacity $q^r(j, i) = f(i, j)$.

A *shortest path algorithm* for MCFP works as follows. Given a minimum cost feasible flow $\mathcal{X}$ transmitting $\widetilde{\zeta}$ ($\widetilde{\zeta} < \zeta$) flow units (possibly $\widetilde{\zeta} = 0$ at the first iteration), we look for a shortest path from s to t in the corresponding incremental graph. If such a path is found, the flow is increased by transmitting along the path an additional flow δ equal to the minimum between $\zeta - \widetilde{\zeta}$ and the minimum residual capacity of an arc in the path. It can be proved that the resulting flow is the minimum cost solution for the transmission of $\widetilde{\zeta} + \delta$ flow units from s to t. The process is iterated until either the prefixed quantity ζ of flow has been obtained or no path from s to t exists (implying that the instance does not have a feasible solution).

This algorithm requires pseudo-polynomial time for solving a general MCFP instance, as ζ shortest path rounds are needed, in the worst case, to obtain the optimal solution. When applied to an LSAP instance, however, its complexity is polynomial as $\zeta = n$. A straightforward implementation of the approach would need a procedure for determining shortest paths in graph D^r that contain arcs with negative costs. Tomizawa [640] and, independently, Edmonds and Karp [250] observed, however, that the shortest path approach remains valid if applied to the incremental graph D^r with costs c^r_{ij} replaced by the corresponding reduced costs. Since, as already observed, at each iteration we have a minimum cost partial flow, the reduced costs are nonnegative and the Dijkstra algorithm can be used to find shortest paths in $O(n^2)$ time.

4.4.2 Basic implementation

In order to obtain a simpler algorithm for LSAP, one can observe that (i) the MCFP induced by an LSAP instance is equivalent to a multi-source multi-sink problem where each vertex of U must transmit a flow unit and each vertex of V must receive a flow unit; hence, (ii), as already proved by Tomizawa [640], an optimal solution can be obtained by considering a source at a time and finding the shortest path emanating from it; (iii) it is not necessary to explicitly build the incremental graph as all computations can be performed on the LSAP bipartite graph G. The resulting algorithm starts with a given (possibly empty) optimal partial assignment and a corresponding optimal dual solution. At each iteration the algorithm

(i) selects an unassigned vertex of U; (ii) considers the incremental graph and finds the shortest path from the selected vertex to an unassigned vertex of V; (iii) augments the partial solution by interchanging the assignments along this path; and (iv) updates the dual variables so that complementary slackness holds. The similarity of this approach to the $O(n^3)$ implementation of the Hungarian algorithm introduced in Section 4.2.2 is thus evident. Indeed, Derigs [226] formally proved that the two methods perform the same augmentations, but the shortest path approach is a more efficient implementation in the sense that a sequence of dual updates performed by Procedure Augment(k) is replaced by a single dual update only performed when augmentation occurs.

The procedure shown in Algorithm 4.9 finds a shortest path arborescence emanating from a given (unassigned) vertex $k \in U$ and terminates execution as soon as an unassigned vertex of V is reached. Sets SU and SV contain the vertices already reached by a shortest path emanating from k (scanned vertices). The value π_j ($j \in V$) is the Dijkstra label, i.e., it stores the cost of the shortest path from k to j that only passes through vertices of $SU \cup SV$. Explicit labels for the vertices of U are not used as they can be handled implicitly (see below). Finally, δ is the cost of the shortest path from k to the last vertex that entered SV, i.e., the largest cost of a path in the arborescence rooted at k.

Algorithm 4.9. Procedure Shortest_path(k).
Find a shortest path in the incremental graph.

for each $j \in V$ **do** $\pi_j := +\infty$;
$SU := SV := \emptyset$;
$sink := 0, \delta := 0, i := k$;
while $sink = 0$ **do**
  $SU := SU \cup \{i\}$;
  for each $j \in V \setminus SV \ : \ \delta + c_{ij} - u_i - v_j < \pi_j$ **do**
    $pred_j := i, \pi_j := \delta + c_{ij} - u_i - v_j$
  endfor;
  $j := \arg\min\{\pi_h \ : \ h \in V \setminus SV\}$;
  $SV := SV \cup \{j\}, \delta := \pi_j$;
  if $row(j) = 0$ **then** $sink := j$
  else $i := row(j)$
endwhile;
comment: dual update;
$u_k := u_k + \delta$;
for each $i \in SU \setminus \{k\}$ **do** $u_i := u_i + \delta - \pi_{\varphi(i)}$;
for each $j \in SV$ **do** $v_j := v_j - \delta + \pi_j$
return $sink$

In the incremental digraph, each unassigned vertex $j \in V$ has no emanating arc, while each assigned vertex $j \in V$ has the unique emanating arc $(j, row(j))$. Due to the optimality of the current partial assignment, the reduced cost of any such arc is zero, so, when j is scanned, the current shortest path is immediately extended to $row(j)$. The label of vertex $row(j) \in U$ would have value π_j, i.e., the minimum label value among unscanned vertices. It follows that it is not necessary to explicitly have labels for the vertices in U as the path is extended by scanning $row(j)$. When $row(j) = 0$, an augmenting path from k to j has been obtained. We next show that Shortest_path(k) produces a new feasible dual solution.

Proposition 4.10. *If the input dual variables satisfy the dual feasibility constraints (4.7) and produce zero reduced costs for the current partial assignment, then Shortest_path(k) returns updated dual variables satisfying (4.7) and a shortest path arborescence whose arcs have an updated reduced cost of value zero.*

Proof. In order to simplify the proof, we explicitly use the Dijkstra labels for the vertices $i \in SU$, namely, $\mu_i = \pi_{\varphi(i)}$ for $i \in SU \setminus \{k\}$ and $\mu_k = 0$. Let $\overline{c}_{ij}$ denote the input reduced costs. Observe that, during the shortest path search, we have

$$\mu_i + \overline{c}_{ij} \geq \pi_j \; \forall \, i \in SU, \; j \in V \tag{4.20}$$

since otherwise $\mu_i + \overline{c}_{ij}$ would be the cost of a path from k to j shorter than the current path of cost π_j.

We first prove that the updated reduced costs are nonnegative, i.e., that (4.7) holds for the updated dual variables. We consider the four pairs i, j whose dual variables are updated in a different way:

(a) $i \in SU, j \in SV$: the updated reduced cost has the value $\overline{c}_{ij} - \delta + \mu_i + \delta - \pi_j$, which is nonnegative by (4.20). Observe in addition that if (i, j) belongs to the shortest path arborescence, then such a cost is zero by definition since $\pi_j = \mu_i + \overline{c}_{ij}$;

(b) $i \in SU, j \notin SV$: the updated reduced cost has the value $\overline{c}_{ij} - \delta + \mu_i$, which is nonnegative by (4.20) and by the fact that $\delta = \min\{\pi_h : h \in V \setminus SV\}$;

(c) $i \notin SU, j \in SV$: the updated reduced cost has the value $\overline{c}_{ij} + \delta - \pi_j$, which is nonnegative since, by construction, we have $\delta \geq \pi_j$ for all $j \in SV$;

(d) $i \notin SU, j \notin SV$: no updating occurs.

Now consider the shortest path arborescence: we have observed that any forward arc (i, j) $(i \in SU, j \in SV)$ has an updated reduced cost equal to zero. The same holds for backward arcs (j, i) $(j \in SV, i \in SU)$ for which the input reduced cost is zero and $\mu_i = \pi_j$. □

Assume that, given a (possibly empty) partial primal solution and a feasible dual solution satisfying complementary slackness, Shortest_path(k) is invoked for an unassigned vertex $k \in U$. The returned shortest path from k to $sink \in V$ can be used to augment the primal solution by removing the assignments corresponding to backward arcs along the path and adding those corresponding assignments to forward arcs. It follows from Proposition 4.10 that this new solution and the updated dual variables satisfy the complementary slackness conditions. We have thus proved correctness of the shortest path method for LSAP given in Algorithm 4.10.

Algorithm 4.10. Hungarian_SP.
Shortest path implementation of the Hungarian algorithm.

initialize u, v, *row*, φ and $\overline{U}$;
while $|\overline{U}| < n$ **do**
  let k be any vertex in $U \setminus \overline{U}$;
  $sink :=$ Shortest_path(k);
  $\overline{U} := \overline{U} \cup \{k\}$, $j := sink$;
  repeat
    $i := pred_j$, $row(j) := i$, $h := \varphi(i)$, $\varphi(i) := j$, $j := h$
  until $i = k$
endwhile

Developing Example 4.3 for Hungarian_SP would not highlight the difference with respect to Hungarian_3, as both algorithms find the optimal solution through a single dual updating (executed by Augment(3) after the scanning of line 1, and by Shortest_path(3) before return). We will instead develop the following example.

Example 4.11. Consider the following input matrix C. By executing the basic preprocessing of Algorithm 4.1, we obtain the dual variables (shown on the left and on the top) and the corresponding reduced cost matrix $\overline{C}$. The partial assignment (underlined in $\overline{C}$) is thus $row = (1,\ 2,\ 3,\ 0)$, and hence, $\varphi = (1,\ 2,\ 3,\ 0)$.

$$
\begin{array}{c}
\begin{matrix} & 0 & 1 & 0 & 0 \end{matrix}\\
\begin{matrix} 6\\4\\1\\2 \end{matrix}
\begin{pmatrix} 6 & 9 & 11 & 10\\ 6 & 5 & 7 & 4\\ 7 & 8 & 1 & 5\\ 3 & 7 & 2 & 9 \end{pmatrix}\\
C
\end{array}
\qquad
\begin{array}{c}
\begin{pmatrix} \underline{0} & 2 & 5 & 4\\ 2 & \underline{0} & 3 & 0\\ 6 & 6 & \underline{0} & 4\\ 1 & 4 & 0 & 7 \end{pmatrix}\\
\overline{C}
\end{array}
$$

Since $\overline{U} = \{1, 2, 3\}$, a single augmentation will produce the optimal solution. The execution of Shortest_path(4) is as follows.

$\pi = (\infty, \infty, \infty, \infty)$, $SU = SV = \emptyset$, $sink = 0$, $\delta = 0$;

$i = 4$: $SU = \{4\}$, $pred = (4, 4, 4, 4)$, $\pi = (1, 4, 0, 7)$;
  $j = 3$: $SV = \{3\}$, $\delta = 0$;
$i = 3$: $SU = \{4, 3\}$, $pred = (4, 4, 4, 3)$, $\pi = (1, 4, 0, 4)$;
  $j = 1$: $SV = \{3, 1\}$, $\delta = 1$;
$i = 1$: $SU = \{4, 3, 1\}$, $pred = (4, 1, 4, 3)$, $\pi = (1, 3, 0, 4)$;
  $j = 2$: $SV = \{3, 1, 2\}$, $\delta = 3$;
$i = 2$: $SU = \{4, 3, 1, 2\}$, $pred = (4, 1, 4, 2)$, $\pi = (1, 3, 0, 3)$;
  $j = 4$: $SV = \{3, 1, 2, 4\}$, $\delta = 3$;
  $sink = 4$, $u = (8, 4, 4, 5)$, $v = (-2, 1, -3, 0)$.

On return, Hungarian_SP produces the optimal solution: $x_{12} = x_{24} = x_{33} = x_{41} = 1$ (and $x_{ij} = 0$ elsewhere). We have thus obtained the required augmentation through a single dual update. It is left as an exercise to check that, in Hungarian_3, the call to Augment(4) produces the same primal and dual solution, but through two dual updates, performed after the scanning of lines 3 and 1, respectively. ■

The time complexity of Shortest_path(k) is $O(n^2)$ time (it is in practice the Dijkstra algorithm for a bipartite graph of $2n$ vertices), so it immediately follows that Hungarian_SP has $O(n^3)$ time complexity.

4.4.3 Efficient implementations

A number of implementations of shortest path algorithms have been proposed in the literature. Engquist [255] gave a shortest path refinement of the Dinic–Kronrod algorithm, and computationally compared it with the primal algorithm by Barr, Glover, and Klingman [68] (see Section 4.5.2) and with the dual algorithms by Hung and Rom [382] (see Section 4.6.1) and Weintraub and Barahona [661]. An algorithm that combines row permutations and shortest paths on a reduced cost matrix was proposed by Bhat and Kinariwala [96]. An efficient labeling technique, especially effective for sparse assignment problems, was proposed by Derigs and Metz [228].

The most efficient shortest path algorithms for LSAP basically differ in three aspects:

(a) the implementation of the procedure that is used for determining the *shortest paths*;

(b) a possible *sparsification technique*, which solves an instance with a reduced number of edges and iteratively adds edges until an optimal solution for the original graph is obtained;

(c) the *preprocessing method*, used to determine a feasible dual solution and a partial primal solution (where less than n rows are matched) satisfying the complementary slackness conditions, such as the $O(n^2)$ time Procedure Basic_preprocessing (Algorithm 4.1), that we have been using so far.

Tarjan [633] and Fredman and Tarjan [278] used special data structures (such as, e.g., Fibonacci heaps) to compute shortest paths so as to obtain algorithms having time complexity $O(nm \log_{(2+m/n)} n)$ and $O(n^2 \log n + nm)$, respectively, that are particularly efficient for sparse matrices. The latter complexity is the best strongly polynomial-time bound known for LSAP.

Sparsification techniques operate in two phases. A *core problem* is first defined by selecting a subset of entries from C, and its primal and dual optimal solutions are determined. If all reduced costs are nonnegative, then these solutions are also optimal for the complete instance. Otherwise, the second phase enlarges the core through additional entries of C and the process is iterated. Carraresi and Sodini [172] and, independently, Glover, Glover, and Klingman [322] developed so-called *threshold algorithms* in which the core problem consists of "short" edges induced by a threshold value that is updated after each augmentation. In Derigs and Metz [229] the core problem is given by the k edges of smallest cost incident to each vertex (where k is a prefixed value depending on n) and the iterative phase enlarges it through post-optimal analysis and out-pricing. Carpaneto and Toth [169] proposed an algorithm in which the core problem is produced by considering the small reduced cost elements. The threshold value is proportional to the average reduced cost value, and the enlargement is performed either by adding the entries with negative reduced cost (when a primal feasible solution has been obtained) or by doubling the threshold (when the sparse problem has no primal feasible solution). A core approach was also used by Lee and Orlin [450] for solving very large random instances of LSAP (including a one million vertex, one trillion edge instance) by generating the edges in ascending order of their costs until a solution can be verified to be optimal for the entire problem. Volgenant [647] modified the Jonker and Volgenant [392] shortest path algorithm (see Section 4.4.4) by initially defining the core through selection of the k lowest cost entries in each row (k a prefixed value) and refining it, for each row, through exchanges between core elements of cost greater than the current average core cost of the row and non-core elements of cost smaller than this average. Diagonal elements are added to ensure the existence of a feasible solution, and the enlargement is produced by negative reduced cost entries.

Many authors have observed that preprocessing is a crucial tool for the implementation of efficient shortest path algorithms for LSAP. An analysis of the expected number of initial assignments produced by a slightly modified version of Procedure Basic_preprocessing (Algorithm 4.1) under the assumption that the costs are a family of independent identically distributed random variables with continuous distribution function can be found in Nawijn and Dorhout [510]. They proved that, for n sufficiently large, the expected number of initial assignments is $(2 - \exp(-1/e) - \exp(-\exp(-1/e)))\, n \approx 0.8073\, n$.

Some of the most effective preprocessing methods are examined in the next section.

4.4.4 Preprocessing

Various preprocessing techniques have been proposed by Carpaneto and Toth [166, 168, 169] and Carpaneto, Martello, and Toth [165]. The most efficient one is shown in Algorithm 4.11.

Initially, a column reduction is performed and a first partial solution is obtained by assigning each column to an unassigned row (if any) corresponding to the minimum column cost. The second phase performs a row reduction and tries to enlarge the current partial assignment. For each unassigned row i, the column j^* corresponding to the minimum reduced row cost is considered. If j^* is not assigned the solution is immediately enlarged, as in Procedure Basic_preprocessing. If instead j^* is currently assigned to row $r = row(j^*)$, an attempt is made to find an alternative assignment for this row, i.e., an unassigned column j having zero reduced cost. If such a column is found, the solution is enlarged by assigning row i to column j^* and row r to column j. Otherwise, the next column j^* having zero reduced cost in row r (if any) is selected and the attempt is iterated. Note that, with respect to the associated bipartite graph $G = (U, V; E)$, this is equivalent to executing, for each unassigned vertex $i \in U$, a modified Procedure Alternate(i) (see Section 4.2.1) which only considers alternating paths composed by three edges or less. In order to preserve an $O(n^2)$ overall time complexity, $next\_j(i)$ stores the first unchecked column in row i ($i = 1, 2, \ldots, n$).

Algorithm 4.11. Procedure Three_edge_preprocessing.
Carpaneto and Toth three edge preprocessing.

for $h := 1$ **to** n **do** $row(h) := \varphi(h) := 0$;
comment: column reduction and partial feasible solution;
for $j := 1$ **to** n **do**
  $i^* := \arg\min\{c_{ij} : i = 1, 2, \dots, n\}$ (breaking ties by $\varphi(i) = 0$);
  $v_j := c_{i^*j}$;
  if $\varphi(i^*) = 0$ **then**
    $\varphi(i^*) := j$, $row(j) := i^*$, $u_{i^*} := 0$, $next\_j(i^*) := j + 1$
  endif
endfor;
comment: row reduction and enlarged partial feasible solution;
for $i := 1$ **to** n **do if** $\varphi(i) = 0$ **then**
  $j^* := \arg\min\{c_{ij} - v_j : j = 1, 2, \dots, n\}$ (breaking ties by $row(j) = 0$);
  $u_i := c_{ij^*} - v_{j^*}$;
  if $row(j^*) = 0$ **then** *found* := `true` **else** *found* := `false`;
  while (not *found* **and** $j^* \le n$) **do**
    if $c_{ij^*} - u_i - v_{j^*} = 0$ **then**
      $r := row(j^*)$, $j := next\_j(r)$;
      while (not *found* **and** $j \le n$) **do**
        if ($row(j) = 0$ **and** $c_{rj} - u_r - v_j = 0$) **then**
          $row(j) := r$, $\varphi(r) := j$, *found* := `true`
        else $j := j + 1$;
        $next\_j(r) := j + 1$
      endwhile
    endif;
    if not *found* **then** $j^* := j^* + 1$
  endwhile;
  if *found* **then** $row(j^*) := i$, $\varphi(i) := j^*$, $next\_j(i) := j^* + 1$
endfor

Example 4.12. Consider the input matrix C below.

$$\begin{array}{cccc} 1 & 6 & 2 & 4 \end{array}$$
$$\underbrace{\begin{pmatrix} 7 & 9 & 10 & 12 \\ 2 & 8 & 7 & 9 \\ 1 & 6 & 7 & 11 \\ 3 & 6 & 2 & 4 \end{pmatrix}}_{C} \qquad \underbrace{\begin{pmatrix} 6 & 3 & 8 & 8 \\ 1 & 2 & 5 & 5 \\ \underline{0} & 0 & 5 & 7 \\ 2 & \underline{0} & 0 & 0 \end{pmatrix}}_{\widetilde{C} = (c_{ij} - v_j)}.$$

Observe that Procedure Basic_preprocessing (Algorithm 4.1) would terminate with $\varphi = (1, 0, 0, 3)$ and $row = (1, 0, 4, 0)$. At the end of the first phase of Three_edge_preprocessing we have $v = (1, 6, 2, 4)$, $\varphi = (0, 0, 1, 2)$, $row = (3, 4, 0, 0)$, $u =$

$(-,\ -,\ 0,\ 0)$, and $next\_j = (-,\ -,\ 2,\ 3)$. Matrix $\widetilde{C} = (c_{ij} - v_j)$ has the current partial assignment shown by the underlined zeroes.

At the first iteration of the second phase we have $i = 1$, $j^* = 2$, $r = 4$, and the inner "while" loop finds an alternative assignment for $j = 3$, thus enlarging the current solution. We obtain $\varphi = (2,\ 0,\ 1,\ 3)$, $row = (3,\ 1,\ 4,\ 0)$, $u = (3,\ -,\ 0,\ 0)$, and $next\_j = (3,\ -,\ 2,\ 4)$. The next iteration sets $u_2 = 1$, but is unable to further enlarge the solution. ∎

In the Jonker and Volgenant [392] algorithm, preprocessing is by far the most important and time-consuming phase. The high computational effectiveness of this algorithm mainly comes from the fact that in many cases the resulting initial partial solution has a number of assignments very close to n, so the shortest path phase obtains the optimal solution within few iterations. The initialization consists of three steps. The first one is a column reduction, performed as in the first part of Algorithm 4.11, but with statement "**for** $j := 1$ **to** n **do**" replaced by "**for** $j := n$ **down to** 1 **do**". (By scanning the columns in reverse order, the low indexed columns are most likely to remain unassigned: During subsequent row scans, in case of ties, the first row minimum is more likely to produce a new assignment.) The second step is a reduction transfer procedure, given in Algorithm 4.12, which updates the dual variables v associated with the currently assigned columns in such a way that each assigned row has the minimum reduced cost in at least two different columns. In this way, in the third step it will be easier to move a row assignment in order to enlarge the solution. Note that dual variables u are not explicitly maintained: For any assignment $(i, \varphi(i))$ the value of u_i is assumed to be $c_{i\varphi(i)} - v_\varphi(i)$.

ALGORITHM 4.12. Procedure Reduction_transfer.
Jonker and Volgenant Reduction transfer.

```
UR := ∅; [comment: unassigned rows]
for i := 1 to n do
    if φ(i) > 0 then
        c* := min{c_ij − v_j : j = 1, 2, ..., n, j ≠ φ(i)};
        v_φ(i) := v_φ(i) − c*
    else UR := UR ∪ {i}
endfor
```

The third step of the Jonker and Volgenant preprocessing is an augmenting row reduction procedure, shown in Algorithm 4.13. For each row $i \in UR$ (the set of unassigned rows), one or more iterations are performed. At each iteration, the minimum and second minimum reduced costs in row i are compared. If their values are different, the dual variable v corresponding to the minimum is updated so that the two resulting reduced costs are equal and row i is assigned to the column, say, j, corresponding to the minimum: if j was previously unassigned the next row of UR is considered; otherwise, a new iteration is performed for the row previously assigned to j. If instead the minimum and second minimum reduced costs in row i are equal, row i is assigned either to the column corresponding to the minimum, if such column is unassigned, or to the one corresponding to the second

$$\begin{matrix} 1 & 6 & 2 & 4 \end{matrix} \qquad \begin{matrix} 1 & 3 & 2 & 4 \end{matrix} \qquad \begin{matrix} -2 & 3 & 2 & 4 \end{matrix}$$

$$\begin{pmatrix} 6 & 3 & 8 & 8 \\ 1 & 2 & 5 & 5 \\ 0 & \underline{0} & 5 & 7 \\ 2 & 0 & 0 & \underline{0} \end{pmatrix} \quad \begin{pmatrix} 6 & \underline{6} & 8 & 8 \\ 1 & 5 & 5 & 5 \\ 0 & 3 & 5 & 7 \\ 2 & 3 & 0 & \underline{0} \end{pmatrix} \quad \begin{pmatrix} 9 & \underline{6} & 8 & 8 \\ 4 & 5 & 5 & 5 \\ \underline{3} & 3 & 5 & 7 \\ 5 & 3 & 0 & \underline{0} \end{pmatrix}$$

(a) (b) (c)

Figure 4.5. *Values $c_{ij} - v_j$ in Augmenting_row_reduction with $\hat{\imath} = 1$.*

minimum otherwise, and in both cases the next row of UR is considered. Note that in the second case no attempt is made to reassign the deassigned row, in order to avoid possible cycling.

ALGORITHM 4.13. Procedure Augmenting_row_reduction.
Jonker and Volgenant Augmenting row reduction.

for each $\hat{\imath} \in UR$ **do**
  $i := \hat{\imath}$;
  repeat
    $f := \arg\min\{c_{ij} - v_j \ : \ j = 1, 2, \dots, n\}$, $u^f := c_{if} - v_f$;
    $s := \arg\min\{c_{ij} - v_j \ : \ j = 1, 2, \dots, n, \ j \neq f\}$, $u^s := c_{is} - v_s$;
    $j := f$;
    if $u^f < u^s$ **then** $v_f := v_f - (u^s - u^f)$
    else if $row(f) > 0$ **then** $j := s$;
    $r := row(j)$, $row(j) := i$, $\varphi(i) := j$, $i := r$;
    if $r > 0$ **then** $\varphi(r) := 0$;
  until ($u^f = u^s$ **or** $r = 0$)
endfor

The effect of Procedure Augmenting_row_reduction is to increase the number of assignments through a series of alternating paths that, within a "for each" iteration, are not necessarily simple paths as usual, but may even reverse direction and be extended by rows and columns visited before. It follows that the procedure is formally non-polynomial, as exit from the "repeat-until" loop only occurs when either an augmenting path has been obtained or the first and second reduced costs are equal. Since there are $O(n)$ "for each" iterations, the time complexity of Augmenting_row_reduction is $O(n^3 C)$, where C denotes the maximum c_{ij} value. This time complexity could be reduced to $O(n^3)$ by limiting to n the number of allowed path extensions, although, as already observed in [392], this number is not even approached in practice. In their original implementation Jonker and Volgenant apply Augmenting_row_reduction twice.

Example 4.13. Consider again Example 4.12. The initial column reduction produces $v = (1, 6, 2, 4)$, $\varphi = (0, 0, 2, 4)$, and $row = (0, 3, 0, 4)$. Figure 4.5(a) shows the reduced matrix (with values v_j in row zero) and the current assignment (underlined).

$$
\begin{array}{cccc}
-3 & 3 & 2 & 4
\end{array}
\qquad
\begin{array}{cccc}
-3 & 2 & 2 & 4
\end{array}
\qquad
\begin{array}{cccc}
-3 & 1 & 2 & 4
\end{array}
\qquad
\begin{array}{cccc}
-4 & 1 & 2 & 4
\end{array}
$$

$$
\begin{pmatrix}
10 & \underline{6} & 8 & 8 \\
\underline{5} & 5 & 5 & 5 \\
4 & 3 & 5 & 7 \\
6 & 3 & 0 & \underline{0}
\end{pmatrix}
\quad
\begin{pmatrix}
10 & 7 & 8 & 8 \\
\underline{5} & 6 & 5 & 5 \\
4 & \underline{4} & 5 & 7 \\
6 & 4 & 0 & \underline{0}
\end{pmatrix}
\quad
\begin{pmatrix}
10 & \underline{8} & 8 & 8 \\
\underline{5} & 7 & 5 & 5 \\
4 & 5 & 5 & 7 \\
6 & 5 & 0 & \underline{0}
\end{pmatrix}
\quad
\begin{pmatrix}
11 & \underline{8} & 8 & 8 \\
6 & 7 & \underline{5} & 5 \\
\underline{5} & 5 & 5 & 7 \\
7 & 5 & 0 & \underline{0}
\end{pmatrix}
$$

(a) (b) (c) (d)

Figure 4.6. *Values $c_{ij} - v_j$ in Augmenting_row_reduction with $\hat{\imath} = 2$.*

Since both assigned rows have zero reduced costs in at least two columns, the only effect of Procedure Reduction_transfer is to define $UR = \{1, 2\}$. The first iteration of Augmenting_row_reduction is thus executed for $\hat{\imath} = 1$. The "repeat-until" loop executes the following iterations:

$i = 1$: $u^f = 3$, $u^s = 6$, $v_2 = 3$, $\varphi = (2,\ 0,\ 0,\ 4)$, $row = (0,\ 1,\ 0,\ 4)$, Figure 4.5(b);

$i = 3$: $u^f = 0$, $u^s = 3$, $v_1 = -2$, $\varphi = (2,\ 0,\ 1,\ 4)$, $row = (3,\ 1,\ 0,\ 4)$, Figure 4.5(c).

The second iteration of Augmenting_row_reduction is executed for $\hat{\imath} = 2$. The "repeat-until" loop executes the following iterations:

$i = 2$: $u^f = 4$, $u^s = 5$, $v_1 = -3$, $\varphi = (2,\ 1,\ 0,\ 4)$, $row = (2,\ 1,\ 0,\ 4)$ Figure 4.6(a);

$i = 3$: $u^f = 3$, $u^s = 4$, $v_2 = 2$, $\varphi = (0,\ 1,\ 2,\ 4)$, $row = (2,\ 3,\ 0,\ 4)$, Figure 4.6(b);

$i = 1$: $u^f = 7$, $u^s = 8$, $v_2 = 1$, $\varphi = (2,\ 1,\ 0,\ 4)$, $row = (2,\ 1,\ 0,\ 4)$, Figure 4.6(c);

$i = 3$: $u^f = 4$, $u^s = 5$, $v_1 = -4$, $\varphi = (2,\ 0,\ 1,\ 4)$, $row = (3,\ 1,\ 0,\ 4)$, Figure 4.6(d);

$i = 2$: $u^f = u^s = 5$, $\varphi = (2,\ 3,\ 1,\ 4)$, $row = (3,\ 1,\ 2,\ 4)$ (optimal assignment).

Note that the alternating path produced by the second iteration is not a simple one, as it visits, in sequence, row vertices 2, 3, 1, 3, 2. ∎

Both Procedures Reduction_transfer and Augmenting_row_reduction are closely related to the original auction method proposed in Bertsekas [86], as is discussed in Section 4.6.3.

Hao and Kocur [362] proposed a preprocessing approach operating in two steps. Step 1 starts by computing reduced costs as in Procedure Basic_preprocessing (Algorithm 4.1), and then finds a maximum cardinality matching in the bipartite partial graph $G^0 = (U, V; E^0)$ that only contains zero cost edges through the algorithm by Chang and McCormick [180]. Step 2 adjusts the costs according to heuristic rules and uses again the maximum cardinality matching algorithm to make new assignments. The approach was computationally tested against the one by Jonker and Volgenant, and it turned out that it is more effective only on random problems with small cost ranges.

Kindervater, Volgenant, de Leve, and van Gijlswijk [417] gave a simple approach to generate alternative dual solutions through shortest path computations.

Computer codes implementing shortest path approaches in various languages are available in the public domain (see Section 4.9). The Algol implementation can be found in Bhat and Kinariwala [96]. The listing of a Fortran implementation is given in Burkard and Derigs [145]. Another Fortran implementation is included in the diskette accompanying

Carpaneto, Martello, and Toth [165]. A Pascal code is given as a listing in Volgenant [647], where it is computationally compared with the algorithms by Carpaneto and Toth [169] and Jonker and Volgenant [392]. Pascal, Fortran, and C++ implementations of the Jonker and Volgenant procedures are discussed in [394].

4.5 Primal algorithms

In this section we consider methods that directly solve the primal problem and are grouped into two categories: Specializations of the primal-simplex algorithm and approaches that cannot be reconducted to pure simplex methods. For both categories, some of the methods we review were developed for the famous Hitchcock-Koopmans transportation problem (see Hitchcock [369]), of which LSAP is a special case. Given a bipartite graph $G = (U, V; E)$ with costs c_{ij} associated with edges $[i, j]$ $(i, j = 1, 2, \ldots, n)$, let the vertices $i \in U$ represent *sources* capable of supplying positive integer amounts a_i and the vertices $j \in V$ represent *sinks* having positive integer demands b_j. Further assume that $\sum_i a_i = \sum_j b_j$. The *transportation problem* is to find the least cost transportation pattern from the sources to the sinks. It follows that the special case where all the supply amounts and the demands are equal to one coincides with LSAP.

Dantzig [211] gave a specialization of the simplex algorithm to network problems, which is the starting point for all simplex-based algorithms for LSAP. Several specializations to the transportation problem were developed in the early 1970s, see, e.g., Glover, Karney, and Klingman [323] and Glover, Karney, Klingman, and Napier [324]. Gavish, Schweitzer, and Shlifer [305] gave computational results on the effect of various pivoting rules on the number of degenerate pivots in the solution of LSAP.

From a practical point of view, the primal algorithms are generally less efficient than other approaches such as, e.g., the shortest path algorithms. They hold, however, considerable theoretical interest for the insight they provide into some relations between LSAP and linear programming. We begin our review with non-simplex algorithms, that have been the first primal algorithms developed since the 1960s for LSAP.

4.5.1 Non-simplex algorithms

Two main primal non-simplex approaches have been developed for LSAP, by Balinski and Gomory [64] and Srinivasan and Thompson [621], respectively.

The first polynomial primal algorithm for LSAP was proposed in 1964 by Balinski and Gomory [64]. The algorithm starts with a primal feasible solution $X = (x_{ij})$ and an (infeasible) dual solution u, v satisfying the complementary slackness conditions (4.8):

$$x_{ij}(c_{ij} - u_i - v_j) = 0 \qquad (i, j = 1, 2, \ldots, n).$$

Given the current pair of solutions $(X, (u, v))$, let $F = \{[i, j] \in E : u_i + v_j \leq c_{ij}\}$ be the set of edges of $G = (U, V; E)$ with nonnegative reduced cost. If $F \equiv E$, the current solution pair is optimal. Otherwise, an edge $[k, l] \in E \setminus F$ is selected and a new pair $(X', (u', v'))$ is obtained such that either

(i) X' has a lower primal solution value than X and (u', v') satisfies

$$u'_i + v'_j \leq c_{ij} \quad \forall\, [i, j] \in F \qquad \text{and} \qquad u'_k + v'_l \leq c_{kl} \tag{4.21}$$

(i.e., the new pair is better both for the primal and the dual) or

(ii) $X' \equiv X$ and (u', v') satisfies

$$u'_i + v'_j \leq c_{ij} \quad \forall\, [i, j] \in F \qquad \text{and} \qquad u'_k + v'_l < u_k + v_l \tag{4.22}$$

(i.e., the new pair is unchanged for the primal and better meets the constraints of the dual).

At each iteration, given the selected edge $[k, l]$ with negative reduced cost, a procedure equivalent to Alternate(k) (Algorithm 4.2) is executed to build an alternating tree rooted at vertex $l \in V$. If vertex $k \in U$ is reached, the assigned and unassigned edges along the path from k to l are interchanged, row k is assigned to column l, and the value of v_l is updated so that the complementary slackness conditions and (4.21) hold. If instead vertex k is not reached, a dual updating similar to that of the Hungarian algorithm is performed so that (4.22) holds while the primal solution is left unchanged.

It is proved in [64] that, with an appropriate choice of edge $[k.l]$, the number of iterations is bounded by $O(n^2)$. The overall time complexity of the Balinski and Gomory approach is thus $O(n^4)$.

Another primal algorithm was proposed by Klein [421]. It operates on the network flow model discussed in Section 4.4.1 by maintaining a feasible flow and iteratively eliminating negative cycles. This is the first *cycle canceling algorithm*, one of the principal techniques for solving minimum cost flow problems (see Ahuja, Magnanti, and Orlin [11]).

In the 1970s Cunningham and Marsh [208] generalized this primal algorithm to solve the minimum weight perfect matching problem in a general graph using the techniques introduced in the 1960s by Edmonds [247, 246].

Srinivasan and Thompson [621] derived from their previous work on an operator theory of parametric programming for the transportation problem (see Srinivasan and Thompson [619, 620]) two different primal algorithms, called the *cell* and the *area* cost operator algorithms. Both start with a dual feasible solution u, v obtained as in Procedure Basic_preprocessing (see Algorithm 4.1) and a primal basic solution X. A new problem is then defined by setting $\hat{c}_{ij} = u_i + v_j$ if $x_{ij} = 1$ and $\hat{c}_{ij} = c_{ij}$ otherwise. In this way, the pair $(X, (u, v))$ satisfies the complementary slackness conditions and, hence, is optimal for the modified problem but not for the original one. After this initialization phase, both algorithms perform a series of iterations in order to change back the costs to their original values. At each iteration the current costs and the current primal and dual solutions are updated in such a way that optimality is preserved. The two approaches basically differ only in the way this updating is performed. The *cell cost operator algorithm* modifies just one cost (a cell) at a time, while the *area cost operator algorithm* simultaneously modifies several costs (an area). Srinivasan and Thompson [621] proved that both algorithms solve LSAP in at most $n(2n + 1) - 1$ iterations. Akgül [18] conjectured that they can be implemented so as to run in $O(n^3)$ time.

4.5.2 Simplex-based algorithms

Let us consider the linear problem associated with LSAP (see Section 4.1.1). It has been shown in Proposition 2.23 that there is a one-to-one correspondence between primal (integer) basic solutions and spanning trees on the associated bipartite graph $G = (U, V; E)$. Given any basic feasible solution, the associated spanning tree T consists of the $2n - 1$ edges $[i, j]$ corresponding to the basic columns (see Proposition 2.22). It is not difficult to obtain a dual solution satisfying the complementary slackness conditions. Starting with a root $r \in U$, we set $u_r = 0$, then $v_j := c_{ij} - u_i$ for all edges $[i, j] \in T$ for which u_i has been defined, then $u_i := c_{ij} - v_j$ for all edges $[i, j] \in T$ for which v_j has been defined, and so on (see also Procedure Compute_Dual(T, r) in Algorithm 4.14 below). If the reduced costs corresponding to the edges of $E \setminus T$ are nonnegative, the basis is optimal. Otherwise, a simplex-based algorithm inserts in the basis an edge $[i, j] \in E \setminus T$ with negative reduced cost and removes from the basis an edge belonging to the unique circuit produced by the addition of $[i, j]$ to T.

It is known from linear programming that the classical *Dantzig rule* [211], which selects the variable with the most negative reduced cost as the entering variable, experimentally proves to be very efficient, but can endlessly cycle when degenerate bases are encountered. Three main techniques can be employed to face this unwanted behavior. Perturbation and lexicographic methods, which however involve additional numerical computation, and the Bland rule [102], which has the drawback of imposing the choice of the new basic variable independently of the magnitude of the corresponding reduced cost. As we will see in the following, for LSAP it is possible to avoid cycling by just keeping specially structured bases.

Remember that the linear problem associated with LSAP has $2n - 1$ rows and n^2 columns and that any primal feasible basic solution has exactly n variables with non-zero value and, hence, is highly degenerate (see Proposition 2.26). This zero pivot phenomenon was studied by Dantzig [211], Balinski [60], and Gavish, Schweitzer, and Shlifer [305]. Degeneracy is the main reason why the first simplex-based algorithms for LSAP required exponential time. Barr, Glover, and Klingman [68] reported computational testings showing that approximately 90 percent of the simplex pivots may be degenerate for LSAP instances.

The main result towards the definition of a polynomial simplex-based algorithm for LSAP was the definition of a subset of bases/trees, which was independently given in the mid-1970s by Cunningham [205] and by Barr, Glover, and Klingman [68]. These are denoted as *strongly feasible trees* in [205] and *alternating path bases* in [68].

Definition 4.14. *Given a tree T in $G = (U, V; E)$, define a vertex $r \in U$ as the root. We will say that an edge $e = [i, j] \in T$ is* odd *(resp.,* even*) if the (unique) path emanating from r and having e as its terminal edge contains an odd (resp., even) number of edges.*

Definition 4.15. *Given a feasible solution X, a tree T in $G = (U, V; E)$ rooted at $r \in U$ is a* strongly feasible tree *if $x_{ij} = 1$ for all odd edges $[i, j] \in T$ and $x_{ij} = 0$ for all even edges $[i, j] \in T$.*

A feasible solution and the corresponding strongly feasible tree are shown by the solid lines of Figure 4.7, where the round vertices are the sources (set U), the square vertices are the sinks (set V), and the root r is source number 3: Thick lines represent the solution ($x_{ii} = 1$ for $i = 1, 2, 3, 4$) and thin lines the other edges of the tree. (Disregard for the moment the

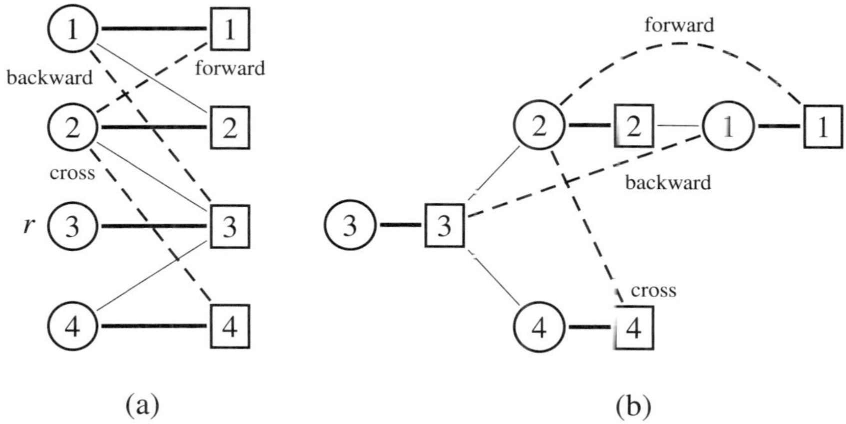

Figure 4.7. (a) *Strongly feasible tree on G;* (b) *an easier view.*

dashed lines.) In Figure 4.7(a), solution and tree are depicted on the usual bipartite graph $G = (U, V; E)$, while in Figure 4.7(b), they are depicted in a way that makes it easier to recognize the tree.

Observe that, from Definition 4.15, we immediately have the following.

Proposition 4.16. *In any strongly feasible tree T the root has degree one and every other vertex of U has degree two.*

Three kinds of non-tree edges are important.

Definition 4.17. *An edge $[i, j] \in E \setminus T$ with $i \in U$ and $j \in V$ is a* forward edge *if i lies on the path that connects r to j in T, is a* backward edge *if j lies on the path that connects r to i in T, and is a* cross edge *if it is neither forward nor backward.*

Examples of forward, backward, and cross edges are depicted as dashed lines in Figure 4.7. Two main results, developed in [205] and [68], hold when pivoting from a strongly feasible tree T.

Theorem 4.18. (Cunningham [205], 1976; Barr, Glover, and Klingman [68], 1977.) *Let T be a strongly feasible tree for an instance of LSAP. Then*

1. *given any edge $[i, j] \in E \setminus T$ with $i \in U$ and $j \in V$, let $C(T, [i, j])$ be the unique circuit in $T \cup \{[i, j]\}$ and let $[i, l]$ be the unique other edge incident to i in this circuit: Then $T \setminus \{[i, l]\} \cup \{[i, j]\}$ is a strongly feasible tree (corresponding to the new solution obtained through a pivot operation involving edges $[i, j]$ and $[i, l]$);*

2. *a pivot performed as in 1 is non-degenerate if and only if $[i, j]$ is a forward edge.*

Suppose for example that backward edge $[i, j] = [1, 3]$ of Figure 4.7 is selected for pivoting. According to 1 above, we have $[i, l] = [1, 2]$ and a degenerate pivoting producing the new strongly feasible tree shown in Figure 4.8 that corresponds to an unchanged solution.

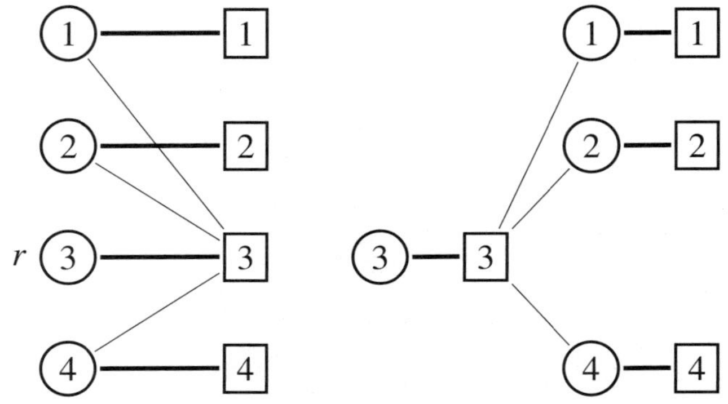

Figure 4.8. *Degenerate pivoting on a backward edge.*

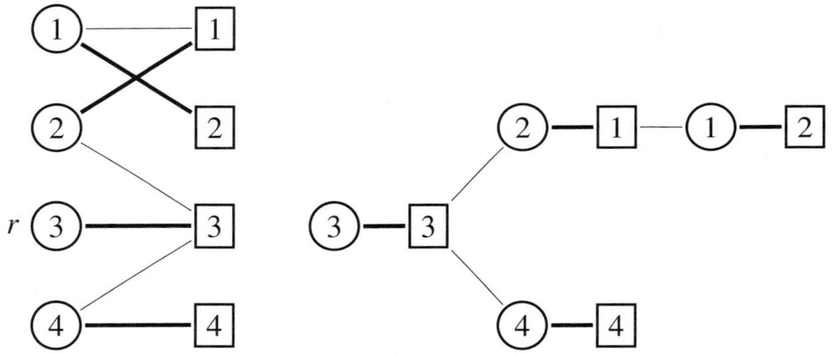

Figure 4.9. *Non-degenerate pivoting on a forward edge.*

If instead forward edge $[i, j] = [2, 1]$ of Figure 4.7 is selected for pivoting, we have $[i, l] = [2, 2]$ and a non-degenerate pivot producing the new solution shown in Figure 4.9.

A simplex algorithm that starts with a strongly feasible tree and only performs (degenerate or non-degenerate) pivots producing strongly feasible trees can thus be outlined as in Algorithm 4.14. Given a feasible primal solution X, a corresponding strongly feasible tree T, and a root $r \in U$, the inner Procedure Compute_Dual(T, r) starts by arbitrarily fixing the value of dual variable u_r to zero. The tree is then traversed and, for each encountered vertex, the corresponding dual variable takes the unique value that produces zero reduced cost. In the "while" loop of Algorithm Primal_simplex, if the reduced costs corresponding to the edges of $E \setminus T$ are nonnegative, we know that the basis corresponding to T is optimal and terminate. Otherwise, a simplex pivot inserts in the basis an edge $[i, j] \in E \setminus T$ such that $c_{ij} - u_i - v_j < 0$ and removes from the basis the unique other edge $[i, l] \in C(T, [i, j])$.

If the pivot is non-degenerate, the primal solution X is updated by complementing the assignments along the alternating cycle $C(T, [i, j])$.

Algorithm 4.14. Primal_simplex.
Strongly feasible trees.

find any feasible primal solution $X = [x_{ij}]$;
let T be a strongly feasible tree corresponding to X, and r any vertex of U;
Compute_Dual(T, r);
while $\overline{E} := \{[i, j] \in E \ : \ c_{ij} - u_i - v_j < 0\} \neq \emptyset$ **do**
    select an edge $[i, j] \in \overline{E}$ (with $i \in U$ and $j \in V$);
    let $[i, l]$ be the unique edge of $C(T, [i, j])$ incident to i with $l \neq j$;
    if $x_{il} = 1$ **then** [**comment:** non-degenerate pivoting]
        for each $[h, k] \in C(T, [i, j])$ **do** $x_{hk} := 1 - x_{hk}$
    endif
    update T as $T := T \setminus \{[i, l]\} \cup \{[i, j]\}$, and let r be any vertex of U;
    Compute_Dual(T, r)
endwhile

Procedure Compute_Dual(T, r)
orient the edges of T away from the root r;
$u_r := 0$;
for each edge $[i, j]$ encountered starting from r and traversing T **do**
    if $i \in U$ **then** $v_j := c_{ij} - u_i$ **else** $u_j := c_{ji} - v_i$

Example 4.19. Given the cost matrix

$$C = \begin{pmatrix} 4 & 3 & 2 & 4 \\ 3 & 3 & 4 & 5 \\ 8 & 4 & 2 & 4 \\ 5 & 6 & 4 & 3 \end{pmatrix},$$

let us start with solution $x_{ii} = 1$ ($i = 1, 2, \ldots, 4$) and the associated strongly feasible tree depicted in Figure 4.7 with root $r = 3$. By executing Procedure Compute_Dual we obtain $u = (2, 2, 0, 2)$, $v = (2, 1, 2, 1)$, $\overline{E} = \{[1, 3], [2, 1]\}$ with $\overline{c}_{13} = -2$ and $\overline{c}_{21} = -1$. By adopting the Dantzig rule, backward edge $[i, j] = [1, 3]$ is selected, so we have $[i, l] = [1, 2]$ and a degenerate pivot producing the tree of Figure 4.8.

The next execution of Compute_Dual gives $u = (0, 2, 0, 2)$, $v = (4, 1, 2, 1)$, $\overline{E} = \{[2, 1], [4, 1]\}$ with $\overline{c}_{21} = -3$ and $\overline{c}_{41} = -1$. We select cross edge $[i, j] = [2, 1]$, so $[i, l] = [2, 3]$ and another degenerate pivot is performed, producing the tree of Figure 4.10. We then obtain $u = (0, -1, 0, 2)$, $v = (4, 4, 2, 1)$, $\overline{E} = \{[1, 2], [4, 1]\}$ with $\overline{c}_{12} = \overline{c}_{41} = -1$.

Forward edge $[i, j] = [1, 2]$ is now selected, so we have $[i, l] = [1, 1]$. A non-degenerate pivot produces the tree of Figure 4.11. Compute_Dual then gives $u = (0, 0, 0, 2)$, $v = (3, 3, 2, 1)$, and $\overline{E} = \emptyset$, so the algorithm terminates with the optimal solution $x_{12} = x_{21} = x_{33} = x_{44} = 1$. Note that this is the same primal solution embedded in the tree of Figure 4.9, for which Compute_Dual would give $u = (3, 2, 0, 2)$, $v = (1, 0, 2, 1)$, and $\overline{E} = \{[1, 3]\}$ with $\overline{c}_{13} = -3$. Hence, a series of degenerate pivots would be necessary to

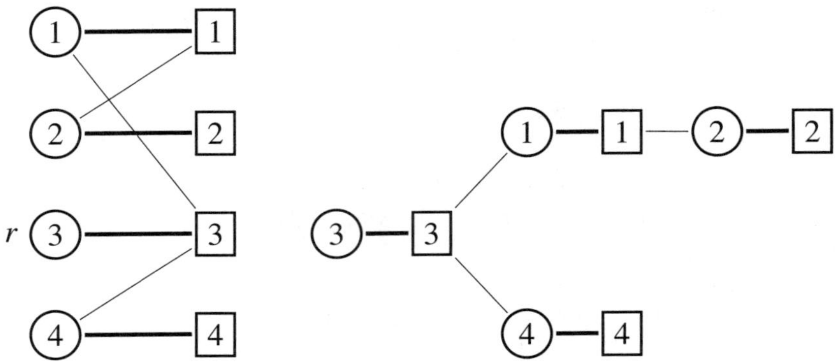

Figure 4.10. *Intermediate solution of Example 4.19.*

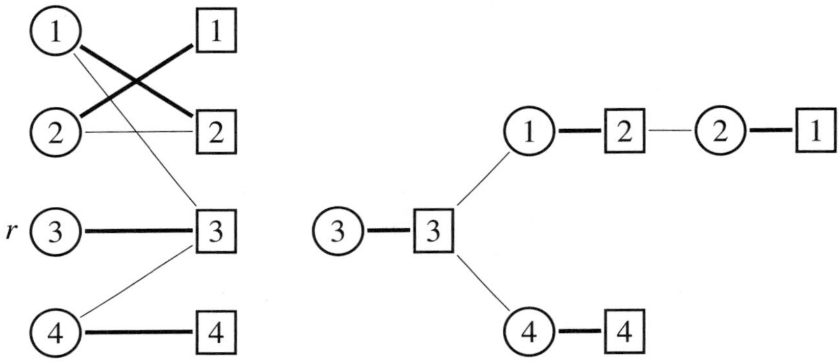

Figure 4.11. *Final solution of Example 4.19.*

prove optimality. It can be seen that even a change of root would not allow one to compute dual variables satisfying $\overline{c}_{13} \geq 0$. ■

It is proved in [205] and [68] that no tree can be repeated in the (degenerate) sequence of strongly feasible trees visited for any feasible solution. Hence, the algorithm will not cycle and it will obtain the optimal solution in a finite number of pivots, regardless of which edge with negative reduced cost is chosen to be the entering edge, and without any reliance on perturbation or lexicographic ordering. However, each feasible solution can correspond to $2\,(2n)^{n-2}$ trees, i.e., the time complexity for any feasible solution is non-polynomial. Cunningham [206] proved that use of specific rules for the selection of the entering edge $[i, j]$ reduces to $O(n^2)$ the number of strongly feasible trees visited at any extreme point. The overall time complexity of the resulting algorithm remains, however, non-polynomial.

A similar result had been obtained a few years earlier by Komáromi [402], who introduced "bases of special structure," which turn out to be strongly feasible trees. He proposed an iterative method that considers cost matrices of increasing size. At each iteration

i ($i = 1, 2, \ldots, n$) an optimal basis of the LSAP associated with the first $i - 1$ rows and columns of matrix C is at hand and the matrix is enlarged by adding the ith row and column. The basis is reoptimized through standard primal simplex pivots. Due to the special structure of the basis, at most $O(n^2)$ degenerate pivots can occur before the objective function decreases. The overall algorithm thus has the same time complexity as the Cunningham [206] method. Unfortunately, the Komàromi paper has been ignored for tens of years.

Polynomial-time primal simplex algorithms were obtained in the 1980s. Hung [381] combined a modified version of the Cunningham rule with the Dantzig rule, ensuring a number of pivots bounded by $O(n^3 \log \Delta)$, where Δ is the difference between the initial and final solution values. The actual time complexity of the overall algorithm depends on the computational effort for implementing the edge selection rule (that amounts to $O(n^3)$ time per pivot in Hung's implementation).

The presence of $\log \Delta$ makes the Hung algorithm weakly polynomial. A strongly polynomial primal simplex algorithm was obtained by Orlin [514], who showed that it is possible to perturbate the given costs to obtain an equivalent problem for which the number of pivots has the same upper bound that it has for the original problem if the Dantzig rule is employed. The resulting algorithm performs at most $O(n^2 \log \Delta)$ pivots, with Δ (defined as above) bounded by $4^n (n!)^2$, i.e., at most $O(n^3 \log n)$ pivots. Note, however, that a strongly polynomial version of the Cunningham [206] algorithm, requiring $O(n^3)$ pivots and $O(n^5)$ time, had been proposed a few years before by Roohy-Laleh [539] in an unpublished Ph.D. thesis.

A different weakly polynomial primal simplex algorithm was obtained by Ahuja and Orlin [12] through a pivot rule that can be regarded as a scaling version of the Dantzig rule. Assuming that the costs are positive integers, they start by defining a threshold value $\varepsilon = 2^{\lceil \log C \rceil}$, where C denotes the maximum c_{ij} value. The pivot rule is to select for pivoting any edge $[i, j]$ having reduced cost $\overline{c}_{ij} \leq -\varepsilon/2$. When there is no edge satisfying this condition, ε is halved and the process is iterated until $\varepsilon < 1$, i.e., the current basis is optimal. The algorithm solves LSAP in $O(n^2 \log C)$ pivots and $O(nm \log C)$ time, where m denotes the number of edges in graph $G = (U, V; E)$.

The primal simplex approach with best time complexity was designed in the 1990s by Akgül [19]. The algorithm operates on primal feasible bases corresponding to strongly feasible trees and adopts the Dantzig rule. Its main peculiarity is the use of a sequence of $2n$ subgraphs of G with an increasing number of edges, the last of which is the original graph G itself. At each iteration a complete basis for the original problem is maintained and a new subgraph is obtained from the previous one by adding a subset of the edges incident to a specified vertex. More precisely, the algorithm starts with a tree T_0 consisting of a single alternating path in $G = (U, V; E)$, rooted at some vertex r. The first subgraph is $G_0 = (U, V; T_0)$, and the starting basis is optimal for it. At each iteration, the current subgraph is enlarged by considering the next vertex, say, i, in path T_0. All edges $[i, j]$ such that j precedes i in the path from r to i along T_0 are added to the subgraph, and a series of pivots transforms the current basis into an optimal one for the new subgraph. The algorithm is based on an analysis of the structure of the set of vertices whose corresponding dual variables are updated during such pivots (*cut-sets*). It is proved that (i) cut-sets are disjoint; and (ii) edges whose source vertex is in a cut-set are dual feasible for all subgraphs that are considered afterwards. In addition, each vertex is subject to at most one dual variable update per iteration and, when the considered vertex i is a sink, the new solution is obtained through at most one pivot. It follows that a pair of consecutive iterations can be performed with at most $k + 2$ pivots, where k is the number of pairs of iterations performed so far, from which a bound $n(n+3)/2 - 4$ on the number of pivots is derived. The algorithm can be implemented with no special data structure to run in $O(n^3)$ time for dense matrices. For sparse matrices, use of the dynamic trees by Sleator and Tarjan [613] and of the Fibonacci heaps by Fredman and Tarjan [278] leads to an $O(n^2 \log n + nm)$ time complexity.

4.6 Dual algorithms

In this section we consider methods that evolve an optimal but infeasible solution until it becomes feasible. Similarly to what we did for the primal algorithms of Section 4.5, these methods will be grouped into two categories: specializations of the dual-simplex algorithm and approaches that cannot be directly reconducted to the simplex method. Non-simplex algorithms have been the first dual algorithms for LSAP.

4.6.1 Non-simplex algorithms

The treatment of the first algorithm that has a dual nature has been anticipated in Section 4.3 due to its relations with the Hungarian algorithm and the shortest path methods. Indeed the 1969 algorithm by Dinic and Kronrod [235] maintains an infeasible solution in which all rows are assigned, while some columns are either unassigned or multiply assigned, and a corresponding optimal dual solution. At each iteration, an unassigned column is assigned and the dual solution is correspondingly updated through a shortest alternating path until the solution becomes feasible.

In 1980 Hung and Rom [382] proposed an algorithm based on the same relaxation of LSAP, namely, the one defined by (4.1), (4.2), and (4.4). Algorithm 4.15 starts by assigning each row to a column with minimum cost in that row. At any iteration (i) a basis is constructed as a shortest path tree spanning all row vertices and column vertices having one or more assignments; (ii) an unassigned column vertex j^* is selected and assigned at minimum reduced cost to a row vertex i^*; and (iii) assigned and unassigned edges along the alternating path in the tree that connects i^* to a multiply assigned column vertex are interchanged.

Algorithm 4.15. Hung_Rom.
Hung–Rom algorithm.

comment: initialization;
define (x_{ij}) by assigning each row to a column with minimum cost in that row;
let V_0, V_1 and V_2 be the sets of column indexes (vertices of V) having, respectively,
    no row assigned, one row assigned, and two or more rows assigned;
for $j := 1$ **to** n **do** $v_j := 0$;
while $V_0 \neq \emptyset$ **do**
    comment: construction of the basis tree;
    select a column $r \in V_2$ as the root, and set $v_r := 0$;

$$\begin{array}{r} \\ 7 \\ 2 \\ 1 \\ (1)\text{-} \end{array} \begin{array}{c} \begin{array}{cccc} 0 & 0(2) & 0(1) & 0 \end{array} \\ \begin{pmatrix} \underline{7} & 9 & 8 & 9 \\ \underline{2} & 8 & 5 & 7 \\ \underline{1} & 6 & 6 & 9 \\ 3 & 6 & \underline{2} & 2 \end{pmatrix} \end{array}$$

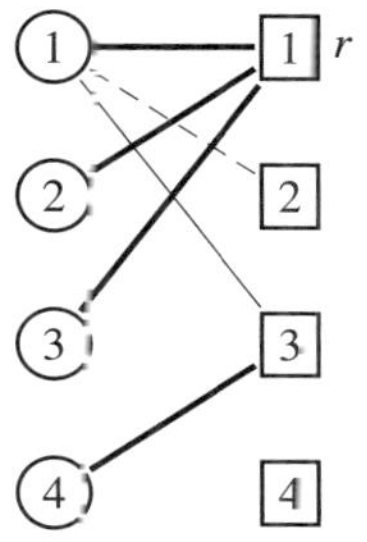

Figure 4.12. *First cycle of Example 4.20.*

for each $i \in U$: $x_{ir} = 1$ **do** $u_i := c_{ir}$;
initialize the basis tree with the edges $[i, r]$ such that $x_{ir} = 1$;
while $\overline{V}_{12} :=$ {columns of $V_1 \cup V_2$ that are not in the tree} $\neq \emptyset$ **do**
  for each column $j \in \overline{V}_{12}$ **do** $\mu_j := \min\{c_{ij} - u_i$: row i is in the tree};
  select $\hat{j} \in \overline{V}_{12}$ with minimum $\mu_j - v_j$ and set $v_{\hat{j}} := \mu_{\hat{j}}$;
  add $[\hat{\imath}, \hat{j}]$ to the tree, where $\hat{\imath}$ is the row vertex determining $\mu_{\hat{j}}$;
  for each $i \in U$: $x_{i\hat{j}} = 1$ **do** $u_i := c_{i\hat{j}} - v_{\hat{j}}$ and add $[i, \hat{j}]$ to the tree
endwhile;
comment: enforcing violated constraints;
select $j^* \in \overline{V}_0$ and set $i^* := \arg\min_{i \in U}\{c_{ij^*} - u_i\}$, $v_{j^*} := c_{i^*j^*} - u_{i^*}$;
let P be the path connecting i^* to a column of V_2 (in the direction of r);
interchange assigned and unassigned edges along P, and set $x_{i^*j^*} := 1$;
update V_0, V_1 and V_2
endwhile

For the sake of simplicity, in the inner "while" loop we have assumed that the μ_j values are computed from scratch at each iteration. This phase can, however, be implemented, in a Dijkstra-like fashion, so that the whole μ_j computation requires $O(n^2)$ time. Since at each *cycle* (outer "while" loop iteration) a new vertex leaves V_0, the overall time complexity of Algorithm Hung_Rom is $O(n^3)$.

Example 4.20. We refer to the numerical instance used for the Dinic–Kronrod algorithm (Example 4.9). Consider the cost matrix shown in Figure 4.12. The u and v values are shown on the left and on the top, respectively, and the figures in brackets refer to subsequent phases of the cycle.

The initialization phase determines the assignment shown by the underlined entries, with $V_0 = \{2, 4\}$, $V_1 = \{3\}$, $V_2 = \{1\}$, and $v = (0, 0, 0, 0)$. In the first cycle, the basis tree is initialized with $r = 1$ and the three edges emanating from it, as shown in the bipartite graph of Figure 4.12, with $u = (7, 2, 1, -)$ and $\overline{V}_{12} = \{3\}$. We then obtain $\mu_3 = 8 - 7$ (row 1), $\hat{j} = 3$, $v_3 = 1$, and edge [1, 3] is added to the tree. We set $u_4 = 1$ and add edge [4, 3], thus completing the basis tree. We now enforce the unassigned column $j^* = 2$: $i^* = 1$, $v_2 = 9 - 7$, $P = \{[1, 1]\}$, $x_{11} = 0$, $x_{12} = 1$, with $V_0 = \{4\}$, $V_1 = \{2, 3\}$, $V_2 = \{1\}$.

$$
\begin{array}{r} \\ (4)7 \\ 2 \\ 1 \\ (\text{-}1)1 \end{array}
\begin{array}{c} \begin{array}{cccc} 0 & 2(5) & 1(3) & 0(3) \end{array} \\
\left(\begin{array}{cccc} 7 & \underline{9} & 8 & 9 \\ \underline{2} & 8 & 5 & 7 \\ \underline{1} & 6 & 6 & 9 \\ 3 & 6 & \underline{2} & 2 \end{array}\right) \end{array}
$$

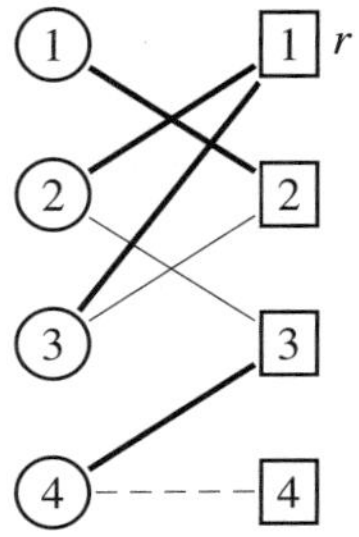

Figure 4.13. *Second cycle of Example 4.20.*

In the second cycle, shown in Figure 4.13, the basis tree is initialized with $r = 1$ and the two edges emanating from it, with $u = (7, 2, 1, 1)$ and $\overline{V}_{12} = \{2, 3\}$. At the first iteration we obtain $\mu_2 = 6 - 1$ (row 3), $\mu_3 = 5 - 2$ (row 2), $\hat{\jmath} = 3$, $v_3 = 3$, and edge $[2, 3]$ is added to the tree. We set $u_4 = -1$ and add edges $[4, 3]$, obtaining $\overline{V}_{12} = \{2\}$. At the second iteration we have $\hat{\jmath} = 2$, $v_2 = 5$, $u_1 = 4$, and edges $[3, 2]$ and $[1, 2]$ complete the basis tree. The unassigned column $j^* = 4$ is enforced: $i^* = 4$, $v_4 = 3$, $P = \{[4, 3], [2, 3], [2, 1]\}$, $x_{43} = 0$, $x_{23} = 1$, $x_{21} = 0$, $x_{44} = 1$. Since we now have $V_0 = \emptyset$, an optimal solution has been obtained. ■

The main difference between the Hung–Rom algorithm and the Dinic–Kronrod algorithm is in the way the alternating path connecting a vertex $j^* \in V_0$ to a vertex of V_2 is obtained. While Hung and Rom compute the whole shortest path tree which spans the vertices of $U \cup V_1 \cup V_2$, Dinic and Kronrod determine only the shortest path from j^* to the first vertex of V_2. In this sense, the Dinic–Kronrod algorithm can be considered more effective than the Hung–Rom algorithm.

4.6.2 Simplex-based algorithms

The *signature method*, proposed in the mid-1980s by Balinski [62], is the most famous dual algorithm for LSAP, although, as is shown later, it cannot be considered a "pure" dual simplex algorithm. The method is related to the strongly feasible trees introduced by Cunningham [205] and by Barr, Glover, and Klingman [68] (see Section 4.5.2) and is based on the following definition of "signature."

Definition 4.21. *Given a spanning tree T in a bipartite graph $G = (U, V; E)$, the* signature *of the tree is the vector $d(T)$ such that $d_i(T)$ is the degree of vertex $i \in U$ in T ($i = 1, 2, \ldots, n$). The number of 1's in $d(T)$ is called the* level *of T.*

We have already seen in Section 4.5.2 that any spanning tree in G admits, for any root $r \in U$, a dual solution u, v satisfying $c_{ij} - u_i - v_j = 0$ for all edges $[i, j] \in T$ obtained, e.g., as in Procedure Compute_Dual of Algorithm 4.14. If, in addition, $c_{ij} - u_i - v_j \geq 0$ for all edges $[i, j] \notin T$, the tree is called a *dual feasible tree*. The signature method is based on the following.

Theorem 4.22. *Consider an instance of LSAP and a dual feasible tree T in the associated bipartite graph $G = (U, V; E)$. If $d_r(T) = 1$ for exactly one root vertex $r \in U$ and $d_i(T) = 2$ for all vertices $i \in U \setminus \{r\}$, then the solution*

$$x_{ij} = 1 \text{ for all odd edges } [i, j] \in T \tag{4.23}$$

(see Definition 4.14), and $x_{ij} = 0$ for all other edges $[i, j] \in E$, is optimal.

Proof. From Definition 4.15 and Proposition 4.16 it is easily seen that (i) T is a strongly feasible tree; and (ii) $X = (x_{ij})$ is a feasible primal solution. It follows that X and the dual variables associated with T satisfy the complementary slackness conditions. □

The signature method, shown in Algorithm 4.16, starts with a dual feasible tree T of level $n - 1$, rooted at $r = 1$, and having signature $d(T) = (n, 1, \dots, 1)$, consisting of edges $[1, j]$ $(j = 1, 2, \dots, n)$ in T, plus $n - 1$ additional edges $[i, j]$ $(i = 2, 3, \dots, n)$ of minimum reduced cost, as the one shown by the solid lines in the bipartite graph of Figure 4.14. A series of cycles is then performed, each of which decreases by one the level of T through a number of pivots, until a dual feasible tree of level 1 is obtained.

The signature of the initial tree of a cycle of level k is $d_1(T) = k + 1$, $d_i(T) = 1$ for exactly k vertices of U, and $d_i(T) = 2$ for the remaining $n - k - 1$ vertices of U. The first pivot decreases the degree of the root vertex $r = 1$ by 1 through removal of an edge $[1, l]$ ($[1, 1]$ in Figure 4.14) and increases that of another row vertex s^* by 1 through addition of an edge $[s^*, l^*]$ ($[2, 3]$ in Figure 4.14). Hence, if $d_{s^*}(T)$ is equal to 2, we know that level $k - 1$ has been reached and the cycle terminates. Otherwise, $d_{s^*}(T)$ is equal to 3 and a new pivot is performed to reduce the degree of s^* by 1, and so on. The final cycle produces a tree whose signature contains exactly one 1 and otherwise 2's from which Theorem 4.22 provides the optimal primal solution.

A pivot removes from T an edge $[s, l]$ such that both s and l have a degree of at least 2. Let T^s and T^l, with $s \in T^s$ and $l \in T^l$, be the two resulting components (see again Figure 4.14, where T^l is shown by the thick edges). The new tree is obtained by computing

$$\delta = \min\{c_{ij} - u_i - v_j \ : \ i \in U(T^l), j \in V(T^s)\}, \tag{4.24}$$

where, for a subtree $\overline{T}$, $U(\overline{T})$, and $V(\overline{T})$ denote the sets of vertices $i \in U$ and $j \in V$, respectively, such that $[i, j] \in \overline{T}$. Subtrees T^s and T^l are then rejoined through an edge $[s^*, l^*]$ for which δ is achieved (the dashed edge in Figure 4.14). The dual variables of the vertices of T^l are updated as

$$\begin{cases} u_i := u_i + \delta & \text{for all } i \in U(T^l), \\ v_j := v_j - \delta & \text{for all } j \in V(T^l). \end{cases} \tag{4.25}$$

Since T is a dual feasible tree, we have $\delta \ge 0$. It can be seen that the choice of δ and (4.25) guarantee that the new tree is dual feasible.

ALGORITHM 4.16. Signature.
Balinski algorithm.

comment: initialization (dual feasible tree of level $n-1$);
$T := \{[1, j] : j = 1, 2, \dots, n\}, u_1 := 0$;
for $j := 1$ **to** n **do** $v_j := c_{1j}$;
for $i := 2$ **to** n **do**
  $j^* := \arg\min\{c_{ij} - v_j : j \in V\}, u_i := c_{ij^*} - v_{j^*}, T := T \cup \{[i, j^*]\}$;
endfor
for $k := n-1$ **down to** 2 **do** [**comment:** k = level of T]
comment: decrease by 1 the level of the current tree;
  select a row vertex $t \in U$ with $d_t(T) = 1$ as the target, and set $s := 1$;
  repeat
    comment: dual pivoting;
    let l be the first column vertex in the path connecting s to t in T;
    $T := T \setminus \{[s, l]\}$, and let T^s, T^l ($s \in T^s, l \in T^l$) be the resulting subtrees;
    $\delta := \min\{c_{ij} - u_i - v_j : i \in U(T^l), j \in V(T^s)\}$;
    let $[s^*, l^*]$ be an edge for which δ is achieved, and set $T := T \cup \{[s^*, l^*]\}$;
    for each $i \in U(T^l)$ **do** $u_i := u_i + \delta$;
    for each $j \in V(T^l)$ **do** $v_j := v_j - \delta$;
    $s := s^*$
  until $d_s(T) = 2$
endfor
comment: determine the optimal primal solution;
let r be the unique row vertex having $d_r(T) = 1$;
for each odd edge $[i, j] \in T$ **do** $x_{ij} := 1$ [**comment:** $x_{ij} = 0$ for all other edges]

It is easy to see that, within a cycle, no vertex $s \in U$ can be involved more than once in a pivot operation in which the leaving edge is incident to s. Hence, at most $n-k$ pivots are performed at level k, leading to an overall bound of $(n-1)(n-2)/2$ pivots. Each pivot, if executed from scratch, requires $O(n^2)$ operations for the computation of δ, hence, producing an overall $O(n^4)$ time complexity. A better implementation, which attaches labels to the vertices of U, was suggested by Cunningham [207] and Goldfarb [334]. In this way the computational effort is reduced to $O(n^2)$ operations per level, thus giving an $O(n^3)$ time complexity for the signature method.

Example 4.23. We refer to the numerical instance used for the Dinic–Kronrod and Hung–Rom algorithms (Examples 4.9 and 4.20). Consider the cost matrix shown in Figure 4.14. The initial dual feasible tree T of level 3, with signature $d(T) = (4, 1, 1, 1)$, is shown by the solid lines in the bipartite graph. Here, too, the u and v values are shown, respectively, on the left and on the top, with figures in brackets referring to subsequent phases of the algorithm. At the beginning of the first cycle, we select row vertex $t = 2$ as the target and we set $s = 1$. Hence $l = 1$ and $[1, 1]$ is the pivoting edge. We have $U(T^l) = \{2, 3\}$ and $V(T^s) = \{2, 3, 4\}$, with subtree T^l highlighted through thick lines. Then $\delta = 2$ is determined by $[s^*, l^*] = [2, 3]$ (dashed edge), the dual variables are updated (values in

$$\begin{array}{r} \\ 0 \\ (-3)\text{-}5 \\ (-4)\text{-}6 \\ -7 \end{array} \begin{array}{cccc} 7(5) & 9 & 8 & 9 \\ \end{array} \left(\begin{array}{cccc} 7 & 9 & 8 & 9 \\ 2 & 8 & 5 & 7 \\ 1 & 6 & 6 & 9 \\ 3 & 6 & 2 & 2 \end{array}\right)$$

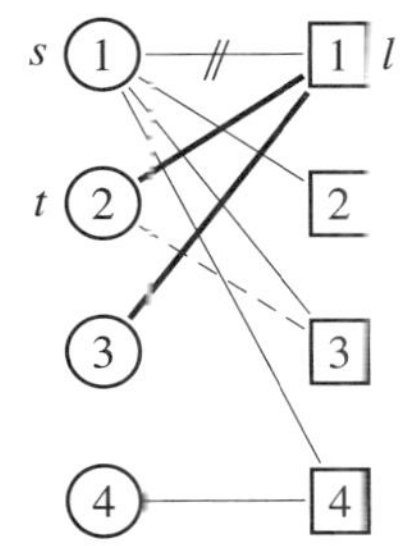

Figure 4.14. *First cycle of Example 4.23.*

$$\begin{array}{r} \\ 0 \\ (-2)\text{-}3 \\ (-3)\text{-}4 \\ -7 \end{array} \begin{array}{cccc} 5(4) & 9 & 8(7) & 9 \\ \end{array} \left(\begin{array}{cccc} 7 & 9 & 8 & 9 \\ 2 & 8 & 5 & 7 \\ 1 & 6 & 6 & 9 \\ 3 & 6 & 2 & 2 \end{array}\right)$$

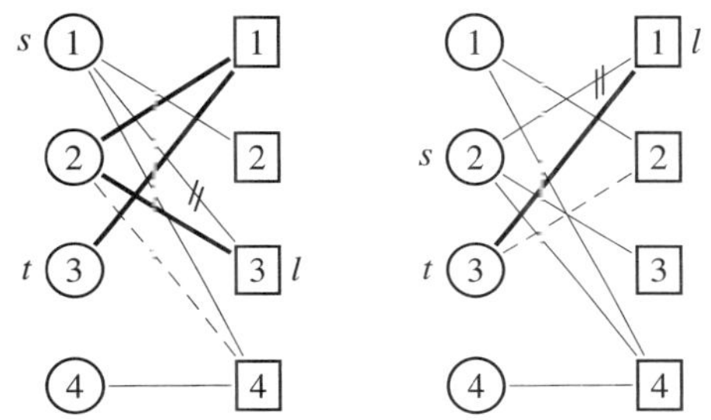

Figure 4.15. *Second cycle of Example 4.23.*

brackets), and the cycle terminates since $d(T) = (3, 2, 1, 1)$ for the new tree T of level 2, shown by the solid lines in the first bipartite graph of Figure 4.15.

At the second cycle we select row vertex $t = 3$ as the target. We set $s = 1$; hence, $l = 3$, so $[1, 3]$ is the pivoting edge. We now have $U(T^l) = \{2, 3\}$ and $V(T^s) = \{2, 4\}$, with subtree T^l shown by the thick lines in the first bipartite graph of Figure 4.15. The value $\delta = 1$ is then determined by $[s^*, l^*] = [2, 4]$ (dashed edge), the dual variables are updated (values in brackets), and we obtain $d(T) = (2, 3, 1, 1)$ for the new tree T shown by the solid lines in the second bipartite graph of Figure 4.15. Since $d_2(T) \neq 2$, the cycle requires an additional pivoting. We have $s = 2$ and $l = 1$, so $[2, 1]$ is the pivoting edge. We obtain $U(T^l) = \{3\}$ and $V(T^s) = \{2, 3, 4\}$, with subtree T^l consisting of the unique edge $[3, 1]$ (the thick line in the second bipartite graph of Figure 4.15). Then $\delta = 0$ is determined by $[s^*, l^*] = [3, 2]$ (dashed edge), the dual variables remain unchanged, and both the cycle and the algorithm terminate since $d(T) = (2, 2, 2, 1)$ for the resulting new tree T of level 1. The optimal primal solution is then computed as $x_{44} = x_{12} = x_{23} = x_{31} = 1$ and $x_{ij} = 0$ elsewhere. ∎

Balinski [61] showed that the signature method may be used to prove that the Hirsch conjecture (see Section 2.2) holds for dual transportation polyhedra.

We have seen in Section 4.6.1 that, some years before the signature method, Hung and Rom [382] had proposed another $O(n^3)$ dual (non-simplex) algorithm in which a series of relaxed problems is solved by updating the current solution through shortest paths until the semi-assignment becomes feasible. In 1987 Kleinschmidt, Lee, and Schannath [422] proved that these two algorithms are equivalent. More specifically, they showed that, if row and column nodes are interchanged in the description of one of the algorithms, then, under some mild restrictions, the following properties hold.

(i) The Balinski algorithm can be started with the Hung–Rom initial tree, and conversely the Hung–Rom algorithm can be started with the Balinski initial tree. Suppose now that both algorithms have reached the same current tree T.

(ii) Any tree T' constructed by the Hung–Rom algorithm (resp., by the Balinski algorithm) in the next cycle can also be reached by the Balinski algorithm (resp., by the Hung–Rom algorithm) in the next cycle.

Goldfarb [334] proposed another signature algorithm which solves a sequence of subproblems, each larger than the preceding one, starting from a 1×1 problem. The kth subproblem is defined by the first k rows and k columns of C. Given the optimal solution to the $k \times k$ subproblem, the corresponding dual feasible tree having signature $(2, \dots, 2, 1)$ is extended to the $(k+1) \times (k+1)$ subproblem by assigning the $(k+1)$th row and column at minimum reduced cost. The new signature either remains optimal or has the form $(2, \dots, 2, 3, 2, \dots, 2, 1, 1)$. In the latter case, the dual pivoting of the Balinski algorithm is executed. Since this step requires $O(k^2)$ pivots instead of $O(n^2)$, the resulting algorithm has a smaller coefficient of n^3.

Variants for solving sparse LSAPs in $O(nm + n^2 \log n)$ time were also proposed by Goldfarb [334].

Balinski [63] observed that the signature method cannot be considered a genuine dual simplex method since a pivot may be made by deleting an edge $[i, j]$ with $x_{ij} = 0$. In addition, the algorithm terminates only when a tree with the wanted signature (containing exactly one 1 and otherwise 2's) is obtained, hence, it may continue to pivot when an optimal solution has been reached. In order to obtain a genuine dual simplex method, Balinski [63] introduced the *strongly dual feasible trees.*

Definition 4.24. *Given a primal solution X, a tree T in $G = (U, V; E)$ rooted at $r \in U$ is a* strongly dual feasible tree *if $x_{ij} \geq 1$ for all even edges $[i, j] \in T$ and $x_{ij} \leq 0$ for all odd edges $[i, j] \in T$ which are not edges $[r, j]$ with $d_j(T) = 1$.*

Note that the initial tree T of the signature method is strongly dual feasible since we can determine the associated primal solution by setting $x_{ij} = 1$ for all edges of $F = \{[i, j] \in T : i \neq 1\}$ (even edges), $x_{1j} = 1 - \sum_{[i,j] \in F} x_{ij}$ for all $j \in V$ (odd edges), and $x_{ij} = 0$ otherwise. For the tree of Figure 4.14 we would obtain $x_{21} = x_{31} = x_{44} = 1$, $x_{11} = -1$, $x_{12} = x_{13} = 1$, $x_{14} = 0$, and $x_{ij} = 0$ otherwise. Balinski [63] gave another pivot rule, obtained by combining the signature method and the Dantzig rule of selecting the edge with the most negative reduced cost, and proved that pivoting with such a rule on a strongly dual feasible tree produces another strongly dual feasible tree. The resulting algorithm only pivots on edges $[i, j]$ with $x_{ij} < 0$; hence, it is a genuine dual simplex method. It converges in at most $(n-1)(n-2)/2$ pivots and $O(n^3)$ time.

Another dual simplex algorithm, which, similarly to the one proposed by Goldfarb [334], solves a sequence of subproblems of increasing size, was later proposed by Akgül [16, 17]. Let $G^k = (U, V^k; E^k)$, with vertex set $V^k = 1, 2, \dots, k \cup \{0\}$ and edge set

$E^k = \{[i, j] \in E : i \in U, j \leq k\} \cup \{[i, 0] : i \in U\}$, denote a graph obtained from the original bipartite graph $G = (U, V; E)$ by (i) removing the last $n - k$ vertices from V and their incident edges; and (ii) adding a fictitious column vertex 0 and all the edges connecting it to the row vertices of U with identical costs $c_{i0} = M$ for some large M. Consider the transportation problem (see Sections 1.2 and 4.5) defined by G^k with supplies $a_i = 1$ at the source vertices of U and demands $b_j = 1$ $(j = 1, 2, \ldots, k)$ and $b_0 = n - k$ at the sink vertices of V^k. The algorithm solves in sequence the transportation problems relative to $G^0, G^1, \ldots, G^n$. At each stage the algorithm finds an optimal strongly dual feasible tree for the kth subproblem and transforms it into an optimal strongly dual feasible tree for the $(k+1)$th subproblem by performing at most $k - 1$ pivots. The overall time complexity is $O(n^3)$ for dense graphs and $O(nm + n^2 \log n)$ for sparse graphs. A similar algorithm was independently obtained by Paparrizos [531].

Another signature approach was proposed by Paparrizos [529]. The algorithm starts with a dual feasible tree and performs a series of stages, each consisting of a number of pivots. These pivots may produce infeasible dual solutions, but each stage terminates with a strongly dual feasible tree. The last tree of the last stage is also primal feasible and hence optimal. The algorithm has time complexity $O(n^4)$. Akgül and Ekin [21] proposed a better implementation of the Paparrizos [529] algorithm that runs on $O(n^3)$ time. Another $O(n^3)$ variant was later presented by Paparrizos [530], a *purely infeasible* algorithm that restores dual feasibility only when it stops with an optimal solution. An improved version of this algorithm was given by Achatz, Kleinschmidt, and Paparrizos [3] while its worst-case behavior was examined by Papamanthou, Paparrizos, Samaras and Stergiou [528].

4.6.3 Auction algorithms

The auction method was introduced in 1981 by Bertsekas [86] for a maximization version of LSAP. For the sake of uniformity with our presentation, we will, however, describe it for the minimization case.

The algorithm maintains a pair $(x, (u, v))$ of a (possibly infeasible) primal solution x (e.g., initially, $x_{ij} = 0$ for all i and j) and a feasible dual solution (u, v) satisfying complementary slackness, obtained as follows. Recall the dual of the continuous relaxation of LSAP (see Section 4.1.2),

$$\begin{aligned} \max \quad & \sum_{i=1}^{n} u_i + \sum_{j=1}^{n} v_j \\ \text{s.t.} \quad & u_i + v_j \leq c_{ij} \quad (i, j = 1, 2, \ldots, n), \end{aligned}$$

and observe that, for a given vector v, the feasible vector u which provides the maximum objective function value is given by

$$u_i = \min\{c_{ij} - v_j : j = 1, 2, \ldots, n\} \tag{4.26}$$

for $i = 1, 2, \ldots, n$. It follows that the dual problem is equivalent to the unconstrained problem $\max q(v)$, where

$$q(v) = \sum_{i=1}^{n} \min_j \{c_{ij} - v_j\} + \sum_{j=1}^{n} v_j. \tag{4.27}$$

For a given vector v, let

$$\varphi(i) = \arg\min\{c_{ij} - v_j \ : \ j = 1, 2, \ldots, n\} \quad (i = 1, 2, \ldots, n) \tag{4.28}$$

denote the column producing the minimum $c_{ij} - v_j$ value. Let UR be a set of unassigned rows and consider the (infeasible) primal solution $x_{i\varphi(i)} = 1$ for $i \in U \setminus UR$ (and $x_{ij} = 0$ otherwise). The pair $(x, (u, v))$, with u given by (4.26), satisfies the complementary slackness conditions (4.8). This solution is primal feasible (and hence optimal) only if $UR = \emptyset$ and $\varphi(i) \neq \varphi(k)$ for $i \neq k$. It is interesting to note that this result coincides with the basic theorem proved in 1969 by Dinic and Kronrod [235] (see Theorem 4.8, Section 4.3, with $\Delta_j = v_j$).

For a given row i, once u_i has been computed as above, let u_i^s denote the second minimum $c_{ij} - v_j$ value of row i, i.e.,

$$u_i^s = \min\{c_{ij} - v_j \ : \ j = 1, 2, \ldots, n, \ j \neq \varphi(i)\}. \tag{4.29}$$

Starting with the empty assignment ($UR = U$), at each iteration the auction algorithm selects an unassigned row $i \in UR$, determines u_i and u_i^s, and considers the following two cases.

Case 1: ($u_i < u_i^s$) or ($u_i = u_i^s$ and $\varphi(i) \neq \varphi(k)\ \forall k \in U \setminus UR$).
Row i is assigned to column $\varphi(i)$ and removed from set UR and the dual variables are updated by setting $v_{\varphi(i)} := v_{\varphi(i)} - (u_i^s - u_i)$ and $u_i := u_i^s$ (*outpricing*) so as to preserve dual feasibility. In addition, if column $\varphi(i)$ was already assigned to a row, such a row reenters UR.
Case 2: ($u_i = u_i^s$ and $\varphi(i) = \varphi(k)$ for some $k \in U \setminus UR$).
The algorithm executes a labeling procedure similar to the Hungarian one which can terminate in two ways:

(a) an augmenting path having zero reduced cost from i to an unassigned column j is found: one more assignment is then obtained by interchanging unassigned and assigned edges along this path, and the dual variables are updated accordingly;

(b) no such path is found: a value δ (the minimum positive column label) is subtracted from the dual variables associated with the columns having a label of value zero and added to the dual variables associated with labeled rows, row i is assigned to column $\varphi(i)$, and row k reenters UR.

The algorithm terminates when $UR = \emptyset$.

Note that the outpricing operation performed in Case 1 produces an alternative assignment for row i, similarly to what happens in the algorithm by Dinic and Kronrod [235] (see Section 4.3). Outpricing was also later used by Jonker and Volgenant [392] in their preprocessing phase (see Section 4.4.4, Algorithms 4.12 and 4.13). Observe in particular the similarity between Case 1 above and Procedure Augmenting_row_reduction, with one relevant difference when $u_i < u_i^s$ and column $\varphi(i)$ was already assigned to a row: while Case 1 just de-assigns such a row, Augmenting_row_reduction tries to reassign it by iterating the process.

The name *auction* comes from an interpretation that is more intuitive for the maximization version of the problem, where we want to maximize the objective function of LSAP under the assignment constraints. Imagine that the columns represent items for sale in an auction and each row i is a customer for whom each item j is worth c_{ij}. The dual variable v_j is the current price of item j, so the difference $c_{ij} - v_j$ is the profit margin of customer i relative to item j. At any iteration some of the items are temporarily assigned to customers k who are willing to pay $v_{\varphi(k)}$ for item $\varphi(k)$. In Case 1, the new customer i (who has no item assigned) chooses the item j, giving him the maximum profit margin, and bids up its price (*bidding* phase) by the difference between this margin and the second maximum profit margin, i.e., by the largest amount for which j will still give him the maximum profit margin. The item j is then assigned to customer i in place of another customer (if any) that had bid earlier for j. The interpretation of Case 2 is less intuitive and involves the concept of *cooperative bidding*, in which several customers simultaneously increase the prices of the corresponding items.

The time complexity of the auction algorithm is pseudo-polynomial, amounting to $O(n^3 + n^2C)$, where C denotes the maximum c_{ij} value. Bertsekas [86] also considers a combined auction–Hungarian algorithm. The combined approach starts with the auction algorithm and switches to the Hungarian algorithm as soon as an internal counter exceeds a prespecified parameter. It can be shown that the resulting approach has time complexity $O(n^3)$.

A polynomial auction algorithm was later obtained by Bertsekas and Eckstein [92] through a relaxed version of the complementary slackness conditions (4.8). A pair $(x, (u, v))$ of a primal and a dual solution is said to satisfy the *ε-complementary slackness* if, given a positive ε value, we have

$$x_{ij}(c_{ij} - u_i - v_j) \le \varepsilon \qquad (i, j = 1, 2, \ldots, n). \tag{4.30}$$

In other words, the assignment of row i to column j is accepted if the corresponding reduced cost does not exceed zero by more than ε.

Consider a feasible primal-dual pair $(x, (u, v))$ satisfying (4.30), and let φ store the primal solution. By summing over all pairs (i, j) we obtain

$$\sum_{i=1}^{n} c_{i\varphi(i)} - \left(\sum_{i=1}^{n} u_i + \sum_{i=1}^{n} v_i\right) \le n\varepsilon, \tag{4.31}$$

i.e., the difference between the value of the primal solution and that of the dual solution is bounded by $n\varepsilon$. Therefore we have the following.

Proposition 4.25. *If the costs c_{ij} are integer and a feasible primal-dual pair satisfies (4.30) for $\varepsilon < 1/n$, then the solution is optimal for LSAP.*

Let us define the *ε-relaxed* problem as that of finding a solution satisfying (4.30). A modified auction algorithm could initially solve the ε-relaxed problem for a large ε value, then decrease this value and reoptimize the solution, iterating until $\varepsilon < 1/n$. In Bertsekas and Eckstein [92] this approach is implemented by multiplying all costs by $n+1$ and solving a sequence of ε-relaxed problems with integer ε values decreasing from an initial large value

$$\begin{pmatrix} 7 & 7 & 8 \\ 2 & 8 & 5 \\ 3 & 6 & 7 \end{pmatrix} \qquad \begin{pmatrix} 28 & 28 & 32 \\ 8 & 32 & 20 \\ 12 & 24 & 28 \end{pmatrix}$$

(a) (b)

Figure 4.16. *Input matrix and scaled matrix for Example 4.26.*

to one. If a primal-dual pair $(x, (u, v))$ is optimal for this 1-relaxed problem, then the pair $(x, (\overline{u}, \overline{v}))$, with $\overline{u}_i = u_i/(n+1)$ for all i and $\overline{v}_j = v_j/(n+1)$ for all j, satisfies $(1/(n+1))$-complementary slackness for the original problem; hence, Proposition 4.25 applies. The ε decrease is such that at most $O(\log(nC))$ ε-relaxed problems are solved. The overall time complexity is $O(nm\log(nC))$.

Each subproblem is solved through an algorithm similar to the original auction algorithm, which alternates a bidding phase and an assignment phase. In the bidding phase an unassigned row i is selected and the bid is determined as

$$v_{\varphi(i)} - (u_i^s - u_i) - \varepsilon, \tag{4.32}$$

where $\varphi(i)$, u_i^s, and u_i are computed through (4.26), (4.28), and (4.29), respectively. In the assignment phase row i is assigned to column $\varphi(i)$ and the corresponding dual variable is updated by setting $v_{\varphi(i)} := v_{\varphi(i)} - (u_i^s - u_i) - \varepsilon$. In addition, if column $\varphi(i)$ was already assigned to a row, such a row becomes unassigned.

Example 4.26. Given the 3×3 input matrix shown in Figure 4.16(a), consider the scaled matrix obtained by multiplying all elements by $n + 1$, shown in Figure 4.16(b).

We start with $\varepsilon = 4$ and $v = (0, 0, 0)$. The first subproblem is solved through the following iterations:

$i = 1$, $(c_{ij} - v_j) = (28, 28, 32)$, $\varphi = (1, -, -)$, $v = (-4, 0, 0)$;

$i = 2$, $(c_{ij} - v_j) = (12, 32, 20)$, $\varphi = (-, 1, -)$, $v = (-16, 0, 0)$;

$i = 3$, $(c_{ij} - v_j) = (28, 24, 28)$, $\varphi = (-, 1, 2)$, $v = (-16, -8, 0)$;

$i = 1$, $(c_{ij} - v_j) = (44, 36, 32)$, $\varphi = (3, 1, 2)$, $v = (-16, -8, -8)$.

The approximate solution is thus $x_{13} = x_{21} = x_{32} = 1$ and $x_{ij} = 0$ otherwise.

We now solve the second subproblem with the final value $\varepsilon = 1$. We de-assign all rows and start with the final v value of the previous subproblem:

$i = 1$, $(c_{ij} - v_j) = (44, 36, 40)$, $\varphi = (2, -, -)$, $v = (-16, -13, -8)$;

$i = 2$, $(c_{ij} - v_j) = (24, 45, 28)$, $\varphi = (2, 1, -)$, $v = (-21, -13, -8)$;

$i = 3$, $(c_{ij} - v_j) = (33, 37, 36)$, $\varphi = (2, -, 1)$, $v = (-25, -13, -8)$;

$i = 2$, $(c_{ij} - v_j) = (33, 45, 28)$, $\varphi = (2, 3, 1)$, $v = (-25, -13, -14)$.

The optimal solution is thus $x_{12} = x_{23} = x_{31} = 1$ and $x_{ij} = 0$ otherwise. ∎

This implementation of the auction algorithm, where one unassigned row bids at a time, is called the "Gauss–Seidel" version. In a different implementation, known as the

"Jacobi" version, all unassigned rows bid simultaneously. This terminology comes from two well-known iterative techniques for solving a system of n linear equations in n variables. The former auction version is generally more efficient, while the latter is more suited for parallel implementations (see Section 4.11).

The auction algorithm can also be implemented by interchanging the role of rows and columns (customers and items). Given a dual vector u and an unassigned column j, we can determine the minimum and the second minimum $c_{ij} - u_i$ values in that column and assign it to the row providing the minimum. In the economic interpretation (referred to the maximization version of the problem) this means that items decrease their prices to a level that is sufficiently low to lure a customer away from his currently selected item. This approach, proposed by Bertsekas, Castañon, and Tsaknakis [91] for the solution of inequality constrained assignment problems, is known as the *reverse* auction algorithm (and conversely the original one can also be termed the *forward* auction algorithm).

Forward and reverse auction algorithms are mathematically equivalent for LSAP. A combination of the two, which frequently switches between the two phases, was proposed by Bertsekas and Castañon [90] (*forward/reverse* auction algorithm). Castañon [174] analyzed the sensitivity of the reverse and forward/reverse auction algorithm to the choice of the scale factor. The computational performance of the forward/reverse implementation was evaluated by Goldberg and Kennedy [328].

Computer codes implementing various auction algorithms (see also Section 4.9) can be downloaded from the home page of Dimitri Bertsekas (see Section 4.9). Schwartz [602] analyzed the expected performance of the forward auction algorithm on the basis of computational experiments performed for a problem of military interest (see Section 5.5).

Orlin and Ahuja [515] used the concept of ε-relaxation to obtain a hybrid algorithm which performs $O(\log(nC))$ scaling phases. Each phase consists of two actions: a preprocessing and a successive shortest path algorithm. The preprocessing phase can be seen both as a modified auction procedure (as in [515]) and as a push-relabel procedure (as in [11], Section 12.4). Each execution of the modified auction procedure terminates in $O(\sqrt{n}\, m)$ time and yields a solution satisfying ε-complementary slackness and having at most $\lceil \sqrt{n}\, \rceil$ unassigned rows. The overall time complexity of the hybrid algorithm is thus $O(\sqrt{n}\, m \log(nC))$, i.e., the same complexity of the Gabow and Tarjan [297] scaling algorithm (see Section 4.2.3). Further, the algorithm uses very simple data structures.

4.6.4 Pseudoflow algorithms

As seen in Section 4.4.1, LSAP can be solved as an MCFP on a unit capacity network obtained from graph $G = (U, V; E)$. An infeasible flow which satisfies the capacity constraints (4.17) but violates the flow conservation constraints (4.16) is called a *pseudoflow*. Pseudoflow algorithms for MCFP operate with ε scaling. For a given ε, they transform a pseudoflow into an ε-optimal flow, i.e., a feasible flow in which the reduced cost of each arc with positive flow does not exceed ε and the reduced cost of each arc with zero flow is nonnegative. The process is iterated for decreasing ε values until $\varepsilon < 1/n$, and, hence, is optimal (see Proposition 4.25). The transformation of a pseudoflow into an ε-optimal flow is obtained through a series of push and relabel operations. A *push* operation sends flow along an arc of the incremental digraph with positive residual capacity. A *relabel* operation handles ε-optimality by modifying the value of the dual variable associated with a vertex.

Pseudoflow algorithms for LSAP have been given by Orlin and Ahuja [515] (see Section 4.6.3), Goldberg, Plotkin, and Vaidya [331, 332], and Goldberg and Kennedy [328]. We next describe the latter algorithm, which is characterized by high computational efficiency. We give a description (Algorithm 4.17, below) specifically oriented to LSAP, different from the one in [328], which is based on the MCFP notation. Let α be a prefixed scale factor ($\alpha = 10$ in the implementation tested in [328]); the algorithm can be outlined as follows.

ALGORITHM 4.17. Pseudoflow.
Goldberg–Kennedy algorithm.

$\varepsilon := \mathcal{C} := \max_{ij}\{c_{ij}\}$;
for $i := 1$ **to** n **do** $u_i := v_i := 0$;
while $\varepsilon \geq 1/n$ **do**
  $\varepsilon := \varepsilon/\alpha$;
  for $i := 1$ **to** n **do for** $j := 1$ **to** n **do** $x_{ij} := 0$;
  for $i := 1$ **to** n **do** $u_i := \min\{c_{ij} - v_j \ : \ [i, j] \in E\}$;
  while x is not a feasible assignment **do**
    select a vertex that corresponds either to an unassigned row $k \in U$
      or to a column $k \in V$ with more than one row assigned;
    $x :=$ *push-relabel* (k, ε)
  endwhile

Algorithm 4.17 can be seen as a "family" of algorithms obtained by designing different strategies to implement the *push-relabel* (k, ε) function.

Since the arcs have unit capacities, each push operation saturates an arc of the incremental graph, which is thus replaced by its reverse arc. The relabel operation sets the dual variable of a vertex to the smallest value that maintains ε-optimality, namely,

$$\begin{cases} u_k := \min_j\{c_{kj} - v_j \ : \ x_{kj} = 0\} & \text{if } k \in U, \\ v_k := \min_i\{c_{ik} - u_i \ : \ x_{ik} = 1\} - \varepsilon & \text{if } k \in V. \end{cases} \tag{4.33}$$

The convergence of the algorithm descends from the following facts (see Goldberg, Plotkin, and Vaidya [331, 332] and Goldberg and Tarjan [333] for a complete description of the push-relabel approach and convergence proofs). During a scaling iteration

(a) the dual variables u monotonically increase;

(b) the dual variables v monotonically decrease;

(c) the absolute variation of any dual variable is $O(n\varepsilon)$.

In their algorithm cost-scaling assignment (CSA) Goldberg and Kennedy [328] adopted for the *push-relabel* (k, ε) function the *double-push* operation, which was independently introduced in Ahuja, Orlin, Stein, and Tarjan [13]. The double-push operation works as follows. An unassigned vertex $k \in U$ is selected to perform the first push operation ($x_{kj} := 1$) on the arc (k, j) with minimum reduced cost among the arcs emanating from k. This operation is immediately followed by a relabel operation (4.33) on k. If j was previously unassigned, the double-push terminates. If instead j now has more than one arc assigned, then it is selected for the second push and for the subsequent second relabel. Note that this second push removes the assignment of a vertex $i \in U$ to j. Therefore, the primal updating of a double-push operation is equivalent to finding either the single-arc augmenting path (k, j) or the two-arc alternating path (k, j)–(j, i).

In order to update the dual variables efficiently, the implementation adopted for algorithm CSA (see Algorithm 4.18 below) starts by finding the two arcs with the smallest and the second smallest reduced cost among those emanating from k, say, (k, j) and (k, z), respectively. After the first push from k to j has been performed, the dual variable u_k is set to $c_{kz} - v_z$. If the second *push* follows, the dual value v_j is set to $c_{kj} - u_k - \varepsilon$.

ALGORITHM 4.18. Procedure Double_push(k).
Double-push function for $k \in U$.

let (k, j) be the arc with the smallest reduced cost emanating from k;
let (k, z) be the arc with the second smallest reduced cost emanating from k;
$x_{kj} = 1$; [**comment:** push (k, j)]
$u_k := c_{kz} - v_z$;
if $x_{ij} = 1$ for some $i \in U$ **then**
    $x_{ij} = 0$; [**comment:** push (j, i)]
    $v_j := c_{kj} - u_k - \varepsilon$
endif

Observe that a vertex $j \in V$ can have more than one vertex of U assigned only in the middle of the double-push operation; hence, the "while" loop in algorithm Pseudoflow always selects vertices $k \in U$. Due to this invariance, the positive flows can be stored, as is usual for LSAP codes, through an array of pointers that associates with each vertex of V its assigned vertex of U, if any (i.e., function *row*(i) defined in (4.11)). Furthermore, it is not necessary to explicitly store the dual variables u since they can be computed as $u_i = \min\{c_{ij} - v_j : x_{ij} = 1, i = 1, 2, \dots, n\}$ (see (4.8)). One can also observe the similarity between the double-push operation and the application of an auction step made by a bidding phase followed by an assignment phase (see Case 1 in Section 4.6.3): the two methods define the same dual value for j and assign/deassign the same elements. In Goldberg and Kennedy [328] several variations of this implementation are proposed and tested. The most efficient one, CSA-Q, implements the double-push strategy through a stack ordering of the unassigned row vertices and use of the implicit value for dual variables u. In addition CSA-Q speeds up the search through the *fourth-best heuristic* (see Bertsekas [87]), which works as follows. Initially, for each row i, the fourth smallest partial reduced cost $c_{ij} - v_j$ is computed and saved in K_i and the three arcs with the three smallest partial reduced costs are stored. When the double-push procedure needs to compute the first and second smallest reduced cost of a row, the search is performed only among these four costs. Since the values of the dual variables v_j monotonically decrease, the partial reduced costs $c_{ij} - v_j$ strictly increase; hence, it is necessary to recompute the four quantities above only when all but possibly one of the stored arcs have partial reduced costs greater than K_i. Goldberg and Kennedy [328] showed the good computational behavior of algorithm CSA-Q by comparing it, through extensive computational experiments, with the algorithms by Jonker and Volgenant [392] (see Section 4.4.4), Castañon [174] (see Section 4.6.3), and Ramakrishnan, Karmarkar, and Kamath [569] (see Section 4.7). The C implementation of this algorithm is available in the public domain (see Section 4.9).

In Goldberg and Kennedy [329] the authors show how the push-relabel algorithm can be implemented so as to achive the best time bound in the cost-scaling context, i.e., $O(\sqrt{n}\, m \log(nC))$.

4.7 Other algorithms

Thompson [637] proposed a recursive non-simplex method which begins by finding an optimal solution to the problem defined by the first row, then finds an optimal solution to the problem defined by rows one and two and so on. (This idea had been previously used by Derigs and Zimmermann [230] for solving linear bottleneck assignment problems; see Section 6.2.) For technical reasons, the algorithm adds a dummy source vertex $n+1$ to U and a dummy sink vertex $n+1$ to V, with $c_{n+1,l} = c_{l,n+1} = 0$ for $l = 1, 2, \dots, n+1$. Let P_k denote the transportation problem consisting of the first k rows, the dummy row $n+1$, and all columns $1, 2, \dots, n+1$, with unit supplies at the first k vertices of U and supply equal to $n-k$ at vertex $n+1$ of U. The demands are always equal to one for all vertices $j \le n$ of V and equal to zero for the dummy vertex $n+1$ of V. The algorithm starts by finding the optimal primal and dual solutions to the trivial problem P_0. At each iteration k ($k = 1, 2, \dots, n$), two recursive steps are performed: Step A constructs the optimal primal solution to P_k from the optimal primal and dual solutions to P_{k-1}, while Step B constructs the optimal dual solution to P_k from its optimal primal solution. The operations performed are similar to analogous operations performed in the algorithms by Derigs and Zimmermann [230] and Barr, Glover, and Klingman [68] (see Section 4.5.2). The algorithm can be implemented so as to require $O(n^3)$ time.

Akgül [18] describes a criss-cross algorithm for LSAP developed by Akgül and Ekin [20]. The term *criss-cross* had been adopted in 1969 by Zionts [668, 669] for a simplex algorithm which (i) can start with any basic solution (possibly both primal and dual infeasible), i.e., with no need for the so-called *Phase 1* of the simplex algorithm; and (ii) at each iteration performs either a primal or a dual pivoting step. However, the original method had no clear guarantee of being finite. The algorithm by Akgül and Ekin is inspired by the variant proposed in 1985 by Terlaky [636], known as the *convergent* criss-cross method: here, termination after finitely many iterations is ensured by the *least-index* pivot rule, which consists of selecting for pivoting the primal or dual infeasible variable with minimum index. (The reader is referred to Fukuda and Terlaky [291] for a more recent overview of these methods.) The LSAP criss-cross algorithm by Akgül and Ekin [20] starts with a specially structured strongly feasible tree (see Section 4.5.2) and alternates dual and primal phases. A *dual phase* destroys the tree by deleting an edge and adds a new edge so that the resulting edge set is dual feasible. A *primal phase* obtains a strongly feasible tree by performing a series of dual simplex pivots. The algorithm has time complexity $O(n^3)$. For sparse graphs, use of the Fibonacci heaps by Fredman and Tarjan [278] reduces the time complexity to $O(n^2 \log n + nm)$.

Ramakrishnan, Karmarkar, and Kamath [569] presented a specialization of a variant of the famous Linear Programming (LP) interior-point method by Karmarkar [406] to LSAP. This variant, known as the *approximate dual projective* (ADP) method, computes discrete approximations to the continuous trajectory of the Karmarkar projective method. The adaptation to LSAP is identical to the general ADP method but for the stopping criterion. The algorithm starts with an LSAP instance formulated as an LP and optionally an initial dual solution. After the ADP method has made sufficient progress towards optimality, a heuristic procedure periodically checks whether an optimal LSAP solution can be obtained from the current LP solution. If this heuristic succeeds, execution terminates; otherwise, the ADP method is resumed. Extensive computational experiments reported in [569] show a good practical behavior of the algorithm on large-size instances.

An important theoretical result comes from a decomposition algorithm by Kao, Lam, Sung, and Ting [403]. Recall that finding the maximum cardinality matching on a bipartite graph $G(U, V; E)$ is equivalent to solving an LSAP (in maximization version) on the same graph with cost $c_{ij} = 1$ for all edges $[i, j] \in E$. The maximum cardinality matching can be obtained in $O(\sqrt{n}m)$ time through the algorithm by Hopcroft and Karp [376] (see Section 3.3). This time complexity was further improved by Ibarra and Moran [384], Alt, Blum, Mehlhorn, and Paul [27], and Feder and Motwani [268] (see Sections 3.6 and 3.4). The time complexity of the Gabow and Tarjan [297] algorithm for LSAP (see Section 4.2.3) is $O(\sqrt{n}m \log(nC))$. It has long been an open question whether the gap between the time complexity of the maximum cardinality matching algorithms and that of LSAP can be closed. In 2001 Kao, Lam, Sung, and Ting [403] proved that it is possible to obtain from G two "lighter" bipartite graphs such that the value of the maximum weight matching of G is equal to the sum of the values of the maximum weight matchings of the lighter graphs. From this result, they derived an algorithm for LSAP whose time complexity may be defined as follows. Let $\kappa(x, y) = \log x / \log(x^2/y)$ and $\mathcal{W} = \sum_{[i,j]\in E} c_{ij}$. Then the algorithm has time complexity $O(\sqrt{n}\mathcal{W}/\kappa(n, \mathcal{W}/\mathcal{C}))$, thus closing the gap for the case $W = o(m \log(n\mathcal{C}))$ (i.e., when W is asymptotically negligible with respect to $m \log(n\mathcal{C})$). When the weights are of order $m\mathcal{C}$, the time complexity of this algorithm can be written as $O(\sqrt{n}\, m\, \mathcal{C} \log_n(n^2/m))$.

4.8 Summary of sequential algorithms

We conclude our review of sequential algorithms for LSAP by summarizing their complexity in Table 4.2, where $\mathcal{C} = \max_{i,j}\{c_{ij}\}$ and $\mathcal{W} = \sum_{[i,j]\in E} c_{ij}$. The most effective computer codes (summarized in Table 4.3 of Section 4.9) have been obtained by implementing algorithms based on shortest paths, auction, and pseudoflow. For the case of sparse instances the complexities of the shortest path implementations can be rewritten by substituting n^2 with m.

4.9 Software

In this section we list computer codes for LSAP that are available in the public domain. Updated links to downloads can be found on the web page associated with the present book, http://www.siam.org/books/ot106/assignmentproblems.html (from now on the *AP web page*). Some of the source codes are directly downloadable, while for others a link is provided.

The paper by Lotfi [465] contains the QuickBasic listing of an $O(n^4)$ time implementation of the Hungarian algorithm (see Section 4.2.1). Two Fortran implementations of the Lawler [448] $O(n^3)$ time version of the Hungarian algorithm (see Section 4.2.2) have been given by Carpaneto and Toth [168] and by Carpaneto, Martello, and Toth [165]. The former code solves sparse LSAPs and is given in [168] as a Fortran listing. Fortran implementations of the latter code for complete and sparse matrices (see Section 4.4.3) are given in [165] as listings and on diskette and can be downloaded from the AP web page, where the C translation of the code for complete matrices is also available.

Table 4.2. *Evolution of algorithms for LSAP.*

Year	Reference	Time complexity	Category
1946	Easterfield [245]	exponential	Combinatorial
1955	Kuhn [438]	$O(n^4)$	Primal-dual
1957	Munkres [502]	$O(n^4)$	Primal-dual
1964	Balinski and Gomory [64]	$O(n^4)$	Primal
1969	Dinic and Kronrod [235]	$O(n^3)$	Dual
1971	Tomizawa [640]	$O(n^3)$	Shortest path
1972	Edmonds and Karp [250]	$O(n^3)$	Shortest path
1976	Lawler [448]	$O(n^3)$	Primal-dual
1976	Cunningham [205]	exponential	Primal simplex
1977	Barr, Glover, and Klingman [68]	exponential	Primal simplex
1978	Cunningham and Marsh [208]	$O(n^3)$	Primal
1980	Hung and Rom [382]	$O(n^3)$	Dual
1981	Bertsekas [86]	$O(n^3 + n^2 C)$	Auction
1985	Balinski [62], Goldfarb [334]	$O(n^3)$	Dual simplex
1985	Gabow [295]	$O(n^{3/4} m \log C)$	Cost scaling
1988	Bertsekas and Eckstein [92]	$O(nm \log(nC))$	Auction + ε-rel.
1989	Gabow and Tarjan [297]	$O(\sqrt{n}\, m \log(nC))$	Cost scaling
1993	Orlin and Ahuja [515]	$O(\sqrt{n}\, m \log(nC))$	Auction + ε-rel.
1993	Akgül [19]	$O(n^3)$	Primal simplex
1995	Goldberg and Kennedy [328]	$O(\sqrt{n}\, m \log(nC))$	Pseudoflow
2001	Kao, Lam, Sung, and Ting [403]	$O(\sqrt{n}\, \mathcal{W} \log(\frac{n^2}{\mathcal{W}/C}) / \log n)$	Decomposition

The listings of Fortran programs implementing a variant of the shortest augmenting path algorithm by Tomizawa [640] (see Section 4.4.1) can be found in the book by Burkard and Derigs [145]. The listing of an Algol implementation of another shortest path algorithm can be found in Bhat and Kinariwala [96]. The listing of the Pascal implementation of an effective shortest path algorithm is given in Jonker and Volgenant [392]. In addition, Pascal, Fortran, and C++ codes can be found on the web page of MagicLogic Optimization, Inc.

Shortest path algorithms based on sparsification techniques are given by Carpaneto and Toth [169] and by Volgenant [647]. A Fortran implementation of the former algorithm can be downloaded from the AP web page, while a Pascal implementation of the latter one is listed in [647].

The book by Bertsekas [88] includes a number of Fortran listings of implementations of auction algorithms for LSAP (see Section 4.6.3). The codes can be downloaded from the author's web page.

The C source code of the pseudoflow algorithm by Goldberg and Kennedy [328] (see Section 4.6.4) can be downloaded from the web page of Andrew Goldberg.

Didactic software is also available. At the AP web page one can execute, by visualizing the most relevant information, Java applets implementing the $O(n^4)$ and $O(n^3)$ Hungarian algorithms of Sections 4.2.1 and 4.2.2, developed by Cheng Cheng Sun [182] and Bergamini [84]. Papamanthou, Paparrizos, and Samaras [527] describe a Java didactic software for visualizing the execution of a primal simplex algorithm (see Section 4.5.2). The applet can

Table 4.3. *The tested codes.*

Acronym	APC APS	CTCS	LAPJV LAPJVsp	LAPm	NAUC	AFLP	AFR	CSA
Reference	[165]	[169]	[392]	[647]	[88]	[88]	[88]	[328]
Year	1988	1987	1987	1996	1991	1991	1991	1995
Method	H	SP	SP	SP	AU+SP	AU	AU	PF
Language	FTN	FTN	PAS	PAS	FTN	FTN	FTN	C

be found on the web page of Nikolaos Samaras. The same page also hosts a Java applet that visualizes the exterior point algorithm by Paparrizos [530] and Achatz, Kleinschmidt, and Paparrizos [3] (see Section 4.6.2) in three possible implementations. The latter software is described in Andreou, Paparrizos, Samaras, and Sifaleras [28].

4.9.1 Computer codes

Among the available software, we have selected and tested eight popular codes. Table 4.3 has one column per code. Each column gives the acronym identifying the code, followed by

(i) the main literature reference;

(ii) the year of publication;

(iii) the nature of the algorithm (H = Hungarian, SP = shortest path, AU = auction, and PF = pseudoflow);

(iv) the language used for the original implementation (FTN = Fortran, PAS = Pascal, C).

All codes work on integer, possibly negative, input costs.

Codes APC and APS: Carpaneto, Martello, and Toth [165]

These codes implement the Lawler $O(n^3)$ version of the Hungarian algorithm. The solution is initialized by executing Procedures Basic_preprocessing of Section 4.1.2 and Three_edge_preprocessing of Section 4.4.4. Both codes consist of a Fortran subroutine that receives the input instance through formal parameters: APC works on a complete cost matrix, while APS works on a sparse instance, provided as a forward star. A very large solution value is returned by APS if the instance does not have a perfect matching. The C translation of APC is also available at the AP web page.

Code CTCS: Carpaneto and Toth [169]

The initialization phase is identical to that of APC. If the number of assignments in the primal partial solution is smaller than $0.6n$, a standard shortest path technique is used to complete the solution. Otherwise, the algorithm constructs a sparse matrix $\widehat{C}$ by heuristically selecting a subset of elements from the original cost matrix C. Using a sparse implementation of the shortest path search, an attempt is made to complete the solution: if a complete assignment does not exist on $\widehat{C}$, some elements from C are heuristically added and the process is iterated. If instead an optimal primal-dual pair for $\widehat{C}$ is found, it is necessary to verify if this is optimal for C: specifically, a check is performed to test if all the reduced costs on C are nonnegative (i.e., if inequalities (4.7) hold). If there are negative reduced costs, the corresponding elements are added to $\widehat{C}$ and the process is iterated. CTCS is a Fortran subroutine that receives the complete cost matrix describing the input instance as a formal parameter.

Codes LAPJV and LAPJVsp: Jonker and Volgenant [392]

The preprocessing phase of both codes, which is the most time consuming part of the algorithm, has been described in Section 4.4.4 and consists of a column reduction followed by Procedures Reduction_transfer and Augmenting_row_reduction. The execution of Procedure Augmenting_row_reduction is repeated twice. The partial solution is then completed through a smart implementation of a Dijkstra-like shortest paths algorithm. Code LAPJV works on a complete cost matrix to be supplied in input. Code LAPJVsp is the sparse version of LAPJV: the user must provide the input instance as a forward star. The experiments have been executed on the Fortran versions of these algorithms. The original LAPJV codes are available at the AP web page.

Code LAPm: Volgenant [647]

Similarly to CTCS, code LAPm (called LAPMOD in the original implementation) constructs a sparse matrix $\widehat{C}$ to contain the main diagonal of C and a subset of heuristically selected small elements of C (using a threshold value depending on n). Note that the presence of the main diagonal ensures that $\widehat{C}$ always contains a feasible assignment. The LSAP instance associated with $\widehat{C}$ is solved through an adaptation of LAPJV to sparse matrices. When an optimal primal-dual pair for $\widehat{C}$ is found, the algorithm performs the same test and update of $\widehat{C}$ as in CTCS and iterates. The experiments have been executed on a Fortran translation, LAPm, of the original Pascal code. LAPm is a subroutine that receives the complete cost matrix describing the input instance as a formal parameter.

Code NAUC: Bertsekas [88]

This is a "Naive AUCtion and sequential shortest path" algorithm without ε-scaling (called NAUCTION_SP in the original implementation). The author [88] describes the code as follows.

> This code implements the sequential shortest path method for the assignment problem, preceded by an extensive initialization using the naive auction algorithm. The code is quite similar in structure and performance to a code of the author [86] and to the code of Jonker and Volgenant [391] and [392]. These codes also combined a naive auction initialization with the sequential shortest path method.

The algorithm performs a prefixed number of auction cycles, each of which is similar to Procedure Augmenting_row_reduction of LAPJV. The number of cycles is defined as a function of the sparsity of the matrix (for dense instances it is equal to two). After the auction phase, the partial solution is completed through shortest paths. The code is a Fortran program that works on a sparse instance provided by the user as a forward star. It solves an LSAP in maximization version. The non-existence of a perfect matching is not checked by the code.

Codes AFLP and AFR: Bertsekas [88]

These two implementations of the auction method with ε-scaling (see Section 4.6.3) differ in the way the scaling technique is implemented. Algorithm AFLP ("Auction with FLoating Point variables," called AUCTION_FLP in the original implementation) uses real variables so it can directly handle the scaled values. Algorithm AFR ("Auction with Forward/Reverse," called AUCTION_FR in the original implementation) uses integer variables and multiplies all data by a constant factor K such that the values assumed by ε are positive integers. As a consequence, AFR can only solve instances where the largest entry of the input matrix is K times smaller than the largest integer that can be stored in a computer word. Both codes are Fortran subroutines working on sparse instances passed by the user through common areas. For AFLP the user is required to provide the input as a forward star, while for AFR a second data structure, equivalent to a backward star, must also be given. This structure is used internally to efficiently implement the reverse phase. In addition, both codes require four values to define the scaling strategy passed through formal parameters. These codes solve an LSAP in maximization version. The non-existence of a perfect matching is not checked.

Code CSA: Goldberg and Kennedy [328]

This is the CSA-Q implementation of the pseudoflow algorithm (see Section 4.6.4). The authors presented several implementations and, on the basis of extensive computational experiments, they concluded that this is best overall. It uses the double-push method and the so-called fourth-best heuristic to speed up the search. CSA comes with a package containing a Makefile that allows one to compile many versions of the code, each with some set of options turned on. The main procedure reads the instance from standard input, prepares the internal data structures, and runs the optimization procedure. The input instance must be provided in DIMACS format (see the DIMACS web page). The code solves an LSAP in maximization version. It is assumed that a perfect matching exists. Differently from the other codes, it is not very easy to use CSA as a subprogram.

4.10 Experimental analysis

In this section we examine the average computational behavior of the eight selected codes on dense and sparse test instances. All the experiments were performed on an Intel Pentium 4 at 2.6 GHz, with four megabytes of RAM memory, running Windows Server 2003. All codes are written in Fortran except for CSA, which is implemented in C. We used compilers Compaq Visual Fortran version 6.6 and Microsoft Visual C++ version 6.0.

In order to have a unique Fortran calling program to run all codes, we implemented two interfaces for CSA. The first one is a Fortran subroutine which receives the cost matrix, stores the costs in a single vector, and calls the second interface, written in C, which prepares the data structures and runs the optimization procedure. The elapsed CPU time was measured for the optimization phase only. The Fortran language stores full matrices by columns, while the C language stores them by rows. Since CSA solves a maximization problem, the first interface stores the opposite of the transposed Fortran cost matrix (i.e., CSA receives cost $-c_{ji}$ for entry (i, j)).

4.10.1 Classes of instances

We used four classes of randomly generated instances from the literature and a deterministic benchmark. Most of these instances were used by Dell'Amico and Toth [220] for evaluating LSAP codes on dense instances. The size n of the cost matrix ranges between 1,000 and 5,000. The random values were generated with the DIMACS completely portable uniform random number generator, universal.c, available at the ftp site of the first DIMACS implementation challenge (see Johnson and McGeoch [390]).

Uniformly random

The costs are uniformly randomly generated integers in the range $[0, K]$ with $K \in \{10^3, 10^4, 10^5, 10^6\}$. This is the most common class of instances used in the literature to test LSAP algorithms (see, e.g., [226], [392], [165], and [328]).

Geometric

We first generate two sets, X and Y, each containing n points with integer coordinates in the square $[1, K] \times [1, K]$ with $K \in \{10^3, 10^4, 10^5, 10^6\}$. Then, for each pair (i, j), c_{ij} takes the truncated Euclidean distance between the ith point of X and the jth point of Y. This class of instances was used by Goldberg and Kennedy [328].

No-Wait Flow-Shop

It is well known (see Papadimitriou and Kanellakis [525]) that an instance of the scheduling problem known as *no-wait flow-shop* can be transformed into an equivalent instance of the *asymmetric traveling salesman problem* (ATSP). We solved LSAPs on cost matrices of ATSP instances derived from no-wait flow-shop scheduling problems with ten machines and up to 5000 jobs having integer processing times uniformly random in the range $[1, 100]$. This class was used by Dell'Amico and Toth [220].

Two-cost

Each entry of the cost matrix has cost 1 with probability 0.5 or cost 10^6 with probability 0.5. This is derived from an analogous class used by Goldberg and Kennedy [328].

Specific benchmark

Machol and Wien [469, 470] defined instances having costs $c_{ij} = (i-1)(j-1)$ $(i, j = 1, 2, \ldots, n)$ that are difficult for LSAP algorithms. We tested five benchmarks with $n \in \{1000, 2000, \ldots, 5000\}$.

Sparse instances

For each of the above classes of dense instances (except for the last one), we obtained sparse instances by randomly selecting, in each row, $2 \log n$ entries.

4.10.2 Experiments

The entries in Tables 4.4–4.12 report the average elapsed CPU time over 10 instances (for all random classes). Each code had a time limit of 500 seconds per instance (except for the Machol–Wien instances of Table 4.8, for which the time limit was set to 3000 seconds).

Table 4.4. *Dense uniformly random instances.*

n	APC	CTCS	LAPJV	LAPm	NAUC	AFLP	AFR	CSA
				$c_{ij} \in [1, 10^3]$				
1000	0.42	0.05	0.05	0.01	0.18	0.31	0.44	0.15
2000	3.70	0.25	0.26	0.05	1.38	4.58	1.75	0.63
3000	21.33	7.62	0.74	0.13	5.84	52.32	7.10	1.95
4000	57.20	14.63	1.75	0.22	13.10	252.19	11.71	5.00
5000	89.78	18.61	2.55	0.30	18.75	tl	37.82	10.42
				$c_{ij} \in [1, 10^4]$				
1000	0.49	0.51	0.05	0.02	0.07	0.34	–	0.14
2000	3.01	3.16	0.26	0.06	0.49	1.53	–	0.61
3000	9.05	9.64	0.75	0.13	1.54	5.19	–	1.41
4000	18.97	20.50	1.85	0.23	3.50	7.19	–	2.58
5000	35.50	38.74	3.78	0.35	7.39	14.27	–	4.28
				$c_{ij} \in [1, 10^5]$				
1000	0.47	0.05	0.05	0.02	0.10	0.29	–	0.14
2000	2.94	0.23	0.28	0.06	0.70	1.01	–	0.60
3000	8.31	8.83	0.76	0.13	2.28	3.03	–	1.39
4000	17.57	18.72	2.05	0.22	5.19	3.79	–	2.60
5000	32.41	34.75	3.86	0.33	10.22	8.15	–	4.29
				$c_{ij} \in [1, 10^6]$				
1000	0.49	0.51	0.05	0.02	0.07	0.34	–	0.14
2000	3.01	3.16	0.26	0.06	0.49	1.53	–	0.61
3000	9.05	9.64	0.75	0.13	1.54	5.19	–	1.41
4000	18.97	20.50	1.85	0.23	3.50	7.19	–	2.58
5000	35.50	38.74	3.78	0.35	7.39	14.27	–	4.28

For the cases where less than ten instances were solved within the time limit, we report in brackets the number of solved instances and compute the average CPU time over them. If no instance was solved within the time limit the entry is "tl." Code AFR cannot handle instances with large costs: for such cases the tables have the symbol "–."

Dense instances

Tables 4.4–4.8 give the computational results for dense instances. Table 4.4 shows that completely random instances are quite easy to solve for all codes except for AFLP. LAPm outperforms all other codes, with LAPJV being the second best. AFR could not be run on instances with large values due to time limits.

Geometric instances (see Table 4.5) are harder, but the codes that can solve the random instances are also able to solve these instances in reasonable CPU times. The winner is again LAPm, followed by CSA.

Table 4.5. *Dense geometric instances.*

n	APC	CTCS	LAPJV	LAPm	NAUC	AFLP	AFR	CSA
				$c_{ij} \in [1, 10^3]$				
1000	1.15	1.19	0.17	0.05	0.70	96.38	135.79(5)	0.19
2000	7.22	7.55	1.07	0.18	5.23	423.59(1)	379.31(1)	0.96
3000	22.76	23.89	3.23	0.37	16.35	tl	tl	2.68
4000	46.09	48.63	6.44	0.60	37.23	tl	tl	4.81
5000	88.51	93.36	12.32	1.00	70.19	tl	tl	8.71
				$c_{ij} \in [1, 10^4]$				
1000	1.17	1.22	0.18	0.06	0.63	155.26	136.09(4)	0.18
2000	7.52	7.85	1.16	0.23	4.76	tl	tl	0.86
3000	23.43	24.54	3.46	0.49	15.05	tl	tl	2.38
4000	48.20	51.13	7.03	0.75	34.41	tl	tl	4.12
5000	93.96	99.87	13.15	1.23	65.07	tl	tl	7.72
				$c_{ij} \in [1, 10^5]$				
1000	1.17	1.21	0.23	0.08	0.57	133.05	–	0.18
2000	7.60	8.00	1.27	0.26	3.80	443.94(1)	–	0.86
3000	23.82	25.30	3.67	0.59	12.53	tl	–	2.36
4000	49.11	52.19	7.16	0.90	28.40	tl	–	4.03
5000	95.25	101.34	13.81	1.50	54.38	tl	–	7.06
				$c_{ij} \in [1, 10^6]$				
1000	1.17	1.22	0.40	0.09	0.65	92.41	–	0.18
2000	7.57	7.89	1.77	0.29	3.64	193.90(2)	–	0.87
3000	23.77	24.68	4.58	0.65	10.91	tl	–	2.29
4000	48.95	51.21	8.82	0.96	23.41	tl	–	4.00
5000	95.12	100.28	16.08	1.63	43.53	tl	–	7.45

Table 4.6. *Dense no-wait flow-shop instances with 10 machines and n jobs.*

n	APC	CTCS	LAPJV	LAPm	NAUC	AFLP	AFR	CSA
1000	4.75	5.54	0.69	0.90	2.13	6.74	5.65	0.13
2000	36.41	45.02	4.83	5.04	15.08	58.79	44.22	0.53
3000	120.81	153.23	15.23	14.14	49.10	177.54	156.99	1.15
4000	290.55	376.28	35.18	30.42	112.45	385.04(5)	117.49(9)	2.04
5000	577.71	753.68	66.40	54.80	213.41	437.78(2)	171.11(9)	3.33

Table 4.7. *Dense two-cost instances.*

n	APC	CTCS	LAPJV	LAPm	NAUC	AFLP	AFR	CSA
1000	0.01	0.01	0.03	0.04	0.02	10.54	–	0.65
2000	0.03	0.03	0.14	0.13	0.06	80.80	–	2.63
3000	0.07	0.07	0.32	0.29	0.15	268.50	–	5.64
4000	0.12	0.12	0.92	0.51	0.26	tl	–	10.28
5000	0.19	0.19	1.83	0.78	0.41	tl	–	16.10

Table 4.8. *Dense Machol and Wien instances. Time limit = 3000 seconds.*

n	APC	CTCS	LAPJV	LAPm	NAUC	AFLP	AFR	CSA
1000	9.61	8.33	4.66	7.91	12.23	188.28	–	2.67
2000	78.51	67.55	37.67	61.12	94.76	tl	–	8.83
3000	315.91	281.66	127.16	203.53	319.44	tl	–	50.56
4000	1239.12	1026.02	302.77	483.31	749.66	tl	–	38.67
5000	2793.93	2544.83	596.41	tl	1486.75	tl	–	82.37

The picture changes in Table 4.6. No-wait flow-shop instances are much harder to solve for all codes, except for CSA, which is the clear winner. The next best codes are LAPJV and LAPm, but their CPU times are one order of magnitude higher. The CPU times are two orders of magnitude higher for the remaining codes.

Two-cost instances (Table 4.7) are very easy to solve for the Hungarian and shortest path algorithms, but are harder for auction and pseudoflow algorithms. APC and CTCS are the fastest codes.

The most difficult instances are the Machol–Wien instances (Table 4.8). All codes have here their highest running times. The only practical code for these cases appears to be CSA.

Table 4.9. *Sparse uniformly random instances.*

n	APS	LAPJVsp	NAUC	AFLP	AFR	CSA
			$c_{ij} \in [1, 10^3]$			
1000	0.02	0.01	0.03	0.04	0.03	0.02
2000	0.09	0.07	0.22	0.07	0.16	0.09
3000	0.22	0.18	0.83	0.30	1.02	0.22
4000	0.44	0.38	1.98	0.92	1.56	0.38
5000	0.75	0.64	4.15	3.28	19.34	0.65
			$c_{ij} \in [1, 10^4]$			
1000	0.03	0.02	0.02	0.07	0.04	0.02
2000	0.12	0.07	0.16	0.18	0.18	0.09
3000	0.32	0.23	0.57	0.58	0.52	0.22
4000	0.59	0.43	1.35	0.78	1.26	0.40
5000	0.92	0.70	2.64	1.50	2.00	0.66
			$c_{ij} \in [1, 10^5]$			
1000	0.03	0.01	0.02	0.08	–	0.02
2000	0.14	0.07	0.12	0.31	–	0.09
3000	0.37	0.21	0.42	0.96	–	0.22
4000	0.70	0.44	0.99	1.26	–	0.39
5000	1.16	0.75	2.08	1.93	–	0.66
			$c_{ij} \in [1, 10^6]$			
1000	0.03	0.02	0.02	0.08	–	0.02
2000	0.14	0.07	0.10	0.35	–	0.09
3000	0.40	0.22	0.33	1.17	–	0.22
4000	0.77	0.41	0.69	1.73	–	0.40
5000	1.28	0.69	1.31	2.95	–	0.67

Sparse instances

Tables 4.9–4.12 give the computational results for sparse instances. Codes CTCS and LAPm are designed to sparsify a dense instance; hence, they were not run on these data sets.

Table 4.9 shows that sparse completely random instances are much easier than their dense counterparts. All codes except AFR can solve them very quickly.

For geometric instances too, the sparse version is considerably easier than the dense version (see Table 4.10). CSA is the fastest code, followed by the shortest path algorithms (being roughly equivalent among them).

No-wait flow-shop instances (Table 4.11) are again more difficult to solve, although they are easier than their dense counterparts. The winner is CSA, followed by APS and LAPJVsp.

Table 4.12 confirms that two-cost instances are very easy to solve. The Hungarian and shortest path algorithms outperform CSA by one order of magnitude and AFLP by three.

Table 4.10. *Sparse geometric instances.*

n	APS	LAPJVsp	NAUC	AFLP	AFR	CSA
			$c_{ij} \in [1, 10^3]$			
1000	0.04	0.03	0.05	0.16	0.24	0.02
2000	0.23	0.23	0.45	1.80	16.63	0.08
3000	0.65	0.66	1.58	5.52	64.99	0.21
4000	1.22	1.35	3.43	25.63	115.51	0.35
5000	2.26	2.59	7.50	46.30	102.88 (3)	0.58
			$c_{ij} \in [1, 10^4]$			
1000	0.05	0.03	0.05	0.18	0.26	0.02
2000	0.29	0.23	0.40	1.44	10.76	0.09
3000	0.78	0.72	1.37	4.23	49.44	0.21
4000	1.44	1.41	2.93	16.47	91.21	0.35
5000	2.68	2.79	6.53	21.10	76.84 (1)	0.58
			$c_{ij} \in [1, 10^5]$			
1000	0.05	0.03	0.04	0.17	–	0.02
2000	0.30	0.24	0.35	1.58	–	0.09
3000	0.87	0.76	1.19	3.51	–	0.21
4000	1.60	1.41	2.49	13.63	–	0.36
5000	2.98	2.79	5.41	40.40	–	0.58
			$c_{ij} \in [1, 10^6]$			
1000	0.05	0.05	0.05	0.18	–	0.02
2000	0.30	0.30	0.35	2.08	–	0.09
3000	0.88	0.84	1.08	3.82	–	0.21
4000	1.63	1.56	2.16	10.92	–	0.35
5000	3.03	3.06	4.58	19.16	–	0.59

Table 4.11. *Sparse no-wait flow-shop instances with 10 machines and n jobs.*

n	APS	LAPJVsp	NAUC	AFLP	AFR	CSA
1000	0.15	0.16	0.39	1.17	0.15	0.03
2000	1.00	1.13	3.48	20.43	1.60	0.13
3000	3.07	3.68	12.00	52.85	4.59	0.28
4000	6.39	7.57	27.90	153.53	9.18	0.46
5000	12.19	14.87	56.40	294.09	24.31	0.73

Table 4.12. *Sparse two-cost instances.*

n	APS	LAPJVsp	NAUC	AFLP	AFR	CSA
1000	0.00	0.00	0.00	1.99	–	0.14
2000	0.01	0.03	0.02	15.36	–	0.52
3000	0.03	0.07	0.05	51.53	–	1.11
4000	0.05	0.12	0.08	119.02	–	1.91
5000	0.10	0.19	0.14	240.38	–	3.10

4.11 Parallel algorithms

Starting in the late 1980s, a number of contributions on parallel algorithms for LSAP appeared in the literature. Some of these papers mainly present theoretical results, while others are more oriented to practical implementations of (i) auction algorithms; (ii) shortest path algorithms; and (iii) primal simplex algorithms.

The theoretical contributions are mostly concerned with implementations on *parallel random access machines* (PRAM). A PRAM is a virtual machine consisting of a set of processors connected by an unbounded shared memory and a common clock. Each processor is a *random access machine* (RAM), and each RAM has an identical program. The RAMs execute the program synchronously, one instruction in one clock cycle, but the RAMs can branch to different parts of the program. Different PRAM models are given by assumptions on the way memory is accessed. In an *exclusive read/exclusive write* (EREW) PRAM every memory cell can be read or written only by one processor at a time. In a *concurrent read/exclusive write* (CREW) PRAM multiple processors can read a memory cell at the same time, but only one can write at a time. In a *concurrent read/concurrent write* (CRCW) PRAM multiple processors can read and write memory cells at the same time.

Practical implementations are mostly tested on *single instruction/multiple data* (SIMD) or *multiple instruction/multiple data* (MIMD) architectures, as defined in the Flynn [275] taxonomy. In a SIMD architecture, multiple processors perform the same operations synchronously on different data, such as, e.g., in an array processor. In a MIMD architecture, multiple processors independently perform different operations on different data, such as, e.g., in a network of workstations. There are two types of MIMD machines:

(i) *shared memory* systems, in which all processors share the same memory;

(ii) *distributed memory* systems, in which each processor has its own memory and constitutes a node of an interconnection network. Some architectures have a fixed interconnection scheme (e.g., the hypercube topology), while in other systems the connection structure may be defined by the user (e.g., transputers with switching/routing devices).

The reader is referred, e.g., to the books by Bertsekas and Tsitsiklis [94], Jan van Leeuwen [451], and Grama, Gupta, Karypis, and Kumar [338] for general introductions to parallel computing and to Duncan [243] for a survey of parallel computer architectures. The results of an extensive experimentation of various LSAP algorithms on several different SIMD and MIMD architectures can be found in Brady, Jung, Nguyen, Raghavan, and Subramonian [113].

4.11.1 Theoretical results

Driscoll, Gabow, Shrairman, and Tarjan [242] introduced new priority queue data structures that can be used to obtain an efficient parallel implementation of the Dijkstra algorithm. Use of this technique in a shortest path algorithm for LSAP gives an $O(nm/p)$ time complexity on an EREW PRAM with $p \le m/(n \log n)$ processors, under the assumption that the costs are nonnegative. Gabow and Tarjan [294] considered a cost scaling algorithm that runs on an EREW PRAM with p processors. The algorithm requires $O(\sqrt{n}m \log(nC) \log(2p)/p)$ time (where C is the maximum c_{ij} value) and $O(m)$ space for $p \le m/(\sqrt{n} \log^2 n)$. For

$p = 1$ this gives the best time bound known for a cost scaling sequential algorithm (see Gabow and Tarjan [297] in Section 4.2.3).

The first deterministic sublinear time algorithm for LSAP instances with integer costs in the range $[-\mathcal{C}, \mathcal{C}]$ was presented by Goldberg, Plotkin, and Vaidya [331, 332]. Their pseudoflow algorithm with ε scaling (see Section 4.6.4) runs on a CRCW PRAM with n^3 processors (at most) and has time complexity $O(n^{2/3} \log^3 n \log(n\mathcal{C}))$. A sequential version has the same time complexity as the Gabow and Tarjan [297] algorithm. Osiakwan and Akl [517] presented a parallel version of the Hungarian algorithm that runs in $O(n^3/p + n^2 p)$ time on an EREW PRAM with $p \le \sqrt{n}$ processors. Goldberg, Plotkin, Shmoys, and Tardos [330] applied interior point techniques in the context of parallel computation. The resulting algorithm solves LSAP instances with integer costs on a CRCW PRAM with m^3 processors in $O(\sqrt{m} \log^2 n \log(n\mathcal{C}))$ time. This time bound compares favorably with the algorithm by Goldberg, Plotkin, and Vaidya [331, 332] in the case of sparse graphs. Fayyazi, Kaeli, and Meleis [267] presented an adjustable linear time parallel algorithm that solves LSAP in $O(n/\omega)$ time using $O(n^{\max(2\omega, 4+\omega)})$ processors, where $\omega \ge 1$ is an integer parameter, i.e., the execution time can be reduced by an unbounded factor at the expense of an increase of the number of processors.

The randomized parallel algorithms by Karp, Upfal, and Wigderson [411] and Mulmuley, Vazirani, and Vazirani [501] solve the minimum-cost matching problem on general graphs using a number of processors that is proportional to $\mathcal{C}$. Orlin and Stein [516] presented a scaling method which, combined with such algorithms, produces parallel algorithms for LSAP in which the number of processors needed is independent of $\mathcal{C}$ (although the time increases proportionally to $\log \mathcal{C}$).

Finally, we mention some results of a different kind, obtained outside the combinatorial optimization "community". Schwiegelshohn and Thiele [605] gave an $O(n^2)$ time parallel implementation of the $O(n^4)$ time sequential shortest path algorithm by Hoffman and Markowitz [373] (see Section 4.4) on an array of $O(n^2)$ computing elements. Megson and Evans [486] designed an architecture for solving LSAP through the Hungarian algorithm. The resulting algorithm runs in $O(n^2)$ time and uses an orthogonally connected array of $(n + 2) \times (n + 2)$ cells consisting of simple adders and control logic. Fayyazi, Kaeli, and Meleis [266] proposed a shortest path algorithm and a related hardware implementation such that the running time is $O(\sqrt{n} \log^2 n)$ using $O(n^3)$ processing elements.

4.11.2 Auction algorithms

Recall that the bidding phase of the Bertsekas [86] auction algorithm (see Section 4.6.3) can be implemented in two ways. One possibility (the Gauss–Seidel version) is to consider a single unassigned row at a time, resulting in the updating of a single dual variable. In a different implementation (the Jacobi version) all unassigned rows are considered at one time, so several dual variables are updated.

Bertsekas [87] and Bertsekas and Castañon [89] proposed different synchronous and asynchronous implementations of the auction algorithm. A *synchronous* parallel algorithm consists of a series of computation phases separated by synchronization points. During each phase, each processor operates independently of the others and waits for the next synchronization point before starting the next phase. In an *asynchronous* parallel algorithm each processor operates independently of the others with data that may be out-of-date if some other processor did complete its task on such data. Occasionally, an asynchronous algorithm can have a synchronization point. Both types of implementation can be adopted both for the Jacobi version, in which all unassigned rows (customers) bid before the dual variables are updated, and the Gauss–Seidel version, in which a single customer bids and a single dual variable is updated. We first discuss synchronous parallelizations.

(i) Jacobi version: each processor performs the bidding of one unassigned customer (or of more than one if there are fewer processors than unassigned customers), and there is a synchronization point when all bids have been completed.

(ii) Gauss–Seidel version: the set of admissible columns (sale items) for the current customer i is partitioned among the processors, and each processor computes the minimum and second minimum $c_{ij} - v_j$ values among the columns assigned to it. When all computations have been completed, there is a synchronization point and one of the processors merges the results in order to compute the bid of i.

(iii) Hybrid version: here, the bidding is performed for a subset S of the unassigned customers and the processors are partitioned among them. Let $P(i)$ be the set of processors dedicated to customer i; the computation proceeds as for the Gauss–Seidel version by partitioning the columns among the processors of $P(i)$.

The assignment phase following each bidding phase is not parallelized, as the potential gain is lower than the associated overhead.

In the asynchronous implementations, the bidding and merging phases are divided into tasks which are stored in a first-in first-out queue. As soon as a processor becomes idle, it starts performing the first task (if any) of this queue. In addition, synchronizations and termination conditions are adopted to guarantee the convergence of the algorithm.

Bertsekas and Castañon [89] present computational experiments on an MIMD computer (the Encore Multimax), showing that the asynchronous implementations outperform the corresponding synchronous implementations and that the best approach is the asynchronous hybrid version. Further details on parallel implementations of the auction algorithm can be found in Bertsekas and Tsitsiklis [94].

Philips and Zenios [546], Wein and Zenios [660], and Li and Zenios [455] computationally compared different implementations of the auction algorithm (Gauss–Seidel, Jacobi, and hybrid versions) on an SIMD computer, the Connection Machine CM-2, with up to 32K processors. The machine was configured as an $N \times N$ grid of processors (where N is n rounded up to the nearest power of 2) with a processor assigned to each entry of the cost matrix. A table in [660] gives comparisons with the codes by Kempka, Kennington, and Zaki [414], Kennington and Wang [415], Balas, Miller, Pekny, and Toth [55], and Zaki [666] (see below). Further computational tests on the Gauss–Seidel and Jacobi implementations were carried out by Kempka, Kennington, and Zaki [414] on an Alliant FX/8 MIMD computer with eight processors and vector-concurrent capabilities, both with and without ε-scaling (see Bertsekas and Eckstein [92], Section 4.6.3).

Schütt and Clausen [601] implemented three distributed algorithms (an Hungarian method, the shortest path algorithm by Balas, Miller, Pekny, and Toth [55], and the auction algorithm by Bertsekas and Castañon [89]) on a cluster of 16 Intel i860 processors, each with 16 MB memory. Another MIMD architecture was used by Buš and Tvrdík [163] to test a variation of the Gauss–Seidel implementation (called a "look-back" auction algorithm) in which each customer is associated with a list of recently assigned items that is used to speed-up the bidding phase. Their experiments were performed on a cluster of 16 Pentium III PCs, each with 256 MB memory, connected with a Myrinet network.

4.11.3 Shortest path algorithms

There are essentially two ways to implement a parallel shortest augmenting path algorithm for LSAP:

(a) to use all processors for finding a single shortest path arborescence;

(b) to find several shortest paths emanating from different unassigned vertices by using one processor per path.

The single-path approach was used by Kennington and Wang [415], who obtained a simple parallel version of the code by Jonker and Volgenant [392] (see Section 4.4.4) by executing the vectorial operations in parallel. The code was tested on a Symmetry S81 with 20 processors. Zaki [666] computationally compared, on an Alliant FX/8 MIMD computer, the Gauss–Seidel implementation of the auction algorithm without ε-scaling (see Kempka, Kennington, and Zaki [414], Section 4.11.2) with a parallel implementation of the Jonker and Volgenant [392] algorithm (see Section 4.4.4) in which the vector capabilities of the machine are used to speed-up the initialization phase and, in the augmentation phase, a single path is determined using all processors jointly. Storøy and Sørevik [624] proposed an implementation where each of n processors is assigned a column of the cost matrix. They implemented all combinations of two initialization routines (the basic preprocessing of Section 4.1.2 and the Jonker and Volgenant [393] preprocessing of Section 4.4.4) and the two augmenting path routines proposed by Carpaneto, Martello, and Toth [165] and Jonker and Volgenant [393] (see Section 4.4). The computational experiments, performed on a MasPar MP2 SIMD computer with 16K processors showed the following.

1. Initialization: for $n/C \geq 3$ the Jonker and Volgenant preprocessing is asymptotically best; otherwise (small n values or $n/C < 3$) the basic preprocessing is best;

2. Augmentation: for $n/C < 2$ the Jonker and Volgenant routine is preferred; otherwise, the Carpaneto, Martello, and Toth routine is more efficient.

Additional experimental results for similar combined implementations are given in Damberg, Storøy, and Sørevik [210].

The implementation of the multi-path approach requires some preliminary theoretical consideration. Indeed, if two or more vertex disjoint augmenting paths are found in parallel, we can immediately update the primal solution using all these paths, as observed in the mid-1980s by Derigs [226]. The dual solution, instead, can be obtained by applying to each shortest path arborescence the standard dual updating (see Algorithm 4.9 in Section 4.4.2) only if all arborescences (and not just the augmenting paths) are disjoint. When the arborescences partially overlap, it is necessary to use an updating procedure that properly considers the fact that one vertex has been separately labeled by more than one processor. Let p be the number of shortest paths searched in parallel. For $h = 1, 2, \dots, p$, let T_h

denote the hth arborescence and $i^h \in U$ (resp., $j^h \in V$) the unassigned vertex where the augmenting path starts (resp., terminates). Moreover, let $SU^h \in U$ and $SV^h \in V$ be the sets of vertices in T^h, π_j^h the label of vertex $j \in SV^h$, and δ^h the cost of the shortest augmenting path from i^h to j^h. Furthermore, let $SU = \bigcup_{h=1}^{p} SU^h$ and $SV = \bigcup_{h=1}^{p} SV^h$. It is shown by Balas, Miller, Pekny, and Toth [55] that a correct dual update can be obtained as follows.

ALGORITHM 4.19. Procedure Multi-path_updating.
Dual update for multiple shortest path arborescences.

for each $i \in SU$ **do**
    if $\exists\, h$ such that $i = i^h$ **then** $u_i := u_i + \delta^h$
    else $u_i := u_i + \max_{h:\varphi(i)\in SV^h}\{\delta^h - \pi^h_{\varphi(i)}\}$;
for each $j \in SV$ **do** $v_j := v_j - \max_{h:j\in SV^h}\{\delta^h - \pi^h_j\}$

where $\varphi(i)$ denotes, as usual, the column assigned to row i. In the Balas, Miller, Pekny, and Toth [55] synchronous implementation all processors work in parallel, each one looking for an augmenting path emanating from a distinct unassigned row vertex. When the number of augmenting paths found exceeds a threshold value, a synchronization point occurs and the processors perform the primal and dual updates. When all updates are completed, another synchronization point is used to allow the processors to start a new search phase. Further computational tricks and use of *d-heaps* to store the reduced costs make the code very effective. Computational experiments were performed on a Butterfly GP1000, an MIMD distributed memory multiprocessor with 14 processors, each directly accessing 4 megabytes of local memory and accessing the local memory of the other processors through a packed switched network. The algorithm solved a dense randomly generated instance with up to 30000 rows and columns and costs in $[0,10^5]$ in 811 seconds.

Bertsekas and Castañon [90] extended the Balas, Miller, Pekny, and Toth [55] algorithm to obtain an asynchronous implementation. In this algorithm, too, each processor looks for a shortest path rooted at a different unassigned vertex, but the coordination is realized by maintaining a "master" copy of a pair of primal and dual solutions in a shared memory. To start an iteration, a processor copies the current master pair. (During this copy operation the master pair is locked, so no other processor can modify it.) The processor finds an augmenting path, then locks the master pair (which in the meantime may have been modified by other processors), checks whether the update is feasible, and, if so, modifies accordingly the master pair. The master pair is then unlocked. Computational testing was carried out on an Encore Multimax computer with 8 processors for three versions of the algorithm (two of which synchronous) by using a simple preprocessing phase in order to highlight the effect of parallelization in the shortest path phase.

4.11.4 Primal simplex algorithms

Recall that primal simplex algorithms for LSAP (see Section 4.5.2) exploit the correspondence between primal basic solutions of the linear problem associated with LSAP and spanning trees on the associated bipartite graph $G = (U, V; E)$. The core of these algorithms, which operate on the special class of strongly feasible trees, consists of two steps:

1. *Pricing*: select an edge $[i, j]$ with negative reduced cost $\overline{c}_j = c_{ij} - u_i - v_j$.

2. *Pivoting*: insert $[i, j]$ in the basis and remove the (unique) other edge $[i, l]$ belonging to the circuit produced by $[i, j]$ (*basis update*); adjust the dual variables (*dual update*).

Step 1 is crucial for the fast convergence of primal simplex algorithms. Indeed it is known that very good results are produced by the Dantzig rule ("select the edge with most negative reduced cost") but, for LSAP, this requires the computation of the reduced costs of all edges, which can be a relatively heavy computational effort for a single iteration. Hence, the parallelization of primal simplex algorithms has mostly concerned the pricing phase.

Miller, Pekny, and Thompson [491] designed a synchronous algorithm in which every processor searches the most negative element in each of k rows of the reduced cost matrix (with $k = 1$ or 2 in the tested implementation). The corresponding row and column indices are stored in a globally accessible pivot queue, and a synchronization point occurs. If the queue is empty, a new search phase is performed. Otherwise, one of the processors performs the pivots in order from the queue, checking before each pivot that the reduced cost remains negative and bypassing pivots where this is not true. The algorithm was tested on a 14 processor BBN Butterfly computer.

An asynchronous implementation was proposed by Peters [542]. In this algorithm there is one *pivot processor*, say, P_1, dedicated to pivoting. While P_1 pivots, the other processors $P_2, \dots, P_p$ (*search processors*) price edges in parallel and deposit candidate edges in a stack that has on top the edge with most negative reduced cost. When a search processor finds an edge $[i, j]$ with negative reduced cost, it locks the stack, checks whether $\overline{c}_{ij}$ is lower than the reduced cost of the top edge, and, if so, deposits $[i, j]$ in the stack and unlocks it. Whenever P_1 has completed a pivoting phase, it acquires the stack and returns a new empty stack to the search processors. If the acquired stack is not empty, P_1 reprices and reorders the first few edges in the stack (since they have been selected while pivoting was taking place) and starts performing pivots. If instead the acquired stack is empty, or it contains no valid entry, P_1 starts to price all edges itself; if it finds no edge with negative reduced cost and the search is not interrupted by any new entry in the stack, the algorithm terminates with the optimal solution. The algorithm was tested on a Sequent Symmetry S81 MIMD shared memory multiprocessor with 10 Intel 80386 processors.

Another asynchronous algorithm was proposed by Barr and Hickman [71], who gave a parallel implementation of the algorithm by Barr, Glover, and Klingman [70]. Here, the parallelization does not concern only the pricing operations since the dual update operations are subdivided so that they can be performed by several processors in parallel. The synchronization is obtained through a monitor, a programming construct invented by Hoare [370]. A *monitor* is a self-scheduled work allocation scheme consisting of

(a) a section of code, controlled by an associated lock, that can be executed by only one processor at a time (*critical section*);

(b) a shared work list, accessible exclusively within the critical section; and

(c) a delay queue for idle processors.

Idle processors enter the critical section one at a time, update the work list, select a task from the list, exit the critical section, perform the task, and return for additional work.

If the monitor has no task available, the processor is placed in the delay queue, to be released when new tasks become available. When all processors are in the delay queue, the execution terminates. In the Barr and Hickman [71] implementation the shared memory includes a list of candidate edges for pivoting (*candidate list*). There are three kinds of tasks:

(a) select an edge from the candidate list and perform a basis update;

(b) perform a specified portion of the dual update; and

(c) price a group of edges and return the one with the most negative reduced cost, if any, to the candidate list.

Computational experiments on a Sequent Symmetry S81 machine with 20 Intel 80386 processors and 32 MB of sharable memory proved that this algorithm is more efficient than the one by Peters [542].

Chapter 5

Further results on the linear sum assignment problem

5.1 Asymptotic analysis

Linear assignment problems with random cost coefficients have found a long standing interest. We assume in this section that the cost coefficients c_{ij} of an LSAP are independent random variables with a common prespecified distribution.

The main question, examined in the next section, concerns the expected optimal value of the problem. Remarkably, the expected optimal value has a well-defined finite limit, denoted as $\mathbb{E}(z^*)$, as n tends to infinity (Aldous [24], 1992).

Further investigations on random problems, reviewed in Section 5.1.2, concern the construction of good approximation algorithms for LSAP.

5.1.1 Expected optimum value

We start by analyzing the average behavior of a simple greedy algorithm for LSAP. We shall use the fact that the expected value of the minimum of n independent random variables uniformly distributed in $[0, 1]$ equals $1/(n+1)$.

Let us execute the following greedy algorithm on an LSAP instance with $n \times n$ cost matrix $C = (c_{ij})$ for which we assume that the cost coefficients c_{ij} are independent random variables uniformly distributed on the interval $[0, 1]$. First, we determine the minimum in the first row of C and delete the first row and the column corresponding to the minimum. The expected value of the first element is $1/(n+1)$. Next, we determine the minimum of the $n-1$ remaining elements in the second row. Its expected value is $1/n$. We delete the second row and the corresponding column of C in which the minimum of the second row was taken and proceed in the same manner with the remaining rows. Thus the expected value of the greedy solution becomes

$$\frac{1}{2} + \frac{1}{3} + \cdots + \frac{1}{n} + \frac{1}{n+1} = H_{n+1} - 1 \sim \log n,$$

where H_n denotes the nth harmonic number and $\log n$ is computed on the natural basis e. Thus we have proven the following.

Proposition 5.1. *If the cost coefficients of a linear sum assignment problem of size n are independent random variables, uniformly distributed in* $[0, 1]$*, then the expected value of the solution provided by the greedy algorithm is* $H_{n+1} - 1$*, which grows with n like* $\log n$.

This is an astonishing result, as in 1969 Donath [238] conjectured that the expected optimal value, when the cost coefficients are independent uniformly distributed random variables in $[0, 1]$, is around $\pi^2/6 \simeq 1.64$ and not increasing like $\log n$. This was also suggested in the mid-1980s by the replica method from statistical physics (see Mézard and Parisi [489, 490]). The replica method often gives correct answers, but is not a rigorous mathematical method. If it is applied to assignment problems, two limits must be interchanged and the analytic continuation of a function must be assumed to exist. However, this cannot be proved to be mathematically correct. Nevertheless, the conjecture could recently be verified. There was a long way up to this result, which we are going to outline below.

In 1979 Walkup [656] showed that 3 is an upper bound on the expected optimal value of LSAP provided the cost coefficients c_{ij} are independent random variables uniformly distributed on $[0, 1]$. His proof is based on Theorem 3.26, which bounds probabilities for the existence of perfect matchings in random bipartite graphs. Walkup's bound was improved to 2 by Karp [409] five years later by conditioning on an optimal basis of the assignment problem. A nice discussion on Karp's bound and its connection to the Dyer-Frieze-McDiarmid [244] inequality can be found in Steele [622]. Walkup and Karp's proofs were not constructive and did not lead to good heuristics. Finally, the upper bound on $\mathbb{E}(z^*)$ was reduced to 1.94 by Coppersmith and Sorkin [196] by analyzing an augmenting path algorithm in the case that the cost coefficients c_{ij} are drawn from an exponential distribution with mean 1.

On the other hand, lower bounds for problems with cost coefficients which are uniformly distributed in $[0, 1]$ were given by Lazarus [449]. This author exploited the weak duality and evaluated the expected value of the dual objective function $\sum_i u_i + \sum_j v_j$ achieved after execution of Procedure Basic_preprocessing of Section 4.1.2. By computations involving first-order statistics, it is shown that the expected value of $\sum_i u_i + \sum_j v_j$, which is a lower bound on the expected optimal value of LSAP, is of order $1 + 1/e + \log n/n$. This yields a bound of 1.368. Moreover, the author evaluated the maximum number of zeroes lying in different rows and columns of the reduced cost matrix after preprocessing. This evaluation implies that the probability of finding an optimal assignment after preprocessing tends to 0 as n tends to infinity.

The lower bound was improved by Goemans and Kodilian [327] to 1.44, and finally to 1.51 by Olin [513] who obtained her bound from a feasible solution of the dual.

From a technical point of view it is much easier to work with random variables drawn from an exponential distribution with mean 1 instead of a uniform distribution on $[0, 1]$. Both distributions have near 0 the density 1; hence, they are indistinguishable in the neighborhood of 0. As when n tends to infinity, only very small values appear in an optimal assignment; since the limit $\mathbb{E}(z^*)$ is finite, both distributions yield the same result. For a rigorous proof of this property see Aldous [24]. Using the exponential distribution with mean 1, Aldous could finally prove that the limit $\mathbb{E}(z^*)$ is indeed $\pi^2/6$. His proof is based on weak convergence arguments started in [24] and on studying matchings in infinite trees. Let us denote by i.i.d. *independent and identically distributed* variables.

Theorem 5.2. (Aldous [25], 2001.) *Let the cost coefficients c_{ij} of a linear sum assignment problem of size n be i.i.d. random variables with an exponential distribution with mean 1. Then the limit $\mathbb{E}(z^*)$ of the expected optimum objective value exists and equals $\pi^2/6$.*

Completely new and different proofs for this theorem have been given by Linusson and Wästlund [461] (see the remarks after Theorem 5.4).

Aldous [25] also proved a conjecture by Houdayer, Bouter de Monvel, and Martin [378] on order ranks. Given a row of the cost matrix he proved that, as n tends to infinity, the probability that an optimal assignment uses the smallest element of this row is $1/2$. The probability that the optimal assignment uses the second smallest element is $1/4$ and so on. More generally, he showed the following result, which is of interest with respect to the greedy algorithm discussed in Proposition 5.1.

Proposition 5.3. *Let the cost coefficients c_{ij} of a linear assignment problem of size n be i.i.d. random variables with an exponential distribution with mean 1. The probability that the optimal assignment uses the kth smallest element of an arbitrary row tends to $1/(2^k)$ as n tends to infinity.*

A similar result for assignment problems with finite size n can be found in Linusson and Wästlund [461].

In 1998 Parisi [538] had conjectured that the expected value $\mathbb{E}(z_n)$ of an optimum assignment of finite size n is

$$\mathbb{E}(z_n) = \sum_{k=1}^{n} \frac{1}{k^2}. \tag{5.1}$$

Note that

$$\mathbb{E}(z^*) = \lim_{n\to\infty} \mathbb{E}(z_n) = \sum_{k=1}^{\infty} \frac{1}{k^2} = \frac{\pi^2}{6}.$$

For $n = 6$ this conjecture has been proven by Alm and Sorkin [26]. In 2003, Linusson and Wästlund [460] gave a computerized proof of this conjecture for $n = 7$. Finally, Linusson and Wästlund [461] and, independently, Nair, Prabhakar, and Sharma [509] proved Parisi's conjecture for arbitrary n by combinatorial arguments. The two proofs are, however, completely different.

Theorem 5.4. *Let the cost coefficients c_{ij} of a linear assignment problem of size n be i.i.d. random variables with an exponential distribution with mean 1. Then the expected cost of the minimum assignment equals*

$$\sum_{k=1}^{n} \frac{1}{k^2}.$$

A new simple proof of this result using a probabilistic setting follows from a more general result shown by Wästlund [658] in 2005. Recently, Wästlund [659] obtained explicit bounds on the expected value of a min cost perfect matching in the complete graph K_n, n even.

Theorem 5.4 directly implies Theorem 5.2 by taking the limit n to ∞. Still another proof for Theorem 5.2 is given in Linusson and Wästlund [460]. For fixed n and p, $p > 1$, they consider a matrix $C = (c_{ij})$ with

$$c_{ij} = 0 \text{ if } \left(\frac{i}{n}\right)^p + \left(\frac{j}{n}\right)^p \geq 1.$$

The non-zero elements are drawn from an exponential distribution with mean 1. The authors can show that

$$\lim_{n\to\infty} \mathbb{E}(z_n) = \left(1 - \frac{1}{p}\right)^2 \cdot \frac{\pi^2}{6}. \tag{5.2}$$

Imagine that a Cartesian coordinate system has its origin in c_{11}, an axe going right along row 1, and the other going down along column 1. In the case $p = 2$ the matrix entries outside the positive quarter of a cycle are zero; hence, (5.2) yields the limit $\pi^2/24$, and if p tends to infinity we get Theorem 5.2 from (5.2).

Frenk, Houweninge, and Rinnooy Kan [280], as well as Olin [513], Aldous [25], and Grundel, Krokhmal, Oliveira, and Pardalos [344], studied other distribution functions for the cost elements c_{ij}. In [280] the authors analyzed the asymptotic behavior of the first-order statistics in the case of distribution functions F defined on $(-\infty, +\infty)$ (with $\lim_{n\to\infty} F^{-1}(1/n) = -\infty$) which fullfill the conditions

$$\int_{-\infty}^{+\infty} |x|F(x)dx < \infty \quad \text{and} \quad \liminf_{x\to+\infty} \frac{F(-x)}{F(-ax)} > 1 \quad \text{for some } a > 1\,,$$

as well as for functions F defined on $(0, \infty)$ (with $\lim_{n\to\infty} F^{-1}(1/n) = 0$). The estimates on the first-order statistics are then used to provide bounds for the expected optimal value of LSAP along the ideas of Walkup [656]. Olin [513] imposed further conditions on F and derived specific bounds which generalize those by Walkup and Lazarus. More precisely, if (i) F admits a continuous density function which is strictly positive in a neighborhood of the origin, (ii) F has finite expected value, and (iii) $F^{-1}(0^+) = \lim_{y\to 0^+} F^{-1}(y)$ exists, then

$$(1 + e^{-1})F^{-1}(0^+) \leq \liminf_{n\to\infty} \mathbb{E}(z^*) \leq \limsup_{n\to\infty} \mathbb{E}(z^*) \leq 3F^{-1}(0^+)\,.$$

Independent and uniformly distributed cost elements c_{ij} on [0, 1] immediately lead to independent and uniformly distributed cost elements $\bar{c}_{ij} = 1 - c_{ij}$ on [0, 1]. Therefore, we can deduce from

$$\max_{\varphi} \sum_{i=1}^{n} c_{i\varphi(i)} = n - \min_{\varphi} \sum_{i=1}^{n} \bar{c}_{i\varphi(i)} \tag{5.3}$$

that the maximum objective function value of a linear assignment problem tends to infinity as the problem size increases. Thus the gap between minimum and maximum objective function values of an LSAP becomes arbitrarily large when the problem size increases. We show later that quadratic assignment problems have a completely different behavior: for them, the gap between the minimum and the maximum value of the feasible solutions tends to 0 as n tends to infinity.

5.1.2 Asymptotic analysis of algorithms

Let us now turn to the probabilistic analysis of algorithms for LSAP. One of the first results in this area was obtained by Karp [408], who analyzed an exact algorithm for LSAP and showed that its expected running time is $O(n^2 \log n)$ in the case of independent costs c_{ij} uniformly distributed on [0, 1]. The analyzed algorithm is a special implementation of an augmenting path algorithm for the assignment problem and uses priority queues to compute a shortest augmenting path in $O(n^2 \log n)$ time. As in an assignment problem n shortest augmenting paths have to be determined, the time complexity of this algorithm is $O(n^3 \log n)$ for general cost matrices. The algorithm reduces the number of insertions in the priority queue at the cost of a slight increase of the number of deletions. This is done by introducing so-called *surrogate items*. Each surrogate item replaces a large number of regular items of the priority queue. In the case that the cost elements c_{ij} are uniformly distributed on [0, 1], the surrogate items do indeed represent and replace many regular items, and thus they reduce the expected overall number of operations associated with the queue to $O(n^2)$. As any of the queue-operations takes $O(\log n)$ times, and since the time for all other operations is $O(n^2)$, this yields the above-mentioned time complexity.

Other results in this area concern *heuristics* and provide worst-case bounds and/or average case bounds in the case that the costs c_{ij} are independently and uniformly distributed on [0, 1]. Avis and Devroye [49] analyzed a heuristic for LSAP proposed by Kurzberg [444]. This heuristic (i) decomposes a large problem of size $n = mk$ into k^2 problems of size $m \times m$ each; (ii) solves the smaller problems; and (iii) combines their solutions to obtain a solution for the original problem. In its space-optimal version ($k = \sqrt{n}$) the heuristics take $O(n)$ space and $O(n^{2.5})$ time. In its time-optimal version ($k = n^{3/4}$) it takes $O(n^{2.25})$ time and $O(n^{1.5})$ space. It is not difficult to show that the worst-case ratio of this heuristic is ∞. (By applying this heuristic to a maximization version of LSAP, one can obtain a worst-case ratio of $\min\{m, k\}$.) However, in the case that the cost coefficients c_{ij} are uniformly distributed in [0, 1], the expected value of the solution produced by the heuristic is smaller than or equal to $k/2$ times the expected optimal value as n tends to infinity.

The first heuristic producing a solution with expected value bounded by a constant was proposed by Avis and Lai [50]. They designed an $O(n^2)$ time algorithm and showed that, for sufficiently large n, the expected value of the total cost of the assignment found is less than 6, provided the cost coefficients are i.i.d. random variables drawn from a uniform distribution on [0, 1]. The heuristic elaborates the idea of Walkup [656] (see Theorem 3.26) and works with a sparse subgraph of the given graph. The sparse subgraph considered by the authors has $10n$ "cheap" edges and contains a perfect matching with high probability. If necessary, the largest matching contained in this graph is completed to a perfect matching of the original graph in a greedy way. The good behavior of the heuristic relies on the fact that the sparse graph contains, with high probability, a cheap perfect matching.

Karp, Rinnooy Kan, and Vohra [410] derived a better heuristic which runs in $O(n \log n)$ time (with $O(n)$ expected time) and provides, with probability $1 - O(n^{-a})$, a solution whose value is smaller than $3 + O(n^{-a})$ for some $a > 0$. The basic idea is similar to that by Avis and Lai: Construct a "cheap" sparse subgraph of the given graph and show that it contains a perfect matching with high probability. Again, if the subgraph does not contain a perfect matching, a solution for the original LSAP instance is determined in a greedy way. The subgraph is introduced in terms of so-called *random 2-out* bipartite graphs. Such a graph contains a perfect matching with probability $1 - O(n^{-a})$ and a maximum matching can be found in $O(n \log n)$ time. Further, how to construct a random 2-out bipartite subgraph with cheap edges for the given LSAP instance is shown. The expected value of a solution which is either obtained as a perfect matching in this subgraph or by a greedy approach to the given LSAP instance, if the former does not exist, equals $3 + O(n^{-a})$.

Schwartz, Steger, and Weissl [603] reported real-world LSAP instances involving dense graphs with $n > 10000$, for which one can be interested in a fast approximation algorithm. The algorithm in [603] selects a small subset of the edges and solves the resulting sparse instance with the algorithm by Fredman and Tarjan [278] (see Section 4.4.3) if the costs are arbitrary or with the algorithm by Goldberg and Kennedy [329] (see Section 4.6.4) if the costs are integer. In the former case the time complexity is $O(n^2 \log n)$, while in the latter case it is $O(n^2)$. The selected edge set is the union of two subsets: E_1, containing the $c \log n$ smallest edges incident with each vertex, and E_2, containing $c'n \log n$ randomly chosen edges (for given parameters c and c'). Set E_1 is aimed at producing a good practical behavior of the algorithm, while set E_2 ensures, on the basis of a result by Erdös and Rényi [257], that the sparse graph has a perfect matching with high probability, i.e., with probability tending to 1 as n tends to infinity. If the input graph is complete and has uniformly distributed edge weights, using a result by Frieze and Sorkin [284], it is proved in [603] that the algorithm finds with high probability the optimal solution. Schwartz, Steger, and Weissl [603] report computational experiments on randomly generated graphs with n ranging from 10 to 1500.

5.2 Monge matrices and the linear sum assignment problem

In certain cases an optimal solution of the linear sum assignment problem is known beforehand, provided the cost matrix of the problem has a special structure.

Definition 5.5. *An $n \times n$ matrix $C = (c_{ij})$ is said to fulfill the* Monge property *(or to be a* Monge matrix*) if (see Figure 5.1)*

$$c_{ij} + c_{kl} \le c_{il} + c_{kj} \tag{5.4}$$

holds for all $1 \le i < k \le n$ and $1 \le j < l \le n$.

Matrix C fulfills the inverse Monge property *(is an* inverse Monge matrix*) if*

$$c_{ij} + c_{kl} \ge c_{il} + c_{kj} \tag{5.5}$$

holds for all $1 \le i < k \le n$ and $1 \le j < l \le n$.

This definition goes back to Hoffman [372], who considered a slightly more general situation, namely, the so-called Monge sequences.

It is easy to see that it is enough to require the Monge property for adjacent rows and adjacent columns. In other words, (5.4) holds if and only if

$$c_{ij} + c_{i+1,j+1} \le c_{i,j+1} + c_{i+1,j} \qquad (i, j = 1, 2, \ldots, n-1). \tag{5.6}$$

$$\begin{array}{c} \\ i \\ \\ k \\ \\ \end{array}\begin{array}{c} \\ \\ \\ \\ \\ \end{array}\begin{pmatrix} & j & & & l & \\ & \vdots & & & \vdots & \\ \cdots & c_{ij} & \cdots & \cdots & c_{il} & \cdots \\ & \vdots & & & \vdots & \\ \cdots & c_{kj} & \cdots & \cdots & c_{kl} & \cdots \\ & \vdots & & & \vdots & \end{pmatrix}$$

Figure 5.1. *Four elements involved in the Monge property.*

An immediate consequence of this observation is that one can test in $O(n^2)$ time whether a given $n \times n$ matrix C is a Monge matrix.

An important subclass of Monge matrices can be generated in the following way: Let $D = (d_{ij})$ be a nonnegative real matrix of order $n \times n$. It is straightforward to prove that the matrix C obtained by

$$c_{ij} = \sum_{k=i}^{n} \sum_{\ell=1}^{j} d_{k\ell} \qquad (i, j = 1, 2, \dots, n) \tag{5.7}$$

is a Monge matrix. In analogy to the notions of distribution and density matrices in probability theory, Gilmore, Lawler, and Shmoys [312] call a matrix C which is given by (5.7) a *distribution matrix* generated by the *density matrix* D.

Monge matrices are in fact only slight generalizations of distribution matrices since the following can be proved (see Burdyuk and Trofimov [124], and Bein and Pathak [80]).

Proposition 5.6. *Every Monge matrix C is the sum of a distribution matrix $\widehat{C}$ and two vectors u and $v \in \mathbb{R}^n$ such that*

$$c_{ij} = \widehat{c}_{ij} + u_i + v_j \qquad (i, j = 1, 2, \dots, n).$$

Examples of Monge matrices can be obtained through the following settings. $C = (c_{ij})$ is a Monge matrix if

- $c_{ij} = a_i + b_j$ for arbitrary real numbers a_i and b_j $(i, j = 1, 2, \dots, n)$;
- $c_{ij} = a_i b_j$ for increasing real numbers $0 \le a_1 \le a_2 \le \cdots \le a_n$ and decreasing real numbers $b_1 \ge b_2 \ge \cdots \ge b_n \ge 0$;
- $c_{ij} = \min(a_i, b_j)$ for increasing real numbers $a_1 \le a_2 \le \cdots \le a_n$ and decreasing real numbers $b_1 \ge b_2 \ge \cdots \ge b_n$;
- $c_{ij} = \max(a_i, b_j)$ for increasing real numbers a_i and b_j $(i, j = 1, 2, \dots, n)$;
- $c_{ij} = |a_i - b_j|^p$ for $p \ge 1$ and decreasing real numbers a_i and b_j $(i, j = 1, 2, \dots, n)$.

Similar examples can be given for inverse Monge matrices. In connection with linear sum assignment problems we can now show the following.

Matrix C fulfills the *weak Monge property* (see Derigs, Goecke, and Schrader [227]) if

$$c_{ii} + c_{kl} \le c_{il} + c_{ki} \qquad \text{for } 1 \le i < k \le n \text{ and } 1 \le i < l \le n. \tag{5.8}$$

Every Monge matrix obviously fulfills (5.8), but not vice versa.

Proposition 5.7. *A linear sum assignment problem whose cost matrix is a weak Monge matrix is solved by the identical permutation.*

If the cost matrix fulfills the inverse Monge property, then the permutation $\varphi(i) = n + 1 - i$ for $i = 1, 2, \ldots, n$ is an optimal solution.

Proof. We prove the first part of this proposition by successive transformations to the identical permutation. Every transformation does not increase the objective function value. The second part can be proved by analogous arguments using the inverse Monge property.

Let φ^* be an optimal solution of the assignment problem whose cost matrix fulfills the Monge property. For $i = 1, 2, \ldots, n$, perform the following step. If $\varphi^*(i) \neq i$, then let $l = \varphi^*(i)$, find k such that $\varphi^*(k) = i$, and interchange the two assignments by setting $\varphi^*(i) = i$ and $\varphi^*(k) = l$. According to (5.8) we have

$$c_{ii} + c_{kl} \le c_{il} + c_{ki}$$

which shows that the new permutation does not have a larger objective function value.

Since at each iteration i we have $\varphi^*(h) = h$ for $h = 1, 2, \ldots, i - 1$, the interchange never involves the preceding rows and columns, so the resulting permutation remains optimal. $\square$

In the case that the cost matrix C is a Monge matrix, there is an intimate connection between minimizing and maximizing $\sum_{i=1}^{n} c_{i\varphi(i)}$. We can use the obvious fact that

if C is a Monge matrix, then $-C$ is an inverse Monge matrix.

Thus Proposition 5.7 tells us that the permutation $\varphi(i) = n + 1 - i$ for $i = 1, 2, \ldots, n$ maximizes the objective function.

The Monge property depends on a proper numbering of the rows and columns of matrix C. A matrix $C = (c_{ij})$ is called a *permuted Monge matrix* if there are permutations φ and ψ of the rows and columns of C, respectively, such that $(c_{\varphi(i)\psi(j)})$ fulfills the Monge property. Deĭneko and Filonenko [214] showed that it can be tested in $O(n^2 + n \log n)$ time if an $n \times n$ matrix fulfills the permuted Monge property (see also Burkard, Klinz, and Rudolf [152] for Deĭneko and Filonenko's recognition algorithm). Their algorithm also provides the permutations φ and ψ, when they exist.

An important special case of LSAP with a permuted Monge matrix as cost matrix arises if the cost coefficients have the form

$$c_{ij} = a_i b_j.$$

Such LSAPs, which arise as subproblems of the quadratic assignment problem (see Section 7.5), can simply be solved in $O(n \log n)$ time by ordering the numbers a_i and b_j. Let

$a_1 \le a_2 \le \cdots \le a_n$ and $b_1 \le b_2 \le \cdots \le b_n$. Then $(a_i - a_k)(b_j - b_l) \ge 0$ for $i < k$ and $j < l$. Thus $a_i b_j + a_k b_l \ge a_i b_l + a_k b_j$, which shows that the matrix $(c_{ij}) = (a_i b_j)$ is an inverse Monge matrix. Therefore Proposition 5.7 implies the following proposition by Hardy, Littlewood, and Pólya [364].

Proposition 5.8. *Let*

$$a_1 \le a_2 \le \cdots \le a_n \quad \textit{and} \quad b_1 \le b_2 \le \cdots \le b_n.$$

Then for any permutation φ

$$\sum_{i=1}^{n} a_i b_{n+1-i} \le \sum_{i=1}^{n} a_i b_{\varphi(i)} \le \sum_{i=1}^{n} a_i b_i$$

In connection with the realization of discrete event systems (see Gaubert, Butkovič, and Cuninghame-Green [303], Burkard and Butkovič [136], and Butkovič and Cuninghame-Green [162]) the problem of a linear sum assignment problem with a symmetric Hankel matrix as cost matrix arises. A Hankel matrix has the form

$$\begin{pmatrix} c_0 & c_1 & \cdots & c_n \\ c_1 & c_2 & \cdots & c_{n+1} \\ \vdots & \vdots & & \vdots \\ c_n & c_{n+1} & \cdots & c_{2n} \end{pmatrix}$$

with arbitrary real numbers $c_0, c_1, \ldots, c_{2n}$. Thus a Hankel matrix has a very special form. It is still an open question whether or not a linear sum assignment problem with such a special cost matrix can be solved in a faster way than by applying a standard algorithm for LSAP. If the sequence c_n is convex (resp., concave), i.e., $c_{r+2} - c_{r+1} \ge c_{r+1} - c_r$ (resp., $c_{r+2} - c_{r+1} \le c_{r+1} - c_r$) for all r, the Hankel matrix is an inverse Monge matrix (resp., a Monge matrix). Therefore Proposition 5.7 yields optimal solutions.

5.3 Max-algebra and the linear sum assignment problem

In max-algebra the conventional arithmetic operations of addition and multiplication of real numbers are replaced by

$$a \oplus b = \max(a, b), \tag{5.9}$$

$$a \otimes b = a + b, \tag{5.10}$$

where $a, b \in \overline{\mathbb{R}} = \mathbb{R} \cup \{-\infty\}$. In particular, $x^{(k)}$ denotes the kth power of x $(k = 0, 1, \ldots)$, in conventional notation $x^{(k)} = kx$. The algebraic system $(\overline{\mathbb{R}}, \oplus, \otimes)$ offers an adequate language for describing the synchronization of production (Cuninghame-Green [203]) and discrete event systems, to mention just a few applications. Though the operation "subtract" does not exist within this system, many notions from conventional linear algebra, like equation systems and eigenvalues, can be developed within this system.

The operations $\oplus$ and $\otimes$ can be extended to vectors and matrices in the same way as in conventional linear algebra. First, we introduce the notation

$$\sum_{1\le i\le n}^{\oplus} a_i = a_1 \oplus a_2 \oplus \cdots \oplus a_n = \max(a_1, a_2, \ldots, a_n)$$

and

$$\prod_{1\le i\le n}^{\otimes} a_i = a_1 \otimes a_2 \otimes \cdots \otimes a_n = \sum_{i=1}^{n} a_i.$$

Thus if $A = (a_{ij})$, $B = (b_{ij})$, and $C = (c_{ij})$ are matrices with elements from $\overline{\mathbb{R}}$ of compatible sizes, we can write $C = A \oplus B$ if $c_{ij} = a_{ij} \oplus b_{ij}$ for all i, j and $C = A \otimes B$ if $c_{ij} = \sum_k^{\oplus} a_{ik} \otimes b_{kj} = \max_k(a_{ik} + b_{kj})$ for all i, j. In max-algebra the unit matrix I is a square matrix of appropriate size whose diagonal elements are all 0 and whose off-diagonal elements are $-\infty$.

Let us consider a linear equation system in max-algebra. Cuninghame-Green [204] showed that the linear equation system $A \otimes x = b$ with an $n \times n$ matrix A has a unique solution $\bar{x}$ if and only if the matrix $C = (c_{ij})$ defined by $c_{ij} = a_{ij} - b_i$ has exactly one maximum in every column and such maxima are in different rows. In this case the solution is given by

$$\overline{x}_j = -\max_{1\le i\le n} c_{ij} \qquad (j = 1, 2, \ldots, n).$$

This means that the corresponding linear sum assignment problem with cost matrix $-C$ has a unique optimal solution.

How can one check whether an assignment problem has a unique optimal solution? We say that matrix B is a *normal form* of matrix C (see Burkard and Butkovič [136]) if

(a) $b_{ij} \ge 0$ $(i, j = 1, 2, \ldots, n)$, and

(b) there is a constant z such that

$$\sum_{i=1}^{n} c_{i\varphi(i)} = z + \sum_{i=1}^{n} b_{i\varphi(i)} \tag{5.11}$$

holds for all permutations φ, and

(c) there exists a permutation φ_0 such that

$$\sum_{i=1}^{n} b_{i\varphi_0(i)} = 0.$$

It is straightforward to see that the two linear sum assignment problems with cost matrices C and B have the same optimal solution. Indeed, the normal form of a matrix C is the matrix of the reduced costs in the optimal solution of the LSAP induced by C; hence, it can be obtained by applying any LSAP algorithm that provides the optimal dual variables.

Normal forms lead to an interesting property of *symmetric* cost matrices. The following proposition shows that we can always achieve a symmetric normal form if the coefficient matrix of the linear assignment problem is symmetric.

Proposition 5.9. *For a symmetric $n \times n$ matrix C, a symmetric normal form can be determined in $O(n^3)$ steps.*

Proof. Given a permutation φ, let us denote by φ^{-1} the permutation such that $\varphi^{-1}(\varphi(i)) = i$. If C is symmetric, then the permutations φ and φ^{-1} yield the same value of the corresponding LSAP. Using (5.11) we get

$$\begin{aligned} 2\sum_{i=1}^{n} c_{i\varphi(i)} &= \sum_{i=1}^{n} c_{i\varphi(i)} + \sum_{i=1}^{n} c_{i\varphi^{-1}(i)} \\ &= 2z + \sum_{i=1}^{n} b_{i\varphi(i)} + \sum_{i=1}^{n} b_{\varphi(i)i} \\ &= 2z + \sum_{i=1}^{n} (b_{i\varphi(i)} + b_{\varphi(i)i}). \end{aligned}$$

Thus

$$\overline{B} = \frac{1}{2}(B + B^T),$$

where B^T denotes the transpose of B, is a symmetric normal form of C. The thesis follows by recalling that B can be found in $O(n^3)$ time by solving the LSAP defined by matrix C. □

The symmetry and property (c) of the normal form imply that every even cycle $(i_1, i_2, \ldots, i_{2k-1}, i_{2k})$ in the cyclic representation of the optimal solution (see Section 1.1) can be split into two-cycles $\langle i_h, i_{h+1} \rangle$ $(h = 1, 3, \ldots, 2k-1)$ without increasing the solution cost, i.e., we have the following.

Corollary 5.10. *If the cost matrix C is symmetric, then there always exists an optimal solution of the linear sum assignment problem consisting only of cycles of length 2 or an odd length in the cyclic representation of φ.*

Returning to general cost matrices, by permuting the rows and columns of B we can achieve that the identical permutation $(1, 2, \ldots, n)$ becomes an optimal solution. Let us define a directed graph $G_B(N, A)$ with node set $N = \{1, 2, \ldots, n\}$ and arcs $(i, j) \in A$ if and only if $i \neq j$ and $b_{ij} = 0$. This shows immediately the following.

Proposition 5.11. *The optimal solution of the assignment problem with cost matrix C is unique if and only if graph G_B is acyclic.*

Further results in this direction can be found in Butkovič [161] and Burkard and Butkovič [136]. In particular, a square matrix C with columns $C_1, C_2, \ldots, C_n$ is called *regular* if it is not possible to find two nonempty, disjoint subsets S and T of $\{1, 2, \ldots, n\}$ and real numbers α_j such that

$$\sum_{j \in S}^{\oplus} \alpha_j \otimes C_j = \sum_{j \in T}^{\oplus} \alpha_j \otimes C_j.$$

The following result has been shown by Butkovič [161].

Proposition 5.12. *Let B be a normal form of C in which (if necessary) the rows or columns have been permuted such that the identical permutation is optimal. Then C is regular if and only if graph G_B does not contain an even cycle.*

The problem of checking if a graph contains an even cycle has been shown to be polynomially solvable by Robertson, Seymour, and Thomas [586]; hence, the same holds for the problem of checking if a given square matrix is regular.

The *max-algebraic permanent* of an $n \times n$ matrix $C = (c_{ij})$ is defined in an analogous way to classical linear algebra by

$$\text{maper}(C) = \sum_{\varphi \in \mathcal{S}_n}^{\oplus} \prod_{1 \leq i \leq n}^{\otimes} c_{i\varphi(i)}, \tag{5.12}$$

where $\mathcal{S}_n$ denotes the set of all permutations of the set $\{1, 2, \ldots, n\}$. In conventional notation,

$$\text{maper}(C) = \max_{\varphi \in \mathcal{S}_n} \sum_{1 \leq i \leq n} c_{i\varphi(i)},$$

which is the solution value of a maximization LSAP with cost matrix C. The max-algebraic permanent of a matrix plays a role in connection with the eigenvalue problem in max-algebra. Cuninghame-Green [204] showed that in max-algebra the characteristic polynomial of a square matrix A is given by

$$\chi_A(x) := \text{maper}(A \oplus x \otimes I).$$

In other words, it is the max-algebraic permanent of the matrix

$$\begin{pmatrix} a_{11} \oplus x & a_{12} & \cdots & a_{1n} \\ a_{21} & a_{22} \oplus x & \cdots & a_{2n} \\ \vdots & \vdots & & \vdots \\ a_{n1} & a_{n2} & \cdots & a_{nn} \oplus x \end{pmatrix}.$$

This means that $\chi_A(x)$ is a max-algebraic polynomial

$$\chi_A(x) = \delta_0 \oplus \delta_1 \otimes x \oplus \cdots \oplus \delta_{n-1} \otimes x^{(n-1)} \oplus x^{(n)}$$

with coefficients $\delta_0, \delta_1, \ldots, \delta_{n-1}$ and $\delta_n = 1$, where $\delta_0 = \text{maper}(A), \ldots, \delta_{n-1} = \max(a_{11}, a_{22}, \ldots, a_{nn})$ (see also Proposition 5.13 below). Written in conventional notation, the characteristic polynomial is

$$\chi_A(x) = \max\left(\delta_0, \delta_1 + x, \ldots, \delta_{n-1} + (n-1)x, nx\right).$$

Thus, viewed as a function in x, the max-algebraic characteristic polynomial of a matrix A is a piecewise linear, convex function. If for some $k \in \{0, 1, \ldots, n\}$ the inequality

$$\delta_k \otimes x^{(k)} \leq \sum_{i \neq k}^{\oplus} \delta_i \otimes x^{(i)}$$

holds for every real x, then the term $\delta_k \otimes x^{(k)}$ is called *inessential*; otherwise, it is called *essential*. Burkard and Butkovič [135] described an $O(n^4)$ method to find all essential terms of the characteristic max-algebraic polynomial by solving a series of linear sum assignment problems. Recently, Gassner and Klinz [302] improved the time-complexity of the algorithm for finding the essential terms of the characteristic max-algebraic polynomial to $O(n^3)$. They showed that these terms can be found by solving the special linear parametric sum assignment problems with cost coefficients

$$c_{ij}^{\lambda} = \begin{cases} c_{ij}, & \text{if } i \neq j, \\ c_{ii} - \lambda, & \text{otherwise.} \end{cases}$$

Cuninghame-Green [204] showed that the max-algebraic characteristic polynomial of a matrix A is closely related to the best principal submatrix assignment problem. Let $A = (a_{ij})$ be an $n \times n$ matrix. Any matrix of the form

$$\begin{pmatrix} a_{i_1 i_1} & a_{i_1 i_2} & \cdots & a_{i_1 i_k} \\ a_{i_2 i_1} & a_{i_2 i_2} & \cdots & a_{i_2 i_k} \\ \vdots & \vdots & & \vdots \\ a_{i_k i_1} & a_{i_k i_2} & \cdots & a_{i_k i_k} \end{pmatrix}$$

with $1 \leq i_1 < i_2 < \cdots < i_k \leq n$ is called a $k \times k$ *principal submatrix*. The *best principal submatrix assignment problem BPSAP(k)* can be stated as follows.

For given k, $1 \leq k \leq n$, find a principal submatrix of size k and a permutation φ of the set $\{1, 2, \ldots, k\}$ such that

$$\sum_{r=1}^{k} a_{i_r \varphi(i_r)}$$

is a minimum.

The following theorem by Cuninghame-Green [204] shows that all (essential and inessential) coefficients δ_k of the characteristic max-algebraic polynomial can be obtained as solutions of the principal submatrix assignment problem.

Proposition 5.13. *Let $\chi_A(x) = \delta_0 \oplus \delta_1 \otimes x \oplus \cdots \oplus \delta_{n-1} \otimes x^{(n-1)} \oplus x^{(n)}$ be the max-algebraic characteristic polynomial of an $n \times n$ matrix A. Then the coefficients δ_k are given by*

$$\delta_k = \sum_{B \in \mathcal{A}_k}^{\oplus} maper(B), \tag{5.13}$$

where $\mathcal{A}_k$ is the set of all principal submatrices of A of size $(n - k) \times (n - k)$.

It is obvious that $\delta_0 = \text{maper}(A) = \max_\varphi \sum_{i=1}^{n} a_{i\phi(i)}$ and that $\delta_{n-1} = \max(a_{11}, a_{22}, \ldots, a_{nn})$. There is, however, no polynomial method known to the authors to solve the best principal submatrix problem in general for a given size k. However, if the entries of matrix A are polynomially bounded, the best principal submatrix assignment problem can be solved, for any k $(1 \leq k \leq n)$, by a randomized polynomial algorithm (see Burkard and Butkovič [136]). In case matrix A has the Monge property, the k smallest entries of its main diagonal are a solution to the best principal submatrix assignment problem *BPSAP(k)*.

5.4 Variants

In this section we discuss four variants of LSAP: the determination of the K best solutions, the k-cardinality assignment problem, the semi-assignment problem, and the assignment problem in the case of rectangular cost matrices. Other variants can be found in Section 5.5.

5.4.1 Ranking solutions

In Chapter 4 we saw several algorithms to find the solution of minimum value to LSAP. In some cases, it may be useful to also determine the second best solution or, more generally, given a value K, the kth minimum cost assignment for $k = 1, 2, \ldots, K$. Indeed, when LSAP is used to model a real-life problem, it may be difficult to represent all the objectives of the problem through a single function. By ranking the first K solutions in nondecreasing order of value we obtain near optimal solutions that may be better than the optimal one for a decision maker looking at all the aspects of the real problem. Brogan [115] (see Section 5.5) used them to solve a radar tracking problem. Applications of ranking algorithms are frequently encountered in multiobjective programming, such as, e.g., in Pedersen, Nielsen, and Andersen [540] and Przybylski, Gandibleux, and Ehrgott [560], where the first K assignments are used inside a two phase method to determine the set of efficient solutions.

Let us consider a bipartite graph $G = (U, V; E)$, with costs $c(e)$ associated with the edges $e \in E$, and denote by $z(M)$ the value of a given perfect matching M (i.e., the value of a feasible solution to the corresponding LSAP). The problem is to find K distinct perfect matchings $M_1, M_2, \ldots, M_K$ such that $z(M_1) \leq z(M_2) \leq \cdots \leq z(M_K)$ and $z(M) \geq z(M_K)$ for any perfect matching M different from $M_1, M_2, \ldots, M_K$. This problem is $\mathcal{NP}$-hard for general K (due to the exponential number of distinct perfect matchings), but polynomially solvable for fixed K.

The first result for ranking the assignments by nondecreasing value was presented by Murty [506] in 1968. His algorithm builds a branch-decision tree where, at each node, an LSAP with additional constraints is solved. The tree exploration is halted as soon as K distinct minimum value assignments have been computed. Let I and O be two disjoint subsets of edges from set E. The LSAP solved at each node, denoted as LSAP(I, O) in the following, computes the minimum value perfect matching M such that $I \subseteq M$ and $M \cap O = \emptyset$. Set I contains the *imposed* edges, while set O stores the *forbidden* edges: M must contain all edges of I and no edges of O. In order to ensure that LSAP(I, O) can have a feasible solution, set I must be a (partial) matching.

At the root node the algorithm sets $I = O = \emptyset$ and finds the optimal solution, say, M, to LSAP(I, O), i.e., to LSAP without additional constraints. Let $M_1 = M$: to compute M_2, $n - 1$ descending nodes are generated as follows. Let the current M consist of the ordered set of edges $\{[i_1, j_1], [i_2, j_2], \ldots, [i_n, j_n]\}$. We consider one edge at a time among the first $n - 1$ edges, and forbid it while imposing all the preceding ones. (Note that no solution exists if $[i_n, j_n]$ is forbidden while imposing all the other edges of M.)

The second best matching is the solution of minimum value among the $n - 1$ solutions of the generated LSAPs. More formally, the computation of M_2 requires one to solve

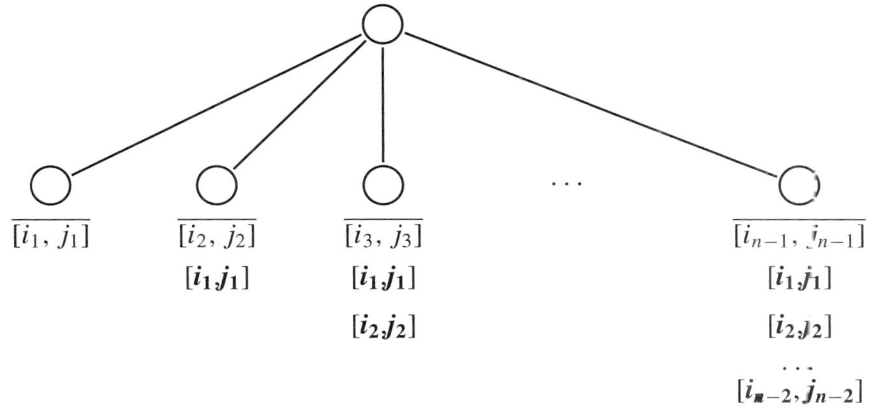

Figure 5.2. *Branching rule of the Murty's algorithm.*

LSAP(I^h, O^h) for $h = 1, 2, \dots, n-1$, where

$$O^h = \{[i_h, j_h]\}, \tag{5.14}$$

$$I^h = \{[i_1, j_1], [i_2, j_2], \dots, [i_{h-1}, j_{h-1}]\}. \tag{5.15}$$

Figure 5.2 depicts the first level of the branch-decision tree: forbidden edges are overlined, while imposed edges are in bold. Problem LSAP(I^1, O^1) is the original problem with a single forbidden edge and can be solved by any LSAP algorithm executed on the original graph with $c([i_1, j_1]) := \infty$. Solving LSAP(I^h, O^h) for $1 < h < n$ requires one to set $c([i_h, j_h]) := \infty$ and to remove from G the vertices of $\{i_1, i_2, \dots, i_{h-1}\}$ and $\{j_1, j_2, \dots, j_{h-1}\}$ together with their incident edges. The LSAP algorithm is thus applied to the resulting subgraph $G' = (U', V'; E')$, which has $|U'| = |V'| = n - h + 1$. Note that this problem could have no feasible solution. By adding the edges of I^h to the solution obtained (if any), we get the optimal solution to LSAP(I^h, O^h).

Let I^* and O^* be the sets I^h and O^h, respectively, that produced M_2. We branch from the corresponding decision node by applying the above scheme to the edges of $M_2 \setminus I^*$ (since the edges of I^* are imposed to all descending nodes). The next matching M_3 is the assignment of minimum value among those computed at the leaves of the resulting branch-decision tree. The procedure is iterated K times to obtain the required matchings. The pseudocode description of the algorithm follows.

ALGORITHM 5.1. Murty.
Algorithm for finding the K minimum value assignments.

Solve LSAP($\emptyset$, $\emptyset$), yielding the perfect matching M of minimum value $z(M)$;
$Q := \{\langle z(M), M, \emptyset, \emptyset \rangle\}$, $k := 1$;

repeat
  comment: select the kth minimum value assignment;
  remove from Q the 4-tuple $\langle z, M, I, O\rangle$ of minimum z value;
  let M consist of the edges in $I \cup \{[i_1, j_1], [i_2, j_2], \ldots, [i_{n-|I|}, j_{n-|I|}]\}$;
  $M_k := M, k := k + 1$;
  if $k \le K$ **then**
    comment: branching phase;
    for $h := 1$ **to** $n - |I| - 1$ **do**
      $O^h := O \cup \{[i_h, j_h]\}, I^h := I \cup \{[i_1, j_1], [i_2, j_2], \ldots, [i_{h-1}, j_{h-1}]\}$;
      solve LSAP(I^h, O^h) to find the perfect matching $\widetilde{M}$ of minimum value;
      if a perfect matching has been found **then** $Q := Q \cup \{\langle z(\widetilde{M}), \widetilde{M}, I^h, O^h\rangle\}$
    endfor
  endif
until $k > K$ or $Q = \emptyset$ [**comment:** if $Q = \emptyset$ less than K assignments exist]

The main repeat-until loop is executed at most K times. Each execution of the branching phase generates $O(n)$ problems, so the overall number of solved LSAPs is $O(Kn)$. Using an $O(n^3)$ routine for LSAP, the algorithm runs in $O(Kn^4)$ time.

Example 5.14. Consider the instance represented in Figure 5.3(a) and let $K = 3$. The $n! = 6$ matchings, shown in Figure 5.3(b), are denoted, by increasing cost, as M_α, M_β, M_γ, M_δ, M_ε, and M_ζ. The initial execution of LSAP($\emptyset$, $\emptyset$) yields M_α, so $Q = \{\langle z(M), M, \emptyset, \emptyset\rangle\} = \{\langle 3, M_\alpha, \emptyset, \emptyset\rangle\}$ and $k = 1$. At the first iteration we empty Q, define the first solution, $M_1 = M_\alpha$, and set $k = 2$. Two children nodes are then generated and solved as LSAP($\emptyset$, $\{[1, 1]\}$) (yielding $\widetilde{M} = M_\gamma$) and LSAP($\{[1, 1]\}$, $\{[2, 2]\}$) (yielding $\widetilde{M} = M_\beta$), and we obtain the new set of tuples, $Q = \{\langle 13, M_\gamma, \emptyset, \{[1, 1]\}\rangle\ \langle 7, M_\beta, \{[1, 1]\}, \{[2, 2]\}\rangle\}$. At the second iteration we remove the second 4-tuple from Q, define the second solution, $M_2 = M_\beta$, and set $k = 3$. A single child node is generated and solved as LSAP($\{[1, 1]\}$, $\{[2, 2], [2, 3]\}$), which yields $\widetilde{M} = \emptyset$. At the third iteration we remove the remaining 4-tuple from Q, define the third solution, $M_3 = M_\gamma$, set $k = 4$, and terminate. ∎

Note that the children nodes generated at each iteration of Murty's algorithm have an increasing number of imposed edges (see (5.14)), so the last LSAPs are easier to solve than the first ones. Pascoal, Captivo, and Clímaco [539] proposed to reverse the order in which the children nodes are generated. In this way the first child has $n - 2$ imposed edges and the corresponding LSAP, which is associated with a 2×2 cost matrix, can be solved in constant time. At each new iteration h, we remove the last edge $[i^*, j^*]$ from the current set of imposed edges and forbid it. It follows that the solution to LSAP(I^h, O^h) can be obtained from the solution to LSAP(I^{h-1}, O^{h-1}) through a single minimum cost augmenting path from i^* to j^* on the corresponding incremental digraph D^r (see Section 4.4.1). It can be proved that D^r has no negative cycle (although it has arcs with negative cost), so the shortest path can be computed through the $O(n^2)$ time algorithm by Bertsekas, Pallottino, and Scutellà [93]. In this way the $O(n)$ children of each decision node can be explored in $O(n^3)$, and the ranking algorithm has time complexity $O(Kn^3)$.

Pedersen, Nielsen, and Andersen [540] obtained the same time complexity by preserving Murty's generation order. At each iteration h, instead of solving LSAP(I^h, O^h) on G' from scratch, they update the dual variables so that the reduced costs define a dual feasible

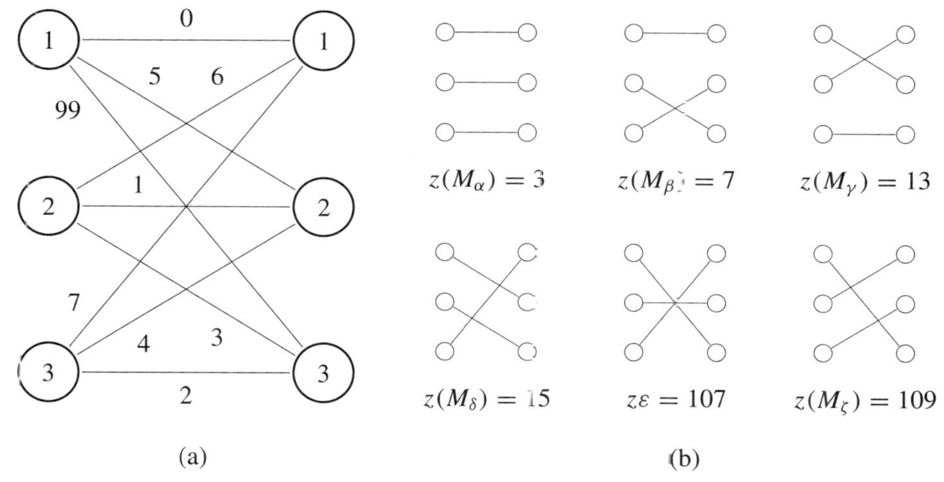

Figure 5.3. (a) *instance for Example 5.14;* (b) *matchings.*

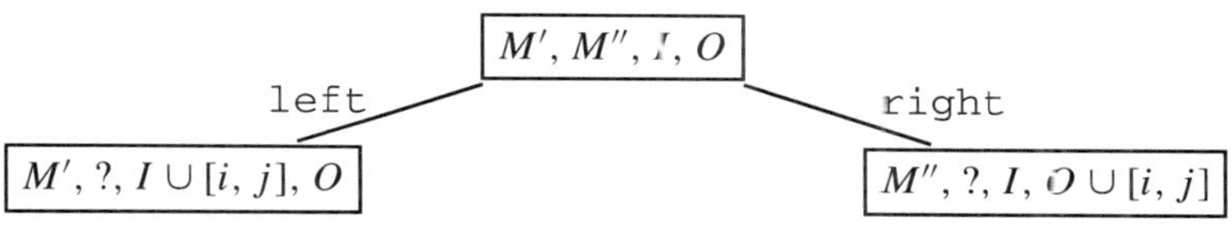

Figure 5.4. *Branching rule of the Chegireddy-Hamacher algorithm.*

solution satisfying complementary slackness for the current partial matching without edge $[i_h, j_h]$. In this way the new matching is determined through a single shortest path augmentation of the standard Dijkstra algorithm. Note that the same result could be obtained by directly applying the Dijkstra algorithm to subgraph G' with no need of updating the dual variables.

We observe that similar techniques were previously used in subtour elimination schemes for the asymmetric traveling salesman problem, in which an LSAP is solved in $O(n^2)$ time at each node of the branch-decision tree (Bellmore and Malone [82]).

Chegireddy and Hamacher [181] proposed an alternative approach to Murty's method that (i) computes the two best assignments associated with each decision node; and (ii) adopts a branching rule that splits each problem into two (instead of $O(n)$) descending nodes. Given a decision node h, let I and O be, respectively, the current sets of imposed and forbidden edges and let M' and M'' denote the first and second minimum value matchings of node h. The algorithm chooses an edge $[i, j] \in M' \setminus M''$ and defines two descending nodes associated with the two pairs of sets $(I \cup \{[i, j]\}, O)$ and $(I, O \cup \{[i, j]\})$. At the first node, s_1, we impose that an edge of the best matching is chosen, so the best matching of s_1 is still M'. At the second node, s_2, we forbid an edge of M' that does not belong to M'', so the best matching of s_2 is M''. The two second best matchings of s_1 and s_2, instead, have to be recomputed. Figure 5.4 describes this branching rule by reporting, for each node, the two best matchings (when known) and the sets of forbidden and imposed edges.

The overall best matching M_1 is the first matching of the root node. The kth best matching M_k is determined at iteration k by selecting the minimum value matching among those that have not yet been stored and are associated with the leaves of the decision tree. When the first descending node s_1, the "`left`" node, of a node h is defined, the best unstored matching that can be associated with it is the second best matching of s_1. On the contrary, the best unstored matching of the second descending node s_2, the "`right`" node, is its best matching. The Chegireddy–Hamacher algorithm can be described as follows. We denote with $LSAP2(I, O)$ the problem of finding the second best perfect matching that uses all the edges in set I and none from set O.

ALGORITHM 5.2. Revised_ranking.
Algorithm for finding the K minimum value assignments.

Solve LSAP(∅, ∅) and LSAP2(∅, ∅) yielding, respectively, the minimum and second
  minimum value perfect matchings M' and M'';
$Q := \{\langle z(M'), M', M'', \emptyset, \emptyset, \texttt{right}\rangle\}$, $k := 1$;
repeat
  comment: select the kth minimum value assignment;
  remove from Q the 6-tuple $\langle z, M', M'', I, O, lr\rangle$ of minimum z value;
  if $lr = \texttt{left}$ **then** $M_k := M''$ **else** $M_k := M'$;
  $k := k + 1$;
  if $k \le K$ **and** $M'' \neq \emptyset$ **then**
    comment: branching phase;
    select an edge $[i, j] \in M' \setminus M''$;
    solve LSAP2($I \cup \{[i, j]\}, O$) yielding the second best perfect matching $\widetilde{M}$;
    if $\widetilde{M} \neq \emptyset$ **then** $Q := Q \cup \{\langle z(\widetilde{M}), M', \widetilde{M}, I \cup \{[i, j]\}, O, \texttt{left}\rangle\}$;
    solve LSAP2($I, O \cup \{[i, j]\}$) yielding the second best perfect matching $\widetilde{M}$;
    $Q := Q \cup \{\langle z(M''), M'', \widetilde{M}, I, O \cup \{[i, j]\}, \texttt{right}\rangle\}$;
    comment: this 6-tuple is stored even if $\widetilde{M} = \emptyset$, in order to preserve M''
  endif
until $k > K$ or $Q = \emptyset$ [**comment:** if $Q = \emptyset$ less than K assignments exist]

Algorithm Revised_ranking executes at most K iterations. At the root node we compute the two best assignments, while at each of the other decision nodes we compute only the second minimum assignment. Chegireddy and Hamacher suggested that one compute the second best assignment by (i) selecting the edges of the best matching, one edge at a time; (ii) forbidding it; and (iii) reoptimizing by means of a single shortest path computation. Using this technique each node can be explored in $O(n^3)$ time. The overall time complexity of the algorithm is $O(Kn^3)$, i.e., the same as that of an efficient implementation of Murty's algorithm.

Example 5.15. We make use of the instance introduced in Example 5.14 (see Figure 5.3) with $K = 3$. The initial execution of LSAP(∅, ∅) and LSAP2(∅, ∅) yields $M' = M_\alpha$ and $M'' = M_\beta$, so we set $Q = \{\langle z(M'), M', M'', I, O, lr\rangle\} = \{\langle 3, M_\alpha, M_\beta, \emptyset, \emptyset, \texttt{right}\rangle\}$ and $k = 1$. At the first iteration we empty Q and set $M_1 = M_\alpha$, $k = 2$. We select $[i, j] = [2, 2]$ and solve LSAP2({[2, 2]}, ∅) (yielding $\widetilde{M} = M_\varepsilon$) and LSAP2(∅, {[2, 2]}) (yielding $\widetilde{M} =$

M_γ). We obtain the new set of 6-tuples $Q = \{\langle 107, M_\alpha, M_\varepsilon, [2, 2], \emptyset, \texttt{left}\rangle, \langle 7, M_\beta, M_\gamma, \emptyset, [2, 2], \texttt{right}\rangle\}$. At the second iteration we remove from Q the second 6-tuple and set $M_2 = M_\beta$, $k = 3$. We select $[i, j] = [1, 1]$ and solve LSAP2($\{[1, 1]\}, \{[2, 2]\}$) (yielding $\widetilde{M} = \emptyset$ so Q is not enlarged) and LSAP2($\emptyset, \{[2, 2], [1, 1]\}$) (yielding $\widetilde{M} = M_\delta$). We now have $Q = \{\langle 107, M_\alpha, M_\varepsilon, \{[2, 2]\}, \emptyset, \texttt{left}\rangle, \langle 13, M_\gamma, M_\delta, \emptyset, \{[2, 2], [1, 1]\}, \texttt{right}\rangle\}$. At the third iteration we remove the second 6-tuple from Q, set $M_3 = M_\gamma$, $k = 4$, and terminate. ∎

Computational experiments performed by Pascoal, Captivo, and Clímaco [539] on C implementations show that their algorithm is faster than that by Chegireddy and Hamacher [181], which in turn is much faster than the original algorithm by Murty [506].

5.4.2 k-cardinality assignment problem

Let $C = (c_{ij})$ be a given $n \times n$ cost matrix and k be a given value not greater than n. The *k-cardinality assignment problem*, introduced by Dell'Amico and Martello [218], asks one to assign exactly k rows to k different columns so that the sum of the corresponding costs is a minimum. The problem can be formulated as

$$\min \sum_{i=1}^{n}\sum_{j=1}^{n} c_{ij}x_{ij} \tag{5.16}$$

$$\text{s.t.} \quad \sum_{j=1}^{n} x_{ij} \le 1 \qquad (i = 1, 2, \dots, n), \tag{5.17}$$

$$\sum_{i=1}^{n} x_{ij} \le 1 \qquad (j = 1, 2, \dots, n), \tag{5.18}$$

$$\sum_{i=1}^{n}\sum_{j=1}^{n} x_{ij} = k, \tag{5.19}$$

$$x_{ij} \in \{0, 1\} \qquad (i, j = 1, 2, \dots, n). \tag{5.20}$$

The linear sum assignment problem is the special case of (5.16)–(5.20) that arises when $k = n$. In this case the model can be simplified by dropping equation (5.19) and replacing the $\le$ signs in (5.17) and (5.18) with the $=$ sign, thus obtaining the LSAP model.

Dell'Amico and Martello [218] proved that the constraint matrix of the k-cardinality assignment problem is totally unimodular; hence, it can be solved by using any LP solver. They also gave a specialized algorithm in which a preprocessing phase determines rows and columns which must be assigned in an optimal solution and constructs a feasible solution, while a primal algorithm obtains the optimal solution through shortest path techniques.

Note that the standard preprocessing techniques adopted for LSAP (see Sections 4.1.2 and 4.4.4) do not extend to the k-cardinality assignment problem, as they are based on the assumption that each row has to be assigned, whereas in this case it is generally impossible to establish that a given row will be assigned in the optimal solution. Consider, e.g., an instance in which $w_{1j} = Q$ for all j, while $w_{ij} \ll Q$ for all j and all $i \neq 1$. The classical Procedure Basic_preprocessing of Section 4.1.2 subtracts from each cost w_{ij} the minimum value in row i and assigns, in turn, each row to one of the columns corresponding to a zero cost, if not yet assigned. Hence, row 1 would be assigned to column 1, but no optimal solution (for $k < n$) assigns row 1.

The preprocessing algorithm by Dell'Amico and Martello [218] finds an optimal solution to (5.16)–(5.20) with k replaced by a smaller value g, i.e., an optimal solution in which g rows are assigned to g columns. It is proved that all rows and columns that are assigned in such optimal solution must also be assigned in an optimal solution to (5.16)–(5.20). Preprocessing is completed by determining additional rows and columns which must be assigned in an optimal solution by computing lower and upper bounds on the optimal solution value and by performing a reduction phase on the cost matrix.

Let $\widetilde{C}$ be the $k \times k$ submatrix of C induced by the k rows and columns that are assigned in the feasible solution found by the preprocessing phase. The optimal solution for the complete cost matrix is then obtained by (i) solving an LSAP for the matrix $\widetilde{C}$; and (ii) executing a series of iterations in which a new row (or a new column) is added to $\widetilde{C}$ and the new optimal solution is obtained through a shortest path computation on a specially structured graph.

The overall algorithm has time complexity $O(n^3)$ for dense matrices. Efficient implementations, both for dense and sparse cost matrices, can be found in Dell'Amico, Lodi, and Martello [216]. Computational experiments reported in [216] show that the code effectively solves very large sparse and dense instances of the problem. Volgenant [649] described a transformation to solve the k-cardinality assignment problem through a standard LSAP algorithm.

5.4.3 Semi-assignment problem

Given an $n \times m$ cost matrix C, with $n \geq m$, and a positive demand vector b of m elements, the *semi-assignment problem* asks one to select one element per row so that the number of elements in each column j is b_j and the sum of the selected elements is a minimum. The problem can be formulated as

$$\min \quad \sum_{i=1}^{n} \sum_{j=1}^{m} c_{ij} x_{ij} \tag{5.21}$$

$$\text{s.t.} \quad \sum_{j=1}^{m} x_{ij} = 1 \qquad (i = 1, 2, \ldots, n), \tag{5.22}$$

$$\sum_{i=1}^{n} x_{ij} = b_j \qquad (j = 1, 2, \ldots, m), \tag{5.23}$$

$$x_{ij} \in \{0, 1\} \qquad (i = 1, 2, \ldots, n;\ j = 1, 2, \ldots, m). \tag{5.24}$$

Note that the problem has a solution only if $\sum_{j=1}^{m} b_j = n$. The semi-assignment problem can be seen as

(i) a generalization of LSAP for which we have $m = n$ and $b_j = 1$ for all j; and

(ii) a special case of the transportation problem (see Section 4.5) in which the right-hand side of (5.22) has a_i instead of 1, with $\sum_{i=1}^{n} a_i = \sum_{j=1}^{m} b_j$.

Barr, Glover, and Klingman [69] described various applications and gave an adaptation of their alternating basis algorithm of Section 4.5.2 to the semi-assignment problem.

Kennington and Wang [416] showed how the various phases of the Jonker and Volgenant [392] shortest path algorithm described in Section 4.4.4 (column reduction, reduction transfer, augmenting row reduction, and shortest path augmentation) can be adapted to handle the right-hand side of (5.23). The resulting algorithm solves the semi-assignment problem in $O(n^2m)$ time, and the computational experiments reported in [416] show that it can efficiently solve large-size sparse and dense instances of the problem. Volgenant [647] discussed the corresponding modifications for the LSAP codes LAPJV and LAPMOD (see Section 4.9.1) and gave a Pascal listing for the resulting LAPMOD code.

5.4.4 Rectangular cost matrix

The generalization of LSAP to rectangular cost matrices can be defined as follows. Given an $n \times m$ cost matrix, and assuming $n \leq m$, assign n rows to n different columns so that the sum of the corresponding elements is a minimum. By adding $m - n$ dummy rows of zero elements, the problem can be solved as an LSAP on the resulting $n \times n$ matrix. Alternatively, one can use the transformation to a minimum cost flow problem of Section 4.4.1, which does not require that vertex sets U and V have equal cardinality.

Specialized algorithms (together with an Algol implementation) were given by Bourgeois and Lassalle [112, 111], who proposed an adaptation of the $O(n^4)$ Munkres [502] algorithm for LSAP (see Section 4.1.3) to the rectangular case.

Volgenant [647] and Bertsekas [88] adapted, respectively, the LAPJV code and the auction method (see Section 4.9.1) to the rectangular case.

5.5 Applications

Linear sum assignment problems occur quite frequently as subproblems in more involved applied combinatorial optimization problems like the quadratic assignment problem (see Chapter 7), asymmetric traveling salesman and vehicle routing problems (see Fischetti, Lodi, and Toth [272] and Toth and Vigo [641] for recent surveys including sections on the use of LSAP in these contexts), or scheduling problems (see, e.g., Section 3.8.2).

Some direct applications can also be found in the literature. The classical personnel assignment problem was first discussed by Votaw and A. Orden [653]. Machol [467] described a practical situation solved through LSAP: the modification of an electromechanical punching typewriter to punch a modified six digit binary code so as to minimize the number of mechanical changes to be made on the code bars.

Machol [468] reported another application of LSAP in which a swimming coach must select four out of n swimmers to form a medley relay team, knowing the time of each swimmer in each of the four strokes (back, breast, butterfly, and free-style). The solution is obtained by solving LSAP on an $n \times n$ cost matrix having a row per swimmer, the first four columns for their times in the four strokes, and the remaining $n - 4$ columns filled by zeroes.

Ewashko and Dudding [263] reported on the use of the Hungarian algorithm for deriving postings for servicemen.

Neng [511] gave an interesting application in railway systems: he considered the problem of assigning engines to trains due to traffic constraints and formulated it as a linear assignment problem.

A relevant application in earth-satellite telecommunication, the time slot assignment problem, was discussed in Section 3.8.2.

Brogan [115] described assignment problems in connection with locating objects in space that could, however, be better modeled as linear bottleneck assignment problems and are therefore discussed in Section 6.2.8.

Schwartz [602] discussed an LSAP arising in military operations, namely, in a scenario of strategic missile exchange between the US and the Soviet Union. The study had been developed during the Cold War, when there was a perceived threat of Soviet attack by thousands of missiles, and the US had to assign thousands of interceptors, each one dedicated to attempt interception of one Soviet missile. The resulting large-size LSAP was solved through the auction algorithm. (The actual experimental results, however, were not given, as this was forbidden by the US Department of Defense.)

Bekker, Braad, and Goldengorin [81] proposed using LSAP to solve a system of two polynomial equations $f(x, y) = 0$, $g(x, y) = 0$ having a finite number of solutions. The solution, consisting of pairs of an x-value and a y-value, is obtained by independently calculating all x- and y-roots and then matching them through a weighted bipartite graph in which the costs represent the errors due to the numerical computations.

Recently, assignment problems found an interesting application in cosmology (Frisch and Sobolevskiĭ [289]). The present distribution of mass in the universe is highly clustered. This distribution arose from minor fluctuations of an almost uniform density field. In a discretized model, the authors propose to solve a linear assignment problem whose cost coefficients are squared distances between primordial positions of fluctuations and recent locations of galaxies.

5.5.1 Mean flow time minimization on parallel machines

Horn [377] showed that the scheduling of n jobs on m machines to minimize the average, and, hence, total, flow time can be solved as an LSAP. Let p_{ij} be the processing time of job i if executed on machine j ($i = 1, 2, \ldots, n$; $j = 1, 2, \ldots, m$). Given the sequence of the n_j jobs executed on machine j, let $[k]$ denote the index of the kth last job in this sequence. Then the total flow time on machine j is

$$F_j = \sum_{k=1}^{n_j} k p_{[k]j}, \tag{5.25}$$

and the mean flow time is F_j/n_j. The problem is then to assign each job to a position k on a machine j so that $\sum_{j=1}^{m} F_j$ is a minimum (with $k = 1$ for the last job, $k = 2$ for the second last, and so on). This can be modeled through a rectangular cost matrix of n rows (one per job) and nm columns (one per (k, j) pair), where entry $(i, (k, j))$ contains value kp_{ij}. By adding $n(m-1)$ dummy rows of zero elements, the problem can be solved as an LSAP on the resulting $nm \times nm$ matrix.

5.5.2 Categorized assignment scheduling

Punnen and Aneja [562] considered two *categorized* assignment problems arising in a scheduling environment in which

(i) n jobs must be assigned to n machines (one job per machine);

(ii) the entries of an $n \times n$ cost matrix C give the processing time c_{ij} required for executing job i on machine j $(i, j = 1, 2, \ldots, n)$; and

(iii) the jobs are partitioned into r sets $S_1, S_2, \ldots, S_r$.

In the first problem one is required to minimize the maximum, over all sets, of the sum of the processing times of the jobs in a set, i.e., the objective function is

$$\min_{\varphi} \max_{1 \le k \le r} \sum_{i \in S_k} c_{i\varphi(i)}, \tag{5.26}$$

where, as usual, φ is a permutation of $\{1, 2, \ldots, n\}$. In the second problem the objective is to minimize the sum, over all sets, of the maximum processing time of a job in the set, i.e.,

$$\min_{\varphi} \sum_{k=1}^{r} \max_{i \in S_k} c_{\,\varphi(i)}. \tag{5.27}$$

The first case arises if the jobs of each set must be processed in sequence but the sets may be processed in parallel, while in the second case the jobs of each set may be processed in parallel but the sets must be processed in sequence. These objective functions are a combination of LSAP and the bottleneck assignment problem treated in Chapter 6.

As shown by Richey and Punnen [583], both (5.26) and (5.27) are $\mathcal{NP}$-hard for general r. The first problem is already $\mathcal{NP}$-hard for $r = 2$, while the second one is polynomially solvable for fixed r by enumerating all possible solutions as follows. A candidate solution is obtained by selecting, for each set S_k, a pair (i, j) and imposing that it is the one that produces $\max_{i \in S_k} c_{i\varphi(i)}$. The solution is tested for feasibility by checking if a complete matching exists in the induced submatrix that only contains, for each set, those entries that do not exceed the candidate's value. Since the number of r-tuples is $O(n^{2r})$, and each check can be performed in $O(n^{2.5})$ time (see Section 3.3), the overall algorithm runs in polynomial time. Punnen and Aneja [562] introduced Tabu search heuristics for both problems (5.26) and (5.27). Aneja and Punnen [29] proposed, for the first problem (also called the *multiple bottleneck assignment problem*), lower bounding techniques based on a decomposition into a series of LSAPs.

5.5.3 Optimal depletion of inventory

Derman and Klein [231] considered a problem of optimal depletion of inventory. A stockpile consists of n items of the same type, each having a known age a_i $(i = 1, 2, \ldots, n)$. A function $f(a)$ gives the expected value of an item of age a when it is withdrawn from the stockpile. A sequence of n values $t_1, t_2, \ldots, t_n$ gives the times at which an item is demanded.

The problem is to determine, for $j = 1, 2, \ldots, n$, the item $i(j)$ to be released at time t_j so that the total expected value,

$$\sum_{j=1}^{n} f(a_{i(j)} + t_j), \tag{5.28}$$

is a maximum. The solution is given by a maximization LSAP on an expected value matrix T with $t_{ij} = f(a_i + t_j)$ $(i, j = 1, 2, \ldots, n)$.

5.5.4 Personnel assignment with seniority and job priority

Caron, Hansen, and Jaumard [164] described a variation of LSAP arising in the daily schedule of nurses that are assigned to various jobs in a hospital due to absence of regular staff or work overload. These nurses belong to the float team and the availability list of the hospital. Every nurse is qualified to perform a subset of jobs. The float team nurses are under contract with the hospital and are assigned to training when no job for which they are qualified is available. Hence, it is desired to assign them as much as possible. Unassigned jobs (if any) are then assigned to nurses on the availability list, in order of seniority. Still unassigned jobs (if any) are done on overtime, at a higher cost. Furthermore, the jobs have different priorities, evaluated by the hospital units.

Given an LSAP in which the rows correspond to persons and the columns to jobs, let the persons be partitioned into s *seniority classes* and the jobs into p *priority classes*. We say that a solution satisfies the *seniority constraints* if and only if no unassigned person can be given a job without a person of the same or higher class becoming unassigned. Similarly, a solution satisfies the *job priority constraints* if and only if no unassigned job can be assigned without a job of the same or higher priority becoming unassigned. In the nurses application above all float team nurses form a unique (highest) class and every nurse of the availability list constitutes a separate class. The entries c_{ij} of the cost matrix give the utility of assigning nurse i to job j, and the problem is in maximization form.

The personnel assignment problem can be solved in $O(n^3)$ time. Caron, Hansen, and Jaumard [164] showed that the problem can be solved as a maximization LSAP if the cost matrix is modified by weighting each seniority and priority class sufficiently heavily with respect to the next one. In this way the solution is the lexicographically highest one. Volgenant [648] proposed a different algorithm that operates on the original costs and determines the optimal solution by iteratively solving LSAPs of increasing size.

5.5.5 Navy personnel planning

Holder [374] considered the following problem arising in the job rotation of the United States Navy personnel. Let U denote a set of sailors that have to be assigned at least one job, and V the set of jobs (with $|V| > |U|$). Define

$$E = \{[i, j] \ : \ \text{sailor } i \text{ can do job } j\}, \tag{5.29}$$

and let c_{ij} be the cost of assigning sailor i to job j. The basic problem is then

$$\min \sum_{[i,j]\in E} c_{ij}x_{ij} \tag{5.30}$$

$$\text{s.t.} \sum_{[i,j]\in E} x_{ij} \geq 1 \quad \text{for all } i \in U, \tag{5.31}$$

$$\sum_{[i,j]\in E} x_{ij} \leq 1 \quad \text{for all } j \in V, \tag{5.32}$$

$$x_{ij} \in \{0, 1\} \quad \text{for all } [i, j] \in E. \tag{5.33}$$

The actual problem, however, is not to make the assignments by determining an optimal solution to (5.30)–(5.33). It is instead to generate, for each sailor, a list of jobs, having a prefixed length ϑ, so that the sailor can choose a job from such list. Holder [374] proposed a model that embeds (5.30)–(5.33), and is parametric in ϑ He showed that solving its continuous relaxation through an interior point method and generating the lists through a parametric analysis on the solution found provides a good heuristic for the overall problem. Recently Volgenant [650] proposed an alternative solution method based on the linear assignment theory and shortest path computations.

Chapter 6
Other types of linear assignment problems

6.1 Introduction

In the two previous chapters we discussed in detail linear assignment problems with a sum objective function of the form

$$\min_{\varphi} \sum_{i=1}^{n} c_{i\varphi(i)}.$$

In various situations, however, it is meaningful to replace the sum objective by a so-called bottleneck objective function. This leads to a *linear bottleneck assignment problem* (LBAP) of the form

$$\min_{\varphi} \max_{1\le i\le n} c_{i\varphi(i)}.$$

We have seen in the introduction that such LBAPs occur in connection with assigning jobs to parallel working machines. The goal is to assign the jobs such that the latest completion time is minimized. We discuss linear bottleneck assignment problems in detail in Section 6.2. In particular, we develop threshold algorithms, a dual method, and a shortest augmenting path method for solving the LBAP. A practically efficient method will be obtained by thinning out the underlying bipartite graph and exploiting the sparsity of the modified problem; see Section 6.2.5. Moreover, we discuss special cases of the LBAP which can be solved in a fast way. Finally, we describe the asymptotic behavior of LBAPs in Section 6.2.7.

Sum and bottleneck assignment problems can be viewed as special cases of a more general model, the so-called *algebraic assignment problem* which is discussed in Section 6.3. In Sections 6.4 and 6.5 we deal with assignment problems where we require to minimize the sum of the k largest cost coefficients in the assignment, or the difference between the largest and the smallest cost coefficient in the assignment. In Section 6.6 we deal with a modified objective: we order the n cost coefficients of an assignment decreasingly and ask for a solution which is lexicographically minimal.

6.2 Bottleneck assignment problem

6.2.1 Background

Linear bottleneck assignment problems were introduced by Fulkerson, Glicksberg, and Gross [292] and occur, e.g., in connection with assigning jobs to parallel machines so as to minimize the latest completion time. Let n jobs and n machines be given. The cost coefficient c_{ij} is the time needed for machine j to complete job i. If the machines work in parallel and we want to assign the jobs to the machines such that the latest completion time is as early as possible, we get an LBAP of the form

$$\min_{\varphi \in \mathcal{S}_n} \max_{1 \le i \le n} c_{i\varphi(i)}. \tag{6.1}$$

If we describe permutations by the corresponding permutation matrices $X = (x_{ij})$, an LBAP can be modeled as

$$\begin{aligned} \min \quad & \max_{1 \le i,j \le n} c_{ij} x_{ij} && && (6.2) \\ \text{s.t.} \quad & \sum_{j=1}^{n} x_{ij} = 1 && (i = 1, 2, \dots, n), \\ & \sum_{i=1}^{n} x_{ij} = 1 && (j = 1, 2, \dots, n), \\ & x_{ij} \in \{0, 1\} && (i, j = 1, 2, \dots, n). \end{aligned}$$

The following lemma comprises two elementary observations concerning LBAPs which are helpful in solving problems in practice.

Lemma 6.1. *Let $C = (c_{ij})$ be the cost matrix of an LBAP. Then the following two statements hold:*

1. *The optimum value of the LBAP is taken by one of the cost coefficients c_{ij}.*
2. *The optimal solution φ^* depends only on the relative order of the cost coefficients and not on their numerical value.*

Example 6.2. Consider an LBAP with cost matrix

$$C = \begin{pmatrix} \pi & \sqrt{3} & 54392 \\ 0 & e^3 & e^3 \\ -2 & \sin \pi/8 & 4 \end{pmatrix}.$$

By ordering the cost coefficients we get

$$-2 < 0 < \sin \pi/8 < \sqrt{3} < \pi < 4 < e^3 = e^3 < 54392.$$

The smallest element is -2 which can be modeled by 0. The second smallest element is 0 which is replaced by 1. Then $\sin \pi/8$ is replaced by 2, and so on. Therefore, we can replace this cost matrix by

$$C = \begin{pmatrix} 4 & 3 & 7 \\ 1 & 6 & 6 \\ 0 & 2 & 5 \end{pmatrix}. \tag{6.3}$$

Thus, if the cost matrix of an LBAP has d different entries, we may model these entries by $0, 1, 2, \ldots, d-1$. The LBAP with cost matrix (6.3) has the optimal solution

$$\varphi^* = (2, 1, 3).$$

The largest value in this solution is $c_{33} = 5$, which corresponds to 4 in the originally given cost matrix. ∎

If one is interested in solving the "opposite" LBAP, i.e., in finding an assignment in which (6.2) is replaced by

$$\max \min_{1 \leq i,j \leq n} c_{ij} x_{ij},$$

according to the second statement of Lemma 6.1, it is enough to model the elements in the opposite way, i.e., with the largest element modeled by 0, the second largest by 1, and so on.

Considering bottleneck assignment problems, Gross [342] proved the following min-max theorem, which was a starting point of the theory on blocking systems; see Edmonds and Fulkerson [249].

Theorem 6.3. (Gross [342], 1959.) *Let $N = \{1, 2, \ldots, n\}$ and let $\mathcal{S}_n$ be the set of all permutations φ on N. Then the following min-max equality holds for an arbitrary $n \times n$ matrix $C = (c_{ij})$ with elements c_{ij} drawn from a totally ordered set:*

$$\min_{\varphi \in \mathcal{S}_n} \max_{i \in N} c_{i\varphi(i)} = \max_{\substack{I,J \subseteq N \\ |I|+|J|=n+1}} \min_{i \in I, j \in J} c_{ij}. \tag{6.4}$$

Note that the Frobenius theorem, Theorem 2.4, can be seen as a special case of Theorem 6.3, arising when C is a 0–1 matrix.

Proof. Let

$$c^* = \min_{\varphi \in \mathcal{S}_n} \max_{i \in N} c_{i\varphi(i)}.$$

We define two 0–1 matrices $\overline{C}$ and $\overline{\overline{C}}$ by

$$\overline{c}_{ij} = \begin{cases} 1 & \text{if } c_{ij} \leq c^*, \\ 0 & \text{otherwise,} \end{cases} \quad \text{and} \quad \overline{\overline{c}}_{ij} = \begin{cases} 1 & \text{if } c_{ij} < c^*, \\ 0 & \text{otherwise.} \end{cases}$$

According to the construction, matrix $\overline{C}$ contains a permutation matrix and matrix $\overline{\overline{C}}$ does not contain any permutation matrix. By applying Theorem 2.4 to matrix $\overline{C}$ we get that any $(k+1) \times (n-k)$ submatrix of C contains an entry $\leq c^*$. By applying Theorem 2.4 to matrix $\overline{\overline{C}}$ we get that there is a $(k+1) \times (n-k)$ submatrix of C which contains only entries $\geq c^*$. Thus we have shown

$$c^* = \max_{\substack{I,J \subseteq N \\ |I|+|J|=n+1}} \min_{i \in I, j \in J} c_{ij}. \quad \square$$

A slight generalization of LBAPs are min-cost maximum matching problems with a bottleneck objective: Let $G = (U, V; E)$ be a bipartite graph with edge set E. Every edge $[i, j]$ has a length c_{ij}. The *bottleneck min-cost maximum matching problem* can be formulated as follows. Find a maximum matching in G such that the maximum length of an edge in this matching is as small as possible:

$$\min\{\max_{[i,j]\in M} c_{ij} \ : \ M \text{ is a maximum matching}\}. \tag{6.5}$$

All solution methods for the LBAP mentioned below can be carried over in a straightforward way to solution methods for the bottleneck min-cost maximum matching problem.

6.2.2 Threshold algorithms

The idea of using threshold methods for LBAP dates back to Garfinkel [301]. A *threshold algorithm* alternates between two phases. In the first phase a cost element c^*—the *threshold value*—is chosen and a *threshold matrix* $\overline{C}$ is defined by

$$\overline{c}_{ij} = \begin{cases} 1 & \text{if } c_{ij} > c^*, \\ 0 & \text{otherwise.} \end{cases} \tag{6.6}$$

In the second phase it is checked whether for the cost matrix $\overline{C}$ there exists an assignment with total cost 0. To check this we construct a bipartite graph $G = (U, V; E)$ with $|U| = |V| = n$ and edges $[i, j] \in E$ if and only if $\overline{c}_{ij} = 0$. (In other words, we have to check whether a bipartite graph with threshold matrix $\overline{C}$ contains a perfect matching or not.) The smallest value c^* for which the corresponding bipartite graph contains a perfect matching is the optimum value of the LBAP (6.1).

There are several ways to implement a threshold algorithm. One possibility is to apply a binary search in the first phase. This leads to an $O(T(n) \log n)$ algorithm, where $T(n)$ is the time complexity for checking the existence of a perfect matching. We can use the matrix algorithm of Ibarra and Moran [384] (see Section 3.6) to decide whether there exists a perfect matching or not. Thus the optimal objective function value of an LBAP can be found by an algorithm of bitwise complexity $O(n^\beta \log^k n)$. For multiplying two $n \times n$ matrices, $O(n^\beta)$ arithmetic operations are needed and k is some integer incurred by long integer arithmetic (hence, it includes the $\log n$ term of the binary search). Coppersmith and Winograd [197] showed that matrix multiplication is possible with $\beta = 2.376$. If we also want the corresponding optimal solution, we can apply the method by Alt, Blum, Mehlhorn, and Paul (see Section 3.4) and get, therefore, a total complexity of $O(n^{2.5}/\sqrt{\log n})$ in the dense case. Here *dense* means that the number of non-zero cost coefficients c_{ij} is $O(n^2)$. This method yields the theoretically best bound for *dense* LBAPs.

Theorem 6.4. *An LBAP with an $n \times n$ cost matrix C can be solved in $O(n^{2.5}/\sqrt{\log n})$ time.*

Let us formulate the threshold method as an algorithm. For a given value c^*, the bipartite graph $G[c^*] = (U, V; E)$ with $|U| = |V| = n$ and has an edge $[i, j] \in E$ if and only if $c_{ij} \le c^*$.

Algorithm 6.1. Threshold.
Threshold algorithm for the LBAP.

let $C = (c_{ij})$ be a given $n \times n$ cost matrix;
$c_0^* := \min_{ij}\{c_{ij}\}$, $c_1^* := \max_{ij}\{c_{ij}\}$;
if $c_0^* = c_1^*$ **then**
  $z := c_0^*$ [**comment:** any permutation of $\{1, 2, \ldots, n\}$ is optimal]
else
  while $C^* = \{c_{ij} \, : \, c_0^* < c_{ij} < c_1^*\} \neq \emptyset$ **do**
    comment: find the median, c^*, of C^*;
    $c^* := \min\{c \in C^* \, : \, |\{c_{ij} \in C^* \, : \, c_{ij} \leq c\}| \geq |C^*|/2\}$;
    Feasibility_check(c^*, c_0^*, c_1^*)
  endwhile;
  if Feasibility_check has not yet been executed for c_0^* **then**
    Feasibility_check(c_0^*, c_0^*, c_1^*)
  endif;
  find a perfect matching in the current bipartite graph $G[c_1^*]$;
  $z := c_1^*$ [**comment:** value of the optimal matching]
endif

Procedure Feasibility_check(c^*, c_0^*, c_1^*)
define the current graph $G[c^*]$;
if $G[c^*]$ contains a perfect matching **then**
  $c_1^* := c^*$
else
  $c_0^* := c^*$
endif

Example 6.5. Let the cost matrix of an LBAP be given as

$$C = \begin{pmatrix} 8 & 2 & 3 & 3 \\ 2 & 7 & 5 & 8 \\ 0 & 9 & 8 & 4 \\ 2 & 5 & 6 & 3 \end{pmatrix}.$$

We find $c_0^* = 0$ and $c_1^* = 9$. The median of the cost coefficients between 0 and 9 is 4. Thus the threshold matrix of $G[4]$ becomes

$$\overline{C}[4] = \begin{pmatrix} 8 & 0 & 0 & 0 \\ 0 & 7 & 5 & 8 \\ 0 & 9 & 8 & 0 \\ 0 & 5 & 6 & 0 \end{pmatrix}.$$

The corresponding bipartite graph is shown in Figure 6.1. The maximum cardinality matching, shown by the thick lines, has 3 edges.

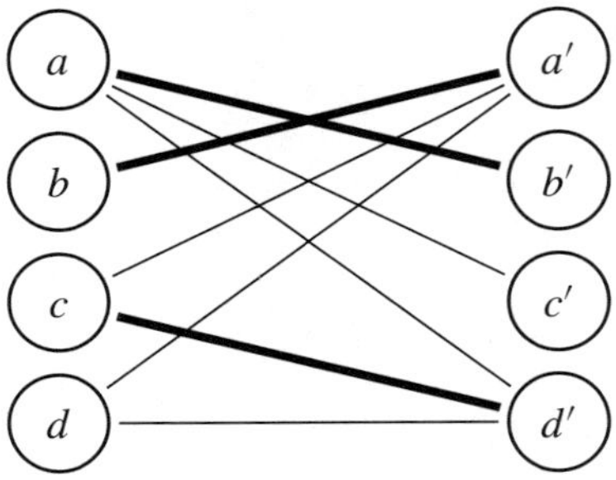

Figure 6.1. *Bipartite graph* $G[4]$.

$$\overline{C}[7] = \begin{pmatrix} 8 & 0 & 0 & 0 \\ 0 & 0 & 0 & 8 \\ 0 & 9 & 8 & 0 \\ 0 & 0 & 0 & 0 \end{pmatrix}$$

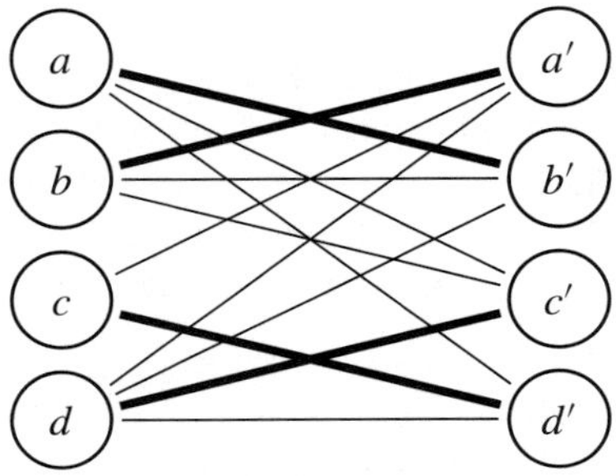

Figure 6.2. *Threshold matrix and bipartite graph* $G[7]$.

$$\overline{C}[5] = \begin{pmatrix} 8 & 0 & 0 & 0 \\ 0 & 7 & 0 & 8 \\ 0 & 9 & 8 & 0 \\ 0 & 0 & 6 & 0 \end{pmatrix}$$

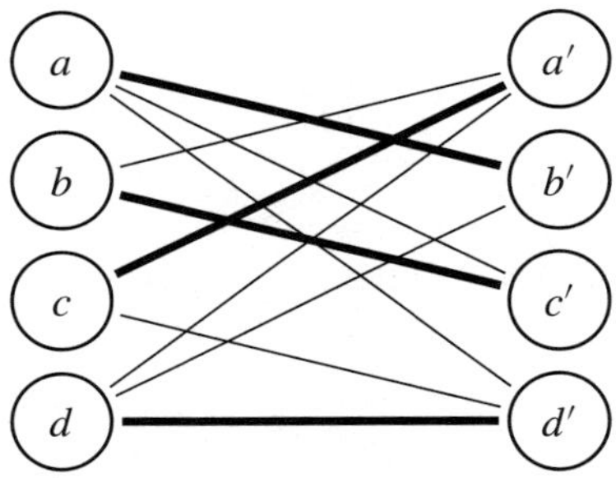

Figure 6.3. *Threshold matrix and bipartite graph* $G[5]$.

Thus we set $c_0^* = 4$ and determine the new value of $c^* = 7$. This leads to the new bipartite graph with complementary adjacency matrix $\overline{C}[7]$ as shown in Figure 6.2. The corresponding maximum matching has cardinality 4. The graph $G[7]$ allows a perfect matching.

Therefore, we set $c_1^* := 7$ and determine $c^* = 5$. The corresponding graph $G[5]$, shown in Figure 6.3, also allows a perfect matching.

Therefore, we set $c_1^* := 5$ and obtain $C^* = \emptyset$. Since the current value $c_0^* = 4$ has already been checked for feasibility, we get the perfect matching in $G[5]$. An optimal solution is thus $\varphi^* = (2, 3, 1, 4)$, with objective function value $z = 5$. ∎

6.2.3 A dual method

Closely related to the threshold method for solving linear bottleneck problems is the following dual method. It uses the fact that one has some information about the threshold value c^*. For example, every row and every column of the cost matrix contains a cost element which contributes to the optimal objective function value. Therefore, the maximum of the row minima and the maximum of the column minima are lower bounds for the optimum objective function value. Thus we may choose as first threshold value c^*

$$c^* = \max_{1 \le k \le n} \left(\min_{1 \le i \le n} c_{ik}, \ \min_{1 \le j \le n} c_{kj} \right). \tag{6.7}$$

This value may be computed in two nested loops. An efficient way to compute it consists in stopping the minimum evaluation as soon as the current minimum gets smaller than the maximum reached up to now. (Another efficient way for computing c^* has been stated by Carpaneto and Toth [167].)

The dual method starts with this value c^* and constructs a minimum row and column cover of all cost elements $c_{ij} \le c^*$. This can be performed by determining a maximum matching in the bipartite graph $G[c^*]$ whose edges $[i, j]$ correspond to cost entries $c_{ij} \le c^*$. The maximum matching immediately leads to a minimum vertex cover of $G[c^*]$ (see the considerations following the labeling algorithm 3.1 of Section 3.2), or, in other terms, to a minimum row and column cover (I, J) of all cost elements $c_{ij} \le c^*$. The index sets I and J contain the indices of the covered rows and columns, respectively. If the matching is not perfect, an additional *uncovered* cost entry contributes to the optimal objective function value. All uncovered cost entries are larger than c^*. Therefore, we replace c^* by

$$c^* = \min_{i \notin I, j \notin J} c_{ij}$$

and define a new graph $G[c^*]$ as before. This graph $G[c^*]$ contains all edges of the previous graph and at least one new edge. Thus one starts from the previously found matching and determines a new maximum matching in $G[c^*]$. This step is repeated until a perfect matching is found. We can summarize this method in the following algorithm.

Algorithm 6.2. Dual_LBAP.
Dual algorithm for the LBAP.

let $C = (c_{ij})$ be a given $n \times n$ cost matrix;
$c^* := \max_{k=1,2,\ldots,n} \ (\min_{i=1,2,\ldots,n} c_{ik}, \ \min_{j=1,2,\ldots,n} c_{kj})$;
$M := \emptyset$;
while $|M| < n$ **do**
    define the bipartite graph $G[c^*]$;
    find a maximum matching M in $G[c^*]$;
    if $|M| < n$ **then**
        let $I \subseteq U$ and $J \subseteq V$ be vertex sets in a minimum vertex cover of $G[c^*]$;
        $c^* := \min_{i \notin I, \ j \notin J} c_{ij}$
    endif
endwhile

Example 6.6. We illustrate the dual method on the same instance used for Example 6.5. We find $c^* = 3$ (the minimum entry of the last two columns). Thus the threshold matrix of $G[3]$ is

$$\overline{C}[3] = \begin{pmatrix} 8 & 0 & 0 & 0 \\ 0 & 7 & 5 & 8 \\ 0 & 9 & 8 & 4 \\ 0 & 5 & 6 & 0 \end{pmatrix}.$$

A maximum matching in the corresponding bipartite graph is given by $M = \{[1,2],[3,1],[4,4]\}$. Since this matching is not perfect, we need a minimum row and column covering of the elements $c_{ij} \leq 3$ and get $I := \{1\}$, $J := \{1,4\}$. The minimum uncovered element has the value 5. Thus we get $c^* := 5$ and obtain graph $G[5]$ of Figure 6.3. An augmentation step leads to the maximum matching $M = \{[1,2],[2,3],[3,1],[4,4]\}$, which is perfect. Thus an optimal solution of this bottleneck assignment problem is $\varphi = (2,3,1,4)$ with value 5. ∎

The threshold method can easily be combined with ideas from the dual method: the smallest uncovered entry is always a lower bound for a feasible threshold value c^*.

6.2.4 Augmenting path methods

Another possibility for solving the LBAP is to mimic the Hungarian method; see Gross [342], Page [521], and Derigs and Zimmermann [230]. The description we give closely follows the implementation proposed by Derigs and Zimmermann [230].

For any matching M in $G = (U, V; E)$, let $U(M) \subseteq U$ be the set of vertices in U which are matched by M. Let the matching M have minimum bottleneck cost among all matchings with fixed set $U(M)$. In order to enlarge a matching M, we can use an augmenting path P with respect to M. But at the same time we want the enlarged matching M' to have minimum cost among all matchings with $U(M')$ fixed. We show in the following proposition that this can be achieved by augmenting matching M with an augmenting path of minimum length. Let S be a set of edges in the underlying graph G; we will denote by $c(S)$ the *bottleneck cost* of S:

$$c(S) = \max_{[i,j]\in S} c_{ij}.$$

Definition 6.7. *Let $P = ([i_1, j_1], [i_2, j_1], [i_2, j_2], \ldots, [i_k, j_k])$ be an augmenting path with respect to a given matching M. The* bottleneck length $\ell(P)$ *of this path is defined by*

$$\ell(P) = \max(c([i_1, j_1]), c([i_2, j_2]), \ldots, c([i_k, j_k])). \tag{6.8}$$

An augmenting path is called b-shortest augmenting path *if its bottleneck length is minimum with respect to all augmenting paths of matching M.*

An explanation is in order concerning the somewhat strange definition of the bottleneck length of an augmenting path. When M is augmented by P, the edges $[i_2, j_1], [i_3, j_2], \ldots, [i_k, j_{k-1}]$ are replaced by the edges $[i_1, j_1], [i_2, j_2], \ldots, [i_k, j_k]$ and thus the length of the replaced edges does not play a role any more (see Part 1 of the proof of Proposition 6.8). For this reason we may skip these edges when defining the bottleneck length of an augmenting path.

Proposition 6.8. *Let C be an $n \times n$ cost matrix of an LBAP and let M be a matching in the corresponding bipartite graph $K_{n,n}$ with minimum cost $c(M)$ among all matchings with the same set $U(M)$. Let P be an augmenting path with respect to M starting in an arbitrary unmatched vertex $i \in U$, which is b-shortest among all augmenting paths starting in this vertex. Then the matching $M' = M \ominus P$ has minimum cost among all matchings in G with the same set $U(M')$. Moreover, the cost $c(M')$ is given by*

$$c(M') = \max(c(M), \ell(P)). \tag{6.9}$$

Proof. 1. We get $c(M) \le c(M')$. Suppose the contrary, namely, $c(M') < c(M)$. In this case there exists an augmenting path

$$P = ([i_1, j_1], [i_2, j_1], [i_2, j_2], \dots, [i_k, j_k])$$

with $i_2, i_3, \dots, i_k$ in $U(M)$ and $\ell(P) < c(M)$. Let us exchange the edges $[i_2, j_1], [i_3, j_2], \dots, [i_k, j_{k-1}]$ of matching M by the edges $[i_1, j_1], [i_2, j_2], \dots, [i_k, j_k]$. Obviously, we get a new matching $\overline{M}$ with $U(\overline{M}) = U(M)$ and $c(\overline{M}) < c(M)$. But this is a contradiction to the assumption that M has minimum cost among all matchings with fixed $U(M)$.

2. Let $\overline{M}$ be any matching with $U(\overline{M}) = U(M')$. Then $c(\overline{M}) \ge c(M')$ holds. Namely, the symmetric difference of the sets M and $\overline{M}$ contains a path $\overline{P}$ which starts in a vertex $i \in U(M') \setminus U(M)$. According to the definition of the augmenting path P which leads to M', this path $\overline{P}$ is an augmenting path with respect to M with a bottleneck length of $\ell(\overline{P}) \ge \ell(P)$. This yields $c(\overline{M}) \ge c(M \ominus \overline{P}) \ge c(M \ominus P) = c(M')$. Therefore, M' has minimum cost among all matchings with the same set $U(M')$.

3. The cost $c(M') = \max(c(M), \ell(P))$ is an immediate consequence of the augmentation step: some edges of M are exchanged by the edges $[i_1, j_1], [i_2, j_2], \dots, [i_k, j_k]$. Due to 1, it cannot happen that the cost decreases during the augmentation step. □

It may well be that $\ell(P) < c(M)$. Moreover, note that Proposition 6.8 becomes wrong if we require that M has minimum cost among all matchings which cover the same vertices in sets U and V as the matching M. For example, consider the cost matrix

$$C = \begin{pmatrix} 5 & 0 \\ 0 & 7 \end{pmatrix}.$$

The matching $M = \{[1, 1]\}$ has cost 5 and obviously has minimum cost among all matchings which match the first vertex in U and the first vertex in V (there is only one such matching). By augmenting M with the path $P = ([2, 1], [1, 1], [1, 2])$ we get the matching $M' = \{[1, 2], [2, 1]\}$ with $c(M') = 0$.

A b-shortest augmenting path can easily be found by adapting Dijkstra's shortest path algorithm to our situation (see Burkard [131]). As mentioned before, the actual values of the cost coefficients do not really play a role in bottleneck problems, only their order is relevant. Therefore, we can assume that all cost coefficients c_{ij} are nonnegative. (We could even assume that all cost coefficients are from the set $\{0, 1, 2, \ldots, n^2 - 1\}$.) Moreover, the cost of the edges of matching M do not play any role during the augmentation step. Therefore, we can assume that during the shortest path computation all edges of M have cost 0.

The b-shortest augmenting path algorithm can be described as follows. Let M be the matching already found in $G = (U, V; E)$ and let c^* be the cost of the matching (i.e., the cost of the most expensive matching edge). Let L be the set of unmatched vertices on the left side of the bipartite graph, i.e., $L \subseteq U$. We choose one vertex of L as the start vertex i_1. During the shortest path algorithm we give every considered vertex $x \in U$ (resp., $x \in V$) a label $(\overline{\alpha}(x), \overline{\beta}(x))$ (resp., $(\alpha(x), \beta(x))$), where $\overline{\alpha}(x)$ (resp., $\alpha(x)$) is the bottleneck length of a b-shortest augmenting path from i_1 to this vertex, found up to this step, and $\overline{\beta}(x)$ (resp., $\beta(x)$) denotes the immediate predecessor on the b-shortest path from i_1 to vertex x. We start by labeling all neighbors of vertex i_1: If the edge $[i_1, j_1]$ has a length not greater than c^*, we label j_1 by (c^*, i_1); otherwise, we label j_1 by $(c_{i_1 j_1}, i_1)$. Those $j \in V$ which are not neighbors of i_1 get the first label $\alpha(j) = \infty$. Now we determine a vertex $j \in V$ with minimum value $\alpha(j)$. If $\alpha(j) = \infty$, then there does not exist an augmenting path starting in i_1. We delete i_1 from L and choose another starting vertex in L. Otherwise, there are two possibilities.

Case 1. Vertex j is unmatched. Then we have found a b-shortest augmenting path from i_1 to j with bottleneck length $\alpha(j)$. This path P can be reconstructed by backtracking on the labels β and $\overline{\beta}$ (alternately), which give the sequence of vertices $(i_1, \ldots, \overline{\beta}(\beta(j)), \beta(j), j)$. In this case, the matching can be augmented to $M' = M \ominus P$ with cost $\max(c^*, \alpha(j))$.

Case 2. Vertex j is matched. We now mark vertex j as scanned. There is a matching edge $[i_2, j]$. We label vertex i_2 by $(\overline{\alpha}(i_2), \overline{\beta}(i_2)) := (\alpha(j), j)$. Moreover, the labels of all unscanned neighbors j_2 of vertex i_2 are updated as follows:

- if $z = \max(\alpha(j), c_{i_2, j_2}) < \alpha(j_2)$, then $\alpha(j_2) := z, \beta(j_2) := i_2$;
- otherwise, keep the label of j_2 unchanged.

Now again an unscanned, labeled vertex, say, j', of V is determined with minimum $\alpha(j')$. If this vertex is unmatched, we have found a b-shortest augmenting path. Otherwise, we continue as above.

The validity of this algorithm is based on the following proposition.

Proposition 6.9. *Let $\overline{X}$ be the set containing vertex i_1 and all labeled vertices of L and X the set containing all scanned vertices of V. Then the following two statements hold in any step of the method described above:*

1. *For any vertex $x \in \overline{X} \setminus \{i_1\}$ (resp., $x \in X$), the value $\overline{\alpha}(x)$ (resp., $\alpha(x)$) is the bottleneck length of a b-shortest path from i_1 to x in G.*
2. *For any labeled, but unscanned, vertex $j \in V$, the value $\alpha(j)$ is the bottleneck length of a b-shortest path from i_1 to j which uses as intermediate vertices only vertices of $\overline{X} \cup X$.*

Proof. We prove the proposition by induction on the set $\overline{X} \cup X$. At the beginning, $\overline{X} = \{i_1\}$, $X = \emptyset$. After labeling all neighbors of i_1, the second statement of Proposition 6.9 is obviously true. Next we determine a vertex $j \in V$ with minimum value $\alpha(j)$. If $\alpha(j) = \infty$, vertex i_1 has no neighbors and we stop. Otherwise, this $\alpha(j)$ is the shortest bottleneck length of a path from i_1 to j, since any path not using the edge $[i_1, j]$ would pass through an edge $[i_1, \bar{j}]$ with length $c_{i_1\bar{j}} \geq \alpha(j)$ and, therefore, has bottleneck length at least $\alpha(j)$. Thus the first statement is true at the beginning, so we can assume that for all steps up to a certain step the two statements of the proposition hold. Let us now analyze the next step.

Case 1. In the next step a vertex $i \in U$ is labeled. Any path to this vertex i passes through the matching edge $[i, j']$ and this edge has bottleneck length 0. Moreover, $j' \in X$ according to the algorithm. This implies by induction that $\alpha(j')$ is the bottleneck length of a b-shortest path from i_1 to i.

Case 2. In the next step a vertex $j \in V$ is labeled from a vertex $i \in \overline{X}$. If $\alpha(j) = \infty$, no path from i_1 to j was up to now possible in graph G. Therefore, the path via i fulfills the second statement of the proposition. If $\alpha(j)$ is finite, then the bottleneck length of the previous path to j and the bottleneck length of the path via vertex i are compared and $\alpha(j)$ becomes the shorter of these two lengths. Therefore, again, the second statement of the proposition holds.

Case 3. We scan a vertex $j \in V$ if $\alpha(j)$ is minimum among all unscanned vertices of V. Any path to j which does not use only vertices in $\overline{X} \cup X$ must pass through a labeled unscanned vertex j' with $\alpha(j') \geq \alpha(j)$. Thus it has at least bottleneck length $\alpha(j)$, showing that after scanning the first property of the proposition remains true. □

Corollary 6.10. *The modified Dijkstra algorithm finds a b-shortest augmenting path.*

As we stop the modified Dijkstra algorithm either when no augmenting path exists or as soon as a first unmatched vertex of V is scanned, we have $j \in X$ and, as j is unmatched, the path from i_1 to j is an augmenting path. Due to the scanning rule, any other augmenting path would have a bottleneck length of at least $\alpha(j)$. Thus we have found a b-shortest augmenting path. As every edge of G is considered at most once, the complexity of this modified Dijkstra method is $O(|E|)$.

Summarizing the ideas above, we can formulate an augmenting path algorithm for the bottleneck assignment problem, or, even more generally, for a maximum matching problem with minimum bottleneck cost, as follows. We start with a lower bound c^* for the optimal objective function value and grow b-shortest augmenting paths (which increase the value of c^*) until a maximum matching is reached.

Algorithm 6.3. Augmenting_LBAP.
Derigs and Zimmermann augmenting path algorithm.

let $G = (U, V; E)$ be a bipartite graph with edge lengths c_{ij} for $[i, j] \in E$;
find a lower bound c^* for the optimum objective function value;
comment: e.g., $c^* := \max(\max_{i\in U} \min_{j\in V} c_{ij},\ \max_{j\in V} \min_{i\in U} c_{ij})$;
find a maximum matching M in the graph $G[c^*]$ having an edge $[i, j]$ iff $c_{ij} \leq c^*$;
if $|M| = |U|$ **then stop** [**comment:** M is an optimum matching of cost c^*]
else let L be the set of unmatched vertices of U
endif

```
while L is nonempty do
    choose an arbitrary vertex i ∈ L;
    L := L \ {i};
    Dijkstra(i); [comment: the procedure returns a path P starting in i]
    if P ≠ nil then
        M := M ⊖ P;
        c* := max(c*, ℓ(P))
    endif
endwhile
comment: M is a maximum matching with minimum cost c*.
```

The b-shortest augmenting paths are found through the following procedure.

ALGORITHM 6.4. Procedure Dijkstra(i).
Modified Dijkstra algorithm for LBAP.

```
label all vertices j ∈ V by (α(j), β(j)) := (∞, nil);
R := V; [comment: R contains the unscanned vertices of V]
ᾱ(i) := c*, P := nil;
Label(i);
while R ≠ ∅ do
    find a vertex j₁ ∈ R with minimum α(j₁);
    if α(j₁) = ∞ then R := ∅
    else
        if j₁ is unmatched then
            find the path P induced by the vertex sequence (i, ..., β̄(β(j₁)), β(j₁), j₁);
            ℓ(P) := α(j₁);
            R := ∅
        else
            let [i₁, j₁] be the matching edge;
            label i₁ with (ᾱ(i₁), β̄(i₁)) = (α(j₁), j₁);
            R := R \ {j₁};
            Label(i₁)
        endif
    endif
endwhile
```

```
Procedure Label(i)
for each neighbor j ∈ R of i do
    if α(j) > max(ᾱ(i), c_ij) then
        α(j) := max(ᾱ(i), c_ij);
        β(j) := i
    endif
endfor
```

Let us illustrate the augmenting path method by means of the following example.

Example 6.11. Let the cost matrix of an LBAP be given as

$$C = \begin{pmatrix} 9 & 1 & 5 & 6 & 3 & 7 \\ 2 & 8 & 1 & 0 & 4 & 2 \\ 6 & 5 & 8 & 4 & 2 & 7 \\ 3 & 6 & 1 & 4 & 2 & 9 \\ 7 & 2 & 6 & 8 & 3 & 4 \\ 5 & 0 & 6 & 7 & 6 & 5 \end{pmatrix}.$$

We find $c^* = 2$ and the corresponding matching $M = \{[2, 1], [3, 5], [4, 3], [5, 2]\}$. Thus $L \subseteq U$ is $L = \{1, 6\}$. We choose the left vertex $i = 1$ and call Procedure Dijkstra (1). We get $R := \{1, 2, 3, 4, 5, 6\} = V$ and the first call to Procedure Label produces the labeling of all vertices of R by

$$\alpha := (9, 2, 5, 6, 3, 7), \quad \beta := (1, 1, 1, 1, 1, 1).$$

We find $j_1 = 2$, which is a matched vertex. Thus we determine $i_1 = 5$ and we label vertex 5 of U by $(\overline{\alpha}(5) := 2, \overline{\beta}(5) := 2)$. Now Label(5) produces, for vertices 1 and 6 of V,

$$(\alpha(1) := 7, \ \beta(1) := 5), \ (\alpha(6) := 4, \ \beta(6) := 5).$$

We get $\min \alpha(j) = 3$ for $j_1 = 5$. Vertex j_1 is again a matched vertex and we find $i_1 = 3$. We label vertex 3 of U by $(\overline{\alpha}(3) := 3, \overline{\beta}(3) := 5)$, and Label(3) gives, for vertices 1 and 4 of V,

$$(\alpha(1) := 6, \ \beta(1) := 3), \ (\alpha(4) := 4, \ \beta(4) := 3).$$

The minimum of $\alpha(j)$, $j \in R = \{1, 3, 4, 6\}$ is attained for $j_1 = 4$ with $\alpha(4) = 4$ and j_1 unmatched. Thus we get as augmenting path $P = ([1, 5], [3, 5], [3, 4])$ with $\ell(P) = 4$. An augmentation step yields the new matching $M = \{[1, 5], [2, 1], [3, 4], [4, 3], [5, 2]\}$ with $c^* := 4$.

We still have an unmatched vertex in U, namely, vertex 6. Thus we call Dijkstra(6) and set

$$\alpha := (5, 4, 6, 7, 6, 5), \quad \beta := (6, 6, 6, 6, 6, 6).$$

Now, j_1 becomes 2. This is a matched vertex. We get $i_1 = 5$ and Label(5) produces, for vertices 5 and 6 of V,

$$(\alpha(5) := 4, \ \beta(5) := 5), \ (\alpha(6) := 4, \ \beta(6) := 5).$$

We determine again $j_1 = 5$, which is matched by $i_1 = 1$, and Label(1) gives, for vertices 3 and 4 of V,

$$(\alpha(3) := 5, \ \beta(3) := 1), \ (\alpha(4) := 6, \ \beta(4) := 1).$$

Now we have $j_1 = 6$ and we found another augmenting path $P([6, 2], [5, 2], [5, 6])$ with $c^* = 4$. An augmenting step yields the perfect matching

$$M = \{[1, 5], [2, 1], [3, 4], [4, 3], [5, 6], [6, 2]\}$$

with $c^* := 4$, which is the optimal solution. ∎

Usually one can find by heuristics a very good matching at the beginning such that only a few augmentation steps are necessary to get a maximum matching with minimum bottleneck cost. Fortran codes for this method can be found in the paper by Derigs and Zimmermann [230], in the book by Burkard and Derigs [145], and in Carpaneto and Toth [167]. The implementations differ in the determination of a starting solution and in the applied data structures. One of the most efficient implementations is described in Derigs [225]. A thorough investigation on computational issues concerning LBAPs can be found in Pferschy [544]. Among others, Pferschy proposes an implementation using sparse subgraphs; see Section 6.2.5.

Gabow and Tarjan [293] as well as Punnen and Nair [563] proposed to combine threshold techniques with shortest augmenting path computations in order to design algorithms with a good running time complexity for the maximum matching problem with minimum bottleneck weight in bipartite graphs. Let a bipartite graph $G = (U, V; E)$ with $n = \min(|U|, |V|)$ and $|E| = m$ be given. Every edge $e \in E$ has a weight $c(e)$. We want to find a maximum matching in G with minimum bottleneck weight. By applying Hopcroft–Karp's procedure to the unweighted graph G, we find the cardinality N of a maximum matching in G. Next, in a first phase we apply a binary search to find a threshold value c^* such that the corresponding graph $G[c^*]$ allows a matching of cardinality at least equal to $N - n/k$. In order to find this threshold value, we test at most $O(\log n)$ graphs $G[c]$ by an approximate version of the Hopcroft–Karp algorithm, which stops if either the matching is maximum or has cardinality at least equal to $N - n/k$. Due to Proposition 3.12, this phase takes $O(km \log n)$ time. In the second phase we grow shortest augmenting paths to get a perfect matching with minimum bottleneck cost. Since at most n/k shortest augmenting paths are necessary to find an optimum matching, and every shortest augmenting path computation takes $O(m)$ time, this second phase can be completed in $O(mn/k)$ time. So we get a total complexity of $O(km \log n + mn/k)$ time for finding an optimal solution to the maximum matching problem with minimum bottleneck weight. Choosing

$$k = \sqrt{\frac{n}{\log n}}$$

yields a total time complexity of $O(m\sqrt{n \log n})$ for the Gabow–Tarjan algorithm.

Punnen and Nair [563] proposed the same technique. However, instead of the Hopcroft–Karp algorithm, they used the method by Alt, Blum, Mehlhorn, and Paul (see Section 3.4) to find an approximate matching M which differs from a maximum matching by at most $\sqrt{n}/k$ edges. Such a matching can be computed in $O(kn^{2.5}/\log n)$ time. So the first phase (binary search for an approximate matching) takes $O(kn^{2.5}/\log n)O(\log n) = O(kn^{2.5})$ time. In the second phase, at most $\sqrt{n}/k$ shortest augmenting paths have to be computed. Thus we get as total complexity $O(kn^{2.5} + m\sqrt{n}/k)$. Choosing

$$k = \sqrt{m}/n$$

yields a total time complexity of $O(n\sqrt{mn})$ for finding a maximum matching with minimum bottleneck cost. Table 6.1 summarizes the complexities of the various algorithms for the bottleneck assignment problem. The matrix threshold method refers to the algorithm described in Section 6.2.2. The table shows that the best time complexities are those of the matrix threshold algorithm for dense graphs and of the Gabow–Tarjan algorithm for sparse graphs. Punnen–Nair's method yields the best time complexity in between.

Table 6.1. *Complexity of threshold-based bottleneck assignment algorithms.*

Method	Complexity	$m = O(n)$	$m = O(n^2)$
Threshold	$O(n^2\sqrt{n/\log n})$	$O(n^2\sqrt{n/\log n})$	$O(n^2\sqrt{n/\log n})$
Gabow-Tarjan	$O(m\sqrt{n\log n})$	$O(n\sqrt{n\log n})$	$O(n^2\sqrt{n\log n})$
Punnen-Nair	$O(n\sqrt{mn})$	$O(n^2)$	$O(n^2\sqrt{n})$

6.2.5 Sparse subgraph techniques

The results on bipartite random graphs (see Section 3.7) already show that very sparse bipartite graphs without isolated vertices allow a perfect matching with high probability. For example, if we assume that the cost coefficients of a bottleneck assignment problem are *independent and identically distributed* (i.i.d.) and we choose only the three smallest entries in every row and column, then there exists a perfect matching in the subgraph consisting of these entries with a probability greater than $1 - (3/n)^4/122$ (see Theorem 3.26). In other words, for $n = 1{,}000$ there is no perfect matching using these entries with a probability less than 10^{-12}. In the following we take advantage of this property by thinning out the given cost matrix and solving a maximum matching problem with minimum bottleneck weight in a sparse graph instead of the originally given problem. Following a suggestion of Pferschy [543], we may describe an algorithm for the bottleneck assignment problem using sparse subgraph techniques as follows.

1. find the smallest entry in every row and column of cost matrix C, and let the corresponding edges form an edge set $\hat{E}$;

2. determine the $(n \log n)$th smallest value c^* of the cost coefficients and add $n \log n$ edges e with $c(e) \le c^*$ to $\hat{E}$;

3. solve a maximum matching problem with minimum bottleneck cost in the sparse bipartite graph with edge set $\hat{E}$, and let k be the cardinality of this matching;

4. if $k < n$ then execute a bottleneck assignment algorithm.

The value c^* in Step 2 can be determined in $O(n^2)$ time by using a linear median algorithm. The sparse graph constructed in Step 2 has $O(n \log n)$ edges. So, when we solve this sparse weighted matching problem by the method of Gabow and Tarjan, it takes $O(n^{3/2} \log^{3/2} n)$ time. Step 4 has to be executed only with a small probability. If we assume that the cost coefficients are i.i.d., then Theorem 3.25 tells us that this step is executed with a probability less than $O(1/\sqrt{n \log n})$. If we use the matrix threshold method for executing Step 4, then the expected running time of Step 4 is $O(n^2/\log n)$. In practice, one may use the solution found in Step 3 as a starting solution. Summarizing, we get the following result by Pferschy [543].

Proposition 6.12. *Let C be an $n \times n$ matrix with i.i.d. entries c_{ij}. Then the bottleneck assignment problem with cost matrix C can be solved in $O(n^2)$ expected time.*

$$
\begin{array}{c}
\\ \\ i \\ \\ k \\ \\
\end{array}
\begin{pmatrix}
 & j & & & l & \\
 & \vdots & & & \vdots & \\
\cdots & c_{ij} & \cdots & \cdots & c_{il} & \cdots \\
 & \vdots & & & \vdots & \\
\cdots & c_{kj} & \cdots & \cdots & c_{kl} & \cdots \\
 & \vdots & & & \vdots &
\end{pmatrix}
$$

Figure 6.4. *Four elements involved in the Monge property.*

The method outlined above not only provides a good bound in terms of expected time complexity but is also very efficient and simple to use. Armstrong and Jin [45] proposed a different algorithm, based on the concept of strong spanning trees (see Balinski [63]), which has worst-case time complexity $O(mn + n^2 \log n)$, where m denotes the number of edges in E.

6.2.6 Special cases

In certain cases an optimal solution of an LBAP is known beforehand, provided the cost matrix of the problem has a special structure.

Definition 6.13. *An $n \times n$ matrix $C = (c_{ij})$ is said to fulfill the* bottleneck Monge property *(or to be a* bottleneck Monge matrix*) if (see Figure 6.4)*

$$\max(c_{ij}, c_{kl}) \le \max(c_{il}, c_{kj}) \tag{6.10}$$

holds for all $1 \le i < k \le n$ and $1 \le j < l \le n$.

Matrix C fulfills the inverse bottleneck Monge property *(is an* inverse bottleneck Monge matrix*) if*

$$\max(c_{ij}, c_{kl}) \ge \max(c_{il}, c_{kj}) \tag{6.11}$$

holds for all $1 \le i < k \le n$ and $1 \le j < l \le n$.

As Klinz, Rudolf, and Woeginger [423] pointed out, it can be determined in $O(n^3)$ steps whether a given matrix fulfills the bottleneck Monge property. But often one knows in advance that a given matrix has the bottleneck Monge property. For example, if $a_1 \le a_2 \le \cdots \le a_n$ and $b_1 \ge b_2 \ge \cdots \ge b_n$, then the matrices $C = (c_{ij})$ defined by

$$
\begin{aligned}
c_{ij} &= a_i + b_j, \\
c_{ij} &= \max(a_i, b_j), \\
c_{ij} &= a_i \cdot b_j, \text{ provided } a_1, b_n \ge 0,
\end{aligned}
$$

satisfy (6.10). If a bottleneck Monge matrix is the cost matrix of an LBAP, we get the identical permutation as an optimal solution.

Proposition 6.14. *If the cost matrix of a bottleneck assignment problem fulfills the bottleneck Monge property, then the identical permutation is an optimal solution.*

If the cost matrix fulfills the inverse bottleneck Monge property, then the permutation $\varphi(i) = n + 1 - i$ *for* $i = 1, 2, \ldots, n$ *is an optimal solution.*

Proof. We show the first part of this proposition by successive transformations to the identical permutation. No transformation increases the objective function value. The second part can be shown by analogous arguments using the inverse bottleneck Monge property.

Let φ^* be an optimal solution of the bottleneck assignment problem whose cost matrix fulfills the bottleneck Monge property. For $i = 1, 2, \ldots, n$, perform the following step. If $\varphi^*(i) \neq i$, then let $l = \varphi^*(i)$, find k such that $\varphi^*(k) = i$, and interchange the two assignments by setting $\varphi^*(i) = i$ and $\varphi^*(k) = l$. According to (6.10) we have

$$\max(c_{ii}, c_{kl}) \leq \max(c_{il}, c_{ki})$$

which shows that the new permutation does not have a larger objective function value.

Since at each iteration i we have $\varphi^*(h) = h$ for $h = 1, 2, \ldots, i - 1$, the interchange never involves preceding rows and columns, so the resulting permutation remains optimal. □

An application of Proposition 6.14 to matrix $C = (a_i b_j)$, where $0 \leq a_1 \leq a_2 \leq \cdots \leq a_n$ and $b_1 \geq b_2 \geq \cdots \geq b_n \geq 0$, yields immediately a bottleneck version for Proposition 5.8.

Proposition 6.15. *Let*

$$0 \leq a_1 \leq a_2 \leq \cdots \leq a_n \quad \textit{and} \quad b_1 \geq b_2 \geq \cdots \geq b_n \geq 0.$$

Then for any permutation φ

$$\max_{1 \leq i \leq n} a_i b_i \leq \max_{1 \leq i \leq n} a_i b_{\varphi(i)} \leq \max_{1 \leq i \leq n} a_i b_{n+1-i}.$$

The bottleneck Monge property depends on a proper numbering of the rows and columns of matrix C. A matrix $C = (c_{ij})$ is called a *permuted bottleneck Monge matrix* if there are permutations φ and ψ of the rows and columns of C, respectively, such that $(c_{\varphi(i)\psi(j)})$ fulfills the bottleneck Monge property. Recognizing permuted bottleneck Monge matrices, however, is more difficult than solving the bottleneck assignment problem directly (see Klinz, Rudolf, and Woeginger [423]), so we do not discuss it here. For further results concerning the bottleneck Monge property see the survey article by Burkard, Klinz, and Rudolf [152].

6.2.7 Asymptotic results

In Section 5.1 we showed that for a linear assignment problem whose cost coefficients are i.i.d. random variables which obey an exponential distribution with mean 1, the expected optimum value equals $\pi^2/6$. Pferschy [543] investigated the asymptotic behavior of LBAPs. He showed that the expected optimal objective function value of an LBAP tends toward the lower end of the range of cost coefficients for any bounded distribution function when the size n of the problem increases. In particular he showed the following.

Proposition 6.16. *Let $F(x)$ be a continuous distribution function such that* $\sup\{x : F(x) < 1\} < \infty$. *The optimal value z_n of a random LBAP whose cost coefficients are i.i.d. according to the distribution function F satisfies*

$$\lim_{n\to\infty} \mathbb{E}(z_n) = \inf\{x : F(x) > 0\}.$$

In the case that the cost coefficients of the LBAP are uniformly distributed in [0, 1], Pferschy derived the following bounds for $\mathbb{E}(z_n)$.

Proposition 6.17. *Let $B(x, y)$ be the Beta function. Then we get for $n > 78$*

$$\mathbb{E}(z_n) < 1 - \left(\frac{2}{n(n+2)}\right)^{2/n} \frac{n}{n+2} + \frac{123}{610n}$$

and

$$\mathbb{E}(z_n) \geq 1 - nB(n, 1 + 1/n) = \frac{\ln n + 0.5749}{n} + O\left(\frac{\ln^2 n}{n^2}\right).$$

6.2.8 Variants and applications

Linear bottleneck assignment problems arise as subproblems in more complex combinatorial problems which model real-life applications.

Bus driver rostering

One of the most important optimization problems for public transport companies is the so-called *bus driver scheduling problem*, which involves the assignment of drivers to a selection of working shifts satisfying the service requirements. The solution methods from the literature usually adopt a decomposition technique which splits the problem into two subproblems: (i) find a set of working shifts that satisfy the service requirements, together with the possible side constraint (certain authors call this problem *the* bus driver scheduling problem); and (ii) assign each shift to one of the available drivers (*rostering problem*). The latter can be difficult to solve when the complex clauses of the union contract have to be taken into account. In most European countries one of the most important clauses requires that the mix of shift types assigned to the drivers in a given finite time horizon are very similar to each other. A simplified version of this rule associates a *weight* with each shift to represent a global measure of the corresponding driving effort, coming from a combination of several factors such as the driving duration (*platform time*) and the total elapsed time (*spread time*). The objective function of the rostering problem can thus be defined as a bottleneck function which requires one to minimize the maximum total weight of the shifts assigned to a driver.

Given an m-day time horizon, we can model the problem through a *layered digraph* (see Section 3.3) having one layer per day. Each layer k contains one vertex for each shift of day k and some additional dummy shifts introduced to give the same number of vertices, say, n, to each layer. The arcs of the digraph join each vertex i of layer k with each vertex j of the adjacent layer $k+1$ if and only if the two corresponding shifts can be performed by the same driver. Each arc starting from shift i of layer k ($k = 1, 2, \ldots, m-2$) is given the weight of shift i. Each arc from shift i of layer $m-1$ to shift j of layer m is given a weight equal to the sum of the weights of the two shifts i and j. A path from the first layer to layer m corresponds to a feasible m-day shifts assignment for a driver. The weight of the path is the weight of the assignment. A feasible solution of the rostering problem is thus given by n paths starting from layer 1 and ending at layer m, and the optimal solution is the one in which the weight of the heaviest path is a minimum. It is easily seen that, for $m = 2$, the problem is exactly an LBAP. It has been proved by Camerini [170] that it is $\mathcal{NP}$-hard for general m. The complexity of the problem is not known for fixed $m > 2$.

Carraresi and Gallo [170] proposed an effective heuristic algorithm for the rostering problem based on the iterative solution of LBAPs obtained by fixing the arc choices for all layers but two adjacent ones.

Product rate variation

Consider a mixed-model assembly line that produces units of n different types. Assume that the setup time when the type changes is negligible and that each unit requires the same production time for any type (the cycle time of the line, adopted, without loss of generality, as the unit of time). We are given the numbers of units u_i of type i ($i = 1, 2, \ldots, n$) to be produced in a time horizon of $U = \sum_{i=1}^{n} u_i$ time units. The general product rate variation problem (PRVP) is then to schedule the production so that the production rate of each type i is as close as possible, over time, to its ideal production rate $r_i = u_i/U$ ($i = 1, 2, \ldots, n$).

Let x_{ij} denote the total production of units of type i in time periods from 1 to j ($i = 1, 2, \ldots, n$; $j = 1, 2, \ldots, U$). Different *discrepancy functions* $f_i(x_{ij}, j)$ can be adopted for measuring the deviation between the actual and the ideal production. If the objective is

$$\min \max_{i,j} f_i(x_{ij}, j), \tag{6.12}$$

the problem is called the min-max PRVP. Bautista, Companys, and Corominas [76] showed how to compute a cost matrix C such that the problem can be solved as an LBAP over C. Moreno and Corominas [498] presented computational results for a number of different implementations of LBAP approaches to the min-max PRVP.

Object tracking

Let us consider n objects which are detected by two passive sensors at geographically different sites. Each sensor measures the horizon elevation angle and the azimuth bearing angle under which the object can be seen, i.e., it provides n lines in the three-dimensional space on which the objects lie. The location of every object is found by intersecting the appropriate lines. Due to small errors during the measurements, these lines might not meet. The pairing of the lines is modeled as follows: let c_{ij} be the smallest distance between the ith line provided by sensor 1 and the jth line provided by sensor 2. Solving an LBAP with cost matrix $C = (c_{ij})$ leads to very good results in practice (see Brogan [115] who used, however, LSAPs instead of the error-minimizing bottleneck problems).

A similar technique can be used for tracking missiles in space. If their locations at two different times t_1 and t_2 are known, we compute the (squared) Euclidean distances between any pair of old and new locations and solve the corresponding LBAP in order to match the points in the right way.

Categorized bottleneck assignment problems

We consider here two scheduling problems in which:

(i) n jobs must be assigned to m machines ($1 \le m \le n$);

(ii) the entries of an $m \times n$ cost matrix C give the processing time c_{ij} required for executing job j on machine i ($i = 1, 2, \dots, m$; $j = 1, 2, \dots, n$); and

(iii) for each machine i, an associated value k_i gives the number of jobs to be assigned to that machine (with $\sum_{i=1}^{m} k_i = n$).

In the first problem the objective is to minimize the maximum, over all machines, of the sum of the processing times of the jobs executed on it:

$$\min \quad \max_{1 \le i \le m} \sum_{j=1}^{n} c_{ij} x_{ij} \tag{6.13}$$

$$\text{s.t.} \quad \sum_{j=1}^{n} x_{ij} = k_i \qquad (i = 1, 2, \dots, m), \tag{6.14}$$

$$\sum_{i=1}^{m} x_{ij} = 1 \qquad (j = 1, 2, \dots, n), \tag{6.15}$$

$$x_{ij} \in \{0, 1\} \qquad (i = 1, 2, \dots, m;\ j = 1, 2, \dots, n). \tag{6.16}$$

In the second problem we consider, for each machine, the maximum processing time of a job executed on it and minimize the sum, over all machines, of such processing times:

$$\min \quad \sum_{i=1}^{m} \max_{1 \le j \le n} \{c_{ij} : x_{ij} = 1\} \tag{6.17}$$

$$\text{s.t.} \quad (6.14) - (6.16). \tag{6.18}$$

For $m = 1$, (6.13)–(6.16) is the LSAP and (6.17)–(6.18) is the LBAP.

Aggarwal, Tikekar, and Hsu [8] presented algorithms for the "exact" solution of both problems. Punnen [561] showed that these algorithms have pseudo-polynomial time complexity. He also proved that both problems are strongly $\mathcal{NP}$-hard for general m. Thus the algorithms in [8] cannot guarantee the optimal solution unless $\mathcal{P} = \mathcal{NP}$, although they may be used as good heuristics.

6.2.9 Software

The AP web page (http://www.siam.org/books/ot106/assignmentproblems.html) associated with this book provides codes for solving the bottleneck assignment problem. Such codes implement a modification of code LAPJV by Jonker and Volgenant [392] (see Section 4.9.1).

At the AP web page one can also execute, by visualizing the most relevant information, a didactic Java applet that implements the threshold algorithm of Section 6.2.2.

6.3 Algebraic assignment problem

Linear sum assignment problems can be solved by only *adding, subtracting,* and *comparing* the cost coefficients. Moreover, a careful reasoning shows that a subtraction $b - a$ occurs only in the case that the cost element a is not greater than b. Thus when we want to determine $b-a$, we ask for an element c such that $a+c = b$. These considerations lead to an algebraic model, originally introduced in Burkard, Hahn, and Zimmermann [150], which allows one to formulate and solve linear assignment problems within a general framework. The underlying algebraic structure is a so-called *d-monoid*, i.e., a totally ordered commutative semigroup $(H, *, \preceq)$ with composition $*$ and linear order relation $\preceq$. We require that the order relation is compatible with the composition $*$, i.e.,

$$a \preceq b \text{ implies } a * c \preceq b * c \text{ for all } c \in H.$$

Moreover, the d-monoid fulfills the following axiom.

Axiom 6.18. *If $a \preceq b$, then there exists an element $c \in H$ such that $a * c = b$.*

Without loss of generality we may always assume that the d-monoid has a neutral element e for which $a * e = a$ holds for all $a \in H$.

Given n^2 cost coefficients $c_{ij} \in H$, the *algebraic linear assignment problem* can be formulated as

$$\min_{\varphi} \ (c_{1\varphi(1)} * c_{2\varphi(2)} * \cdots * c_{n\varphi(n)}). \tag{6.19}$$

The objective function value of permutation φ with respect to the cost matrix C is denoted by

$$z[C, \varphi] = c_{1\varphi(1)} * c_{2\varphi(2)} * \cdots * c_{n\varphi(n)} \,.$$

Special examples for d-monoids are as follows.

- $H = \mathbb{R}$ with the addition as composition and the usual order relation. This model leads to LSAPs:

$$\min_{\varphi} \ (c_{1\varphi(1)} + c_{2\varphi(2)} + \cdots + c_{n\varphi(n)}).$$

- H is the set of extended real numbers $\overline{\mathbb{R}}$ (including $-\infty$) with the usual order relation. The composition is defined by $a * b = \max\ (a, b)$. This model leads to LBAPs:

$$\min_{\varphi} \max\{c_{1\varphi(1)}, c_{2\varphi(2)}, \ldots, c_{n\varphi(n)}\}.$$

- $H = \mathbb{R}^n$, the composition is the vector addition, and the order relation is the lexicographical order. This leads to lexicographical sum assignment problems.
- $H = \overline{\mathbb{R}} \times \mathbb{R}$, the order relation is the lexicographical order, and the composition is defined by

$$(a,b) * (\overline{a},\overline{b}) = \begin{cases} (a,b) & \text{if } a > \overline{a}, \\ (\overline{a},\overline{b}) & \text{if } a < \overline{a}, \\ (a, b+\overline{b}) & \text{if } a = \overline{a}. \end{cases}$$

This leads to the so-called *time-cost assignment problem*, in which a pair (a, b) represents the (time, cost) pair of an assignment. We want to find an assignment which first minimizes the maximum occurring time. Second, under all solutions which yield this time, a solution with minimum cost is to be found. Such problems occur if n customers have to be served as fast a possible (e.g., under emergency aspects), and then a cost minimal optimal solution should be found.

For solving the algebraic assignment problem we shall successively transform the cost matrix until we find a "zero" cover which has n elements (see Proposition 2.9). To realize this we have to describe suitable transformations of the cost matrix C (which we call *admissible transformations*), we have to define *zero elements* in H, and we have to explain in which way the admissible transformations can be applied in order to solve the algebraic assignment problem. We start with the definition of admissible transformations.

Definition 6.19. (Admissible transformation.) *A transformation T of the $n \times n$ matrix $C = (c_{ij})$ to the matrix $\overline{C} = (\overline{c}_{ij})$ is called* admissible *with* index $z(T)$ *if*

$$z[C,\varphi] = z(T) * z[\overline{C},\varphi]$$

for all $\varphi \in \mathcal{S}_n$.

When we perform an admissible transformation T after an admissible transformation S, we get again an admissible transformation. If S and T have the indices $z(S)$ and $z(T)$, respectively, their composition has index $z(S) * z(T)$.

The definition above states a property of admissible transformations, but says nothing about what an admissible transformation looks like. This is provided by the following theorem by Burkard, Hahn, and Zimmermann [150].

Theorem 6.20. (Admissible transformations for assignment problems.) *Let $I, J \subseteq \{1, 2, \ldots, n\}$, $m = |I| + |J| - n \geq 1$, and $c = \min\{c_{ij} : i \in I, j \in J\}$. Then the transformation $T = T(I, J)$ of the cost coefficients c_{ij} to new cost coefficients $\overline{c}_{ij}$ defined by*

$$\overline{c}_{ij} * c = c_{ij}, \quad \text{for } i \in I, j \in J, \tag{6.20}$$

$$\overline{c}_{ij} = c_{ij} * c, \quad \text{for } i \notin I, j \notin J, \tag{6.21}$$

$$\overline{c}_{ij} = c_{ij}, \quad \text{otherwise}, \tag{6.22}$$

is admissible with $z(T) = c * c * \cdots * c$, where the expression on the right-hand side contains m factors.

Note that we make use of Axiom 6.18 in the first line of the definition of $\overline{c}_{ij}$!

Proof. Let φ be an arbitrary permutation of $\{1, 2, \dots, n\}$ and let n_0 be the number of pairs $(i, \varphi(i))$ with $i \in I$ and $\varphi(i) \in J$. Similarly, let n_1 be the number of pairs $(i, \varphi(i))$ with $i \in I$ and $\varphi(i) \notin J$, or with $i \notin I$ and $\varphi(i) \in J$. Let n_2 be the number of pairs $(i, \varphi(i))$ with $i \notin I$ and $\varphi(i) \notin J$. Obviously, $n_0 + n_1 + n_2 = n$ and $2n_0 + n_1 = |I| + |J|$. This implies

$$n_0 - n_2 = |I| + |J| - n = m. \tag{6.23}$$

As the right-hand side in (6.23) does not depend on the particular permutation φ, (6.23) holds for all permutations on $\{1, 2, \dots, n\}$.

Let $C[i \in I]$ denote the composition of all cost coefficients $c_{i\varphi(i)}$ with $i \in I$, and let c^k denote $c * c * \cdots * c$ (k times). Using this notation, (6.20) yields for any permutation φ

$$z(C, \varphi) = C[i \in I] * C[i \notin I] = c^{n_0} * \overline{C}[i \in I] * C[i \notin I].$$

Now, (6.21) and (6.22) yield

$$c^{n_2} * C[i \notin I] = \overline{C}[i \notin I].$$

Thus we get for all permutations φ

$$z(C, \varphi) = c^m * z(\overline{C}, \varphi)$$

which shows that the transformation in Theorem 6.20 is feasible with index c^m. $\square$

In the semigroup H the role of 0-elements is replaced by so-called *dominated elements*. An element $a \in H$ is *dominated* by an element $z \in H$ if $a * z = z$. Thus in $(\mathbb{R}, +, \leq)$ the 0 is dominated by any other number. In $(\mathbb{R}, \max, \leq)$ every element a, $a \leq z$, is dominated by z. Now we can formulate the following optimality criterion.

Theorem 6.21. *Let T be an admissible transformation such that there exists a permutation $\hat{\varphi}$ with the following properties:*

1. $z(T) * \overline{c}_{ij} \geq z(T)$ *for all i, j;*
2. $z[\overline{C}, \hat{\varphi}] * z(T) = z(T)$.

Then $\hat{\varphi}$ is an optimal assignment with value $z(T)$.

The first property in Theorem 6.21 says that all cost coefficients $\overline{c}_{ij}$ are "nonnegative" (with respect to $z(T)$). The second property says that the current objective function value is already dominated by $z(T)$, i.e., it has value "0".

Proof. Let φ be an arbitrary permutation. According to Definition 6.19 and properties 1 and 2 of the thesis we get

$$z[C, \varphi] = z(T) * z[\overline{C}, \varphi] \geq z(T) = z(T) * z[\overline{C}, \hat{\varphi}] = z[C, \hat{\varphi}].$$

Therefore, $\hat{\varphi}$ is optimal. $\square$

Now we have to specify in which way the admissible transformations should be applied. This is stated in the following algorithm.

ALGORITHM 6.5. Algebraic_assignment.
Algorithm for solving linear algebraic assignment problems.

$z := e$, where e denotes the neutral element of semigroup H;
comment: row reduction in matrix C;
for $k := 1$ **to** n **do**
  perform an admissible transformation $T(I, J)$ with $I = \{k\}$ and $J = \{1, 2, \ldots, n\}$;
  replace matrix C by the transformed matrix $\overline{C}$;
  $z := z * z(T(I, J))$
endfor;
comment: column reduction;
for $k := 1$ **to** n **do**
  perform an admissible transformation $T(I, J)$ with $I = \{1, 2, \ldots, n\}$ and $J = \{k\}$;
  replace matrix C by the transformed matrix $\overline{C}$;
  $z := z * z(T(I, J))$
endfor;
find a minimum cover of the elements c_{ij} for which $c_{ij} * z = z$;
$I := \{i : \text{row } i \text{ is uncovered}\}$;
$J := \{j : \text{column } j \text{ is uncovered}\}$;
while $|I| + |J| > n$ **do**
  perform an admissible transformation $T(I, J)$;
  replace matrix C by the transformed matrix $\overline{C}$;
  $z := z * z(T(I, J))$;
  find a minimum cover of the elements c_{ij} for which $c_{ij} * z = z$;
  $I := \{i : \text{row } i \text{ is uncovered}\}$;
  $J := \{j : \text{column } j \text{ is uncovered}\}$
endwhile
comment: the current cover corresponds to a maximum matching of value z.

After performing the row reduction, all elements in the transformed matrix are "nonnegative" with respect to z, namely, $\overline{c}_{ij} * z \geq z$. After the column reduction, every row and column in the transformed cost matrix contains at least one element which is dominated by z. All other elements remain nonnegative with respect to z. Then a bipartite graph $G = (U, V; E)$ is defined, where set U corresponds to the rows of matrix C and set V corresponds to the columns of matrix C. Graph G contains an edge (i, j) if and only if the element $\overline{c}_{ij}$ is dominated by the current objective function value z. A minimum cover of the dominated elements is obtained by determining a maximum matching in G (see König's theorem, Theorem 2.7). If the size of this minimum cover is less than n (i.e., if the maximum matching has cardinality less than n), a further admissible transformation is performed. The definition of sets I and J in the "while" loop guarantees that after each admissible transformation at least one further cost coefficient is dominated by the current objective function value.

It is rather straightforward to show that this algorithm yields an optimal solution of the algebraic assignment problem after at most $O(n^2)$ admissible transformations.

If the composition $*$ is specialized to "+", Algorithm 6.5 becomes a variant of the Hungarian method. If the composition is specialized to the *max* operation, then we obtain the bottleneck assignment problem and the above algorithm is a variant of the threshold method.

If the semigroup $(H, *, \preceq)$ fulfills the *weak cancellation rule*, namely,

$$\text{if } a * c = b * c, \text{ then either } a = b \text{ or } a * c = b, \tag{6.24}$$

then an algebraic assignment problem can be solved in $O(n^3)$ time (see Burkard and Zimmermann [158] as well as Frieze [283]). For further results in this direction, consult the survey on algebraic optimization by Burkard and Zimmermann [159].

Finally, we address a case where the solution of an algebraic assignment problem can be stated explicitly. We say that a cost matrix $C = (c_{ij})$ fulfills the *algebraic Monge property* if it fulfills the following conditions:

$$c_{ij} * c_{kl} \preceq c_{il} * c_{kj}, \quad \text{for } 1 \le i < k \le n, \ 1 \le j < l \le n. \tag{6.25}$$

As in the sum and bottleneck cases of Monge matrices, the following can be shown.

Theorem 6.22. *If the cost matrix C of an algebraic linear assignment problem fulfills the algebraic Monge property (6.25), then the identical permutation is an optimal solution.*

The problem of recognizing *permuted* algebraic Monge matrices is rather subtle in general. It can be shown that this problem is $\mathcal{NP}$-hard if $n \ge 3$ and the ordered semigroup fulfills no additional property. It becomes, however, polynomially solvable if, for instance, a weak cancellation rule (6.24) is fulfilled. For details, see Burkard, Klinz, and Rudolf [152].

6.4 Sum-k assignment problem

Given an $n \times n$ cost matrix $C = (c_{ij})$ and a value k not greater than n, the *sum-k assignment problem* is to assign each row to a different column so that the sum of the k largest selected costs is a minimum. The two extreme cases, $k = n$ and $k = 1$, coincide, respectively, with LSAP and with the bottleneck assignment problem (see Section 6.2). This problem cannot be modeled as an algebraic assignment problem. Grygiel [349] investigated it within an algebraic framework and designed an $O(n^5)$ algorithm in the case of real cost coefficients.

6.5 Balanced assignment problem

Let $C = (c_{ij})$ be a given real $n \times n$ matrix and φ be an arbitrary permutation of the set $\{1, 2, \dots, n\}$. We call

$$\mathrm{sp}(\varphi) = \max_i \{c_{i\varphi(i)}\} - \min_i \{c_{i\varphi(i)}\} \tag{6.26}$$

the *spread* of solution φ. The *balanced assignment problem*, introduced by Martello, Pulleyblank, Toth, and de Werra [480] in the more general framework of balanced optimization problems, minimizes the spread of an assignment solution. The balanced assignment problem is closely related to bottleneck assignment problems. Given a real $n \times n$ matrix $C = (c_{ij})$, the balanced assignment problem can be formulated as

$$\min_{\varphi} \ \mathrm{sp}(\varphi). \tag{6.27}$$

For solving this problem, the following algorithm can be used, which is a modification of the original proposal by Martello, Pulleyblank, Toth, and de Werra.

ALGORITHM 6.6. Balanced_assignment.
Algorithm for the balanced assignment problem.

let an $n \times n$ cost matrix $C = (c_{ij})$ be given;
let φ be an optimal solution of the LBAP with cost matrix C;
$l := \min_i\{c_{i\varphi(i)}\}$;
$u := \max_i\{c_{i\varphi(i)}\}$;
$\mathrm{sp}(\varphi) := u - l$;
$Q := \{(i, j) : l < c_{ij}\}$;
fail := `false`;
while $\mathrm{sp}(\varphi) > 0$ **and** *fail* = `false` **do**
    $E := \{(i, j) \in Q : c_{ij} \le u\}$;
    find a minimum vertex cover VC in the bipartite graph with edge set E;
    if $|VC| = n$ **then**
        let φ be the corresponding perfect matching;
        $l := \min_i\{c_{i\varphi(i)}\}$;
        $\mathrm{sp}(\varphi) := u - l$
    else
        $\overline{Q} := \{(i, j) \in Q : (i, j) \text{ is uncovered}\}$;
        if $\overline{Q} \neq \emptyset$ **then**
            $u := \min\{c_{ij} : (i, j) \in \overline{Q}\}$;
            $l := u - \mathrm{sp}(\varphi)$
        else *fail* := `true`
    endif;
    $Q := Q \setminus \{(i, j) \in Q : c_{ij} \le l\}$
endwhile

The correctness of the algorithm follows from the following facts. The optimal solution of the corresponding linear bottleneck problem is the first candidate for an optimal solution of the balanced assignment problem. Observe that any feasible assignment must use elements of value at least equal to u and that no lower spread can be obtained by using elements of value less than or equal to l. In the next steps the value u is kept fixed and the spread is minimized by iteratively increasing the value of l. Only when this is not further possible, i.e., with the current l and u values no perfect matching exists, the value of u is increased. As in this case every feasible solution of the assignment problem must contain an uncovered element, we can increase u to the smallest uncovered element. At the same time we may forbid all elements which would lead to the same or a larger spread than that already found. An optimal solution is reached if either $\mathrm{sp}(\varphi) = 0$ or u cannot be increased any more.

With this implementation the algorithm breaks ties in the solution value by selecting the balanced solution that minimizes the minimum (and hence the maximum) cost. For cases where the solution that maximizes these values is preferred in case of a tie, it is enough to execute the algorithm on a transformed instance having costs $\widetilde{c}_{ij} = -c_{ij}$.

The following example illustrates algorithm Balanced_assignment.

Example 6.23. Let the cost matrix of a balanced assignment problem be given as

$$C = \begin{pmatrix} 8 & 9 & 3 & 2 \\ 4 & 7 & 3 & 8 \\ 0 & 8 & 10 & 4 \\ 2 & 5 & 8 & 3 \end{pmatrix}.$$

The bottleneck assignment problem with this cost matrix has the optimum value 5. An optimal solution is $\varphi = (4, 3, 1, 2)$ with $l = 0$, $u = 5$, and $\text{sp}(\varphi) = 5$. We forbid c_{31} (set Q contains all elements but $(3, 1)$) so the resulting bipartite graph has a vertex cover of cardinality n corresponding to the matching $\varphi = (3, 1, 4, 2)$ with $l = 3$ and $\text{sp}(\varphi) = 2$. We remove from Q all elements of value not greater than 3, so the nonforbidden elements are

$$C = \begin{pmatrix} 8 & 9 & - & - \\ 4 & 7 & - & 8 \\ - & 8 & 10 & 4 \\ - & 5 & 8 & - \end{pmatrix}.$$

The next bipartite graph only contains the three edges with value 4 or 5. The last three rows provide a minimum vertex cover, and thus $\overline{Q} = \{(1, 1), (1, 2)\}$. We increase u to 8 and l to 6, and we forbid all elements with a value not greater than 6. The edges of the new bipartite graph correspond to the values 7 and 8 in the following matrix:

$$C = \begin{pmatrix} 8 & 9 & - & - \\ - & 7 & - & 8 \\ - & 8 & 10 & - \\ - & - & 8 & - \end{pmatrix}.$$

This graph has a vertex cover of cardinality n corresponding to a perfect matching $\varphi = (1, 4, 2, 3)$ with $l = 8$ and $\text{sp}(\varphi) = 0$. Therefore, we have reached the optimal solution. ■

If, at each iteration, the vertex cover is determined from scratch, using the $O(n^{2.5})$ Hopcroft–Karp algorithm of Section 3.3, the overall time complexity of Balanced_assignment is $O(n^{4.5})$. This can be improved if, at each iteration, the new vertex cover is obtained from the current partial solution by completing the corresponding nonperfect matching through augmenting path techniques (see Section 3.2). In the worst case this implementation will perform an augmentation for each element that is forbidden when l is increased and an augmentation for each new edge that enters the bipartite graph when u is increased. Since a single augmentation requires $O(n^2)$ time, the resulting time complexity is $O(n^4)$.

Balanced linear assignment problems can be used in a heuristic for decomposing traffic matrices arising from TDMA systems (see Section 3.8.2) in at most n switch modes (see Balas and Landweer [51], who used LBAPs instead of balanced linear assignment problems in this context). Given the traffic matrix T, let φ^* be an optimal solution of the balanced assignment problem with coefficient matrix T. We set $\lambda_1 = \max_{1\leq i\leq n} t_{i\varphi^*(i)}$ and forbid the elements $t_{i\varphi^*(i)}$, $i = 1, 2, \ldots, n$. With this new matrix we solve the next balanced assignment problem and determine λ_2. We continue in this way until all elements of T are forbidden. The rational behind this approach is that during the application of a fixed switch mode all involved stations should have about the same workload.

6.6 Lexicographic bottleneck assignment problem

Let $C = (c_{ij})$ be a given real $n \times n$ matrix and let φ be a permutation of the set $\{1, 2, \ldots, n\}$. By ordering the n elements $c_{i\varphi(i)}$ decreasingly, we get a vector $w(\varphi) = (w_1, w_2, \ldots, w_n)$ where w_k corresponds to the kth largest cost element in solution φ.

Example 6.24. Consider the matrix

$$C = \begin{pmatrix} 8 & 9 & 3 & 2 \\ 4 & 7 & 3 & 8 \\ 0 & 8 & 10 & 4 \\ 2 & 5 & 8 & 3 \end{pmatrix}$$

and the permutation $\varphi = (3, 1, 4, 2)$, which yields

$$c_{1\varphi(1)} = 3, \ c_{2\varphi(2)} = 4, \ c_{3\varphi(3)} = 4, \ c_{4\varphi(4)} = 5.$$

Thus $w(\varphi)$ is the vector

$$w(\varphi) = (5, 4, 4, 3).$$

When we consider the permutation $\psi = (4, 3, 1, 2)$, we get

$$w(\psi) = (5, 3, 2, 0).$$

Note that both φ and ψ are optimal solutions of the bottleneck assignment problem with cost matrix C. Solution ψ, however, yields a lexicographically smaller vector w and is therefore preferred, in this context, to the solution described by φ. ∎

The problem of finding a permutation φ which yields a lexicographic minimal vector $w(\varphi)$ is known as the *lexicographic bottleneck assignment problem (LexBAP)*. Discrete optimization problems with a lexicographic bottleneck objective were originally introduced by Burkard and Rendl [155]. Their approach was later improved by Sokkalingam and Aneja [615], who showed that linear programming duality helps in solving this problem. We mainly follow here the approach of the latter two authors. For the sake of clarity we first give a simplified version of the algorithm.

As in the case of bottleneck assignment problems, we can assume that C has the entries $0, 1, 2, \ldots, m-1$. First, we solve a bottleneck assignment problem with cost matrix C. Its optimal value z defines the largest (first) entry of the vector $w(\varphi)$. Since in a lexicographic minimal solution as few entries as possible have the value z, we next solve an LSAP with the cost matrix $D[z]$ defined by

$$d_{ij} = \begin{cases} \infty & \text{if } c_{ij} > z, \\ 1 & \text{if } c_{ij} = z, \\ 0 & \text{otherwise.} \end{cases} \quad (6.28)$$

The set of optimal solutions of this problem contains the set of the optimal solutions of the LexBAP, and possibly other solutions. Since LSAPs can be written as a linear program, we can characterize the optimal solutions by complementary slackness conditions. Consider the linear assignment problem

$$\begin{aligned} \min \quad & \sum_{i=1}^{n} \sum_{j=1}^{n} d_{ij} \\ \text{s.t.} \quad & \sum_{j=1}^{n} x_{ij} = 1 \quad (i = 1, 2, \ldots, n), \\ & \sum_{i=1}^{n} x_{ij} = 1 \quad (j = 1, 2, \ldots, n), \\ & x_{ij} \geq 0 \qquad (i, j = 1, 2, \ldots, n). \end{aligned}$$

The dual linear program reads as follows:

$$\begin{aligned} \max \quad & \sum_{i=1}^{n} u_i + \sum_{j=1}^{n} v_j \\ \text{s.t.} \quad & u_i + v_j \leq d_{ij} \qquad (i, j = 1, 2, \ldots, n). \end{aligned}$$

According to the theory of complementary slackness (see Section 4.1.2), the pair $(x, (u, v))$ of optimal solutions of these two linear programs is characterized by the condition

$$d_{ij} > u_i + v_j \text{ implies } x_{ij} = 0. \quad (6.29)$$

To guarantee that in the following subproblems we only consider solutions which are optimal for cost matrix $D[z]$, we forbid all entries which fulfill $d_{ij} > u_i + v_j$. Let F be the set of forbidden entries: at this stage, F contains all elements (i, j) for which $c_{ij} > z$ or $d_{ij} > u_i + v_j$.

At the next step of our solution procedure we set up a new cost matrix $D[z-1]$ defined by

$$d_{ij} = \begin{cases} \infty & \text{if } (i, j) \in F, \\ 1 & \text{if } c_{ij} = z - 1, \\ 0 & \text{otherwise.} \end{cases} \quad (6.30)$$

Note in particular that all previous entries $d_{ij} = 1$ with $(i, j) \notin F$ become $d_{ij} = 0$. We again determine an optimal dual solution of the LSAP with cost matrix $D[z-1]$, forbid new entries according to the complementary slackness conditions, and set up a new cost matrix $D[z-2]$. We proceed in this way until we find an optimal solution for the LSAP with cost matrix $D[0]$, which is the optimal solution of the LexBAP. Summarizing, we get the following algorithm.

Algorithm 6.7. Lex_BAP.
Algorithm for the LexBAP.

let an $n \times n$ cost matrix $C = (c_{ij})$ with entries $0, 1, \dots, n^2 - 1$ be given;
let z be the optimal objective function value of the LBAP with cost matrix C;
$F := \{(i, j) : c_{ij} > z\}$;
while $z \geq 0$ **do**
  if $(i, j) \in F$ **then** $d_{ij} := \infty$
  else
    if $c_{ij} = z$ **then** $d_{ij} := 1$ **else** $d_{ij} := 0$
  endif;
  find an optimal dual solution (u, v) for the LSAP with cost matrix (d_{ij});
  $F := F \cup \{(i, j) : d_{ij} > u_i + v_j\}$;
  $z := z - 1$
endwhile
let φ be the optimal primal solution of the last LSAP.
comment: φ is also an optimal solution of the given LexBAP.

Example 6.25. Consider the matrix

$$C = \begin{pmatrix} 6 & 6 & 4 & 2 & 1 \\ 0 & 6 & 5 & 1 & 1 \\ 2 & 5 & 6 & 3 & 1 \\ 5 & 2 & 3 & 4 & 6 \\ 6 & 6 & 2 & 0 & 1 \end{pmatrix}.$$

The bottleneck assignment problem with this cost matrix has an optimal solution $\varphi = (4, 5, 1, 2, 3)$, with optimum value $z = 2$. We set $F = \{(i, j) : c_{ij} > 2\}$ and obtain as the cost matrix for the first LSAP:

$$D[2] = \begin{pmatrix} \infty & \infty & \infty & 1 & 0 \\ 0 & \infty & \infty & 0 & 0 \\ 1 & \infty & \infty & \infty & 0 \\ \infty & 1 & \infty & \infty & \infty \\ \infty & \infty & 1 & 0 & 0 \end{pmatrix}.$$

An LSAP algorithm yields an optimal dual solution

$$u = (1, 0, 1, 1, 0) \text{ and } v = (0, 0, 1, 0, -1).$$

As $d_{25} = d_{55} = 0 > u_2 + v_5 = u_5 + v_5 = -1$, we add the elements (2, 5) and (5, 5) to the set F of forbidden entries. Now we set $z = 1$ and determine an optimal dual solution of the

LSAP with cost matrix $D[1]$ defined as

$$D[1] = \begin{pmatrix} \infty & \infty & \infty & 0 & 1 \\ 0 & \infty & \infty & 1 & \infty \\ 0 & \infty & \infty & \infty & 1 \\ \infty & 0 & \infty & \infty & \infty \\ \infty & \infty & 0 & 0 & \infty \end{pmatrix}.$$

Note that $c_{55} = 1$, but $d_{55} = \infty$ since $(5, 5) \in F$. An optimal dual solution is given by

$$u = (0, 0, 0, 0, 0) \text{ and } v = (0, 0, 0, 0, 1).$$

Since $d_{24} = 1 > u_2 + v_4 = 0$, we add element $(2, 4)$ to the set F of forbidden entries. Now $z = 0$ is reached, so we solve an LSAP with matrix

$$D[0] = \begin{pmatrix} \infty & \infty & \infty & 0 & 0 \\ 1 & \infty & \infty & \infty & \infty \\ 0 & \infty & \infty & \infty & 0 \\ \infty & 0 & \infty & \infty & \infty \\ \infty & \infty & 0 & 1 & \infty \end{pmatrix}.$$

The optimal primal solution of this problem is the assignment $\varphi = (4, 1, 5, 2, 3)$, which yields the optimal vector $w(\varphi) = (2, 2, 2, 1, 0)$. ∎

The reduction of z by 1 in the "while" loop of the algorithm is simple, but it is not the fastest implementation. For the cost matrix

$$C = \begin{pmatrix} 24 & 23 & 22 & 21 & 20 \\ 19 & 18 & 17 & 16 & 15 \\ 14 & 13 & 12 & 11 & 10 \\ 9 & 8 & 7 & 6 & 5 \\ 4 & 3 & 2 & 1 & 0 \end{pmatrix}$$

we get as optimal value for the bottleneck assignment problem $z = 20$ (the smallest entry in the first row). Then we would have to solve four linear assignment problems, namely, one each for $z = 20, 19, 18$, and 17, where we would add the elements $(2, 1)$, $(2, 2)$, and $(2, 3)$ to the set F of forbidden elements. Instead, one can solve an LBAP for the matrix

$$C = \begin{pmatrix} \infty & \infty & \infty & \infty & 0 \\ 19 & 18 & 17 & 16 & 15 \\ 14 & 13 & 12 & 11 & 10 \\ 9 & 8 & 7 & 6 & 5 \\ 4 & 3 & 2 & 1 & 0 \end{pmatrix},$$

which immediately yields the next relevant value for z, namely, $z = 16$. All entries with values $16 < c_{ij} < 20$ can now be forbidden, i.e., added to set F. With this modification, the algorithm has time complexity $O(n^4)$, since at each of the n iterations at least one entry of vector $w(\varphi)$ is fixed. In one iteration a bottleneck assignment and a sum assignment problem have to be solved, which can be done in $O(n^3)$ time.

Della Croce, Paschos, and Tsoukias [201] described another approach for lexicographic bottleneck problems. Let k be the number of different cost values. Their algorithm replaces the original cost coefficients by 0-1 vectors with k components: If c is the rth largest cost coefficient among the k different values, then c is replaced by the vector $(\gamma_1, \dots, \gamma_k)$ with

$$\gamma_i = \begin{cases} 1 & \text{if } i = r, \\ 0 & \text{otherwise.} \end{cases}$$

Then the problem is solved as a lexicographic sum problem. In the case of assignment problems, the lexicographic sum assignment problem can be solved by the methods described in Section 6.3. Moreover, for LexBAPs we can choose k as the number of different cost coefficients less than or equal to the optimal value of the bottleneck assignment problem. This leads to an algorithm of time complexity $O(n^3k)$ for the LexBAP.

Chapter 7

Quadratic assignment problems: Formulations and bounds

7.1 Introduction

Quadratic assignment problems (QAPs) belong to the most difficult combinatorial optimization problems. Because of their many real-world applications, many authors have investigated this problem class. For a monograph on QAPs, see the book by Çela [176]. A volume with selected papers on this topic was edited by Pardalos and Wolkowicz [536]. Some of the more recent surveys are Burkard [130], Pardalos, Rendl, and Wolkowicz [535], Burkard, Çela, Pardalos, and Pitsoulis [142], Rendl [574], and Loiola, Maia de Abreu, Boaventura-Netto, Hahn, and Querido [464]. The Quadratic Assignment Problem Library (QAPLIB), set up by Burkard, Karisch, and Rendl [151] and currently maintained by P. Hahn at http://www.seas.upenn.edu/qaplib/, contains not only many test examples with computational results, but also a detailed bibliography on this topic and a survey on the latest results.

7.1.1 Models and applications

The QAP was introduced by Koopmans and Beckmann [427] in 1957 as a mathematical model for the location of indivisible economical activities. We want to assign n facilities to n locations with the cost being proportional to the flow between the facilities multiplied by the distances between the locations plus, eventually, costs for placing the facilities at their respective locations. The objective is to allocate each facility to a location such that the total cost is minimized. We can model this assignment problem by means of three $n \times n$ matrices:

$A = (a_{ik})$, where a_{ik} is the flow from facility i to facility k;

$B = (b_{jl})$, where b_{jl} is the distance from location j to location l;

$C = (c_{ij})$, where c_{ij} is the cost of placing facility i at location j.

The QAP in Koopmans–Beckmann form can now be written as

$$\min_{\varphi \in \mathcal{S}_n} \left(\sum_{i=1}^{n} \sum_{k=1}^{n} a_{ik} b_{\varphi(i)\varphi(k)} + \sum_{i=1}^{n} c_{i\varphi(i)} \right), \tag{7.1}$$

where, as usual, $\mathcal{S}_n$ is the set of all permutations of the integers $1, 2, \ldots, n$. Each individual product $a_{ik} b_{\varphi(i)\varphi(k)}$ is the transportation cost caused by assigning facility i to location $\varphi(i)$ and facility k to location $\varphi(k)$. Thus each term $c_{i\varphi(i)} + \sum_{k=1}^{n} a_{ik} b_{\varphi(i)\varphi(k)}$ is the total cost given, for facility i, by the cost for installing it at location $\varphi(i)$, plus the transportation costs to all facilities k, if installed at locations $\varphi(1), \varphi(2), \ldots, \varphi(n)$.

An instance of the QAP with input matrices A, B, and C is denoted by $QAP(A, B, C)$. If there is no linear term (hence, no matrix C), we just write $QAP(A, B)$.

In many cases matrix B fulfills the triangle inequality $b_{jl} + b_{lr} \geq b_{jr}$ for all j, l, and r. In these cases $QAP(A, B)$ is called *metric QAP*.

A more general version of the QAP was considered by Lawler [445]. Lawler introduced a four-index cost array $D = (d_{ijkl})$ instead of the three matrices A, B, and C and obtained the general form of a QAP as

$$\min_{\varphi \in \mathcal{S}_n} \sum_{i=1}^{n} \sum_{k=1}^{n} d_{i\varphi(i)k\varphi(k)}. \tag{7.2}$$

The relationship with the Koopmans–Beckmann problem is

$$\begin{aligned} d_{ijkl} &= a_{ik} b_{jl} \qquad (i, j, k, l = 1, 2, \ldots, n;\ i \neq k \text{ or } j \neq l);\\ d_{ijij} &= a_{ii} b_{jj} + c_{ij} \quad (i, j = 1, 2, \ldots, n). \end{aligned}$$

Example 7.1. We consider a Koopmans–Beckmann problem $QAP(A, B, C)$ with $n = 3$ and input matrices

$$A = \begin{pmatrix} 1 & 2 & 4 \\ 3 & 4 & 5 \\ 5 & 6 & 1 \end{pmatrix}, \quad B = \begin{pmatrix} 2 & 3 & 6 \\ 1 & 4 & 7 \\ 5 & 6 & 2 \end{pmatrix}, \quad \text{and } C = \begin{pmatrix} 9 & 7 & 9 \\ 6 & 5 & 7 \\ 8 & 9 & 8 \end{pmatrix}.$$

Given a permutation, say, $\varphi = (2, 1, 3)$, we can easily compute the corresponding objective function value by first permuting the rows and columns of B according to φ, as in

$$B_\varphi = (b_{\varphi(i)\varphi(k)}) = \begin{pmatrix} 4 & 1 & 7 \\ 3 & 2 & 6 \\ 6 & 5 & 2 \end{pmatrix},$$

and then deriving from (7.1)

$$z = (4 + 2 + 28) + (9 + 8 + 30) + (30 + 30 + 2) + (7 + 6 + 8) = 164.$$

In order to obtain the equivalent Lawler's form, we need to define the four-index matrix D. Let us represent it through n^2 square matrices D^{ij} of order n. Matrix D^{ij} is formed by the elements d_{ijkl} with fixed indices i and j and variable indices $k, l = 1, 2, \dots, n$, for example,

$$D^{11} = \begin{pmatrix} 11 & 3 & 6 \\ 4 & 6 & 12 \\ 8 & 12 & 24 \end{pmatrix}.$$

In order to compute the objective function value corresponding to the same permutation $\varphi = (2, 1, 3)$, we need the matrices D^{12}, D^{21}, and D^{33}:

$$D^{12} = \begin{pmatrix} 1 & 11 & 7 \\ 2 & 8 & 14 \\ 4 & 16 & 28 \end{pmatrix}, D^{21} = \begin{pmatrix} 6 & 9 & 18 \\ 14 & 12 & 24 \\ 10 & 15 & 30 \end{pmatrix}, D^{33} = \begin{pmatrix} 25 & 30 & 10 \\ 30 & 36 & 12 \\ 5 & 6 & 10 \end{pmatrix}.$$

Note that each matrix D^{ij} has the cost $c_{i\varphi(i)}$ added to the element stored in row i and column j. We obtain

$$z = (11 + 2 + 28) + (9 + 14 + 30) + (30 + 30 + 10) = 164. \quad \blacksquare$$

It is astonishing how many real-life applications can be modeled as QAPs. An early natural application in location theory was used by Dickey and Hopkins [233] in a *campus planning* model. The problem consists of planning the sites of n buildings on a campus, where b_{jl} is the distance from site j to site l and a_{ik} is the traffic intensity between building i and building k. The objective is to minimize the total weekly walking distance between the buildings. Another early application in this area was described by Steinberg [623] who minimized the number of connections in a backboard wiring problem, nowadays an outdated technology of historical interest only. Elshafei [251] used QAPs in hospital planning. Bos [109] described a related problem for forest parks.

In addition to facility location, QAPs appear in a variety of applications such as computer manufacturing, scheduling, process communications, and turbine balancing. In the field of ergonomics Pollatschek, Gershoni, and Radday [551] as well as Burkard and Offermann [153] showed that QAPs can be applied to *typewriter keyboard design*. The problem is to arrange the keys on a keyboard so as to minimize the time needed to write texts. Let the set of integers $N = \{1, 2, \dots, n\}$ denote the set of symbols to be arranged. Then a_{ik} denotes the frequency of the appearance of the ordered pair of symbols i and k. The entries of the distance matrix b_{jl} are the times needed to press the key in position l after pressing the key in position j. A permutation $\varphi \in \mathcal{S}_n$ describes an assignment of symbols to keys. An optimal solution φ^* for the QAP minimizes the average time for writing a text. A similar application related to ergonomic design is the development of control boards in order to minimize eye fatigue by McCormick [484].

The *turbine runner problem* was originally studied by Bolotnikov [106] and Stoyan, Sokolovskii, and Yakovlev [625]. The blades of a turbine, which due to manufacturing have slightly different masses, should be welded on the turbine runner such that the center of gravity coincides with the axis of the runner. Mosevich [499], Schlegel [598], and Laporte and Mercure [446] applied the QAP model to this problem. It has been shown that the minimization of the distance between the center of gravity and the axis of the runner is $\mathcal{NP}$-hard, whereas the maximization can be obtained in polynomial time (see Burkard, Çela, Rote, and Woeginger [143] and Section 8.4).

Other applications concern the ranking of archeological data (Krarup and Pruzan [430]), the ranking of a team in a relay race (Heffley [366]), the scheduling of parallel production lines (Geoffrion and Graves [309]), the analysis of chemical reactions for organic compounds (Ugi, Bauer, Brandt, Friedrich, Gasteiger, Jochum, and Schubert [644]), the arrangement of numbers around a dartboard under risk maximization (Eiselt and Laporte [254]), the arrangement of probes on microarray chips (de Carvalho Jr. and Rahmann [173]), and combinatorial data analysis (Hubert [380]). Winter and Zimmermann used a QAP for modeling the shunting of trams in a storage yard [665].

A number of the problems studied in combinatorial optimization are special cases of QAPs. We shortly describe some of them in the next sections.

7.1.2 Traveling salesman problem

The most prominent special case of QAP is the *traveling salesman problem* (TSP): given n cities and a matrix $A = (a_{ik})$ of distances between the cities, find a cyclic permutation ψ (i.e., a permutation representing a single cycle, see Section 1.1) which leads to a shortest tour through these n cities. Thus the TSP can be stated as

$$\min_{\psi \text{ cyclic}} \sum_{i=1}^{n} a_{i\psi(i)}. \tag{7.3}$$

We can model the traveling salesman problem as a Koopmans–Beckmann problem with the matrices A and B, where $B = (b_{jl})$ describes any cyclic permutation. For example, matrix B can be chosen as

$$B = \begin{pmatrix} 0 & 1 & 0 & 0 & 0 \\ 0 & 0 & 1 & 0 & 0 \\ 0 & 0 & 0 & 1 & 0 \\ 0 & 0 & 0 & 0 & 1 \\ 1 & 0 & 0 & 0 & 0 \end{pmatrix}.$$

Thus the Koopmans–Beckmann problem

$$\min_{\varphi \in S_n} \sum_{i=1}^{n} \sum_{k=1}^{n} a_{ik} b_{\varphi(i)\varphi(k)} \tag{7.4}$$

is an equivalent formulation to (7.3). The required permutation is obtained by setting $\psi(i) = k$ if $b_{\varphi(i)\varphi(k)} = 1$.

7.1.3 Minimum weight feedback arc set problem

In the *minimum weight feedback arc set problem* (FASP) we are given a weighted digraph $D = (N; \hat{A})$ with vertex set N and arc set $\hat{A}$. The goal is to remove a set of arcs from $\hat{A}$ with minimum total weight such that the new graph is *acyclic*, i.e., it does not contain directed cycles. Obviously the minimum weight feedback arc set problem is equivalent to the problem of finding an acyclic subgraph of D with maximum weight. The unweighted version of the FASP, that is an FASP where the arc weights of the underlying digraph are equal to 0 or 1, is called the *acyclic subdigraph problem* and has been thoroughly studied by Jünger [395]. Both versions of the FASP are known to be $\mathcal{NP}$-hard (see Karp [407], Garey and Johnson [300]).

In an acyclic digraph the nodes can be labeled topologically, i.e., we can give a label $\varphi(i)$ to each vertex i such that each arc (i, j) fulfills $\varphi(i) < \varphi(j)$. Therefore, the FASP can be formulated as a QAP. Let A be the weighted adjacency matrix of digraph D. Since the sum of weights is constant, we maximize the sum of the arc weights of an acyclic subgraph of D if we consider

$$\min_{\varphi \in \mathcal{S}_n} \sum_{i=1}^{n} \sum_{k=1}^{n} a_{ik} b_{\varphi(i)\varphi(k)},$$

where

$$b_{jl} = \begin{cases} 1 & \text{if } j > l, \\ 0 & \text{otherwise.} \end{cases}$$

Therefore, the FASP can be modeled as $QAP(A, -B)$ (since we minimize the objective function in (7.1)). The resulting feedback arc set is $\{(i, j) \in \hat{A} : \varphi(i) > \varphi(j)\}$.

A special case of the minimum weight feedback arc problem, and therefore also of the QAP, is the so-called *triangulation problem of input-output matrices* which plays an important role in econometrics as it is used to forecast the development of industries (see Leontief [452]). In this problem the rows and columns of a given $n \times n$ matrix A have to be permuted simultaneously such that the sum of elements in the permuted matrix above the main diagonal is minimized. For modeling this problem as a QAP, let us introduce a matrix $B = (b_{jl})$ with

$$b_{jl} = \begin{cases} 1 & \text{if } l > j, \\ 0 & \text{otherwise.} \end{cases}$$

Thus the triangulation problem becomes the QAP

$$\min_{\varphi \in \mathcal{S}_n} \sum_{i=1}^{n} \sum_{k=1}^{n} a_{ik} b_{\varphi(i)\varphi(k)}.$$

Concerning the triangulation of input-output matrices, we refer the interested reader to the monograph by Reinelt [570] in which not only the polyhedral structure of the underlying linear ordering polytope is described, but also economic implications of the problem are discussed, and a powerful solution algorithm is given. Also note that the first monograph on QAPs by Conrad [195] was devoted to the QAP, the TSP, and the triangulation of input-output matrices.

7.1.4 Graph partitioning and maximum clique problem

Two well studied $\mathcal{NP}$-hard combinatorial optimization problems, which are special cases of the QAP, are the graph partitioning problem and the maximum clique problem.

In the *graph partitioning problem* (GPP) we are given a weighted graph $G = (V; E)$ with n vertices and a number r which divides n. We want to partition the vertex set V into r subsets $V_1, V_2, \dots, V_r$ of equal cardinality such that the total weight of the edges between

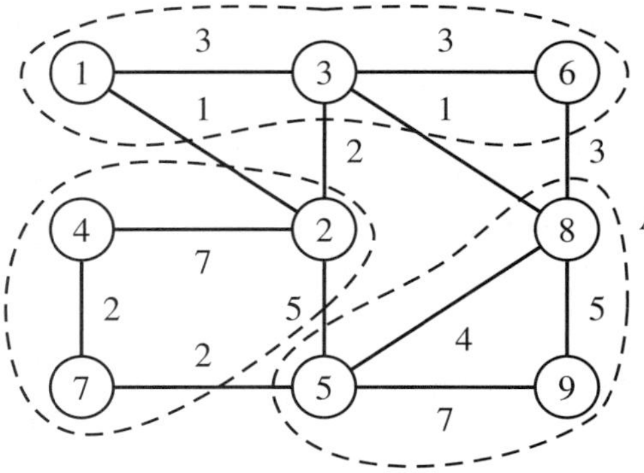

$$A = \begin{pmatrix} 0 & 1 & 3 & 0 & 0 & 0 & 0 & 0 & 0 \\ 1 & 0 & 2 & 7 & 5 & 0 & 0 & 0 & 0 \\ 3 & 2 & 0 & 0 & 0 & 3 & 0 & 1 & 0 \\ 0 & 7 & 0 & 0 & 0 & 0 & 2 & 0 & 0 \\ 0 & 5 & 0 & 0 & 0 & 0 & 2 & 4 & 7 \\ 0 & 0 & 3 & 0 & 0 & 0 & 0 & 3 & 0 \\ 0 & 0 & 0 & 2 & 2 & 0 & 0 & 0 & 0 \\ 0 & 0 & 1 & 0 & 4 & 3 & 0 & 0 & 5 \\ 0 & 0 & 0 & 0 & 7 & 0 & 0 & 5 & 0 \end{pmatrix}$$

Figure 7.1. *Graph of Example 7.2 and its weighted adjacency matrix.*

all sets V_i and V_j $(i, j = 1, 2, \dots, r;\ i \neq j)$ is minimized. Since the sum of all edge weights is a constant, this task is equivalent to maximizing the weights on edges within all components defined by $V_1, \dots, V_r$. Therefore, we can formulate this problem in the following way as a *QAP*(A, B): matrix A is the weighted adjacency matrix of G and matrix B aims at maximizing the sum of edge weights within the single components. It contains in the diagonal n/r matrices of size $r \times r$ with constant entries -1, while the remaining elements are 0 (see the example below). The resulting partition is $V_h = \{i \in V : \lceil \varphi(i)/r \rceil = h\}$ $(h = 1, 2, \dots, r)$. For more information on graph partitioning problems the reader is referred to Lengauer [451].

Example 7.2. We want to partition the graph shown in Figure 7.1 into three components so that the sum of weights of edges between different components is minimized.

Since G should be partitioned into three components, we choose as matrix B the matrix

$$B = \begin{pmatrix} -1 & -1 & -1 & 0 & 0 & 0 & 0 & 0 & 0 \\ -1 & -1 & -1 & 0 & 0 & 0 & 0 & 0 & 0 \\ -1 & -1 & -1 & 0 & 0 & 0 & 0 & 0 & 0 \\ 0 & 0 & 0 & -1 & -1 & -1 & 0 & 0 & 0 \\ 0 & 0 & 0 & -1 & -1 & -1 & 0 & 0 & 0 \\ 0 & 0 & 0 & -1 & -1 & -1 & 0 & 0 & 0 \\ 0 & 0 & 0 & 0 & 0 & 0 & -1 & -1 & -1 \\ 0 & 0 & 0 & 0 & 0 & 0 & -1 & -1 & -1 \\ 0 & 0 & 0 & 0 & 0 & 0 & -1 & -1 & -1 \end{pmatrix}.$$

The GPP can now be stated as $QAP(A, B)$. An optimal solution is $\varphi = (1, 4, 2, 5, 7, 3, 6, 8, 9)$, hence, the optimal partition (shown by the dashed lines in Figure 7.1) is $V_1 = \{1, 3, 6\}$, $V_2 = \{2, 4, 7\}$, $V_3 = \{5, 8, 9\}$. ∎

In the *maximum clique problem* (MCP) we are given an unweighted simple graph $G = (V; E)$ with n vertices and we wish to find a subset $V' \subseteq V$ of maximum cardinality $|V'| = r$ which induces a clique in G (i.e., such that each pair of vertices of V' is connected by an edge of G). For a given tentative value r, the problem of deciding if G has a clique of size r can be modeled as a QAP in the following form. Matrix A is the (unweighted) adjacency matrix of graph G. Matrix B models a clique of size r and $n-r$ isolated vertices. This matrix contains in its upper left part an $(r \times r)$ matrix with constant entries -1 and, otherwise, 0 entries, as in the example below. There exists a clique of size r in G if and only if the optimal value of the corresponding QAP is $-r(r-1)$. The corresponding clique is $V' = \{i \in V : \varphi(i) \leq r\}$. For an extensive survey on clique problems see Pardalos and Xue [537].

Example 7.3. We want to check whether the graph shown in Figure 7.1 has a clique of size $r = 4$. The adjacency matrix A is given by

$$A = \begin{pmatrix} 0 & 1 & 1 & 0 & 0 & 0 & 0 & 0 & 0 \\ 1 & 0 & 1 & 1 & 1 & 0 & 0 & 0 & 0 \\ 1 & 1 & 0 & 0 & 0 & 1 & 0 & 1 & 0 \\ 0 & 1 & 0 & 0 & 0 & 0 & 1 & 0 & 0 \\ 0 & 1 & 0 & 0 & 0 & 0 & 1 & 1 & 1 \\ 0 & 0 & 1 & 0 & 0 & 0 & 0 & 1 & 0 \\ 0 & 0 & 0 & 1 & 1 & 0 & 0 & 0 & 0 \\ 0 & 0 & 1 & 0 & 1 & 1 & 0 & 0 & 1 \\ 0 & 0 & 0 & 0 & 1 & 0 & 0 & 1 & 0 \end{pmatrix}. \tag{7.5}$$

We choose as matrix B the matrix

$$B = \begin{pmatrix} -1 & -1 & -1 & -1 & 0 & 0 & 0 & 0 & 0 \\ -1 & -1 & -1 & -1 & 0 & 0 & 0 & 0 & 0 \\ -1 & -1 & -1 & -1 & 0 & 0 & 0 & 0 & 0 \\ -1 & -1 & -1 & -1 & 0 & 0 & 0 & 0 & 0 \\ 0 & 0 & 0 & 0 & 0 & 0 & 0 & 0 & 0 \\ 0 & 0 & 0 & 0 & 0 & 0 & 0 & 0 & 0 \\ 0 & 0 & 0 & 0 & 0 & 0 & 0 & 0 & 0 \\ 0 & 0 & 0 & 0 & 0 & 0 & 0 & 0 & 0 \\ 0 & 0 & 0 & 0 & 0 & 0 & 0 & 0 & 0 \end{pmatrix}.$$

If the corresponding QAP returns -12 as the optimal value, then G has a clique with 4 vertices. This is obviously not the case in this example. ∎

7.1.5 Graph isomorphism and graph packing problems

Let two undirected graphs $G_1 = (V_1; E_1)$ and $G_2 = (V_2; E_2)$ be given. We assume that $|V_1| = |V_2| = n$. The two graphs are said to be *isomorphic* if there exists a bijective mapping from V_1 to V_2 with the property that $(i, j) \in E_1$ if and only if $(\varphi(i), \varphi(j)) \in E_2$. Let A be the adjacency matrix of graph G_1 and let B be the adjacency matrix of graph G_2. Then we get the following.

Proposition 7.4. *The graphs G_1 and G_2 are isomorphic if and only if the QAP*

$$z^* = \max_{\varphi \in \mathcal{S}_n} \sum_{i=1}^{n} \sum_{k=1}^{n} a_{ik} b_{\varphi(i)\varphi(k)}$$

has the optimal objective function value $z^* = |E_1| (= |E_2|)$.

A permutation π of $\{1, 2, \dots, n\}$ is called a *packing of G_2 into G_1* if $(i, j) \in E_1$ implies $(\pi(i), \pi(j)) \notin E_2$, for $1 \le i, j \le n$ (see Bollobás [104]). In other words, a packing of G_2 into G_1 is an embedding of the vertices of G_2 into the vertices of G_1 such that no pair of edges coincides. The *graph packing problem* consists of finding a packing of G_2 into G_1, if one exists, or proving that no packing exists.

The graph packing problem can be formulated as a Koopmans–Beckmann problem where A is the adjacency matrix of G_1 and B is the adjacency matrix of G_2. There exists a packing of G_2 into G_1 if and only if the optimal value of this QAP is equal to 0. The formulation of the packing problem as a QAP also shows that the problem is symmetric: if there is a packing of G_2 into G_1, then there is also a packing of G_1 into G_2. In general, the packing problem is $\mathcal{NP}$-hard. Polynomially solvable special cases have been discussed by Çela [176].

7.1.6 Quadratic bottleneck assignment problem

By replacing the sums in the objective function of a QAP by the maximum, we get the so-called *quadratic bottleneck assignment problem* (QBAP). A QBAP (in Koopmans–Beckmann form) can be formulated as

$$\min_{\varphi} \max_{1 \le i,k \le n} a_{ik} b_{\varphi(i)\varphi(k)}. \tag{7.6}$$

Such problems occur in a variety of applications such as very large-scale integration (VLSI) design, numerical analysis (bandwidth minimization), and quadratic assignment under time aspects, to mention just a few. Basically, all QAP applications give rise to a QBAP model as well, because it often makes sense to minimize the largest cost instead of the overall cost incurred by some decision.

The QBAP will be examined in detail in Chapter 9.

7.1.7 Complexity issues

In contrast to linear assignment problems, QAPs are very hard to solve. Currently it is only possible to find an optimal solution (and to prove optimality) for QAPs up to the size of about $n = 30$, and even this requires much computational effort (see Section 8.1). The inherent difficulty for solving the QAP is also reflected by its computational complexity. It is easy to see that the QAP belongs to the class of $\mathcal{NP}$-hard problems since it contains the TSP as a special case (see Section 7.1.2). Sahni and Gonzalez [593] showed that even finding an approximate solution for the QAP within some constant factor from the optimum value cannot be done in polynomial time, unless $\mathcal{P} = \mathcal{NP}$. These results hold even for Koopmans–Beckmann QAPs with coefficient matrices fulfilling the triangle inequality, as proved by Queyranne [568]. However, for a maximization version of the QAP in which matrix B satisfies the triangle inequality, Arkin, Hassin, and Sviridenko [44] provided an $O(n^3)$ time algorithm that guarantees a solution value within 1/4 from the optimum. In addition, the linear dense arrangement problem, which is a special Koopmans–Beckmann QAP, admits a polynomial-time approximation scheme (see Arora, Frieze, and Kaplan [46]). In the *linear dense arrangement problem*, matrix A is the distance matrix of n points which are regularly spaced on a line, i.e., points with abscissae given by $x_p = p$, $p = 1, \ldots, n$ and matrix B is a dense 0-1 matrix, i.e., the number of 1-entries in B is at least kn^2 for some constant k.

Even local search is hard in the case of the QAP. It can be shown (see Pardalos, Rendl, and Wolkowicz [535] for details) that the QAP is $\mathcal{PLS}$-complete with respect to the pairwise exchange neighborhood structure, as well as with respect to a Lin–Kernighan-like neighborhood structure (see Murthy, Pardalos, and Li [505]). This implies that the time complexity of a local search method for the QAP using either of the two mentioned neighborhood structures is exponential in the worst case. Moreover, from results of Papadimitriou and Wolfe [526] it follows that deciding whether a given local optimal solution of the QAP is also globally optimum is $\mathcal{NP}$-complete.

The QBAP is $\mathcal{NP}$-hard as well, since it contains the Hamiltonian cycle problem as a special case. To see this, consider a bottleneck $QAP(A, B)$ where A is the complementary adjacency matrix of the given graph (see equation (2.3)) and B is the permutation matrix of a cyclic permutation.

7.2 Formulations

There exist different but equivalent mathematical formulations for the QAP which stress different structural characteristics of the problem and lead to different solution approaches.

It is immediate that we can write the Koopmans–Beckmann QAP (7.1) as an integer quadratic program of the form

$$\min \quad \sum_{i=1}^{n}\sum_{j=1}^{n}\sum_{k=1}^{n}\sum_{l=1}^{n} a_{ik}b_{jl}x_{ij}x_{kl} + \sum_{i=1}^{n}\sum_{j=1}^{n} c_{ij}x_{ij} \tag{7.7}$$

$$\text{s.t.} \quad \sum_{j=1}^{n} x_{ij} = 1 \qquad (i = 1, 2, \ldots, n), \tag{7.8}$$

$$\sum_{i=1}^{n} x_{ij} = 1 \qquad (j = 1, 2, \ldots, n), \tag{7.9}$$

$$x_{ij} \in \{0, 1\} \qquad (i, j = 1, 2, \ldots, n), \tag{7.10}$$

where

$$x_{ij} = \begin{cases} 1 & \text{if facility } i \text{ is assigned to location } j; \\ 0 & \text{otherwise.} \end{cases}$$

Recall that every permutation φ can be represented by a permutation matrix $X_\varphi = (x_{ij})$ with $x_{ij} = 1$ if $j = \varphi(i)$ and $x_{ij} = 0$ otherwise. Let $\mathbf{X}_n$ denote the set of all $n \times n$ permutation matrices, i.e., all matrices defined by (7.9)–(7.10). Lawler's general form (7.2) can be written as

$$\min \sum_{i=1}^{n}\sum_{j=1}^{n}\sum_{k=1}^{n}\sum_{l=1}^{n} d_{ijkl}x_{ij}x_{kl} \tag{7.11}$$

$$\text{s.t. } X \in \mathbf{X}_n. \tag{7.12}$$

We can write the Koopmans–Beckmann QAP in a more compact way by defining an inner product between matrices. The *inner product* of two real $n \times n$ matrices A and B is defined by

$$\langle A, B \rangle = \sum_{i=1}^{n} \sum_{j=1}^{n} a_{ij} b_{ij}.$$

Given an $n \times n$ matrix B and a permutation $\varphi \in \mathcal{S}_n$, with the associated permutation matrix $X_\varphi \in \mathbf{X}_n$, we have

$$X_\varphi B X_\varphi^T = (b_{\varphi(i)\varphi(k)}), \tag{7.13}$$

where X^T is the transposed matrix X. Thus a Koopmans–Beckmann QAP can be written as

$$\begin{aligned} &\min \langle A, XBX^T \rangle + \langle C, X \rangle \\ &\text{s.t. } X \in \mathbf{X}_n. \end{aligned} \tag{7.14}$$

7.2.1 Trace formulation

The last formulation leads immediately to the so-called *trace formulation* of a Koopmans–Beckmann problem. The *trace* of an $n \times n$ matrix $A = (a_{ik})$ is the sum of its diagonal elements:

$$\text{tr}(A) = \sum_{i=1}^{n} a_{ii}.$$

Some simple properties of the trace operator are

$$\begin{aligned} \text{tr}(A) &= \text{tr}(A^T); \\ \text{tr}(A + B) &= \text{tr}(A) + \text{tr}(B); \\ \text{tr}(AB) &= \text{tr}(A^T B^T). \end{aligned}$$

Therefore, the inner product

$$\sum_{i=1}^{n} \sum_{k=1}^{n} a_{ik} b_{ik}$$

can be written as the trace of the product AB^T of the two matrices $A = (a_{ik})$ and $B = (b_{ik})$. Moreover, the matrix $(b_{\varphi(i)\varphi(k)})$ can be written as $X_\varphi B X_\varphi^T$ (see (7.13)). Since $\text{tr}(CX^T) = \sum_{i=1}^{n} c_{i\varphi(i)}$, the QAP can be formulated as

$$\begin{aligned} &\min \text{tr}((AXB^T + C)X^T) \\ &\text{s.t. } X \in \mathbf{X}_n. \end{aligned} \tag{7.15}$$

The trace formulation of the QAP appeared first in Edwards [252] and was used by Finke, Burkard, and Rendl [270] to introduce eigenvalue bounds for QAPs (see Section 7.7).

Example 7.5. We consider the same instance used in Example 7.1. We first illustrate the quadratic program formulation. The permutation matrix corresponding to $\varphi = (2, 1, 3)$ is

$$X_\varphi = \begin{pmatrix} 0 & 1 & 0 \\ 1 & 0 & 0 \\ 0 & 0 & 1 \end{pmatrix};$$

hence,

$$X_\varphi B X_\varphi^T = \begin{pmatrix} 0 & 1 & 0 \\ 1 & 0 & 0 \\ 0 & 0 & 1 \end{pmatrix} \begin{pmatrix} 2 & 3 & 6 \\ 1 & 4 & 7 \\ 5 & 6 & 2 \end{pmatrix} \begin{pmatrix} 0 & 1 & 0 \\ 1 & 0 & 0 \\ 0 & 0 & 1 \end{pmatrix} = \begin{pmatrix} 4 & 1 & 7 \\ 3 & 2 & 6 \\ 6 & 5 & 2 \end{pmatrix}.$$

The objective function value is then computed, through the inner products, as $z = \langle A, X_\varphi B X_\varphi^T \rangle + \langle C, X_\varphi \rangle$:

$$\begin{aligned} z &= \left\langle \begin{pmatrix} 1 & 2 & 4 \\ 3 & 4 & 5 \\ 5 & 6 & 1 \end{pmatrix}, \begin{pmatrix} 4 & 1 & 7 \\ 3 & 2 & 6 \\ 6 & 5 & 2 \end{pmatrix} \right\rangle + \left\langle \begin{pmatrix} 9 & 7 & 9 \\ 6 & 5 & 7 \\ 8 & 9 & 8 \end{pmatrix}, \begin{pmatrix} 0 & 1 & 0 \\ 1 & 0 & 0 \\ 0 & 0 & 1 \end{pmatrix} \right\rangle \\ &= (4+2+28+9+8+30+30+30+2) + (7+6+8) = 164. \end{aligned}$$

In order to compute the same objective function value through the trace formulation we need to compute

$$A X_\varphi B^T = \begin{pmatrix} 1 & 2 & 4 \\ 3 & 4 & 5 \\ 5 & 6 & 1 \end{pmatrix} \begin{pmatrix} 0 & 1 & 0 \\ 1 & 0 & 0 \\ 0 & 0 & 1 \end{pmatrix} \begin{pmatrix} 2 & 1 & 5 \\ 3 & 4 & 6 \\ 6 & 7 & 2 \end{pmatrix} = \begin{pmatrix} 31 & 34 & 24 \\ 47 & 51 & 48 \\ 33 & 33 & 62 \end{pmatrix},$$

from which

$$\begin{aligned} (A X_\varphi B^T + C) X_\varphi^T &= \left(\begin{pmatrix} 31 & 34 & 24 \\ 47 & 51 & 48 \\ 33 & 33 & 62 \end{pmatrix} - \begin{pmatrix} 9 & 7 & 9 \\ 6 & 5 & 7 \\ 8 & 9 & 8 \end{pmatrix} \right) \begin{pmatrix} 0 & 1 & 0 \\ 1 & 0 & 0 \\ 0 & 0 & 1 \end{pmatrix} \\ &= \begin{pmatrix} 41 & 40 & 33 \\ 55 & 53 & 55 \\ 42 & 41 & 70 \end{pmatrix}. \end{aligned}$$

The trace of the resulting matrix gives $z = 41 + 53 + 70 = 164$. ∎

7.2.2 Kronecker product formulation

Let U and V be two real $n \times n$ matrices. The *Kronecker product* of matrices U and V is defined as

$$U \otimes V = \begin{pmatrix} u_{11}V & u_{12}V & \cdots & u_{1n}V \\ \vdots & \vdots & \ddots & \vdots \\ u_{n1}V & u_{n2}V & \cdots & u_{nn}V \end{pmatrix}.$$

For $U = (u_{ij})$ and $V = (v_{kl})$ the Kronecker product $U \otimes V$ is the $n^2 \times n^2$ matrix formed by all possible products $u_{ij}v_{kl}$.

Let X be a permutation matrix. Using the four index cost matrix D introduced in equation (7.2), we can write the cost coefficients of a general QAP as an $n^2 \times n^2$ matrix $D = (D^{ij})$, where every $n \times n$ matrix D^{ij} is formed by the elements d_{ijkl} with fixed indices i and j and variable indices $k, l = 1, 2, \ldots, n$. That is, the entry d_{ijkl} lies in the $((i-1)n+k)$th row and $((j-1)n+l)$th column of matrix D. For $n = 3$, for example, the cost matrix D has the form

$$D = \begin{pmatrix} D^{11} & D^{12} & D^{13} \\ D^{21} & D^{22} & D^{23} \\ D^{31} & D^{32} & D^{33} \end{pmatrix}.$$

Using this notation, the objective function (7.11) can be rewritten as

$$\sum_{i=1}^{n}\sum_{j=1}^{n} x_{ij}\langle D^{ij}, X\rangle. \tag{7.16}$$

This leads to the Kronecker product formulation of a QAP (see Lawler [445])

$$\min \langle D, Y\rangle \tag{7.17}$$
$$\text{s.t. } Y = X \otimes X, \tag{7.18}$$
$$X \in \mathbf{X}_n. \tag{7.19}$$

This formulation is used for computing the so-called Gilmore–Lawler bound for general QAPs (see Section 7.6).

7.2.3 Convex and concave integer models

For any $n \times n$ matrix U, let $\mathrm{vec}(U) \in \mathbb{R}^{n^2}$ be the vector formed by the columns of U. (In the following, if no confusion arises we write just x instead of $\mathrm{vec}(X)$.) A well-known identity from matrix analysis states that

$$\mathrm{vec}(UXV) = (V^T \otimes U)\,\mathrm{vec}(X). \tag{7.20}$$

Using this identity and $\mathrm{tr}(AB) = \mathrm{tr}(B^T A^T)$ the trace formulation of a QAP can be rewritten as

$$\mathrm{tr}(AXB^T X^T) = x^T\,\mathrm{vec}(AXB^T) = x^T(B \otimes A)x.$$

Therefore, a Koopmans–Beckmann problem can be formulated as

$$\min x^T(B \otimes A)x + \mathrm{vec}(C)^T x \tag{7.21}$$
$$\text{s.t. } X \in \mathbf{X}_n.$$

The element $a_{ik}b_{jl}$ lies in the $((j-1)n+i)$th row and $((l-1)n+k)$th column of matrix $B \otimes A$. Therefore, we can arrange the n^4 cost coefficients d_{ijkl} in a new way so that the element d_{ijkl} lies in row $(j-1)n+i$ and column $(l-1)n+k$ of an $n^2 \times n^2$ matrix $\tilde{D}$. The general QAP can then be written as

$$\begin{aligned} &\min x^T \tilde{D} x \\ &\text{s.t. } X \in \mathbf{X}_n. \end{aligned} \tag{7.22}$$

Since for any vector x we have $x^T \tilde{D} x = x^T[\frac{1}{2}(\tilde{D} + \tilde{D}^T)]x$, we can assume that $\tilde{D}$ is symmetric. The addition of a constant to the entries of the main diagonal of $\tilde{D}$ does not change the optimal solutions of the corresponding QAP since it simply adds a constant to the objective function value. By adding a large positive constant to the diagonal elements of $\tilde{D}$, we can achieve that all eigenvalues of $\tilde{D}$ become positive, i.e. that $\tilde{D}$ is positive definite. If we add a sufficiently small constant to the diagonal elements of $\tilde{D}$, we can achieve that all eigenvalues of $\tilde{D}$ become negative, i.e., that $\tilde{D}$ is negative definite. Thus we can write a QAP as a quadratic convex program or as a quadratic concave program.

Example 7.6. We consider again the instance used in Example 7.1. In order to compute the solution value corresponding to permutation $\varphi = (2, 1, 3)$ through the Kronecker product formulation, we need to define matrices D and $Y_\varphi = X_\varphi \otimes X_\varphi$:

$$D = \left(\begin{array}{rrr|rrr|rrr} 11 & 3 & 6 & 1 & 11 & 7 & 5 & 6 & 11 \\ 4 & 6 & 12 & 2 & 8 & 14 & 10 & 12 & 4 \\ 8 & 12 & 24 & 4 & 16 & 28 & 20 & 24 & 8 \\ \hline 6 & 9 & 18 & 3 & 12 & 21 & 15 & 18 & 6 \\ 14 & 12 & 24 & 4 & 21 & 28 & 20 & 24 & 15 \\ 10 & 15 & 30 & 5 & 20 & 35 & 25 & 30 & 10 \\ \hline 10 & 15 & 30 & 5 & 20 & 35 & 25 & 30 & 10 \\ 12 & 18 & 36 & 6 & 24 & 42 & 30 & 36 & 12 \\ 10 & 3 & 6 & 1 & 13 & 7 & 5 & 6 & 10 \end{array}\right);$$

$$Y_\varphi = \left(\begin{array}{ccc|ccc|ccc} 0 & 0 & 0 & 0 & 1 & 0 & 0 & 0 & 0 \\ 0 & 0 & 0 & 1 & 0 & 0 & 0 & 0 & 0 \\ 0 & 0 & 0 & 0 & 0 & 1 & 0 & 0 & 0 \\ \hline 0 & 1 & 0 & 0 & 0 & 0 & 0 & 0 & 0 \\ 1 & 0 & 0 & 0 & 0 & 0 & 0 & 0 & 0 \\ 0 & 0 & 1 & 0 & 0 & 0 & 0 & 0 & 0 \\ \hline 0 & 0 & 0 & 0 & 0 & 0 & 0 & 1 & 0 \\ 0 & 0 & 0 & 0 & 0 & 0 & 1 & 0 & 0 \\ 0 & 0 & 0 & 0 & 0 & 0 & 0 & 0 & 1 \end{array}\right).$$

Then we obtain

$$z = \langle D, Y_\varphi \rangle = 11 + 2 + 28 + 9 + 14 + 30 + 30 + 30 + 10 = 164.$$

Finally, we consider the convex formulation. Let $x_\varphi = \text{vec}(X_\varphi)$. For permutation $\varphi = (2, 1, 3)$ we have

$z = x_\varphi^T (B \otimes A) x_\varphi + \text{vec}(C)^T x_\varphi$

$$= \begin{pmatrix} 0\\1\\0\\1\\0\\0\\0\\0\\1 \end{pmatrix}^T \begin{pmatrix} 2 & 4 & 8 & 3 & 6 & 12 & 6 & 12 & 24\\ 6 & 8 & 10 & 9 & 12 & 15 & 18 & 24 & 30\\ 10 & 12 & 2 & 15 & 18 & 3 & 30 & 36 & 6\\ 1 & 2 & 4 & 4 & 8 & 16 & 7 & 14 & 28\\ 3 & 4 & 5 & 12 & 16 & 20 & 21 & 28 & 35\\ 5 & 6 & 1 & 20 & 24 & 4 & 35 & 42 & 7\\ 5 & 10 & 20 & 6 & 12 & 24 & 2 & 4 & 8\\ 15 & 20 & 25 & 18 & 24 & 30 & 6 & 8 & 10\\ 25 & 30 & 5 & 30 & 36 & 6 & 10 & 12 & 2 \end{pmatrix} \begin{pmatrix} 0\\1\\0\\1\\0\\0\\0\\0\\1 \end{pmatrix} + \begin{pmatrix} 9\\6\\8\\7\\5\\9\\9\\7\\8 \end{pmatrix}^T \begin{pmatrix} 0\\1\\0\\1\\0\\0\\0\\0\\1 \end{pmatrix} = 164. \quad \blacksquare$$

7.2.4 Mean objective value of feasible solutions

The mean and the variance of the objective function values of all feasible solutions can be computed directly from the input data in a more or less straightforward way. They play a role in heuristics for finding solutions of good quality (as in the algorithm by Graves and Whinston [339] discussed in Section 8.2.2).

Let a Koopmans–Beckmann problem (7.1) be given. By redefining $C = (c_{ij})$ as $c_{ij} := c_{ij} + a_{ii}b_{jj}$ $(i, j = 1, 2, \ldots, n)$, we can assume that the diagonal entries of the input matrices A and B are 0. We denote the probability that an index i is assigned to an index j by $P(i \to j)$. It is easy to see that

$$P(i \to j) = \frac{1}{n} \quad \text{for all } i,\ j \tag{7.23}$$

and that

$$P((i \to j) \wedge (k \to l)) = \frac{1}{n(n-1)} \quad \text{for } i \neq k,\ j \neq l. \tag{7.24}$$

The mean objective value $\mu(A, B, C)$ of (7.1) is given by

$$\mu(A, B, C) = \sum_{i=1}^{n} \sum_{j=1}^{n} c_{ij} P(i \to j) + \sum_{i=1}^{n} \sum_{j=1}^{n} \sum_{k=1}^{n} \sum_{l=1}^{n} a_{ik} b_{jl} P((i \to j) \wedge (k \to l)). \tag{7.25}$$

This immediately yields the following.

Proposition 7.7. *The mean of the objective function values of a Koopmans–Beckmann problem* (7.1) *is given by*

$$\mu(A, B, C) = \frac{1}{n} \sum_{i=1}^{n} \sum_{j=1}^{n} c_{ij} + \frac{1}{n(n-1)} \left(\sum_{i=1}^{n} \sum_{k=1}^{n} a_{ik} \right) \left(\sum_{j=1}^{n} \sum_{l=1}^{n} b_{jl} \right). \tag{7.26}$$

In the case of a general QAP (7.2) one gets

$$\mu(D) = \frac{1}{n}\sum_{i=1}^{n}\sum_{j=1}^{n} d_{ijij} + \frac{1}{n(n-1)}\sum_{i=1}^{n}\sum_{j=1}^{n}\sum_{\substack{k=1\\k\neq i}}^{n}\sum_{\substack{l=1\\l\neq j}}^{n} d_{ijkl}. \tag{7.27}$$

These formulae are easy to evaluate and can be extended to the case where some of the indices are already fixed (see Graves and Whinston [339]).

The computation of the variance $\sigma^2(A, B, C)$ is more involved and needs the evaluation of partial sums of the given data.

7.3 Linearizations

In all formulations seen so far the objective function of the QAP is a quadratic function. It was observed that the quadratic objective function can be linearized by introducing new variables and requiring additional constraints. Lawler [445] replaced the quadratic terms $x_{ij}x_{kl}$ in the objective function (7.11) by n^4 variables

$$y_{ijkl} = x_{ij}x_{kl} \quad (i, j, k, l = 1, 2, \ldots, n)$$

and obtained in this way an integer program with $n^4 + n^2$ binary variables and $n^4 + 2n + 1$ constraints. The QAP (7.11)–(7.12) can be rewritten as an integer program in the following form:

$$\min \quad \sum_{i=1}^{n}\sum_{j=1}^{n}\sum_{k=1}^{n}\sum_{l=1}^{n} d_{ijkl}y_{ijkl}$$

$$\text{s.t.} \quad \sum_{i=1}^{n}\sum_{j=1}^{n}\sum_{k=1}^{n}\sum_{l=1}^{n} y_{ijkl} = n^2, \tag{7.29}$$

$$x_{ij} + x_{kl} - 2y_{ijkl} \geq 0 \qquad (i, j, k, l = 1, 2, \ldots, n), \tag{7.30}$$

$$y_{ijkl} \in \{0, 1\} \qquad (i, j, k, l = 1, 2, \ldots, n), \tag{7.31}$$

$$X \in \mathbf{X}_n. \tag{7.32}$$

If X is a feasible solution of (7.11)–(7.12) and we define $y_{ijkl} = x_{ij}x_{kl}$, then y_{ijkl} obviously fulfills constraints (7.29)–(7.31) and yields the same objective function value. Conversely, let $(x_{ij}) = X$ and (y_{ijkl}) be a feasible solution of (7.29)–(7.32). Observe that y_{ijkl} can take the value 1 only if $x_{ij} = x_{kl} = 1$. Since X is a permutation matrix, the number of pairs $((i, j), (k, l))$ such that $x_{ij} = x_{kl} = 1$ is exactly n^2. Therefore, we have

$$y_{ijkl} = 1 \text{ if and only if } x_{ij} = x_{kl} = 1,$$

which implies $y_{ijkl} = x_{ij}x_{kl}$. Thus the two objective function values are equal.

For computational purposes this linearization is rather inconvenient due to the many additional binary variables and constraints. Thus efforts have been made to find linearizations with fewer variables and constraints. Bazaraa and Sherali [78] showed that the QAP can be written as a mixed integer linear program with n^2 binary variables, $n^2(n-1)^2/2$ real variables, and $2n^2$ linear constraints. A linearization with a smaller number of additional variables and constraints has been described by Kaufman and Broeckx [412], who derived a mixed integer linear programming formulation of the QAP using n^2 new real variables and $2n^2$ additional constraints.

7.3.1 Kaufman–Broeckx

Recall that by using the matrix D^{ij} defined by

$$d_{kl}^{ij} = (a_{ik}b_{jl})$$

(see Section 7.2.2), the objective function of a Koopmans–Beckmann problem $QAP(A, B)$ can be written as

$$\sum_{i=1}^{n}\sum_{j=1}^{n}\sum_{k=1}^{n}\sum_{l=1}^{n} a_{ik}b_{jl}x_{ij}x_{kl} = \sum_{i=1}^{n}\sum_{j=1}^{n} x_{ij}\langle D^{ij}, X\rangle,$$

where X is the matrix $X = (x_{kl})$.

Kaufman and Broeckx defined n^2 real variables w_{ij} as

$$w_{ij} = x_{ij}\langle D^{ij}, X\rangle = x_{ij}\sum_{k=1}^{n}\sum_{l=1}^{n} a_{ik}b_{jl}x_{kl}. \tag{7.33}$$

Using these variables, the objective function of a Koopmans–Beckmann problem can be rewritten as

$$\sum_{i=1}^{n}\sum_{j=1}^{n} w_{ij}.$$

Kaufman and Broeckx [412] proved the following.

Proposition 7.8. *The Koopmans–Beckmann problem $QAP(A, B)$ with nonnegative cost coefficients a_{ik} and b_{jl}*

$$\min \sum_{i=1}^{n}\sum_{j=1}^{n}\sum_{k=1}^{n}\sum_{l=1}^{n} a_{ik}b_{jl}x_{ij}x_{kl}$$
$$\text{s.t. } X \in \mathbf{X}_n$$

is equivalent to the following mixed-integer linear program:

$$\min \sum_{i=1}^{n}\sum_{j=1}^{n} w_{ij} \tag{7.34}$$

$$\text{s.t.} \quad f_{ij}x_{ij} + \langle D^{ij}, X\rangle - w_{ij} \le f_{ij} \quad (i, j = 1, 2, \ldots, n), \tag{7.35}$$

$$w_{ij} \ge 0 \quad (i, j = 1, 2, \ldots, n), \tag{7.36}$$

$$X \in \mathbf{X}_n, \tag{7.37}$$

where

$$f_{ij} = \sum_{k=1}^{n}\sum_{l=1}^{n} a_{ik}b_{jl}. \tag{7.38}$$

Note that we can assume nonnegative cost coefficients without loss of generality, since adding a constant to all cost coefficients does not change the optimal solution. The Kaufman–Broeckx formulation can also be obtained by applying a general linearization strategy proposed by Glover [317].

Proof. If (x_{ij}) is a feasible solution of the QAP, then we obtain a feasible solution of (7.34)–(7.37) with the same objective function value by defining

$$w_{ij} = x_{ij}\langle D^{ij}, X\rangle. \tag{7.39}$$

Conversely, let (x_{ij}) and (w_{ij}) be an optimal solution of (7.34)–(7.37): we have to prove that (7.39) holds. If $x_{ij} = 0$, then w_{ij} must be 0 since $\langle D^{ij}, X\rangle < f_{ij}$ by definition and the sum of the nonnegative variables w_{ij} is minimized. If $x_{ij} = 1$, then constraints (7.35) imply $\langle D^{ij}, X\rangle \leq w_{ij}$. Again the argument that we minimize the sum of variables w_{ij} yields (7.39). □

7.3.2 Balas–Mazzola

A similar linearization was proposed by Balas and Mazzola [53] (see also Burkard and Bönniger [134]). Let g_{ij} be an upper bound for $\{\langle D^{ij}, X\rangle : X \in \mathbf{X}_n\}$. Moreover, we define for every $Y \in \mathbf{X}_n$ a matrix $H(Y) = (h_{kl}(Y))$ and a real number $h(Y)$ by

$$h_{kl}(Y) = g_{kl}y_{kl} + \sum_{i=1}^{n}\sum_{j=1}^{n} d_{ijkl}y_{ij} \tag{7.40}$$

and

$$h(Y) = \langle G, Y\rangle. \tag{7.41}$$

Then we get the following.

Proposition 7.9. *Every optimal solution X^* of a QAP(A, B) with nonnegative cost coefficients d_{ijkl} uniquely corresponds to an optimal solution (z^*, X^*) of the integer program*

$$\min z \tag{7.42}$$

$$\text{s.t. } z \geq \langle H(Y), X\rangle - h(Y) \quad \text{for all } Y \in \mathbf{X}_n. \tag{7.43}$$

Proof. Definition (7.40) yields, for any $Y \in \mathbf{X}_n$,

$$\begin{aligned}
\langle H(Y), X\rangle - h(Y) &= \sum_{k=1}^{n}\sum_{l=1}^{n}\left(g_{kl}y_{kl} + \sum_{i=1}^{n}\sum_{j=1}^{n} a_{ik}b_{jl}y_{ij}\right) x_{kl} - \langle G, Y\rangle \\
&= \sum_{i=1}^{n}\sum_{j=1}^{n}\left(\sum_{k=1}^{n}\sum_{l=1}^{n} a_{ik}b_{jl}x_{kl}\right) y_{ij} \\
&\qquad + \sum_{k=1}^{n}\sum_{l=1}^{n} g_{kl}x_{kl}y_{kl} - \sum_{k=1}^{n}\sum_{l=1}^{n} g_{kl}y_{kl} \\
&= \sum_{i=1}^{n}\sum_{j=1}^{n}\Big(\langle D^{ij}, X\rangle + g_{ij}(x_{ij}-1)\Big) y_{ij} \\
&= \langle W(X), Y\rangle,
\end{aligned}$$

where $W(X) = (w_{ij}(X))$ is defined by

$$w_{ij}(X) = \langle D^{ij}, X\rangle + g_{ij}(x_{ij}-1).$$

Therefore, (7.42)–(7.43) can be rewritten as

$$\min\{z : z \geq \langle W(X), Y\rangle\} = \min_{X\in\mathbf{X}_n}\max_{Y\in\mathbf{X}_n}\langle W(X), Y\rangle.$$

For fixed $X = \overline{X}$ an optimal solution of

$$\max_{Y\in\mathbf{X}_n}\langle W(\overline{X}), Y\rangle$$

is given by $Y = \overline{X}$, since

$$\begin{aligned}
&\text{(i) } \overline{x}_{ij} = 1 \text{ implies } w_{ij}(\overline{X}) = \langle D^{ij}, \overline{X}\rangle \geq 0; \\
&\text{(ii) } \overline{x}_{ij} = 0 \text{ implies } w_{ij}(\overline{X}) = \langle D^{ij}, \overline{X}\rangle - g_{ij} \leq 0.
\end{aligned}$$

Note that Y is feasible since $\overline{X} \in \mathbf{X}_n$. Now we get

$$\begin{aligned}
\langle W(\overline{X}), Y\rangle &= \langle W(\overline{X}), \overline{X}\rangle \\
&= \sum_{i=1}^{n}\sum_{j=1}^{n}\sum_{k=1}^{n}\sum_{l=1}^{n} a_{ik}b_{jl}\overline{x}_{ij}\overline{x}_{kl} + \sum_{i=1}^{n}\sum_{j=1}^{n} g_{ij}(\overline{x}_{ij}-1)\overline{x}_{ij}.
\end{aligned}$$

Since $\overline{x}_{ik} \in \{0, 1\}$ implies $(\overline{x}_{ik}-1)\overline{x}_{ik} = 0$ for all i and k, we obtain

$$\langle W(\overline{X}), \overline{X}\rangle = \sum_{i=1}^{n}\sum_{j=1}^{n}\sum_{k=1}^{n}\sum_{l=1}^{n} a_{ik}b_{jl}\overline{x}_{ij}\overline{x}_{kl}.$$

Therefore,

$$\min_{X\in\mathbf{X}_n}\max_{Y\in\mathbf{X}_n}\langle W(X), Y\rangle = \min_{X\in\mathbf{X}_n}\langle W(X), X\rangle = \min_{X\in\mathbf{X}_n}\sum_{i=1}^{n}\sum_{j=1}^{n}\sum_{k=1}^{n}\sum_{l=1}^{n} a_{ik}b_{jl}x_{ij}x_{kl}. \qquad \square$$

7.3.3 Frieze–Yadegar

Frieze and Yadegar [287] replaced the products $x_{ij}x_{kl}$ of the binary variables by continuous variables y_{ijkl} ($y_{ijkl} = x_{ij}x_{kl}$). They obtained the following mixed integer linear programming formulation for the QAP (7.11)–(7.12) using n^4 real variables, n^2 binary variables, and $n^4 + 4n^3 + 2n$ constraints, plus the nonnegativity constraints on the continuous variables:

$$
\begin{aligned}
\min \quad & \sum_{i=1}^{n}\sum_{j=1}^{n}\sum_{k=1}^{n}\sum_{l=1}^{n} d_{ijkl}y_{ijkl} && && (7.44)\\
\text{s.t.} \quad & \sum_{i=1}^{n} y_{ijkl} = x_{kl} && (j,k,l = 1,2,\ldots,n), && (7.45)\\
& \sum_{j=1}^{n} y_{ijkl} = x_{kl} && (i,k,l = 1,2,\ldots,n), && (7.46)\\
& \sum_{k=1}^{n} y_{ijkl} = x_{ij} && (i,j,l = 1,2,\ldots,n), && (7.47)\\
& \sum_{l=1}^{n} y_{ijkl} = x_{ij} && (i,j,k = 1,2,\ldots,n), && (7.48)\\
& X \in \mathbf{X}_n, && && (7.49)\\
& 0 \le y_{ijkl} \le 1 && (i,j,k,l = 1,2,\ldots,n). && (7.50)
\end{aligned}
$$

Proposition 7.10. *The QAP* (7.11)–(7.12) *is equivalent to the mixed-integer linear program* (7.44)–(7.50).

Proof. Given an $x_{ij} \in \mathbf{X}_n$ and defining $y_{ijkl} = x_{ij}x_{kl}$, it is straightforward that the constraints (7.45)–(7.48) and (7.50) are fulfilled, and that (7.44) and the objective function of the QAP yield the same value. Conversely, let (x_{ij}) and (y_{ijkl}) be a solution of (7.44)–(7.50) and observe that

$$
\begin{aligned}
x_{ij} = 0 \quad &\text{implies} \quad y_{ijkl} = 0 \ \text{ for all } \ k,l, && (7.51)\\
x_{kl} = 0 \quad &\text{implies} \quad y_{ijkl} = 0 \ \text{ for all } \ i,j. && (7.52)
\end{aligned}
$$

We now have only to show that $x_{ij} = x_{kl} = 1$ implies $y_{ijkl} = 1$. Let $x_{kl} = 1$ and observe that, for a fixed j, constraints (7.45) require that $\sum_{i=1}^{n} y_{ijkl}$ has total value 1. But (7.51) imposes that only one of the y variables in the summation can take a positive value, namely, y_{ijkl}, which must take value 1, thus concluding the proof. □

7.3.4 Adams–Johnson

Adams and Johnson [5] presented a 0-1 linear integer programming formulation for the QAP which resembles to a certain extent the linearization of Frieze and Yadegar. It is based on a linearization technique for general 0-1 polynomial programs introduced by Adams and Sherali [6, 7]. The QAP with a coefficient array $D = (d_{ijkl})$ is proven to be equivalent to the following mixed 0-1 linear program:

$$\min \sum_{i=1}^{n}\sum_{j=1}^{n}\sum_{k=1}^{n}\sum_{l=1}^{n} d_{ijkl}y_{ijkl} \tag{7.53}$$

$$\text{s.t.} \sum_{i=1}^{n} y_{ijkl} = x_{kl} \qquad (j,k,l = 1,2,\dots,n), \tag{7.54}$$

$$\sum_{j=1}^{n} y_{ijkl} = x_{kl} \qquad (i,k,l = 1,2,\dots,n), \tag{7.55}$$

$$y_{ijkl} = y_{klij} \qquad (i,j,k,l = 1,2,\dots,n), \tag{7.56}$$

$$X \in \mathbf{X}_n \tag{7.57}$$

$$y_{ijkl} \geq 0 \qquad (i,j,k,l = 1,2,\dots,n), \tag{7.58}$$

where each y_{ijkl} represents the product $x_{ij}x_{kl}$. The above formulation contains n^2 binary variables x_{ij}, n^4 continuous variables y_{ijkl}, and $n^4 + 2n^3 + 2n$ constraints, excluding the nonnegativity constraints on the continuous variables. The constraint set (7.54)–(7.58) describes a solution matrix Y which is the Kronecker product of two permutation matrices X (see (7.17)–(7.19)). Hence, this formulation of the QAP is equivalent to (7.11)–(7.12).

Another simple way to prove that this is a valid linearization is to observe that $x_{ij}x_{kl} = x_{kl}x_{ij}$, i.e., matrix (y_{ijkl}) has the symmetry $y_{ijkl} = y_{klij}$. The Frieze–Yadegar linearization can thus be rewritten by substituting (7.47)–(7.48) with (7.56).

Adams and Johnson [5] noted that a significantly smaller formulation, both in terms of variables and constraints, could be obtained, but the structure of the continuous relaxation above is favorable for solving the problem approximately by means of the Lagrangean dual (see Section 7.5). The theoretical strength of this linearization lies in the fact that it comprises the previous linearizations, which are just linear combinations of its constraints.

7.3.5 Higher level linearizations

Recently, Adams, Guignard, Hahn, and Hightower [4] refined the Adams–Johnson approach by using a *reformulation-linearization technique* developed by Sherali and Adams [606, 607, 608] for general 0-1 programming problems. In particular, they implemented a *level-2 reformulation* which can be obtained from the Adams–Johnson formulation as follows: (i) multiply (7.54)–(7.58) by the binary variable x_{pq}; (ii) linearize all nonlinear terms using new variables

$$z_{ijklpq} = x_{ij}x_{kl}x_{pq};$$

and (iii) append the resulting equations to the Adams–Johnson model. Thus the following new constraints are added to (7.53)–(7.58):

$$\sum_{\substack{i=1\\ i\neq k,p}}^{n} z_{ijklpq} = y_{klpq} \quad (j,k,l,p,q = 1,2,\ldots,n,\ j\neq l\neq q,\ k\neq p), \tag{7.59}$$

$$\sum_{\substack{j=1\\ j\neq l,q}}^{n} z_{ijklpq} = y_{klpq} \quad (i,k,l,p,q = 1,2,\ldots,n,\ i\neq k\neq p,\ l\neq q). \tag{7.60}$$

$$z_{ijklpq} = z_{klijpq} = z_{ijpqkl} = z_{klpqij} = z_{pqijkl} = z_{pqklij}$$
$$(i,j,k,l,p,q = 1,2,\ldots,n,\ i<k<p,\ j\neq l\neq q), \tag{7.61}$$

$$z_{ijklpq} \geq 0 \quad (i,j,k,l,p,q = 1,2,\ldots,n,\ i\neq k\neq p,\ j\neq l\neq q). \tag{7.62}$$

The resulting linear program (7.53)–(7.62) yields considerably tighter bounds than the Adams–Johnson linearization. Adams, Guignard, Hahn, and Hightower proposed a relaxation of the symmetry constraints (7.56) and (7.61) in a Lagrangean fashion and proved that the resulting program can be efficiently solved by computing one LSAP of size n, n^2 LSAPs of size $n-1$, and $n^2(n-1)^2$ LSAPs of size $n-2$. A dual ascent method is used to find an optimal set of Lagrangean multipliers.

The above technique can be extended to obtain level-3 reformulations, which can produce better lower bound values but require very extensive computations (Hahn [356]).

7.4 Quadratic assignment polytopes

Polyhedral combinatorics has proven to be a powerful tool for attacking hard combinatorial optimization problems. The main idea of polyhedral combinatorics is the study of the polytope which occurs as the convex hull of all feasible solutions of a combinatorial optimization problem. Knowing a linear description of a polytope, i.e., its defining hyperplanes, an optimal solution of the underlying combinatorial optimization problem can be found by applying linear programming techniques, for example, interior point methods. A nice monograph on geometric algorithms and combinatorial optimization was written by Grötschel, Lovász, and Schrijver [343].

A direct application of polyhedral combinatorics to the QAP is not possible, since the objective function is quadratic. So, first one has to find a suitable linearization for the QAP and in the next step one can study the convex hull of all (integral) feasible points of the linearized problem. Studies on the QAP polytope were begun by Barvinok [73] and Padberg and Rijal [519] (see also Rijal's Ph.D. thesis [584]). A few years later Jünger and Kaibel [396, 398, 397] contributed further interesting results (see also Kaibel's Ph.D. thesis [400]).

The linearizations discussed in the previous sections can be used as a starting point for the definition of the QAP polytope. For example, we can define the *QAP polytope* denoted by QAP_n as convex hull of all 0-1 vectors (x_{ij}, y_{ijkl}), $1\leq i,j,k,l\leq n$, which are feasible solutions of the mixed integer linear program (MILP) formulation of Adams and Johnson [5]. Jünger and Kaibel [396] chose another way of describing the QAP polytope. Their formulation provides more insight into the problem. For each $n\in\mathbb{N}$ they considered the graph $G_n = (V_n; E_n)$ with vertex set $V_n = \{(i,j) : 1\leq i,j\leq n\}$ and edge set $E_n = \{((i,j),(k,l)) : i\neq k, j\neq l\}$. Obviously, every permutation matrix $X\in\mathbf{X}_n$

corresponds to a maximal clique in G_n with cardinality n and vice versa. Given an instance of the Lawler QAP with coefficients d_{ijkl}, we introduce d_{ijij} as vertex weights and d_{ijkl} for $i \neq k,\ j \neq l$ as edge weights in G_n. Finding an optimal solution of the QAP is equivalent to finding a maximal clique with minimum total vertex and edge weights. For each clique C in G_n with n vertices we denote its incidence vector by (x^C, y^C), where $x^C \in \mathbb{R}^{n^2}$, $y^C \in \mathbb{R}^{\frac{n^2(n-1)^2}{2}}$:

$$x_{ij} = \begin{cases} 1 & \text{if } (i,j) \in C, \\ 0 & \text{otherwise,} \end{cases} \qquad y_{ijkl} = \begin{cases} 1 & \text{if } (i,j),(k,l) \in C, \\ 0 & \text{otherwise.} \end{cases}$$

The QAP polytope QAP_n is the convex hull of all vectors (x^C, y^C) where C is a clique with n vertices in G_n. It turns out (see Kaibel [400]) that the traveling salesman polytope and the linear ordering polytope are projections of QAP_n. Moreover, QAP_n is a face of the so-called *Boolean quadric polytope*. (For a definition and study of the Boolean quadric polytope see, e.g., Padberg [518]).

Barvinok [73], Padberg and Rijal [519], and Jünger and Kaibel [396] independently computed the dimension of QAP_n and showed that the inequalities $y_{ijkl} \geq 0, i \neq k, j \neq l$, are facet defining. The corresponding facets are usually called *trivial facets* of QAP_n. Moreover, Padberg and Rijal as well as Jünger and Kaibel independently showed that the affine hull of QAP_n is described by the following equations:

$$\sum_{j=1}^{n} x_{ij} = 1 \qquad (i = 1, 2, \ldots, n), \tag{7.63}$$

$$\sum_{i=1}^{n} x_{ij} = 1 \qquad (j = 1, 2, \ldots, n), \tag{7.64}$$

$$-x_{kl} + \sum_{\substack{i=1 \\ i \neq k}}^{n} y_{ijkl} = 0 \qquad (i, j, k = 1, 2, \ldots, n; i \neq k), \tag{7.65}$$

$$-x_{ij} + \sum_{l=1}^{j-1} y_{ijkl} + \sum_{l=j+1}^{n} y_{ijkl} = 0 \qquad (i, j, l = 1, 2, \ldots, n; j \neq l). \tag{7.66}$$

This result was strengthened by Barvinok [73], Padberg and Rijal [519], and Jünger and Kaibel [396].

Theorem 7.11.

(i) *The affine hull of the QAP polytope QAP_n is given by the linear equations (7.63)–(7.66). Redundant equations can be eliminated, resulting in a set of equations which are linearly independent. The rank of such an equation system is $2n(n-1)^2 - (n-1)(n-2)$ for $n \geq 3$.*

(ii) *The dimension of QAP_n is equal to $1 + (n-1)^2 + n(n-1)(n-2)(n-3)/2$ for $n \geq 3$.*

(iii) *The inequalities $y_{ijkl} \geq 0$, $i < k$, $j \neq l$, define facets of QAP_n.*

Padberg and Rijal [519] identified two additional classes of valid inequalities for QAP_n, the so-called *clique* inequalities and the *cut* inequalities. They gave conditions under which the cut inequalities are not facet defining. A larger class of valid inequalities, the so-called *box inequalities*, was described by Jünger and Kaibel [397]. These inequalities are obtained by exploiting the relationship between the Boolean quadric polytope and the QAP polytope. A nice feature of the box inequalities is that it can be decided efficiently whether they are facet defining or not. In the latter case some facet defining inequalities which dominate the corresponding box inequality can be derived.

The case of symmetric coefficient matrices A and B of a Koopmans–Beckmann problem leads to another linear description of the QAP which takes the symmetry into account. The corresponding polytope is known as the *symmetric QAP polytope* $SQAP_n$. Padberg and Rijal [519] modified the system of equations that describes QAP_n to obtain a minimal linear description of the affine hull of $SQAP_n$. Jünger and Kaibel [396] proved that the dimension of $SQAP_n$ is $(n-1)^2 + n^2(n-3)^2/4$ (which was already conjectured by Padberg and Rijal). Moreover, they introduced a class of facet defining inequalities, the so-called *curtain inequalities*. The separation problem for these inequalities has been shown to be $\mathcal{NP}$-hard.

7.5 Gilmore–Lawler bound and reduction methods

Since the QAP is $\mathcal{NP}$-hard, good lower bounds are of great importance for solving it by implicit enumeration procedures like branch-and-bound. A good lower bound for a combinatorial optimization problem should be tight and easy to compute compared with solving the problem itself. In this and in the next sections we describe lower bounds based on linearizations and eigenvalues, as well as more recent approaches that use semidefinite programming and ideas from convex analysis.

7.5.1 Gilmore–Lawler bound

The term *Gilmore–Lawler bound* (GLB) comes from folklore: Gilmore [311] introduced this bound in 1962 for Koopmans–Beckmann problems, and the next year Lawler [447] extended it to general QAPs.

Let us consider a Koopmans–Beckmann problem $QAP(A, B, C)$. Without loss of generality we can assume that all entries in matrices A and B are nonnegative. Since the diagonal entries of A and B may enter the linear part of the QAP by defining

$$c_{ij} := c_{ij} + a_{ii}b_{jj},$$

we do not consider these entries in the following. For each row index i let $\hat{a}_i$ be the $(n-1)$-dimensional vector obtained from the ith row of A by deleting the element a_{ii}. Similarly, define $\hat{b}_j$ for every row j of matrix B. According to Proposition 5.8 we get the minimum scalar product

$$\langle a, b\rangle^- = \min_{\varphi} \sum_{i=1}^{n-1} a_i b_{\varphi(i)}$$

of two vectors $a, b \in \mathbb{R}^{n-1}$ by sorting the elements of a nondecreasingly and the elements of b nonincreasingly.

Let us suppose that a solution φ maps i to $j = \varphi(i)$. We can fix the indices i and j and get a lower bound for

$$\sum_{k=1}^{n} a_{ik} b_{j\varphi(k)}$$

by computing the minimum scalar product $\langle \hat{a}_i, \hat{b}_j \rangle^-$ and adding the term $a_{ii}b_{jj}$. So, in order to find a lower bound on the optimum value of a QAP instance, we first compute the n^2 minimum scalar products $\langle \hat{a}_i, \hat{b}_j \rangle^-$ and define a new cost matrix $L = (l_{ij})$ by

$$l_{ij} = c_{ij} + a_{ii}b_{jj} + \langle \hat{a}_i, \hat{b}_j \rangle^- \quad (i, j = 1, 2, \ldots, n). \tag{7.67}$$

We obtain the Gilmore–Lawler lower bound for the Koopmans–Beckmann QAP by solving the LSAP with cost matrix L. The appropriate sorting of the rows and columns of A and B can be done in $O(n^2 \log n)$ time. The computation of all l_{ij} values takes $O(n^3)$ time, and the same amount of time is needed to solve the last LSAP. Thus the Gilmore–Lawler bound for Koopmans–Beckmann QAPs can be computed in $O(n^3)$ time. This bound is easy to compute, but it deteriorates quickly with increasing size n of the problem.

Example 7.12. Let us consider a Koopmans–Beckmann problem $QAP(A, B, C)$ with the following data:

$$A = \begin{pmatrix} 7 & 2 & 1 & 7 \\ 0 & 0 & 4 & 2 \\ 3 & 0 & 5 & 6 \\ 2 & 3 & 4 & 0 \end{pmatrix} \quad B = \begin{pmatrix} 1 & 7 & 3 & 1 \\ 2 & 1 & 0 & 6 \\ 0 & 4 & 2 & 6 \\ 0 & 0 & 3 & 7 \end{pmatrix} \quad \text{and } C = 0.$$

Now we get

$$\begin{aligned} \hat{a}_1 &= (2, 1, 7), \quad \hat{b}_1 = (7, 3, 1); \\ \hat{a}_2 &= (0, 4, 2), \quad \hat{b}_2 = (2, 0, 6); \\ \hat{a}_3 &= (3, 0, 6), \quad \hat{b}_3 = (0, 4, 6); \\ \hat{a}_4 &= (2, 3, 4), \quad \hat{b}_4 = (0, 0, 3). \end{aligned}$$

Therefore, the matrix

$$(\langle \hat{a}_i, \hat{b}_j \rangle^-) = \begin{pmatrix} 20 & 10 & 14 & 3 \\ 10 & 4 & 8 & 0 \\ 15 & 6 & 12 & 0 \\ 27 & 18 & 24 & 6 \end{pmatrix},$$

together with matrices C and

$$(a_{ii}b_{jj}) = \begin{pmatrix} 7 & 7 & 14 & 49 \\ 0 & 0 & 0 & 0 \\ 5 & 5 & 10 & 35 \\ 0 & 0 & 0 & 0 \end{pmatrix},$$

yields

$$L = \begin{pmatrix} 27 & 17 & 28 & 52 \\ 10 & 4 & 8 & 0 \\ 20 & 11 & 22 & 35 \\ 27 & 18 & 24 & 6 \end{pmatrix}.$$

The optimal solution of the linear assignment problem with cost matrix L is the permutation $\varphi = (2, 3, 1, 4)$ with optimum value $z = 51$. Therefore, we get the lower bound 51 for the optimum value of this QAP instance.

We can also easily compute the exact objective function value for the permutation $\varphi = (2, 3, 1, 4)$ by permuting the rows and columns of matrix B according to φ,

$$B_\varphi = (b_{\varphi(i)\varphi(k)}) = \begin{pmatrix} 1 & 0 & 2 & 6 \\ 4 & 2 & 0 & 6 \\ 7 & 3 & 1 & 1 \\ 0 & 3 & 0 & 7 \end{pmatrix},$$

and computing the objective function value $\langle A, B_\varphi \rangle$, which for this feasible solution is equal to 104. ■

Given a Koopmans–Beckmann problem, it is an $\mathcal{NP}$-complete problem to decide whether the corresponding GLB equals the optimum value, as shown by Li, Pardalos, Ramakrishnan, and Resende [457].

Proposition 7.13. *Let $z^*(A, B)$ be the optimum value of an instance of $QAP(A, B)$, and let $GLB(A, B)$ be the corresponding Gilmore–Lawler bound. It is $\mathcal{NP}$-complete to decide whether $GLB(A, B) = z^*(A, B)$.*

Proof. We reduce the Hamiltonian cycle problem to our decision problem. Let $G = (V; E)$ be an undirected simple graph with n nonisolated vertices. We define A to be the adjacency matrix of a cycle of length n. Let $B = (b_{jl})$ be defined by

$$b_{jl} = \begin{cases} 0 & \text{for } j = l, \\ 1 & \text{for } (j, l) \in E, \\ 2 & \text{otherwise.} \end{cases} \tag{7.68}$$

Every row of matrix A has $n - 1$ entries 0 and just one nondiagonal entry 1. The smallest nondiagonal element in every row of matrix B is 1. Therefore, the GLB has the value n due to the fact that all ordered products are 1. On the other hand, $z^*(A, B) = n$ if and only if graph $G = (V; E)$ contains a Hamiltonian cycle. □

Other authors proposed bounding strategies similar to those by Gilmore and Gomory. The basic idea relies again on decomposing the given QAP into a number of subproblems which can be solved efficiently: first solve each subproblem, then build a matrix with the optimal values of the subproblems, and solve a linear assignment problem with such matrix as cost matrix to obtain a lower bound for the given QAP. For example, Christofides and Gerrard [188] decomposed the Koopmans–Beckmann $QAP(A, B)$ based on isomorphic subgraphs of graphs whose weighted adjacency matrices are A and B. The GLB is obtained

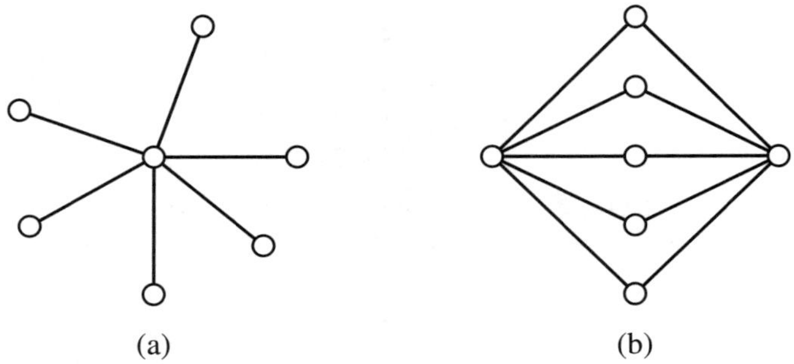

Figure 7.2. (a) *A star graph;* (b) *a double star graph.*

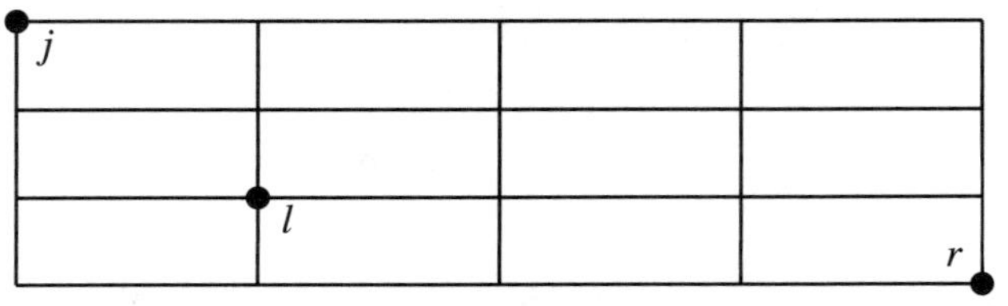

Figure 7.3. *A shortest path triple* (j, l, r) *in a rectangular grid.*

as a special case if these subgraphs are stars (see Figure 7.2(a)), and it generally outperforms the bounds obtained by employing other subgraphs, such as single edges or double stars (see Figure 7.2(b)) introduced by Gavett and Plyter [304].

In the case of a metric QAP the GLB can be strengthened by decomposing the flow matrix A. Let $z^*(A, B)$ be the optimal value of $QAP(A, B)$. For $A = A_1 + A_2$ we get

$$\min_{\varphi}\langle A_1 + A_2, B_\varphi\rangle \geq \min_{\varphi}\langle A_1, B_\varphi\rangle + \min_{\varphi}\langle A_2, B_\varphi\rangle.$$

Therefore,

$$z^*(A, B) \geq z^*(A_1, B) + z^*(A_2, B).$$

We call a triple (j, l, r) of locations a *shortest path triple* if l lies on a shortest path from j to r, i.e.,

$$b_{jl} + b_{lr} = b_{jr}. \tag{7.69}$$

Figure 7.3 shows a shortest path triple in the case where the locations are grid points in a rectangular grid in which the distances are measured by the l_1-metric. Now let us associate the following symmetric matrix T to the shortest path triple (j, l, r):

$$t_{jl} = t_{lj} = t_{lr} = t_{rl} = 1, \quad t_{jr} = t_{rj} = -1, \quad \text{and } t_{ik} = 0 \text{ otherwise.}$$

Chakrapani and Skorin-Kapov [179] showed the following.

Lemma 7.14. *If T is the matrix associated with a shortest path triple, then we get* $z^*(\alpha T, B) = 0$ *for all* $\alpha \geq 0$.

Proof. For any permutation φ we have

$$\langle \alpha T, B_\varphi \rangle = 2\alpha(b_{\varphi(j)\varphi(l)} + b_{\varphi(l)\varphi(r)} - b_{\varphi(j)\varphi(r)}) \geq 0$$

due to the triangle inequality. If φ is the identical permutation we get, by (7.69), $\langle \alpha T, B \rangle = 0$, which shows the lemma. $\square$

By choosing matrix $A_1 = A - \alpha T$ and $A_2 = \alpha T$, where matrix T corresponds to a shortest path triple, Lemma 7.14 guarantees that the Gilmore–Lawler bound (GLB) of $QAP(A_1, B)$ yields a lower bound for $z^*(A, B)$. For improving the GLB of $QAP(A, B)$, one can choose a set $\mathcal{T}$ of shortest path triples together with suitable positive constants and compute the GLB of the problem $QAP(A - \sum \alpha_s T_s, B)$ where the matrices T_s correspond to shortest path triples in $\mathcal{T}$. This technique has been used by Chakrapani and Skorin-Kapov [179] to generate improved Gilmore–Lawler bounds. Palubetskis [524] used such a decomposition to generate QAPs with known optimal objective function values. A further improvement was suggested in Palubeckis [544] where simple graphs other than triangles are used for the decomposition.

7.5.2 Reduction methods

The quality of the Gilmore–Lawler bound may be improved if the given problem is transformed so that the contribution of the quadratic term in the objective function is decreased by moving costs to the linear term. This is the aim of the so-called *reduction methods*. The reduction method was introduced by Conrad [195] and has been subsequently investigated by many researchers, such as Burkard [126], Roucairol [591], Christofides, Mingozzi, and Toth [189], Edwards [252], and Frieze and Yadegar [287]. The general idea is to decompose each quadratic cost coefficient into several terms so as to guarantee that some of them end up being linear cost coefficients and can be moved to the linear term of the objective function.

In the case of a Koopmans–Beckmann QAP the general decomposition scheme is

$$a_{ik} = \overline{a}_{ik} + r_k + u_i \quad (i, k = 1, 2, \ldots, n; i \neq k), \tag{7.70}$$

$$b_{jl} = \overline{b}_{jl} + s_l + v_j \quad (j, l = 1, 2, \ldots, n; j \neq l), \tag{7.71}$$

with real vectors $r, s, u, v \in \mathbb{R}^n$. Frieze and Yadegar [287] have shown that the inclusion of vectors u and v does not affect the value of lower bound GLB. Therefore, these vectors are redundant and will no longer be considered. Since we can add the terms $a_{ii}b_{jj}$ to the linear term c_{ij} by redefining $c_{ij} = c_{ij} + a_{ii}b_{jj}$, we only need to consider a reduction of a_{ik} for $i \neq k$ and of b_{jl} for $j \neq l$. In the following we assume that $a_{ii} = \overline{a}_{ii} = 0$ and $b_{jj} = \overline{b}_{jj} = 0$.

From (7.70) and (7.71) we have, for $i \neq k$ and $j \neq l$,

$$\overline{a}_{ik}\overline{b}_{jl} = (a_{ik} - r_k)(b_{jl} - s_l) = a_{ik}b_{jl} - r_k b_{jl} - s_l c_{ik} + r_k s_l.$$

Therefore, we get

$$\sum_{i=1}^{n}\sum_{k=1}^{n} a_{ik}b_{\varphi(i)\varphi(k)} + \sum_{k=1}^{n} c_{k\varphi(k)} = \sum_{i=1}^{n}\sum_{k=1}^{n} \overline{a}_{ik}\overline{b}_{\varphi(i)\varphi(k)} + \sum_{k=1}^{n} \overline{c}_{k\varphi(k)}$$

with

$$\overline{c}_{kl} = c_{kl} + r_k \sum_{j=1}^{n} b_{jl} + s_l \sum_{i=1}^{n} a_{ik} - (n-1) r_k s_l. \tag{7.72}$$

Several rules have been suggested on how to choose r_k and s_l. A choice which guarantees that, after reduction, all reduced elements $\overline{a}_{ik}$ and $\overline{b}_{jl}$ are nonnegative is

$$\begin{aligned} r_k &= \min\{a_{ik} : 1 \leq i \leq n; i \neq k\} \quad (k = 1, 2, \dots, n), \\ s_l &= \min\{b_{jl} : 1 \leq j \leq n; j \neq l\} \quad (l = 1, 2, \dots, n). \end{aligned}$$

Example 7.15. Let us consider the same Koopmans–Beckmann problem as in Example 7.12. First we split the objective function of the problem into a linear and a quadratic term by redefining

$$c_{kl} = c_{kl} + a_{kk} b_{ll} \quad (k, l = 1, 2, \dots, n)$$

and setting $a_{ii} = b_{ii} = 0$ for $i = 1, 2, \dots, n$. We obtain

$$C = \begin{pmatrix} 7 & 7 & 14 & 49 \\ 0 & 0 & 0 & 0 \\ 5 & 5 & 10 & 35 \\ 0 & 0 & 0 & 0 \end{pmatrix}.$$

The reduction of matrix A yields

$$r_1 = r_2 = 0, \ r_3 = 1, \ \text{and } r_4 = 2$$

and the reduced matrix

$$\overline{A} = \begin{pmatrix} 0 & 2 & 0 & 5 \\ 0 & 0 & 3 & 0 \\ 3 & 0 & 0 & 4 \\ 2 & 3 & 3 & 0 \end{pmatrix}.$$

The reduction of matrix B yields

$$s_1 = s_2 = s_3 = 0 \text{ and } s_4 = 1$$

and the reduced matrix

$$\overline{B} = \begin{pmatrix} 0 & 7 & 3 & 0 \\ 2 & 0 & 0 & 5 \\ 0 & 4 & 0 & 5 \\ 0 & 0 & 3 & 0 \end{pmatrix}.$$

The linear part becomes, according to (7.72),

$$\overline{C} = \begin{pmatrix} 7 & 7 & 14 & 49 \\ 0 & 0 & 0 & 0 \\ 5 & 5 & 10 & 35 \\ 0 & 0 & 0 & 0 \end{pmatrix} + \begin{pmatrix} 0 & 0 & 0 & 5 \\ 0 & 0 & 0 & 5 \\ 2 & 11 & 6 & 19 \\ 4 & 22 & 12 & 35 \end{pmatrix} = \begin{pmatrix} 7 & 7 & 14 & 54 \\ 0 & 0 & 0 & 5 \\ 7 & 16 & 16 & 54 \\ 4 & 22 & 12 & 35 \end{pmatrix}.$$

Now we can apply the Gilmore–Lawler bound to the reduced matrices $\overline{A}$ and $\overline{B}$. We get

$$\begin{aligned} \hat{a}_1 &= (2,0,5), \quad \hat{b}_1 = (7,3,0), \\ \hat{a}_2 &= (0,3,0), \quad \hat{b}_2 = (2,0,5), \\ \hat{a}_3 &= (3,0,4), \quad \hat{b}_3 = (0,4,5), \\ \hat{a}_4 &= (2,3,3), \quad \hat{b}_4 = (0,0,3), \end{aligned}$$

from which

$$(\langle \hat{a}_i, \hat{b}_j \rangle^-) = \begin{pmatrix} 6 & 4 & 8 & 0 \\ 0 & 0 & 0 & 0 \\ 9 & 6 & 12 & 0 \\ 23 & 16 & 22 & 6 \end{pmatrix}.$$

Adding this matrix to $\overline{C}$ yields

$$L = \begin{pmatrix} 13 & 11 & 22 & 54 \\ 0 & 0 & 0 & 5 \\ 16 & 22 & 28 & 54 \\ 27 & 38 & 34 & 41 \end{pmatrix}.$$

The optimal solution of an LSAP with cost matrix L is the permutation $\varphi = (2,4,1,3)$, which yields a lower bound value equal to 66. This is a slight improvement upon the Gilmore–Lawler bound of 51. The objective function value of the QAP for this permutation is 97. ■

Li, Pardalos, Ramakrishnan, and Resende [457] proposed choosing the reduction terms r_k and s_l so as to minimize the variance of all off-diagonal entries of matrix $\overline{A}$ and $\overline{B}$. Let $m(A_j)$ be the arithmetic mean of the off-diagonal entries in the jth column of matrix A:

$$m(A_j) = \frac{1}{n-1} \sum_{\substack{i=1 \\ i \neq j}}^{n} a_{ij}. \tag{7.73}$$

The value $m(B_j)$ is defined in an analogous way. The choice proposed in [457] is

$$r_k = m(A_n) - m(A_k), \tag{7.74}$$

$$s_l = m(B_n) - m(B_l). \tag{7.75}$$

Computational tests showed that this choice leads, in most of the tested cases, to a better value than the Gilmore–Lawler bound and that its performance increases with increasing values of n. Pardalos, Ramakrishnan, and Resende also developed another reduction scheme, where matrices A and B are replaced by $A + \Delta(A)$ and $B + \Delta(B)$. Matrices $\Delta(A)$ and $\Delta(B)$ do not have constant row entries as above. Computational tests showed, however, that their choice of $\Delta(A)$ and $\Delta(B)$ does not lead to an essential improvement of the bound computed by (7.74) and (7.75), although the bounds based on $\Delta(A)$ and $\Delta(B)$ are much more complicated to compute. Therefore, we refrain here from a detailed description of these bounds and refer the interested reader to [457].

Frieze and Yadegar [287] showed an interesting connection between the Gilmore–Lawler bound and a Lagrangean relaxation of their linearization. Let us recall the Frieze–Yadegar linearization (7.44)–(7.50). By relaxing constraints (7.45) and (7.46) via Lagrangean multipliers α_{jkl} and β_{ikl}, respectively, we obtain

$$
\begin{aligned}
L(\alpha, \beta) = \min \quad & \sum_{i=1}^{n}\sum_{j=1}^{n}\sum_{k=1}^{n}\sum_{l=1}^{n} d_{ijkl} y_{ijkl} + \sum_{i=1}^{n}\sum_{j=1}^{n} e_{ij} x_{ij} \\
\text{s.t.} \quad & \sum_{k=1}^{n} y_{ijkl} = x_{ij} && (i, j, l = 1, 2, \dots, n), \\
& \sum_{l=1}^{n} y_{ijkl} = x_{ij} && (i, j, k = 1, 2, \dots, n), \\
& 0 \le y_{ijkl} \le 1 && (i, j, k, l = 1, 2, \dots, n), \\
& X \in \mathbf{X}_n,
\end{aligned}
$$

where

$$
\begin{aligned}
d_{ijkl} &= a_{ik} b_{jl} - \alpha_{jkl} - \beta_{ikl}, \\
e_{ij} &= \sum_{l=1}^{n} \alpha_{lij} + \sum_{k=1}^{n} \beta_{kij}.
\end{aligned}
$$

Frieze and Yadegar [287] showed the following.

Proposition 7.16. *Any lower bound obtained by the Gilmore–Lawler method with reduction cannot be larger than*

$$L^* = \max_{\alpha,\beta} L(\alpha, \beta).$$

According to a result by Geoffrion [308] the value L^* equals the minimum objective function value of the linear relaxation of the Frieze–Yadegar linearization. Frieze and Yadegar proposed a subgradient method for approximately computing the value L^*.

In branch-and-bound algorithms and in some heuristics, we have to compute lower bounds on the objective function value of a QAP under the additional constraint that some of the indices are already assigned and therefore fixed. Let $M \subset \{1, 2, \dots, n\}$ and let $\varphi(i)$ be fixed for $i \in M$. We call (M, φ) a *partial permutation.*

Given a partial permutation (M, φ), how can we find a lower bound for

$$\min_{\varphi} \sum_{i=1}^{n}\sum_{k=1}^{n} a_{ik} b_{\varphi(i)\varphi(k)}$$

under the side constraint that $\varphi(i)$ is already fixed to $\overline{\varphi}(i)$ for $i \in M$? We can rewrite the objective function as

$$\sum_{i\in M}\sum_{k\in M} a_{ik} b_{\overline{\varphi}(i)\overline{\varphi}(k)} + \sum_{i\in M}\sum_{k\notin M} \left(a_{ik} b_{\overline{\varphi}(i)\varphi(k)} + a_{ki} b_{\varphi(k)\overline{\varphi}(i)}\right) + \sum_{i\notin M}\sum_{k\notin M} a_{ik} b_{\varphi(i)\varphi(k)}. \quad (7.76)$$

The first term in (7.76) is a constant, since $\overline{\varphi}(i)$ and $\overline{\varphi}(k)$ are fixed. For the second term in (7.76) a lower bound can be derived by solving an LSAP with cost coefficients $\overline{c}_{kl}$ for $k \notin M$ and $l \notin \overline{\varphi}(M) = \{\overline{\varphi}(i) \;:\; i \in M\}$ defined by

$$\overline{c}_{kl} = a_{kk}b_{ll} + \sum_{i \in M} \left(a_{ik}b_{\overline{\varphi}(i)l} + a_{ki}b_{l\overline{\varphi}(i)}\right).$$

The third term in (7.76) can be viewed as the objective function of a QAP with problem size $n - |M|$. Therefore, the previously derived Gilmore–Lawler and reduction bounds can be applied to this term.

7.6 Admissible transformations and other bounds

Let us consider a general QAP (7.2) and denote the objective function value of a feasible solution X by $z(D, X)$. In analogy to Definition 6.19 we have the following.

Definition 7.17. *The transformation T which transforms the cost coefficients D of* (7.2) *into $\overline{D}$ is* admissible *with index $z(T)$ if for all $X \in \mathbf{X}_n$ the equation*

$$z(D, X) = z(T) + z(\overline{D}, X) \tag{7.77}$$

holds.

When we perform an admissible transformation T with index $z(T)$ after an admissible transformation S with index $z(S)$, we get again an admissible transformation. It has the index $z(S) + z(T)$. Theorem 6.21 can now be expressed as follows.

Proposition 7.18. *Let T be an admissible transformation of D into $\overline{D}$ and let $X^* \in \mathbf{X}_n$ be such that*

1. $\overline{D} \geq 0$;
2. $z(\overline{D}, X^*) = 0$.

Then X^* is an optimal solution of (7.2) *with objective function value $z(T)$.*

Proof. Let X be any feasible solution of (7.2). According to Definition 7.17 and assumptions 1 and 2 above we get

$$z(D, X) = z(T) + z(\overline{D}, X) \geq z(T) = z(T) + z(\overline{D}, X^*) = z(D, X^*),$$

which means that X^* is an optimal solution with value $z(T)$. □

7.6.1 Admissible transformations

We start from the Kronecker product formulation (7.17)–(7.19) where the cost coefficients d_{ijkl} are arranged in an $n^2 \times n^2$ matrix D such that the entry d_{ijkl} lies in row $(i-1)n + k$ and column $(j-1)n + l$. Remember that we can write D as a matrix (D^{ij}) where every

D^{ij} is the matrix (d_{ijkl}) with fixed indices i and j. The coefficients d_{ijil} with $j \neq l$ and the coefficients d_{ijkj} with $i \neq k$ can never occur in the objective function since φ is a one-to-one mapping. Therefore, we can assume

$$d_{ijkl} = \infty \quad \text{for } (i = k \text{ and } j \neq l) \text{ or } (i \neq k \text{ and } j = l). \tag{7.78}$$

For $n = 3$, for example, the cost matrix has the form (using "$*$" for ∞)

$$D = \begin{pmatrix} D^{11} & D^{12} & D^{13} \\ D^{21} & D^{22} & D^{23} \\ D^{31} & D^{32} & D^{33} \end{pmatrix}$$

$$= \left(\begin{array}{ccc|ccc|ccc} d_{1111} & * & * & * & d_{1212} & * & * & * & d_{1313} \\ * & d_{1122} & d_{1123} & d_{1221} & * & d_{1223} & d_{1321} & d_{1322} & * \\ * & d_{1132} & d_{1133} & d_{1231} & * & d_{1233} & d_{1331} & d_{1332} & * \\ \hline * & d_{2112} & d_{2113} & d_{2211} & * & d_{2213} & d_{2311} & d_{2312} & * \\ d_{2121} & * & * & * & d_{2222} & * & * & * & d_{2323} \\ * & d_{2132} & d_{2133} & d_{2231} & * & d_{2233} & d_{2331} & d_{2332} & * \\ \hline * & d_{3112} & d_{3113} & d_{3211} & * & d_{3213} & d_{3311} & d_{3312} & * \\ * & d_{3122} & d_{3123} & d_{3221} & * & d_{3223} & d_{3321} & d_{3322} & * \\ d_{3131} & * & * & * & d_{3232} & * & * & * & d_{3333} \end{array}\right).$$

Now we get the following.

Proposition 7.19. (Admissible transformations for the QAP.) *Let $D = (D^{ij})$ be the cost matrix of a general QAP* (7.2). *We will consider two types of transformation:*

I. *Let D^{ij} be a fixed submatrix of D. Adding arbitrary constants to the rows and columns of D^{ij} such that the constants sum up to 0 yields an admissible transformation T with index $z(T) = 0$.*

II. *Adding arbitrary constants to the rows and columns of D yields an admissible transformation T whose index $z(T)$ equals the negative sum of the constants.*

Remark: Since every submatrix D^{ij} has only one entry $d_{ijij} \neq \infty$ in row i and column j (*leading element*), the first part of Proposition 7.19 tells that we can add or subtract arbitrary constants to the rows $k \neq i$ and columns $l \neq j$, provided that we, respectively, subtract or add the same constants to the leading element d_{ijij}. The second part of Proposition 7.19 tells us that when we can collect the leading elements d_{ijij} in an $n \times n$ matrix $L = (l_{ij})$ with $l_{ij} = d_{ijij}$ and apply any admissible transformations to this LSAP, we get an admissible transformation for the QAP.

Proof.

I. From the Hungarian method (see Section 4.2.1) we know that adding constants u_k to the rows and v_l to the columns of matrix D^{ij} yields, for $\overline{d}_{ijkl} = d_{ijkl} + u_k + v_l$,

$$\langle D^{ij}, X \rangle = -\left(\sum_{k=1}^{n} u_k + \sum_{l=1}^{n} v_l \right) + \langle \overline{D}^{ij}, X \rangle \quad \text{for all } X \in \mathbf{X_n}.$$

Since we assume $\sum_k u_k + \sum_l v_l = 0$, we can rewrite the objective function of (7.2) as

$$\sum_{i=1}^{n}\sum_{j=1}^{n}\sum_{k=1}^{n}\sum_{l=1}^{n} d_{ijkl}x_{ij}x_{kl} = \sum_{i=1}^{n}\sum_{j=1}^{n}\langle D^{ij}, X\rangle x_{ij} = \sum_{i=1}^{n}\sum_{j=1}^{n}\langle \overline{D}^{ij}, X\rangle x_{ij}, \tag{7.79}$$

thus getting $z(T) = 0$.

II. Applying the same argument to the $n^2 \times n^2$ matrix D yields

$$\langle D, Y\rangle = -\left(\sum_{r=1}^{n^2} u_r + \sum_{s=1}^{n^2} v_s\right) + \langle \overline{D}, Y\rangle \quad \text{for all } Y \in \mathbf{X}_{n^2};$$

hence, in particular for $Y = X \otimes X$ with $X \in \mathbf{X_n}$, we get an admissible transformation with $z(T) = -(\sum_{r=1}^{n^2} u_r + \sum_{s=1}^{n^2} v_s)$. □

7.6.2 General Gilmore–Lawler bound

Lawler [445] suggested the following way to compute a lower bound for (7.2). For every pair (i, j) we solve the LSAP with cost matrix D^{ij} and, according to (7.73), we need only to solve an LSAP of size $n - 1$. Let u_k $(k \neq i)$ and v_l $(l \neq j)$ be the corresponding dual variables. By setting

$$\overline{d}_{ijij} = d_{ijij} + \sum_{\substack{k=1\\k\neq i}}^{n} u_k + \sum_{\substack{l=1\\l\neq j}}^{n} v_l;$$

$$\overline{d}_{ijkl} = d_{ijkl} - u_k - v_l \qquad (k, l = 1, 2, \ldots, n;\ k \neq i, l \neq j),$$

we get an admissible transformation. The dual variables u_k and v_l can be chosen such that $\overline{d}_{ijkl} \geq 0$. Now we collect the leading elements in an $n \times n$ matrix and solve the linear assignment problem with the cost coefficients $l_{ij} = \overline{d}_{ijij}$. This again produces an admissible transformation whose index is denoted as z_L. Since $\overline{D} \geq 0$, the value z_L is a lower bound on all feasible solutions of the QAP. Let φ^* and the corresponding permutation matrix X^* be the optimal solution of the LSAP with cost matrix $L = (l_{ij})$. According to Proposition 7.18 the permutation φ^* is an optimal solution of the QAP if

$$\sum_{i=1}^{n}\langle \overline{D}^{i\varphi^*(i)}, X^*\rangle = 0. \tag{7.80}$$

As several authors noted (Burkard [126], Frieze and Yadegar [287], Hahn and Grant [358]), a further strengthening of the bound is possible by taking into account that $x_{ij}x_{kl} = x_{kl}x_{ij}$. Due to such a relation we can set

$$d_{ijkl} = \begin{cases} d_{ijkl} + d_{klij} & \text{for } k > i, \\ 0 & \text{for } k < i, \end{cases} \tag{7.81}$$

and apply afterward the bounding procedure described above.

Example 7.20. Let the cost coefficients of a general QAP be given by

$$D = \left(\begin{array}{ccc|ccc|ccc} 10 & \ast & \ast & \ast & 2 & \ast & \ast & \ast & 3 \\ \ast & 5 & 3 & 7 & \ast & 4 & 4 & 1 & \ast \\ \ast & 1 & 2 & 3 & \ast & 2 & 8 & 9 & \ast \\ \hline \ast & 1 & 2 & 1 & \ast & 5 & 4 & 2 & \ast \\ 2 & \ast & \ast & \ast & 2 & \ast & \ast & \ast & 3 \\ \ast & 5 & 5 & 8 & \ast & 9 & 1 & 8 & \ast \\ \hline \ast & 7 & 8 & 9 & \ast & 1 & 4 & 3 & \ast \\ \ast & 1 & 4 & 1 & \ast & 8 & 1 & 2 & \ast \\ 9 & \ast & \ast & \ast & 9 & \ast & \ast & \ast & 0 \end{array} \right).$$

The symmetrization (7.81) yields the new cost matrix

$$\left(\begin{array}{ccc|ccc|ccc} 10 & \ast & \ast & \ast & 2 & \ast & \ast & \ast & 3 \\ \ast & 6 & 7 & 8 & \ast & 6 & 6 & 6 & \ast \\ \ast & 10 & 6 & 10 & \ast & 5 & 16 & 10 & \ast \\ \hline \ast & 0 & 0 & 0 & \ast & 0 & 0 & 0 & \ast \\ 2 & \ast & \ast & \ast & 2 & \ast & \ast & \ast & 3 \\ \ast & 6 & 6 & 9 & \ast & 11 & 5 & 16 & \ast \\ \hline \ast & 0 & 0 & 0 & \ast & 0 & 0 & 0 & \ast \\ \ast & 0 & 0 & 0 & \ast & 0 & 0 & 0 & \ast \\ 9 & \ast & \ast & \ast & 9 & \ast & \ast & \ast & 0 \end{array} \right).$$

Solving the nine LSAPs with cost matrices D^{ij}, we get the following matrix, which contains the optimum value of each subproblem in its leading element:

$$\left(\begin{array}{ccc|ccc|ccc} 22 & \ast & \ast & \ast & 15 & \ast & \ast & \ast & 19 \\ \ast & 0 & 1 & 0 & \ast & 0 & 0 & 0 & \ast \\ \ast & 4 & 0 & 3 & \ast & 0 & 6 & 0 & \ast \\ \hline \ast & 0 & 0 & 0 & \ast & 0 & 0 & 0 & \ast \\ 8 & \ast & \ast & \ast & 11 & \ast & \ast & \ast & 8 \\ \ast & 0 & 0 & 0 & \ast & 2 & 0 & 11 & \ast \\ \hline \ast & 0 & 0 & 0 & \ast & 0 & 0 & 0 & \ast \\ \ast & 0 & 0 & 0 & \ast & 0 & 0 & 0 & \ast \\ 9 & \ast & \ast & \ast & 9 & \ast & \ast & \ast & 0 \end{array} \right).$$

In order to find a lower bound we collect the leading elements in a cost matrix $L = (l_{ij})$:

$$L = \begin{pmatrix} 22 & 15 & 19 \\ 8 & 11 & 8 \\ 9 & 9 & 0 \end{pmatrix}.$$

The optimal solution for the LSAP with cost matrix L is given by

$$x_{12} = x_{21} = x_{33} = 1 \tag{7.82}$$

with value 23. Therefore, 23 is a lower bound for the optimal objective function value of the given QAP. Moreover, for all permutation matrices X, the equation

$$\langle L, X \rangle = 23 + \langle \overline{L}, X \rangle$$

with

$$\overline{L} = \begin{pmatrix} 7 & 0 & 4 \\ 0 & 3 & 0 \\ 9 & 9 & 0 \end{pmatrix}$$

holds. If we replace the leading entries in the transformed cost matrix of the QAP by the new values $\overline{l}_{ij}$, we get a new cost matrix $\overline{D}$:

$$\left(\begin{array}{ccc|ccc|ccc} 7 & * & * & * & 0 & * & * & * & 4 \\ * & 0 & 1 & 0 & * & 0 & 0 & 0 & * \\ * & 4 & 0 & 3 & * & 0 & 6 & 0 & * \\ \hline * & 0 & 0 & 0 & * & 0 & 0 & 0 & * \\ 0 & * & * & * & 3 & * & * & * & 0 \\ * & 0 & 0 & 0 & * & 2 & 0 & 11 & * \\ \hline * & 0 & 0 & 0 & * & 0 & 0 & 0 & * \\ * & 0 & 0 & 0 & * & 0 & 0 & 0 & * \\ 9 & * & * & * & 9 & * & * & * & 0 \end{array}\right).$$

Now it is easy to verify that (7.80) holds. Therefore, the solution $Y = X \otimes X$ with X defined by (7.82) yields the objective function value 0 for the reduced QAP with objective function $\sum_{i=1}^{n} \sum_{j=1}^{n} \langle \overline{D}^{ij}, X \rangle x_{ij}$, which shows that (7.82) is an optimal solution of the given QAP. ∎

Hahn and Grant [358] suggested the following procedure to further strengthen this bound in the case $\overline{L} \neq 0$. They distribute the leading values $\overline{l}_{ij}$ equally to the other elements of the transformed matrices $\overline{D}^{ij}$, for example,

$$\overline{d}_{ijkl} = \begin{cases} \overline{d}_{ijkl} + \overline{l}_{ij}/(n-1) & \text{for } i \neq k \text{ and } j \neq l, \\ 0 & \text{for } i = k \text{ or } j = l. \end{cases}$$

This is an admissible transformation of the first type in Proposition 7.19. Then, a symmetrization step is again applied and the new leading elements are collected as coefficients of a new LSAP (admissible transformation of the second type). If the solution of this LSAP has a positive objective function value, this value can be added to the previous lower bound, which is thereby improved. The procedure can be iterated until no further improvement occurs.

Similar approaches which apply alternately admissible transformations of types I and II were suggested by Assad and Xu [47] as well as by Carraresi and Malucelli [171]. They differ only in the choice of the parameters of the transformation.

The bounds obtained in this section are closely related to a Lagrangean relaxation obtained from the Adams–Johnson [5] linearization (7.53)–(7.58). Adams and Johnson added the complementarity constraints

$$y_{ijkl} = y_{klij} \quad (i, j, k, l = 1, 2, \ldots, n) \tag{7.83}$$

via Lagragean parameters α_{ijkl} to the objective function (7.53), thus obtaining the Lagrangean function $L(\alpha)$. Again due to Geoffrion [308] we get that $\max_\alpha L(\alpha)$ equals the optimum value of the continuous relaxation of (7.53)–(7.58). For maximizing $L(\alpha)$ Adams and Johnson proposed an iterative dual ascent method. At each iteration $L(\alpha)$ is evaluated for the fixed value α. Clearly, $L(\alpha)$ is a lower bound for the considered QAP. Then the multipliers α_{ijkl} are updated by using the information contained in the dual variables of the LSAPs solved during the previous iteration. The algorithm stops after having performed a prespecified number of iterations. Adams and Johnson showed that $L(0)$ equals the Gilmore–Lawler bound, whereas GLBs obtained after applying reductions as well as the bounds of Carraresi and Malucelli [171] and Assad and Xu [47] equal $L(\alpha)$ for special settings of the Lagrangean multipliers α_{ijkl}.

Karisch, Çela, Clausen, and Espersen [404] considered the dual of the continuous relaxation of the mixed-integer linear program (7.53)–(7.58) proposed by Adams and Johnson. They developed an iterative algorithm for approximately solving this dual and showed that the Hahn-Grant bound, the Adams–Johnson bound $L(\alpha)$, and all other Gilmore–Lawler-like bounds can be obtained by applying this algorithm with specific settings for the control parameters. Moreover, they identified a setting of parameters which produces a bound that is competitive with the Hahn-Grant bound, but provides a better time/quality trade-off.

Resende, Ramakrishnan, and Drezner [578] used an interior point approach to solve the continuous relaxation of the Adams–Johnson linearization, which turns out to be highly degenerate. This method yields rather strong bounds, but it is computationally expensive. It has been reported that the bounding algorithm by Karisch, Çela, Clausen, and Espersen yields bounds of about the same quality with an effort which is at least one order of magnitude less than the interior point method.

7.7 Eigenvalue bounds

The trace formulation of Koopmans–Beckmann problems is the basis for a new class of bounds, the so-called *eigenvalue bounds* which were introduced in 1987 by Finke, Burkard, and Rendl [270]. We consider first the case of a Koopmans–Beckmann problem *QAP*(A, B) with symmetric matrices A and B. Later, we comment on how to proceed in the non-symmetric case. When implemented carefully, eigenvalue bounds are quite strong in comparison with Gilmore–Lawler bounds, but they are usually expensive to compute. Moreover, they deteriorate quickly when lower levels in a branch-and-bound tree are searched (see Clausen, Karisch, Perregaard, and Rendl [191]).

In this section we make extensive use of the properties of the trace operator (see Section 7.2.1).

7.7.1 Symmetric Koopmans–Beckmann problems

Let A and B be real symmetric matrices. In this case all their eigenvalues are real. We collect the eigenvalues $\lambda_1, \lambda_2, \ldots, \lambda_n$ of matrix A in a vector λ and the eigenvalues $\mu_1, \mu_2, \ldots, \mu_n$ of matrix B in a vector μ. For any permutation matrix X_φ, the matrix $B_\varphi^T = X_\varphi B^T X_\varphi^T$ has the same eigenvalues as B. Let $\text{diag}(\lambda)$ denote a square matrix whose elements are all zero, except for those on the main diagonal, which have the values $\lambda_1, \lambda_2, \ldots, \lambda_n$. The matrices A and B_φ^T possess diagonalizations of the form $A = P\Lambda P^T$ and $B_\varphi^T = QMQ^T$ with orthogonal matrices P and Q and diagonal matrices $\Lambda = \text{diag}(\lambda)$ and $M = \text{diag}(\mu)$.

Let $p_1, p_2, \dots, p_n$ denote the columns of P and $q_1, q_2, \dots, q_n$ the columns of Q. Then

$$A = \sum_{i=1}^{n} \lambda_i p_i p_i^T \quad \text{and} \quad B_\varphi^T = \sum_{i=1}^{n} \mu_i q_i q_i^T.$$

Therefore,

$$\langle A, B_\varphi \rangle = \operatorname{tr}(AB_\varphi^T) = \operatorname{tr}\left(\sum_{i=1}^{n} \sum_{j=1}^{n} \lambda_i \mu_j p_i p_i^T q_j q_j^T \right) = \sum_{i=1}^{n} \sum_{j=1}^{n} \lambda_i \mu_j \langle p_i, q_j \rangle^2.$$

Since $p_1, \dots, p_n$ and $q_1, \dots, q_n$ are orthonormal bases of $\mathbb{R}^n$, we get

$$\sum_{j=1}^{n} \langle p_i, q_j \rangle^2 = \langle p_i, p_i \rangle = 1 \quad \text{and} \quad \sum_{i=1}^{n} \langle p_i, q_j \rangle^2 = \langle q_j, q_j \rangle = 1.$$

This shows that matrix $S_\varphi = (\langle p_i, q_j \rangle^2)$ is a doubly stochastic matrix. According to Birkhoff's theorem 2.18, matrix S_φ can be written as a convex combination of permutation matrices X_r:

$$S_\varphi = \sum_r \alpha_r X_r.$$

Thus we get

$$\operatorname{tr}(AB_\varphi^T) = \sum_r \alpha_r \langle \lambda, X_r \mu \rangle. \tag{7.84}$$

Let us denote by $\langle \lambda, \mu \rangle^-$ the minimum scalar product of the vectors λ and μ,

$$\langle \lambda, \mu \rangle^- = \min_\varphi \sum_{i=1}^{n} \lambda_i \mu_{\varphi(i)}$$

and let

$$\langle \lambda, \mu \rangle^+ = \max_\varphi \sum_{i=1}^{n} \lambda_i \mu_{\varphi(i)}.$$

Proposition 5.8 tells us how to compute $\langle \lambda, \mu \rangle^-$ and $\langle \lambda, \mu \rangle^+$. Equation (7.84) yields

$$\langle \lambda, \mu \rangle^- \le \sum_r \alpha_r \langle \lambda, X_r \mu \rangle \le \langle \lambda, \mu \rangle^+.$$

Thus we have shown (see Finke, Burkard, and Rendl [270]) the following.

Proposition 7.21. *All objective function values of a Koopmans–Beckmann problem with symmetric matrices A and B lie in the interval*

$$\langle \lambda, \mu \rangle^- \le \sum_{i=1}^{n} \sum_{k=1}^{n} a_{ik} b_{\varphi(i)\varphi(k)} \le \langle \lambda, \mu \rangle^+. \tag{7.85}$$

The bound given by Proposition 7.21 is not very strong, as A and B may have negative eigenvalues. One way to improve the bound is to apply reductions to matrices A and B. The *spread* $\operatorname{sp}(A)$ of a symmetric matrix A is defined as the difference between its largest and smallest eigenvalue

$$\operatorname{sp}(A) = \max_{i,j} |\lambda_i - \lambda_j|. \tag{7.86}$$

Unfortunately, there is no simple formula to compute the spread. Mirsky [495] suggested the following approximation for $\operatorname{sp}(A)$:

$$\operatorname{sp}(A) \leq \left(2\sum_{i=1}^{n}\sum_{k=1}^{n} a_{ik}^2 - \frac{2}{n}\left(\sum_{i=1}^{n} a_{ii}\right)^2 \right)^{1/2} = m(A). \tag{7.87}$$

Minimizing $m(A)$ leads to a system of linear equations from which the coefficients of the reduction can be computed explicitly. The reduction of $A = (a_{ik})$ to $\overline{A} = (\overline{a}_{ik})$ has the form

$$a_{ik} = \overline{a}_{ik} + \alpha_i + \alpha_k + \gamma_{ik},$$

where the coefficients α_i and γ_{ik} are computed by

$$z = \frac{1}{2(n-1)}\left(\sum_{i=1}^{n}\sum_{k=1}^{n} a_{ik} - \sum_{i=1}^{n} a_{ii}\right),$$

$$\alpha_i = \frac{1}{n-2}\left(\sum_{k=1}^{n} a_{ik} - a_{ii} - z\right),$$

$$\gamma_{ik} = \begin{cases} a_{ii} - 2\alpha_i & \text{for } i = k, \\ 0 & \text{otherwise.} \end{cases}$$

The reduced matrix $\overline{A}$ is again symmetric and has zeroes in the diagonal. Moreover, every row and column sum of $\overline{A}$ equals 0. Analogous values z, β_j, and δ_{jl} can be computed, for reducing $B = (b_{ij})$ to $\overline{B} = (\overline{b}_{ij})$, by replacing a_{ik} with b_{ik} in the equations above.

Using these properties we can derive the following lower bound for QAPs. We reduce the matrices A and B by

$$a_{ik} = \overline{a}_{ik} + \alpha_i + \alpha_k + \gamma_{ik},$$
$$b_{jl} = \overline{b}_{jl} + \beta_j + \beta_l + \delta_{jl}.$$

Then we get

$$\sum_{i=1}^{n}\sum_{k=1}^{n} a_{ik} b_{\varphi(i)\varphi(k)} = \sum_{i=1}^{n}\sum_{k=1}^{n} \overline{a}_{ik}\overline{b}_{\varphi(i)\varphi(k)} + \sum_{i=1}^{n}\sum_{k=1}^{n} \overline{b}_{\varphi(i)\varphi(k)}(\alpha_i + \alpha_k + \gamma_{ik})$$
$$+ \sum_{i=1}^{n}\sum_{k=1}^{n} \overline{a}_{ik}(\beta_{\varphi(i)} + \beta_{\varphi(k)} + \delta_{\varphi(i)\varphi(k)}).$$

Since all row and column sums of the symmetric matrix $\overline{A}$ are zero, $\overline{a}_{ii} = 0$ and $\delta_{ik} = 0$ for $i \neq k$, and we get

$$\sum_{i=1}^{n}\sum_{k=1}^{n} \overline{a}_{ik}(\beta_{\varphi(i)} + \beta_{\varphi(k)} + \delta_{\varphi(i)\varphi(k)}) = 0.$$

Moreover, since B is symmetric, we have

$$\sum_{i=1}^{n}\sum_{k=1}^{n} b_{\varphi(i)\varphi(k)}(\alpha_i + \alpha_k + \gamma_{ik}) = \sum_{i=1}^{n} a_{ii} b_{\varphi(i)\varphi(i)} + 2\sum_{i=1}^{n} \alpha_i \sum_{\substack{k=1\\k\neq i}}^{n} b_{\varphi(i)\varphi(k)}.$$

We can collect the linear terms in a cost matrix $C = (c_{ij})$ of a linear assignment problem:

$$c_{ij} = a_{ii} b_{jj} + 2\alpha_i \sum_{\substack{k=1\\k\neq j}}^{n} b_{jk}.$$

Let us denote the eigenvalues of $\overline{A}$ by $\overline{\lambda}_1, \dots, \overline{\lambda}_n$ and the eigenvalues of $\overline{B}$ by $\overline{\mu}_1, \dots, \overline{\mu}_n$. Then we get the new lower bound

$$\langle \overline{\lambda}, \overline{\mu} \rangle^- + \min_{\varphi} \sum_{i=1}^{n} c_{i\varphi(i)}. \tag{7.88}$$

Computational experiments showed that for large-size problems the bound (7.88) is stronger than the Gilmore–Lawler bound.

Example 7.22. (Rendl [572]). Let a Koopmans–Beckmann problem be given by

$$A = \begin{pmatrix} 0 & 1 & 1 & 3 \\ 1 & 0 & 4 & 4 \\ 1 & 4 & 0 & 5 \\ 3 & 4 & 5 & 0 \end{pmatrix} \quad \text{and} \quad B = \begin{pmatrix} 0 & 1 & 3 & 1 \\ 1 & 0 & 4 & 2 \\ 3 & 4 & 0 & 2 \\ 1 & 2 & 2 & 0 \end{pmatrix}.$$

The reduction of A yields

$$z = 6; \alpha_1 = -\tfrac{1}{2}, \alpha_2 = \tfrac{3}{2}, \alpha_3 = 2, \alpha_4 = 3; \gamma_{11} = 1, \gamma_{22} = -3, \gamma_{33} = -4, \gamma_{44} = -6$$

and

$$\overline{A} = \frac{1}{2}\begin{pmatrix} 0 & 0 & -1 & 1 \\ 0 & 0 & 1 & -1 \\ -1 & 1 & 0 & 0 \\ 1 & -1 & 0 & 0 \end{pmatrix}$$

with the eigenvalues

$$\overline{\lambda}_1 = -1, \quad \overline{\lambda}_2 = \overline{\lambda}_3 = 0, \quad \overline{\lambda}_4 = 1.$$

The reduction of B yields the values

$$z = \tfrac{13}{3}; \beta_1 = \tfrac{1}{3}, \beta_2 = \tfrac{4}{3}, \beta_3 = \tfrac{7}{3}, \beta_4 = \tfrac{1}{3}; \delta_{11} = -\tfrac{2}{3}, \delta_{22} = -\tfrac{8}{3}, \delta_{33} = -\tfrac{14}{3}, \delta_{44} = -\tfrac{2}{3}$$

and

$$\overline{B} = \frac{1}{3} \begin{pmatrix} 0 & -2 & 1 & 1 \\ -2 & 0 & 1 & 1 \\ 1 & 1 & 0 & -2 \\ 1 & 1 & -2 & 0 \end{pmatrix}$$

with the eigenvalues

$$\overline{\mu}_1 = \overline{\mu}_2 = \tfrac{2}{3}, \ \ \overline{\mu}_3 = 0, \ \ \overline{\mu}_4 = -\tfrac{4}{3}.$$

Since $a_{ii} = 0$ for all $i = 1, 2, \ldots, n$, the linear assignment problem

$$\min_{\varphi} \sum_{i=1}^{n} c_{i\varphi(i)}$$

can be solved just by taking the minimum scalar product of the vectors $2\alpha = (-1, 3, 4, 6)$ and $b = (5, 7, 9, 5)$, which yields the value 62. Therefore, we get the lower bound

$$\langle \overline{\lambda}, \overline{\mu} \rangle^- + \langle 2\alpha, b \rangle^- = -2 + 62 = 60.$$

The Gilmore–Lawler bound yields for this example the value 59. The optimal value of this Koopmans–Beckmann instance is known to be 60. ∎

Rendl and Wolkowicz [576] proposed a different reduction scheme. Obviously, the eigenvalue bound given by (7.88) is a function on the $4n$ parameters

$$(\alpha_1, \alpha_2, \ldots, \alpha_n;\ \beta_1, \beta_2, \ldots, \beta_n;\ \gamma_{11}, \gamma_{22}, \ldots, \gamma_{nn};\ \delta_{11}, \delta_{22}, \ldots, \delta_{nn}).$$

Maximizing this function yields a best possible bound using reductions. Rendl and Wolkowicz showed that this function is nonlinear, nonsmooth, and nonconcave. They applied a steepest ascent method for approximately maximizing the bound. The new bound produces good results, but is expensive to compute.

It is well known that the set of permutation matrices can be characterized as $\mathbf{X}_n = \mathcal{O}_n \cap \mathcal{E}_n \cap \mathcal{N}_n$, where

$\mathcal{O}_n = \{X \ : \ X^T X = I\}$	is the set of *orthogonal* $n \times n$ *matrices,*
$\mathcal{E}_n = \{X \ : \ Xe = X^T e = e\}$	is the set of $n \times n$ *matrices with row and column sums equal to one, and*
$\mathcal{N}_n = \{X \ : \ X \geq 0\}$	is the set of *nonnegative* $n \times n$ *matrices.*

Here, I denotes the $n \times n$ identity matrix and $e = (1, 1, \ldots, 1)^T$ is the n-dimensional vector of all ones. Therefore, the QAP can be relaxed by deleting one or two of the matrix sets $\mathcal{O}_n$, $\mathcal{E}_n$, and $\mathcal{N}_n$ in the intersection $\mathbf{X}_n = \mathcal{O}_n \cap \mathcal{E}_n \cap \mathcal{N}_n$. Rendl and Wolkowicz [576] have shown that the eigenvalue bound is obtained by relaxing the feasible set to the set of orthogonal matrices:

$$\min_{X \in \mathcal{O}_n} \operatorname{tr}(AXB^T X^T) = \langle \lambda, \mu \rangle^-,$$

$$\max_{X \in \mathcal{O}_n} \operatorname{tr}(AXB^T X^T) = \langle \lambda, \mu \rangle^+.$$

For a short proof of this result see Rendl [574].

A tighter relaxation was proposed in Hadley, Rendl, and Wolkowicz [353, 354], who relaxed the set of permutation matrices to $\mathcal{O}_n \cap \mathcal{E}_n$. They observed that the solution of the equation systems

$$Xe = e \text{ and } X^T e = e \tag{7.89}$$

can explicitly be written as

$$X = \frac{1}{n}ee^T + VYV^T, \tag{7.90}$$

where Y is an $(n-1) \times (n-1)$ matrix, while V is an $n \times (n-1)$ matrix having rank $n-1$ and satisfying $V^T e = 0$. Thus V is a basis of the orthogonal complement of vector e. Since X is supposed to be orthogonal, it is assumed that $V^T V = I_{n-1}$ holds, where I_{n-1} denotes the $(n-1) \times (n-1)$ identity matrix. In this case we have

$$X^T X = I \text{ if and only if } Y^T Y = I_{n-1}.$$

Substituting X in the objective function $\operatorname{tr}(AXB^T X^T)$ and exploiting the symmetry of matrices A and B yields

$$\operatorname{tr}(AXBX^T) = \operatorname{tr}\left(A\left(\frac{1}{n}ee^T + VYV^T\right)B\left(\frac{1}{n}ee^T + VYV^T\right)^T\right)$$

$$= \operatorname{tr}\left(\frac{1}{n^2}Aee^TBee^T + \frac{1}{n}Aee^TBVY^TV^T + \frac{1}{n}AVYV^TBee^T + AVYV^TBVY^TV^T\right)$$

$$= \frac{(e^TAe)(e^TBe)}{n^2} + \operatorname{tr}\left(\frac{2}{n}Aee^TBVY^TV^T + AVYV^TBVY^TV^T\right).$$

Since

$$VYV^T = X - \frac{1}{n}ee^T,$$

we have

$$\operatorname{tr}\left(\frac{2}{n}Aee^TBVY^TV^T\right) = \operatorname{tr}\left(\frac{2}{n}Aee^TBX - \frac{2(e^TAe)(e^TBe)}{n^2}\right).$$

Moreover,

$$\operatorname{tr}(AVYV^TBVY^TV^T) = \operatorname{tr}\left((V^TAV)Y(V^TBV)Y^T\right).$$

So, let $\tilde{A} = V^TAV$ and $\tilde{B} = V^TBV$. The matrices $\tilde{A}$ and $\tilde{B}$ are $(n-1) \times (n-1)$ matrices. We collect their eigenvalues in the vectors $\lambda_{\tilde{A}}$ and $\mu_{\tilde{B}}$, respectively. Thus we get the *projection bound* by Hadley, Rendl, and Wolkowicz [353, 354].

Proposition 7.23. *Let*

$$D = \frac{2}{n}Aee^TB.$$

The projection bound for the symmetric QAP(A, B) is given by

$$\langle \lambda_{\tilde{A}}, \mu_{\tilde{B}} \rangle^- + \min_{\varphi} d_{i\varphi(i)} - \frac{(e^TAe)(e^TBe)}{n^2}. \tag{7.91}$$

Karisch and Rendl [405] studied the projection bound in connection with metric QAPs by applying a triangle reduction as discussed at the end of Section 7.5.1.

7.7.2 Nonsymmetric Koopmans–Beckmann problems

All bounds considered above rely on the fact that the eigenvalues of a real, symmetric matrix are real. This is no longer the case when we consider nonsymmetric matrices A and B. Let us first consider the case where only one of the matrices A and B is nonsymmetric. In the case that matrix A is symmetric, we get the following.

Lemma 7.24. *Let A and B be real $n \times n$ matrices, matrix A be symmetric, and $\hat{B} = 1/2(B + B^T)$. Then we have, for any permutation matrix X,*

$$\operatorname{tr}(AXB^TX^T) = \operatorname{tr}(AX\hat{B}^TX^T). \tag{7.92}$$

In particular, if B is skew symmetric, then $\operatorname{tr}(AXB^TX^T) = 0$.

Proof. The proof relies on some simple properties of the trace operator (see Section 7.2.1). We have

$$\begin{aligned}\operatorname{tr}(AX\hat{B}^TX^T) &= \frac{1}{2}\left(\operatorname{tr}(AXB^TX^T) + \operatorname{tr}(AXBX^T)\right)\\ &= \frac{1}{2}\left(\operatorname{tr}(AXB^TX^T) + \operatorname{tr}(AXB^TX^T)\right) = \operatorname{tr}(AXB^TX^T).\end{aligned}$$

The second part of the lemma follows from $\hat{B} = 0$, if B is skew symmetric. □

Thus the lemma tells us that we can symmetrize the second matrix if one matrix is already symmetric, and apply the methods discussed in the previous section.

In the case that both matrices A and B are nonsymmetric, Hadley, Rendl, and Wolkowicz [352] found a nice way to derive eigenvalue bounds for the problem. They transformed the given matrices to certain complex Hermitian matrices which yield the same objective function value. Recall that a matrix H is *Hermitian* if H equals the transposed, conjugate complex matrix H^*. If H is real, then $H^* = H^T$. Every Hermitian matrix has real eigenvalues. Note that in the remainder of this section "i" will denote $\sqrt{-1}$. Let us define

$$A_+ = \frac{1}{2}(A + A^T), \quad A_- = \frac{1}{2}(A - A^T), \tag{7.93}$$

$$\tilde{A}_+ = \frac{1}{2}(A_+ + iA_-), \quad \tilde{A}_- = \frac{1}{2}(A_+ - iA_-). \tag{7.94}$$

Note that A_+ is symmetric, A_- is skew symmetric, and $\tilde{A}_+$ as well as $\tilde{A}_-$ are Hermitian, i.e., $\tilde{A}_+ = \tilde{A}_+^*$ and $\tilde{A}_- = \tilde{A}_-^*$. $\tilde{A}_+$ is the positive Hermitian part of A and $\tilde{A}_-$ is the negative Hermitian part of A. In particular, $\tilde{A}_+$ and $\tilde{A}_-$ have only real eigenvalues. Hadley, Rendl, and Wolkowicz [352] have shown the following.

Proposition 7.25. *Let A, B, and X be any real $n \times n$ matrices. Then*

$$\operatorname{tr}(AXB^TX^T) = \operatorname{tr}(\tilde{A}_+X\tilde{B}_+^*X^T). \tag{7.95}$$

Proof. Since A_+ and B_+ are symmetric and A_- and B_- are skew symmetric, we can apply Lemma 7.24 and get

$$\begin{aligned}\operatorname{tr}(AXB^TX^T) &= \operatorname{tr}\left((A_+ + A_-)X(B_+ + B_-)^TX^T\right)\\ &= \operatorname{tr}\left((A_+ + A_-)X(B_+ - B_-)X^T\right)\\ &= \operatorname{tr}\left(A_+XB_+X^T\right) - \operatorname{tr}\left(A_-XB_-X^T\right).\end{aligned}$$

On the other hand, by applying Lemma 7.24 to the skew symmetric matrices A_- and B_- we again get

$$\begin{aligned}\operatorname{tr}(\tilde{A}_+X\tilde{B}_+^*X^T) &= \operatorname{tr}\left((A_+ + iA_-)X(B_+ + iB_-)^*X^T\right)\\ &= \operatorname{tr}\left((A_+ + iA_-)X(B_+ + iB_-)X^T\right)\\ &= \operatorname{tr}\left(A_+XB_+X^T\right) - \operatorname{tr}\left(A_-XB_-X^T\right),\end{aligned}$$

which establishes (7.95). □

An immediate consequence of this proposition is the following.

Proposition 7.26. *Let a Koopmans–Beckmann problem with real matrices A and B be given. Let* $\lambda = (\lambda_1, \lambda_2, \ldots, \lambda_n)$ *be the vector of the eigenvalues of the Hermitian matrix* $\tilde{A}_+$ *and let* $\mu = (\mu_1, \mu_2, \ldots, \mu_n)$ *be the vector of the eigenvalues of the Hermitian matrix* $\tilde{B}_+$*. Then we have, for any permutation* φ*,*

$$\langle \lambda, \mu \rangle^- \le \sum_{i=1}^{n}\sum_{k=1}^{n} a_{ik} b_{\varphi(i)\varphi(k)} \le \langle \lambda, \mu \rangle^-. \tag{7.96}$$

Up until now, not many computational experiments have been performed with this bound. In particular it seems that a sharpening of this bound by some kind of reduction methods has not yet been investigated.

7.8 Bounds by semidefinite programming

In this section we consider again only symmetric Koopmans–Beckmannn problems, i.e., we suppose that both matrices A and B in $QAP(A, B)$ are symmetric. Recall that an $n \times n$ matrix U is called *positive semidefinite* if

$$x^TUx \ge 0 \text{ for all } x \in \mathbb{R}^n.$$

We shall express the fact that U is positive semidefinite by $U \succeq 0$. (Indeed, positive semidefiniteness leads to an order relation, the so-called Löwner ordering.) In Section 7.2.3 we formulated the quadratic term of the Koopmans–Beckmann QAP in the form

$$\min x^T(B \otimes A)x$$

$$\text{s.t. } X \in \mathbf{X}_n,$$

where $x = \mathrm{vec}(X)$. Now,

$$x^T(B \otimes A)x = \mathrm{tr}\left((B \otimes A)(xx^T)\right).$$

By defining the $n^2 \times n^2$ matrix Y by xx^T we can write the QAP in the form

$$\begin{aligned} &\min \mathrm{tr}\,((B \otimes A)Y) \\ &\text{s.t. } Y = xx^T, \\ &\qquad X \in \mathbf{X}_n. \end{aligned}$$

When we want to linearize this problem we have to choose Y in the convex hull

$$P = \mathrm{conv}\{xx^T : x = \mathrm{vec}(X),\ X \in \mathbf{X}_n\}.$$

It can be shown that the matrix $Y - yy^T$ is positive semidefinite for $Y \in P$ and $y = \mathrm{diag}(Y)$. This immediately leads to a first semidefinite relaxation of $QAP(A, B)$. Zhao, Karisch, Rendl, and Wolkowicz [667] showed that the constraint $Y \succeq 0$ can be strenghtened by exploiting the fact that all row and column sums of a permutation matrix X are equal to 1 (similar to the projection bound discussed in Section 7.7). They defined an $n \times (n-1)$ matrix V by

$$V = \begin{pmatrix} I_{n-1} \\ -e_{n-1}^T \end{pmatrix}$$

and an $n^2 \times ((n-1)^2 + 1)$ matrix W by

$$W = \left(\frac{1}{n} e \otimes e,\ V \otimes V\right)$$

and proved the following.

Lemma 7.27. *Let $Y \in P$. Then there exists a symmetric, positive semidefinite matrix R of order $(n-1)^2 + 1$, indexed from 0 to $(n-1)^2$ such that*

$$r_{00} = 1 \quad \textit{and} \quad Y = WRW^T.$$

This leads to the following semidefinite relaxation of the QAP:

$$\begin{aligned} &\min \mathrm{tr}(B \otimes A)Y \\ &\text{s.t. } Y = WRW^T, \\ &\qquad Y - \mathrm{diag}(Y)\,\mathrm{diag}(Y)^T \succeq 0, \\ &\qquad r_{00} = 1. \end{aligned} \tag{7.97}$$

Though this semidefinite program is formulated in the variable Y, which represents matrices from P, the actual degrees of freedom are given by matrix R. Thus during computations the variable Y is eliminated by the substitution $Y = WRW^T$. The nonlinear constraint

$$Y - \mathrm{diag}(Y)\,\mathrm{diag}(Y)^T \succeq 0$$

can be rewritten as

$$\begin{pmatrix} 1 & r^T \\ r & WRW^T \end{pmatrix} \succeq 0 \quad \text{and} \quad r = \mathrm{diag}(WRW^T).$$

This bound can be further strengthened by taking into account that all entries in matrix Y are nonnegative and the symmetry conditions $y_{ijkl} = y_{klij}$ hold (recall that $y_{ijkl} = x_{ij}x_{kl} = x_{kl}x_{ij}$). Moreover, one can impose the so-called *pattern constraints* (see (7.78)):

$$y_{ijkl} = 0 \quad \text{for } (i = k \text{ and } j \neq l) \text{ or } (i \neq k \text{ and } j = l). \tag{7.98}$$

Finally, one can impose the orthogonality constraints $X^T X = XX^T = I$ on X, which are quadratic in X but linear in Y.

Rendl and Sotirov [575] showed that interior point methods are not well suited to solving a semidefinite program with nonnegativity constraints $y_{ijkl} \geq 0$ and the pattern constraints (7.98). They developed a bundle method based on subgradient optimization for handling these additional constraints in a semidefinite program and obtained good results. For details, see Rendl and Sotirov [575] as well as the recent research report by Sotirov and Wolkowicz [616] where the authors study several semidefinite programming relaxations of the QAP, exploit their structure, and develop efficient solution techniques.

A lift-and-project method for solving the semidefinite relaxation problem was recently proposed by Burer and Vandenbussche [125]. A cutting plane algorithm based upon a semidefinte relaxation of the QAP has been developed by Faye and Roupin [265].

7.9 Bounds by convex quadratic programming

Anstreicher and Brixius [34] developed a novel powerful bound for the symmetric Koopmans–Beckmann problem which is based on convexity arguments. With the help of this bound, it was possible for the first time to solve the hard test example nug30 [512] with size $n = 30$ to optimality (see Section 8.1.4). The idea of Anstreicher and Brixius is based on the identity

$$0 = \mathrm{tr}\left(S(I - XX^T)\right) = \mathrm{tr}(S) - \mathrm{tr}\left((I \otimes S)xx^T\right)$$

which holds for any orthogonal matrix X and any symmetric matrix S. As in the previous section, x denotes $\mathrm{vec}(X)$. This identity immediately yields, for any orthogonal matrix X and any symmetric matrices S and T,

$$\begin{aligned} \mathrm{tr}\left(AXB^T X^T\right) &= \mathrm{tr}(S) + \mathrm{tr}(T) + \mathrm{tr}\left((B \otimes A - I \otimes S - T \otimes I)xx^T\right) \\ &= \mathrm{tr}(S) + \mathrm{tr}(T) + x^T Q x \end{aligned} \tag{7.99}$$

with $Q = (B \otimes A - I \otimes S - T \otimes I)$. If one fixes S and T, the right-hand side of (7.99) becomes a convex, quadratic program in x.

Using diagonalizations of the symmetric matrices A and B, Anstreicher and Brixius [34] showed the following.

Proposition 7.28. *Let A and B be symmetric $(n \times n)$ matrices and let X be an orthogonal matrix. Then*

$$\min_X \operatorname{tr}\left(AXB^TX^T\right) = \max\{\operatorname{tr}(S) + \operatorname{tr}(T) \;:\; B \otimes A - I \otimes S - T \otimes I \succeq 0; \quad S, T \text{ symmetric}\}. \tag{7.100}$$

Anstreicher and Brixius used the optimal dual solutions $s = (s_1, s_2, \dots, s_n)$ and $t = (t_1, t_2, \dots, t_n)$ of (7.100), which correspond to the diagonal entries of the optimal symmetric matrices S and T in the program on the right-hand side of (7.100), and obtained as a new bound for the optimal objective function value $z(A, B)$ of $QAP(A, B, C)$

$$\sum_{i=1}^{n} s_i + \sum_{i=1}^{n} t_i + \min\left\{x^TQx + c^Tx \;:\; X \text{ doubly stochastic}\right\}. \tag{7.101}$$

This bound was used by Anstreicher, Brixius, Goux, and Linderoth [35] in a parallel branch-and-bound algorithm for the QAP, which solved some well-known previously unsolved test problems.

Chapter 8

Quadratic assignment problems: Algorithms

8.1 Exact algorithms

All main algorithmic techniques for the exact solution of $\mathcal{NP}$-hard problems have been used for attacking the QAP: decomposition, branch-and-bound, and branch-and-cut. The QAP, however, bravely resisted. After decades of attacks, the results obtained on its optimal solution are far from being satisfactory with respect to those obtained for most combinatorial optimization problems.

The set of test instances proposed in the late 1960s by Nugent, Vollmann, and Ruml [512] and enlarged in the following years (*Nugent instances*) are widely considered to be "stubborn" QAP instances and have become a standard challenge for exact QAP algorithms. Up until the mid-1990s the largest such instance ever solved to optimality had size equal to 22. Better results came from technological progress. Parallel algorithms implemented on supercomputers could solve the Nugent instances up to size 25 (see Brüngger, Clausen, Marzetta, and Perregaard [121, 122] and Clausen and Perregaard [192]). In recent years, the use of very large scale distributed computation (*grid computing*) allowed for the first time the solution of Nugent instances of size 27, 28, and 30 (see Anstreicher, Brixius, Goux, and Linderoth [35] and Adams, Guignard, Hahn, and Hightower [4]).

QAPs have also been tackled with some success by global optimization methods (see Enkhbat and Javzandulam [256]).

8.1.1 Benders' decomposition

Traditional cutting plane algorithms for the QAP were developed by different authors, such as Bazaraa and Sherali [78, 79], Balas and Mazzola [52, 53, 54], and Kaufman and Broeckx [412]. These algorithms are based on mixed integer linear programming linearizations of the problem (see Section 7.3) which are suitable for Benders' decomposition.

Consider a linearization of the QAP having the form

$$\begin{aligned} &\min d^T y \\ &\text{s.t. } Ex + Fy \geq e, \\ &\quad X \in \mathbf{X}_n, \\ &\quad y \geq 0, \end{aligned}$$

y being continuous variables and x being binary variables such that $x = \text{vec}(X)$. In algorithms based on the decomposition invented by Benders [83], this formulation is decomposed into a master problem and a slave problem. For a fixed assignment $\overline{x}$, we obtain the *slave problem*

$$\begin{aligned} &\min d^T y \\ &\text{s.t. } Fy \geq e - E\overline{x}, \\ &\quad y \geq 0, \end{aligned}$$

which contains only the variables introduced for linearizing the objective function. Hence, the slave is a linear program whose optimal primal and dual solutions can be obtained in polynomial time. The optimal dual variables induce a cut in the master problem. The algorithm starts with any feasible assignment $\overline{x}$. Then it iteratively solves the slave problem, adds the resulting cut to the master problem, and solves the master to obtain a new solution $\overline{x}$, hence, a new slave. Let u_k denote the dual solution of the slave at iteration k: the *master problem* is

$$\begin{aligned} &\min z \\ &\text{s.t. } z \geq u_i(e - Ex) \quad (i = 1, 2, \ldots, k), \\ &\quad X \in \mathbf{X}_n, \end{aligned}$$

where $x = \text{vec}(X)$. The solution of the master problem is a lower bound for the original QAP. On the other hand, the objective function value of the QAP corresponding to any feasible solution $\overline{x}$ is an upper bound. The algorithm terminates when these bounds coincide. For the QAP, the convergence rate is generally poor, except for toy instances. However, heuristics derived from this approach produce good suboptimal solutions in early stages of the search (see, e.g., Burkard and Bönniger [134] and Bazaraa and Sherali [79]). Miranda, Luna, Mateus, and Ferreira [493] used Benders' decomposition for QAP instances arising in a heuristic for electronic components placement problems.

8.1.2 Branch-and-bound

Branch-and-bound algorithms have been successfully used to solve many $\mathcal{NP}$-hard combinatorial optimization problems, and they appear to be the most effective exact algorithms for solving QAPs. All successful methods for the QAP belong to this class.

A branch-and-bound algorithm dynamically subdivides the feasible solutions space into smaller regions, thus generating a series of subproblems represented through the nodes of a *branch-decision tree*. For each node, a bound on the value of the best solution it can produce is computed by solving a relaxation (frequently the LP relaxation) of the corresponding subproblem (*bounding phase*). If this value is not better than that of the best feasible solution found so far for the original problem (*incumbent solution*), the node is fathomed and another node is explored. Otherwise, the region corresponding to the current node is further subdivided, producing new subproblems with smaller feasible regions (*branching phase*). Whenever a new (better) feasible solution for the original problem is obtained at a node, the incumbent solution is updated. The process terminates when no further branching is possible, i.e., when the entire feasible solutions space has been implicitly searched. The basic ingredients of a branch-and-bound algorithm are

- bounding technique;
- branching methodology;
- exploration strategy.

Bounding techniques

We have discussed in the previous sections the main lower bounds developed for the QAP. For more than 30 years all of the most efficient branch-and-bound algorithms for this problem employed the Gilmore–Lawler bound (GLB) discussed in Section 7.5.1. The main reason for this choice was that other bounds, which outperform it in terms of bound quality, are too expensive in terms of computation time.

Only in the last decade have other bounds been successfully incorporated into branch-and-bound algorithms. Pardalos, Ramakrishnan, Resende, and Li [534] solved some previously unsolved instances from the QAPLIB through a branch-and-bound algorithm which employs the variance reduction lower bound proposed in Li, Pardalos, Ramakrishnan, and Resende [457] (see Section 7.5.2). The bound by Hahn and Grant [358] (see Section 7.6.2), based on a dual procedure similar to the Hungarian algorithm for the LSAP (see Section 4.2), was used with good results in the algorithms by Hahn, Grant, and Hall [357] and Hahn, Hightower, Johnson, Guignard, and Roucairol [359] (see also Hahn [355]). Results of comparable quality were obtained by Brixius and Anstreicher [114] with a branch-and-bound algorithm incorporating the convex quadratic programming bound developed in Anstreicher and Brixius [34] (see Section 7.9). Very good results were recently obtained by Adams, Guignard, Hahn, and Hightower [4] with lower bounds computed through the 2-level reformulation-linearization technique (see Section 7.3.4) and a Lagrangean relaxation solved by means of a number of LSAPs.

A special case of the QAP where matrices A and B are symmetric and the non-zero flows form a tree in the graph induced by A (*Tree QAP*) arises in practice in assembly line layout and in pipeline design. The Tree QAP is $\mathcal{NP}$-hard and contains the TSP as a special case. Christofides and Benavent [186] proposed a branch-and-bound algorithm in which the bounds are computed using a Lagrangean relaxation that is solved through dynamic programming.

Branching methodologies

Various types of branching rules have been used for producing the branch-decision tree. In the *single assignment* branching (Gilmore [311], Lawler [447]), which is probably the most efficient methodology, each problem generates subproblems by fixing the location of one of the facilities which is not yet assigned. The selection of the facility-location pair usually depends on the bounding technique: if the GLB bound is adopted, the choice is frequently formulated in terms of the reduced costs of the last assignment problem solved to bound the subproblem which is currently being branched (see, e.g., Bazaraa and Kirca [77], Burkard [130], Mautor and Roucairol [483]).

The *pair assignment* algorithms (Gavett and Plyter [304], Land [445], Nugent, Vollmann, and Ruml [512]) at a branching step assign a pair of facilities to a pair of locations.

In *relative positioning* algorithms (Mirchandani and Obata [494]) the levels of the branch-decision tree do not correspond to the number of facilities already assigned to locations, but the fixed assignments within each subproblem are determined in terms of distances between facilities. Numerical results show that pair assignment or relative positioning algorithms are generally outperformed by single-assignment algorithms.

Another interesting methodology was proposed by Roucairol [592]. *Polytomic* branching does not produce a binary branch-decision tree, as most of the other approaches, but each decision node generates $n+1$ children nodes as follows. Let φ denote the permutation corresponding to the solution of the last linear assignment problem solved to compute the Gilmore–Lawler bound for the current node of the search tree. Let $\mathcal{S}_n^{(i)}$ denote the subset of $\mathcal{S}_n$ (the set of all permutations of $\{1, 2, \ldots, n\}$) consisting of those permutations π such that $\pi(i) = \varphi(i)$ and $\overline{\mathcal{S}}_n^{(i)}$ the set of those permutations such that $\pi(i) \neq \varphi(i)$. The $n+1$ descendant nodes correspond then to the sets of feasible solutions given by

$$
\begin{aligned}
&\overline{\mathcal{S}}_n^{(1)},\\
&\mathcal{S}_n^{(1)} \bigcap \overline{\mathcal{S}}_n^{(2)},\\
&\ldots,\\
&\mathcal{S}_n^{(1)} \bigcap \mathcal{S}_n^{(2)} \bigcap \cdots \bigcap \mathcal{S}_n^{(n-1)} \bigcap \overline{\mathcal{S}}_n^{(n)},\\
&\mathcal{S}_n^{(1)} \bigcap \mathcal{S}_n^{(2)} \bigcap \cdots \bigcap \mathcal{S}_n^{(n)}.
\end{aligned}
$$

The term *polytomic* is also used for the branching scheme adopted by Mautor and Roucairol [483] in which either one facility is assigned to all the available locations (*row branching*) or all available facilities are assigned to one location (*column branching*). This strategy has been used in the most effective recent algorithms for the exact solution of the QAP.

Exploration strategy

This issue concerns the so-called selection rule, which determines the choice of the subproblem to be branched. The adopted strategies range from problem-independent search (such as depth-first or breadth-first) to instance dependent criteria related to the maximization of lower bounds or reduced costs. There seems to be no clear winner among the different strategies.

8.1.3 Branch-and-cut

Branch-and-cut algorithms, developed in the late 1980s (see, e.g., Padberg and Rinaldi [520]), cleverly combine branch-and-bound and traditional cutting plane methods. Similarly to traditional branch-and-bound methods for integer linear programming methods, they are based on the linear relaxation of the problem, but additionally they make use of valid (possibly facet defining) inequalities known to hold for all feasible solutions of the original problem.

Consider any node of the branch-decision tree. If the solution of the relaxation of the current subproblem is feasible for the original problem, the exploration of the node is complete. Otherwise, if some of the above-mentioned inequalities are violated, a *cutting* phase is performed: one or more of the violated inequalities are added to the subproblem relaxation and a new solution is determined. When none of the valid inequalities is violated, but some integrality constraint is, the algorithm performs a branching step.

All the related elements of branch-and-bound algorithms (bounding techniques, branching methodology, exploration strategy) play a role in branch-and-cut algorithms. The key ingredient is, however, the use of cuts which are valid for the whole polytope of the feasible solutions, while traditional cutting plane algorithms frequently use cuts which are only valid for the current node (so the whole computation has to be done from scratch for different variable fixings). In addition, cuts produced by facet defining inequalities can be considerably "stronger" than traditional cuts, thus considerably accelerating the convergence of the algorithm.

As seen in Section 7.4, some properties and few facet defining inequalities of the QAP polytope are known, but stronger theoretical developments are needed for obtaining successful branch-and-cut algorithms for the QAP. Some efforts in this direction have been made by Padberg and Rijal [519], who tested their algorithm on sparse QAP instances from QAPLIB, and by Kaibel [400, 401], who used branch-and-cut to compute lower bounds for QAP instances from QAPLIB. A large family of valid inequalities inducing facets was proposed by Blanchard, Elloumi, Faye, and Wicker [101]. Recently, Erdoğan and Tansel [259] proposed two new integer programs derived from a flow-based linearization technique and reported on computational experiments with a branch-and-cut algorithm.

8.1.4 Parallel and massively parallel algorithms

The extraordinary difficulty of exactly solving QAP instances made this problem a natural candidate for high performance parallel computations (see Section 4.11 for a general introduction to parallel computing). As previously mentioned, classical benchmarks for exact QAP algorithms are the Nugent [512] instances, usually denoted as Nugxx ("xx" indicating that the instance has size $n = xx$). Parallel branch-and-bound algorithms could exactly solve, for the first time, "large" Nugent instances of size 20 or more.

Traditional parallel computers have been used for solving, in the years shown in parentheses:

- nug20 (1997): Clausen and Perregaard [192], in 16 hours, on a MEIKO system of 16 Intel i860 processors, each with 16 MB of memory;

- nug21 and nug22 (1997, 1998): Brüngger, Clausen, Marzetta, and Perregaard [121, 122], respectively, in 2 and 12 days, on a NEC Cenju-3 system using up to 96 processors;

- nug25 (1999): Marzetta and Brüngger [482], in 30 days, on between 64 and 128 processors of an Intel Paragon and 9 DEC Alpha processors.

In recent years, branch-and-bound algorithms for the QAP have been used for grid computing implementations. A *computational grid* is a collection of a large number of geographically distributed heterogeneous computers that are dynamically available during their idle times. The grid is managed through a *resource management software* which detects the available processors and assigns them jobs to execute. This system is perfectly suitable to a parallel branch-and-bound algorithm: a *master machine* handles the branch-decision tree and assigns task (nonexplored nodes) to the *worker machines*, which report the results of their computations back to the master (*Master-Worker paradigm*).

In 2002 Anstreicher, Brixius, Goux, and Linderoth [35] built up a federation of over 2,500 CPUs distributed around the globe, handled through the MW Master-Worker class library (see Goux, Kulkarni, Linderoth, and Yoder [336]) utilizing the Condor [463] resource management software. Using an adaptation of the branch-and-bound algorithm by Brixius and Anstreicher [114], which incorporates the bound by Anstreicher and Brixius [34](see Section 7.9), they could solve the three surviving Nugent instances. The total CPU times were normalized to times on a single HP9000 C3000 workstation:

- nug27 was solved in 65 CPU days;

- nug28 was solved in 10 CPU months;

- nug30 was solved in 7 CPU years.

Another test instance of size 30 required more than 17 years (C3000 equivalent). Together with computations performed for large-size TSPs (see the TSP page at Georgia Tech), these are among the largest CPU times ever spent for solving single instances of combinatorial optimization problems.

In 2007 Adams, Guignard, Hahn, and Hightower [4] computed lower bounds through a reformulation-linearization technique, developed by Sherali and Adams [606, 607, 608] for general 0-1 programming problems, which provides different levels of representation that give increasing strength to the bound (see Section 7.3.4). Bounds from the level-1 form had been implemented in Hahn, Grant, and Hall [357] and Hahn, Hightower, Johnson, Guignard, and Roucairol [359] (see also Hahn [355]). By implementing level-2 bounds the following results were obtained (CPU times are again normalized to times on a single HP9000 C3000):

- nug27 was solved in 20 CPU days;

- nug28 was solved in 5 CPU months;

- nug30 was solved in 2.5 CPU years.

To our knowledge, these are currently the best computational results obtained for the exact solution of these instances.

8.2 Heuristics

We have shown in the previous sections that using the best exact algorithms currently available, and allowing high running times, it is only possible to solve QAPs of small size. The development of heuristic algorithms to find quality solutions in short computing times is thus very reasonable. Due to its practical and theoretical relevance, the QAP has been attacked with several constructive heuristics and tens of pure or hybrid local search methods. For each of the main approaches we give a short description of the method and we present a selection of the most relevant implementations and results.

8.2.1 Construction algorithms

These heuristics are greedy approaches which usually start with an empty solution and iteratively select a facility and assign it to a free location. More formally, at a generic iteration h the current partial solution is given by the set $M \subset \{1, 2, \dots, n\}$ of the h assigned facilities and by the corresponding *partial permutation* (M, φ) (i.e., $\varphi(i)$ is fixed for $i \in M$ and $\varphi(i) = \texttt{null}$ for $i \notin M$). The algorithm starts with $M = \emptyset$ and, at each iteration h, selects a facility i_h, a location j_h and increases the partial solution by setting $\varphi(i_h) = j_h$ and adding i_h to M. The differences among the various greedy approaches are in the rules used to select the current facility i_h and its location j_h

In 1962 Gilmore [311] proposed a basic constructive algorithm whose refined version is due to Burkard [130]. At each iteration h the refined algorithm computes the Gilmore–Lawler bound by solving an LSAP on a $(n - h + 1) \times (n - h + 1)$ matrix (l_{ij}) computed as in Section 7.5.1, but considering only the indices of the facilities and the locations not yet assigned. Let $\widetilde{\varphi}$ be the corresponding optimal solution: the algorithm selects the facility i_h such that $c_{i_h, \widetilde{\varphi}(i_h)} = \max\{c_{i, \widetilde{\varphi}(i)} : i \notin M\}$ and assigns it to location $j_h = \widetilde{\varphi}(i_h)$. The time complexity is $O(n^4)$.

Another construction method which yields good results has been proposed by Müller-Merbach [500]. At each iteration h the facility i_h is simply chosen as the hth element of a prefixed ordering $i_1, i_2, \dots, i_n$ of the indices $1, 2, \dots, n$. To chose the location j_h we compute the increase $\Delta z(j)$ of the objective function value due to the assignment $\varphi(i_h) = j$, for $j = 1, 2, \dots, n$, and select j_h so that $\Delta z(j_h) = \min_j \Delta z(j)$. If the tentative location j has not been assigned to a facility of M, the computation of $\Delta z(j)$ is immediate. Otherwise, if there is a facility $i_\ell \in M$ with $\varphi(i_\ell) = j$, we need to assign i_ℓ to a new location. To compute $\Delta z(j)$ we assign i_ℓ, in turn, to each of the free locations and we select the location, say, l, that minimizes the overall cost increase induced by the assignments $\varphi(i_h) = j$ and $\varphi(i_\ell) = l$. The time complexity of the method is $O(n^3)$ since we perform n iterations and the number of reassignments to be evaluated at a generic iteration h is $n - h - 1$ for each of the $h - 1$ facilities in M. Finally, we observe that the initial order of the facilities is arbitrary, so we can obtain different solutions by running this algorithm several times with different ordering rules.

8.2.2 Limited exact algorithms

It is well known that exact methods are often able to find good solutions at early stages of the search, but employ a lot of time to marginally improve that solution or to prove its optimality.

This behavior suggests a classical way to design a heuristic algorithm: reduce the computing time by imposing some limit to the exact method. To this end we can use several methods: a simple time limit, an upper bound on the number of iterations, a restriction on the choices we can adopt at some decision point, etc.

Graves and Whinston [339] proposed an implicit enumeration algorithm consisting of a branch-decision tree in which the exact lower bound to be computed at each node is substituted by a probabilistic bound which is stronger than the exact one but, in some cases, is not valid. It follows that the algorithm implicitly enumerates more nodes and reduces the overall computing time, but it may disregard some subtree containing the optimal solution. Given a partial solution, the associated probabilistic bound is $\mu - t\sigma$ where μ and σ are, respectively, the mean and standard deviation of the possible completions of such solution (see Section 7.2.4) and t is a parameter related to the confidence level of the bound.

Burkard and Bönniger [134] considered the Balas–Mazzola linearization (7.42)–(7.43), relaxed it by generating a limited number of constraints, and iteratively solved the problem $\min\{z : z \geq \langle H(Y), X\rangle - h(Y)\}$ with Y fixed and corresponding to a single constraint. The algorithm starts with a random permutation matrix Y and a search direction (ℓ^0_{ij}) (possibly empty). At each iteration r the new direction (ℓ^r_{ij}) is obtained from (ℓ^{r-1}_{ij}) by adding matrix H (see (7.40)) normalized with a term that decreases when the current solution has a large value. The solution of an LSAP is involved in the construction of the new permutation. Algorithm 8.1 formally describes the procedure.

ALGORITHM 8.1. Procedure BB_QAP.
Burkard and Bönniger heuristic for QAP.

compute $g_{kl} = (\max_{1\leq p\leq n} a_{kp})(\max_{1\leq q\leq n} b_{lq})$ and set $\ell^0_{kl} := 0$ for $k, l = 1, 2, \ldots, n$;
randomly generate a permutation matrix $Y = (y_{ij})$;

$\overline{X} := Y, \overline{z} := \sum_{k=1}^{n}\sum_{l=1}^{n}\sum_{i=1}^{n}\sum_{j=1}^{n} a_{ik}b_{jl}y_{ij}$;

for $r := 1$ **to** ρ **do**

  $h_{kl} := g_{kl}y_{kl} + \sum_{i=1}^{n}\sum_{j=1}^{n} d_{ijkl}y_{ij}$ for $k, l = 1, 2, \ldots, n$ [**comment:** see (7.40)];
  $h := \langle G, Y\rangle$ [**comment:** see (7.41)];
  solve an LSAP with cost matrix (h_{ij}) and let z be its optimum value;
  $\delta := \max(1, |z - h|)$;
  $\ell^r_{ij} := \ell^{r-1}_{ij} + h_{ij}/\delta$ for $i, j = 1, 2, \ldots, n$;
  solve an LSAP with cost matrix (ℓ^r_{ij}) yielding solution Y and optimum value z^r;
  if $z^r < \overline{z}$ **then** $\overline{z} := z^r$, $\overline{X} := Y$

endfor;
return the solution $\overline{X}$ and its value $\overline{z}$.

The input parameter ρ is used to determine the number of constraints and solutions to be generated. Due to the random nature of the starting solution, the entire procedure can be run several times.

8.2.3 Basic elements of metaheuristic methods

A *metaheuristic* is a generic scheme (or template) for organizing a search in the solution space of an optimization problem in order to find good solutions. The literature distinguishes between *single solution* methods and *population based* methods. The first class includes, e.g., *Simulated Annealing* (SA), *Tabu Search* (TS), and *Greedy Randomized Adaptive Search Procedure* (GRASP), while in the second class we find *Genetic Algorithms* (GA), *Scatter Search* (SS), and *Ant Colony Optimization* (ACO).

The trajectory followed by a solution during the search is often guided by a *neighborhood* function $\mathcal{N}$ which associates with any solution s a portion $\mathcal{N}(s)$ of the solution space containing solutions which are "close" to s. We say that two solutions s and s' are close to each other if s' can be obtained by applying some "simple" operator to s. The term *move* denotes the transformation of s into s'.

A typical metaheuristic iteration explores the neighborhood $\mathcal{N}(s)$ of the current solution s, looking for a solution s' that meets certain requirements, to be accepted as new current solution for the next iteration. The most characteristic feature of these algorithms is that *improving solutions* are not the only ones that can be accepted. In certain situations (for example, when no improving solution exists in $\mathcal{N}(s)$) a move that transforms s into a *worsening solution* s' can be accepted in the hope that the subsequent iterations will drive the trajectory out of a local minimum of the solution space.

The simplest neighborhood for the QAP is the *pairwise exchange*, which associates with a permutation φ all the permutations obtained by swapping the location assignment of two different facilities. This neighborhood is generalized by the *k-exchange* neighborhood, which can be formally described as follows. Let φ and φ' denote two different solutions and define their *distance* as

$$d(\varphi, \varphi') = |\{i : \varphi(i) \neq \varphi'(i)\}|. \tag{8.1}$$

The k-exchange neighborhood of a solution φ is then

$$\mathcal{N}_k(\varphi) = \{\varphi(i)' \in \mathcal{S} : d(\varphi, \varphi') \leq k\} \quad (\text{with } 2 \leq k \leq n). \tag{8.2}$$

A *Lin–Kernighan-like* neighborhood structure (see Murthy, Pardalos, and Li [505]) consists in a sequence of applications of neighborhood $\mathcal{N}_2$. The search starts by finding the best solution in the neighborhood of the current permutation φ and performing the corresponding exchange. The two facilities involved in the exchange are then "frozen," and a new neighborhood search is performed by only allowing swaps within the remaining $n-2$ facilities. The procedure is iterated h times, until either $h = n/2 - 1$ or the best solution of the restricted neighborhood $\mathcal{N}_2$ is nonimproving. Let $\varphi_1, \varphi_2, \dots, \varphi_h$ denote the solutions obtained in the h iterations and compute $\varphi' = \arg\min\{z(\varphi_i) : i = 1, 2, \dots, h\}$, where $z(\varphi)$ denotes the value of solution φ. Permutation φ' is then selected as the neighbor of the starting solution φ. A similar but wider neighborhood was proposed by Li, Pardalos, and Resende [458]. They adopted the same scheme, but instead of "freezing" the swapped locations they avoided the swaps in which *both* facilities have been swapped in a previous iteration.

An *iterative improvement* algorithm starts with a tentative permutation and iteratively moves to the best neighbor if it is strictly better than the current solution. If no improving solution exists in the current neighborhood the algorithm terminates by returning a local optimum (i.e., a solution φ such that $z(\varphi) \leq z(\varphi')$ for all $\varphi' \in \mathcal{N}(\varphi)$). The CRAFT method introduced by Buffa, Armour, and Vollmann [123] is one of the first examples of iterative improvement algorithm for QAP (although in the literature it is often erroneously described as a constructive algorithm). In order to get better results, improvement methods can be performed several times starting with different initial solutions. We call this method a *multistart* algorithm.

The implementation of a neighborhood exploration is an important issue in the design of efficient heuristic algorithms as it greatly affects the overall performance. Consider the pairwise exchange neighborhood $\mathcal{N}_2$ and observe that, given a solution φ, the size of $\mathcal{N}_2(\varphi)$ is $\binom{n}{2}$. The value $z(\varphi')$ of each solution $\varphi' \in \mathcal{N}_2(\varphi)$ can be evaluated in $O(n)$ (see Heider [367] and Burkard and Rendl [154]) by computing the difference $z(\varphi') - z(\varphi)$ which is affected only by the terms involved in the exchange. More precisely, let φ' be the solution obtained by exchanging the locations of the two facilities r and s. Then for symmetrical cost matrices we have

$$\begin{aligned}\Delta(\varphi, r, s) = z(\varphi') - z(\varphi) = \quad & (a_{rr} - a_{ss})(b_{\varphi(r)\varphi(r)} - b_{\varphi(s)\varphi(s)}) \\ & +2\sum_{\substack{k=1\\k\neq r,s}}^{n}(a_{rk} - a_{sk})(b_{\varphi(s)\varphi(k)} - b_{\varphi(r)\varphi(k)}), \qquad (8.3)\end{aligned}$$

and the exploration from scratch of the entire neighborhood of $\mathcal{N}_2(\varphi)$ requires $O(n^3)$ time. (A slightly more complicated formula has to be used for asymmetric cost matrices.) Frieze, Yadegar, El-Horbaty, and Parkinson [288] observed that if we store the values $\Delta(\varphi, r, s)$ the evaluation of $\mathcal{N}_2(\varphi')$ can be done in $O(n^2)$ time. Observe that indeed $\Delta(\varphi', u, v)$, with u and v different from r or s, can be evaluated in $O(1)$ using

$$\begin{aligned}&\Delta(\varphi', u, v) = \Delta(\varphi, u, v) \\ &+ 2(a_{ur} - a_{vr} - a_{us} + a_{vs})(b_{\varphi'(s)\varphi'(u)} - b_{\varphi'(s)\varphi'(v)} + b_{\varphi'(r)\varphi'(v)} - b_{\varphi'(r)\varphi'(u)}) \qquad (8.4)\end{aligned}$$

while each of the $O(n)$ solutions with u or v equal to r or s can be computed in $O(n)$ time by means of (8.3). It follows that each neighborhood search except for the first one can be implemented to run in $O(n^2)$ time.

The main elements that differentiate the various metaheuristic templates are (a) the method for selecting a neighboring solution; (b) the *intensification* method, i.e., a strategy to perform a deep search in areas of the solution space where it is expected to find high quality solutions; and (c) the *diversification* method, i.e., a strategy to escape from areas with poor solutions. In the next sections we present a selection of the metaheuristic algorithms that have been applied to solve the QAP. A survey of several metaheuristic approaches to the solution of the QAP has been presented by Drezner, Hahn, and Taillard [241], who also introduced new benchmark instances that are difficult for standard metaheuristic algorithms since the variance of the local optima computed with neighborhood $\mathcal{N}_2$ is very high. On the other side these instances are relatively easy for modern exact methods that can solve them up to size $n = 75$. A statistical analysis to compare nondeterministic heuristic methods has been also proposed in [241]. Angel and Zissimopoulos [30, 31, 32, 33] investigated the ruggedness of the search space explored by a metaheuristic for the QAP and the quality of the local minima obtainable from a given neighborhood.

8.2.4 Simulated annealing

Simulated annealing (SA) is one of the first metaheuristic approaches, going back to the seminal papers by Kirkpatrick, Gelatt, and Vecchi [418] and Černý [177]. The algorithm gets inspiration from a method used by physicians to obtain a state of minimum energy of a multiparticle physical system. It is a thermal process that first increases the temperature of a solid in a *heat bath* since it melts, then carefully decreases it until the particles of the solid arrange themselves in a low energy state. The process is repeated until the *ground state* of the solid is reached. Given a combinatorial optimization problem, we associate the solutions with the states of the physical system and the corresponding objective function values to the energy.

SA starts with a high "temperature" value T and performs a search in the solution space by randomly selecting neighboring solutions and accepting them accordingly to a probability function. After a number of iterations, depending on the so-called *cooling scheme*, the temperature is reduced, thus inducing a decrease in the probability of accepting a worsening solution. More precisely, given two solutions φ and φ' with $\varphi' \in \mathcal{N}(\varphi)$, the probability of accepting the transition from φ to φ' is

$$P(\varphi' \text{ accepted}) = \begin{cases} 1 & \text{if } z(\varphi') < z(\varphi), \\ e^{-(z(\varphi')-z(\varphi))/(k_B T)} & \text{if } z(\varphi') \geq z(\varphi), \end{cases} \tag{8.5}$$

where k_B is the Boltzmann constant. SA can be mathematically modeled by a Markov chain, and using this model it is possible to show that, under certain conditions on the slowness of the cooling process, the algorithm asymptotically converges to a global optimal solution. For a detailed discussion of the convergence and other theoretical aspects of SA the reader is referred to the books by Aarts and Korst [2] and Aarts and Lenstra [1].

Burkard and Rendl [154] applied, for the first time, SA to the QAP. They used the pairwise exchange neighborhood $\mathcal{N}_2$ and a simple cooling scheme. In each of the $2n$ iterations the temperature is kept constant, and then it is halved at the next iteration. The resulting SA was tested on some problem instances with $n \leq 36$, comparing it, in particular, with the Burkard and Bönniger [134] limited exact algorithm. SA proved to be superior to the other methods, but sensitive to the starting solution, which was randomly generated.

Other SA algorithms for the QAP were proposed by different authors, such as, e.g., Wilhelm and Ward [664] and Connolly [194]. A C++ implementation of the Connolly algorithm has been given by Taillard [629] (see the companion web page for this volume, http://www.siam.org/books/ot106/assignmentproblems.html). All of these algorithms employ the pairwise exchange neighborhood and differ in the way the cooling process and the termination criterion are defined. The numerical experiments show that the performance of SA strongly depends on the values of the control parameters and on the choice of the cooling schedule.

8.2.5 Tabu search

The *tabu search* (TS) method introduced by Glover [319, 320] uses a simple and effective rule to avoid being trapped in a local minimum: when no improving solution exists in the current neighborhood, it selects a worsening solution that was not visited in the past. It is obviously impractical to store all the visited solutions and to check, for each neighboring solution, if it is identical to a previous one. Therefore, Glover proposed the adoption of a *tabu list* that only stores some *attribute* of each solution. In most cases the attribute is the *move* performed to transform a solution into a neighboring one. For example, if we use the pairwise exchange neighborhood, we can store in the tabu list the pair of facilities that have been swapped to obtain the new solution. At the next iterations we will not allow this pair of facilities to be swapped again. Due to this use of the memory, the trajectory of the search in a TS algorithm is guided by two elements: the local investigation of the neighborhood of the current solution and the past evolution. The reader is referred to Glover, Taillard, and de Werra [326] and to Glover and Laguna [325] for a comprehensive introduction to tabu search algorithms.

Several other ingredients have been proposed to design effective TS algorithms. First of all the tabu list length (known as *tabu tenure*) has a finite value, so old moves exit from the list and their application is permitted again. Moreover the value of the tabu tenure can either be fixed or it can vary according to some strategy. Basically the length is decreased to intensify the search (i.e., to give a higher degree of freedom for the choice of the next moves) and is increased to diversify the search. Other ingredients are (i) an *aspiration criterion*, i.e., a condition that, when fulfilled, cancels a tabu status, and (ii) a *long-term* memory, generally based on the frequency of the application of particular moves, which is used for a better guiding of the search. Finally, observe that, due to the limited length of the tabu list and to the fact that we store attributes instead of complete solutions, there is no guarantee that the method does not enter a loop where it iteratively visits the same solutions. Therefore, a stopping criterion based on some simple rule is always adopted (limitations on the running time, on the number of iterations, etc.).

Due to the high and rich variety of tools and strategies, there is a lot of freedom in the implementation of a tabu search. The resulting performance strongly depends on the implementation chosen, but these algorithms are usually capable of finding, on average, good solutions.

A first TS algorithm for QAP was presented by Skorin-Kapov [611]. Her algorithm uses the pairwise exchange neighborhood $\mathcal{N}_2$, considers as tabu a move that swaps a pair of facilities already swapped and stored in the tabu list, and adopts a fixed tabu tenure. It also adopts some restarting mechanism to continue the search when all the solutions in a neighborhood are tabu. In particular, a long-term memory is used to modify the distance matrix B so as to penalize the assignment of facilities to pairs of locations that are often involved in a swap.

The *robust tabu search* by Taillard [631] is another TS implementation which uses the pairwise neighborhood, but it implements its evaluation with the efficient method proposed by Frieze, Yadegar, El-Horbaty, and Parkinson [288]. The tabu list consists of a matrix TL where the facilities are associated with the rows and the locations to the columns. When an exchange assigns facility i to location j and facility i' to location j' such that $\varphi(i) = j'$ and $\varphi(i') = j$, the algorithm stores in TL_{ij} the number of the iteration in which the swap is performed. At the following iterations a swap is declared tabu if both facilities are assigned to locations they had occupied in the last tabu tenure iterations. The tabu tenure is randomly chosen between prefixed minimum and maximum values and is frequently updated.

Battiti and Tecchiolli [74] introduced the *reactive tabu search* (RTS), which involves a sophisticated mechanism for adapting the tabu tenure and an original diversification scheme. The neighborhood and the tabu status are the same as in Skorin-Kapov [611] and Taillard [631]. The algorithm *reacts* during the evolution of the search by increasing the tabu tenure when a solution is repeated along the search, and decreasing it if no repetition occurs for a certain number of iterations. Hashing functions, binary trees, and bucket lists are used to store the solutions and to check if a neighbor solution was already visited. If a solution is repeated more than once, the algorithm performs a diversification phase based on a *random walk*, i.e., on the execution of a number of random swaps which are also stored in the tabu list to avoid an immediate return to the region of the walk's starting solution. The numerical results show that RTS is competitive with robust tabu search in terms of number of iterations performed to reach the best solution. In [75] the same authors compared their RTS with an implementation of the SA by Burkard and Rendl [154]. They showed that if short computing times are allowed, then SA beats RTS, but the latter needs less CPU time than SA to reach average results which are as close as 1% to the best-known solution.

A number of parallel implementations of TS have been proposed. Taillard [631] presented two parallel version of his robust TS. In the first one the neighborhood search is decomposed by giving a subset of the exchanges to each processor and leaving the overall coordination to a master processor. In the second implementation each processor executes an independent run starting from different solutions. Algorithms based on the parallelization scheme used in the first implementation by Taillard have been proposed, e.g., by Chakrapani and Skorin-Kapov [178] and Talbi, Hafidi, and Geib [632]. Recently James, Rego, and Glover [389] proposed an implementation of the second strategy of parallelization in which, in addition, some reference solutions are stored in a global memory to guide the search of the single TS algorithms. The results of computational experiments with a large number of benchmark instances, using almost all of the most recent metaheuristic algorithms, are provided in [389].

8.2.6 Genetic algorithms

The *genetic algorithms* (GAs) were introduced by Holland [375] in 1975. The basic idea is to mimic the evolutionary mechanisms acting in the selection process in nature. GA starts with a set of feasible solutions (*initial population*), which are generated either randomly or by using some heuristic, and are encoded through a string. The algorithm performs a number of iterations that evolve the population through the following steps:

1. select some pairs of individuals in the population (*parents*);
2. apply a *crossover* operator to the strings encoding each pair of parents in order to generate one or more new solutions (*children*);
3. possibly apply a *mutation* operator to randomly modify a child;
4. possibly apply *immigration* to generate individuals not related to the current population;

5. select some individuals to be added to the population and some individuals to be removed from it.

This basic scheme was applied to the QAP by Tate and Smith [635], who encoded the individuals as permutations and initialized the population with random solutions. The crossover operator they adopted is a technique which generates one child for each pair of parents as follows: (i) each facility assigned to the same location in both parents remains assigned to such location; (ii) unassigned locations are scanned from left to right and, for each location, one of the facilities assigned to it in the two parents is randomly selected and assigned to the location; and (iii) the remaining facilities are assigned to the remaining location in a greedy way. The resulting algorithm has a poor performance, frequently failing in generating the best-known solution even for QAPs of small or moderate size.

Better results can be obtained with *hybrid algorithms*, which apply some local optimization tool to improve the quality of each individual. Fleurent and Ferland [273] obtained promising results on some benchmark problems by applying TS to optimize the individuals. They allowed a large number of TS iterations (up to $4{,}000n$) for the initial population, but much shorter runs (up to $4n$ TS iterations) for the population of the remaining individuals generated by GA. Ahuja, Orlin, and Tivari [14] obtained good results on large scale QAPs from QAPLIB with a hybrid GA which uses the following ingredients: (i) the initial population is produced by the construction phase of a GRASP heuristic (see Section 8.2.7); (ii) a simple implementation of the *path relinking* strategy (see Glover [321]) is selected, among three different crossovers, through computational experiments; and (iii) a cooperative multistart technique, called *tournamenting,* is introduced. The tournamenting method starts by generating an even number $h > 1$ of initial populations and applies GA with a limited number of iterations to each population. It then reduces the number of populations to $h/2$ by merging pairs of populations and eliminating 50% of the resulting individuals. The process is iterated until $h = 1$. Another crossover tested in [14] is the *optimized crossover,* an application to the QAP of the merging operator used by Balas and Niehaus [56] for the maximum clique problem. Let φ_1 and φ_2 denote two solutions, and construct a bipartite graph $G = (U, V; E_1 \cup E_2)$ where the vertex set U (resp., V) contains one vertex for each facility (resp., location) and E_1 (resp., E_2) is the set of the edges $(i, \varphi_1(i))$ (resp., $(i, \varphi_2(i))$). The optimized child solution φ_3 is given by the minimum cost perfect matching of G. Note that this matching contains all the edges contained in $E_1 \cap E_2$ and either the odd or the even edges of each cycle of G.

Lim, Yuan, and Omatu [459] implemented a hybrid GA with a local search based on the k-exchange neighborhood, while Drezner [239, 240], Rodríguez, MacPhee, Bonham, and Bhavsar [588], and Misevicious [496] tested a GA with different versions of a TS as local optimizer.

8.2.7 Greedy randomized adaptive search procedure

The *greedy randomized adaptive search procedure* (GRASP) introduced by Feo and Resende [269] consists of a randomized construction phase and a local improvement phase.

Li, Pardalos, and Resende [458] proposed a GRASP implementation for the solution of the QAP. Their construction phase is divided into two steps: in the first step two facility-location pairs are selected and fixed, while in the second step $n-2$ iterations are performed, each one fixing a single pair. The first step considers all the $O(n^4)$ couples $((i, j), (k, l))$ of facility-location pairs (where i and k refer to the facilities and j, l to the locations) and sorts them by nonincreasing $a_{ik}b_{jl}$ values. A couple $((i, j), (k, l))$ is randomly selected among the first α ($\alpha \geq 1$) smallest cost couples and its two facility-location pairs are fixed. In the second step each iteration assigns a single facility-location pair by randomly selecting this pair among the β ($\beta > 1$) pairs which give the minimum increment to the objective function value of the current partial solution. The improvement phase consists of a local search using the 2-exchange neighborhood $\mathcal{N}_2$ described in Section 8.2.3. The Fortran listing of this code has been presented in Pardalos, Pitsoulis, and Resende [533].

A straightforward parallel implementation can be obtained by executing a GRASP on each processor (see Pardalos, Pitsoulis, and Resende [532]). Due to the random choices of the first phase different runs evolve in different ways.

8.2.8 Ant colony optimization

The ant colony optimization (ACO), originally introduced by Colorni, Dorigo, and Maniezzo [193], is a metaheuristic approach which takes inspiration from the behavior of an ant colony in search of food. An *ant* is a simple computation agent, which iteratively constructs a solution basing its decisions on its status, i.e., the partial solution it has constructed so far, and some information on the solutions constructed by other ants. More precisely, at each step an ant defines a set of feasible expansions of its partial solution (for the QAP an expansion is a facility-location assignment) and chooses one with a probabilistic method. The attractiveness of assigning a facility i to a location j depends on the so-called *pheromone trail*, i.e., a value stored in a global array (τ_{ij}) accessible to all ants. The value τ_{ij} is computed using the objective function value of the previous solutions that assign i to j and is decreased when proceeding from one iteration to the next.

More effective algorithms are obtained by applying a local search to the solutions constructed by the ants, so giving a *hybrid ACO* method. Maniezzo and Colorni [479] proposed a hybrid ANT algorithm which uses a local search optimization based on the 2-exchange neighborhood $\mathcal{N}_2$. Gambardella, Taillard, and Dorigo [299] proposed a hybrid method which works with complete solutions instead of partial ones. They start by generating a random solution for each ant and then perform some optimization by means of a fast local search procedure based on a restriction of $\mathcal{N}_2$, which, however, does not guarantee termination with a local minimum. Next, they iteratively modify all the solutions using the pheromone information and optimize each of them with the restricted local search. In addition, two further techniques are used: (i) an intensification mechanism that updates (τ_{ij}) when a new globally better solution is found; and (ii) a diversification technique that deletes the pheromone information and restarts the search if no improvement has been found in the last $n/2$ iterations.

Taillard [629] proposed the so-called *fast ANT* (FANT) method which is inspired by the hybrid ACO method, but differs from it in two main respects: (i) it does not use a population, but constructs one solution at a time; (ii) when a new solution φ is generated, FANT reinforces each value $\tau_{i\varphi(i)}$, but at the same time it gives a strong reinforcement to the values $\tau_{i\varphi^*(i)}$, where φ^* is the best solution so far. This memory updating provides a fast convergence toward the best solutions, but it also restricts the search to a small area of the search space. Therefore, a diversification method has been introduced which resets matrix (τ_{ij}) when the current solution is identical to the best solution found.

Tseng and Liang [642] used a hybrid ACO method to generate the initial population of a hybrid GA.

Stützle and Hoos [626] presented the *max-min* ACO method, which differs from a classical ant method in three respects: (i) in order to exploit the best solutions found, at each iteration only one ant adds pheromone (such ant may be the one which found the best solution in the current iteration or the one which found the best solution from the beginning of the trail); (ii) in order to avoid stagnation of the search, the τ_{ij} values are limited to an interval $[\tau_{\min}, \tau_{\max}]$; and (iii) pheromone τ_{ij} is initialized to $\tau_{\max}$ to achieve a higher exploration of solutions at the start of the algorithm.

Computational experiments show that ACO methods are competitive heuristics for real-life instances where there are a few good solutions, clustered together. For instances which have many good solutions distributed somewhat uniformly in the search space, they are outperformed by the other heuristics.

8.2.9 Scatter search and path relinking

The *scatter search* (SS) method was introduced in 1997 by Glover [318]. It shares the concept of population with the genetic algorithms, but differs from them in three main respects: (i) the population is controlled with clustering techniques in order to maintain the diversity of the solutions while keeping a set of high quality solutions; (ii) the number of solutions that can take part in the construction of a new solution is not restricted to two; and (iii) the size of the population is usually much smaller than in GAs.

An SS algorithm starts by generating a set of solutions (*reference set*) ensuring that it contains both high quality solutions and solutions with high diversity from each other. The *diversity* of two solutions φ, φ' is a measure of the effort required to transform φ into φ' (the distance $d(\varphi, \varphi')$ defined in equation (8.1) is a possible diversity index). The reference set is then iteratively updated through the following steps:

i. select subsets of solutions;
ii. generate new solutions by combining the solutions of each subset;
iii. locally optimize the new solutions;
iv. replace some solution of the reference set with new ones by preserving diversity and quality.

The *combination method* used by SS algorithms for the QAP is as follows. Given a subset $\mathcal{S}$ of solutions, we define a score matrix (s_{ij}) where each entry (i, j) is a linear combination of the solutions of $\mathcal{S}$ which assign i to j (the multipliers used for the combination are often related to the objective function value of the corresponding solutions). An unassigned facility i is then assigned to an unassigned location j chosen with a probability proportional to the scores.

The scores play the role of an *adaptive memory* system which drives the evolution using in a strategic way the information gathered in the past by the algorithm. Note that this use of memory information resembles the pheromone concept devised some years later for the ACO method.

In the SS implementation by Cung, Mautor, Michelon, and Tavares [202] two combination methods are used. Both use linear combinations with multipliers equal to 1, but the first one adopts for the score the linear combination of the solutions in the subset $\mathcal{S}$, while the second one uses the inverse of the combination of all solutions in the reference set. In the first method the score is aimed at generating high quality solutions, whereas in the second one s_{ij} is the inverse of the frequency of appearance of the assignment $\varphi(i) = j$, thus inducing the creation of solutions with high diversity from the previous ones. The procedure used to locally optimize a solution is a simple TS method.

Using the concept of reference set and adaptive memory, but disregarding the concept of diversity of the solutions, Fleurent and Glover [274] proposed a constructive multistart heuristic enhanced with local optimization and diversification techniques.

Path relinking (PR) relies on the same principles as SS but generalizes it by replacing the linear combination operator used to generate new solutions with an operator working in the neighborhood space, thus providing a framework for local search algorithms to explore adaptive memory in an evolutionary fashion. Given a pair of solutions, PR marks the best one as a *guiding solution* and considers a path in the neighborhood space which goes from the worst solution to the guiding one, i.e., a sequence of moves that transforms the worst solution into the guiding one. All the solutions of the path are candidates to enter the reference set. James, Rego, and Glover [388] implemented sequential and parallel versions of a PR method that combines two solutions φ_1 and φ_2 as follows. Assume that φ_1 is the guiding solution and iteratively compare $\varphi_1(i)$ with $\varphi_2(i)$ for $i = 1, 2, \ldots, n$. If $\varphi_1(i) \neq \varphi_2(i)$, tentatively perform a 2-exchange in φ_2 which assigns facility i to location $\varphi_1(i)$. If the new solution has an objective function value not worse than the value it had before the exchange, then the assignment of i to $\varphi_1(i)$ is confirmed, otherwise it is discarded and $\varphi_2(i)$ remains unchanged. The local improvement of the solutions is obtained by applying the Taillard robust TS (see Section 8.2.5). A parallel version of the algorithm is obtained by applying the improvement method concurrently on each solution generated by the combination method.

8.2.10 Large scale and variable neighborhood search

We defined in Section 8.2.3 the k-exchange neighborhood, for general k values, but we have also seen that almost all the practical implementations use it with $k = 2$. Indeed, for increasing values of $k > 2$, the number of neighbor solutions becomes so huge that it is impractical to explicitly search the neighborhood. Such neighborhoods fall under the category of *very large-scale neighborhoods* (VLSN), structures with a size so large that it is necessary to use implicit enumeration to find improving solutions. One of the tools used to perform a search in a VLSN is the *improvement graph*, which associates with each feasible solution a directed graph such that there is a one-to-one correspondence between the neighbor of this solution and some cycle (possibly satisfying a set of constraints). For a survey on algorithms based on VLSN structures see, e.g, Ahuja, Ergun, Orlin, and Punnen [9].

An application to the solution of the QAP has been presented by Ahuja, Jha, Orlin, and Sharma [10]. Given a starting solution φ, they proposed using an improvement digraph $D = (N; A)$ where the vertex set N contains one vertex for each facility and the arc set A contains all ordered pairs (i, h), with $i \neq h$. Each cycle $C = (i_1, i_2, \ldots, i_k, i_1)$ is associated with a neighbor φ' of φ obtained by setting $\varphi'(i_l) = \varphi(i_{l+1})$ for $l = 1, 2, \dots, k-1$, $\varphi'(i_k) = \varphi(i_1)$ and $\varphi'(i) = \varphi(i)$ for i not in C. Note that the set of all cycles of D with length smaller or equal to k exactly corresponds to the k-exchange neighborhood. Each arc (i, j) is associated with a cost such that $z(\varphi) + z(C)$ (where $z(C)$ denotes the cost of cycle C) approximates $z(\varphi')$. Several implementations of a generic procedure to evaluate neighbors have been proposed and computationally tested using a multistart algorithm.

Another strategy that involves the use of complex neighborhoods is the *variable neighborhood search* (VNS) introduced by Mladenović and Hansen [497]. The basic idea is to use several neighborhood structures and to explore them in a systematic way, by increasing complexity.

Concerning QAP, an application of VNS was proposed by Taillard and Gambardella [630]. The neighborhoods used are the k-exchange neighborhoods with k smaller or equal to a parameter $k_{\max}$. The exploration strategy consists of a number, say, n_I, of iterations which use one neighborhood at a time in a cyclic manner, as follows.

randomly generate a solution φ^* and set $k := 1$;
for $i := 1$ **to** n_I **do**
  randomly select $\varphi' \in \mathcal{N}_k(\varphi^*)$;
  apply a fast improving procedure to φ';
  if $(z(\varphi') < z(\varphi^*))$ **then** set $\varphi^* := \varphi'$, $k := 1$;
  else set $k := k \bmod k_{\max} + 1$
endfor

8.3 Computer codes

The main web resource for the QAP is the QAPLIB by Burkard, Karisch, and Rendl [151], available at http://www.seas.upenn.edu/qaplib/. QAPLIB contains instances, references, news, links, and software codes for the QAP. The instances are due to different authors and range from real-life applications to randomly generated test problems. The most celebrated instances are those by Nugent, Vollmann, and Ruml [512].

From the software section of QAPLIB one can download (directly or through links to other web pages) the following free source codes.

qapglb.f A Fortran code which computes the Gilmore-Lawler bound (see Section 7.5.1) as implemented by Burkard and Derigs [145]. This code is dimensioned to compute the bound for instances of size up to 256.

qapbb.f A Fortran code which implements the branch-and-bound algorithm by Burkard and Derigs [145].

qapeli.f A Fortran implementation of the projection bound (see Proposition 7.23) developed by Karisch and Rendl [405] for symmetric metric QAPs.

GRASP A link redirects the user to the home page of Mauricio Resende, from which one can download the source files (compressed tar-files) of two Fortran implementations of GRASP for dense and sparse QAPs (see Section 8.2.7).

qapsim.f A Fortran implementation of the SA algorithm by Burkard and Rendl [154].

Ro-TS, FANT, SA Three links redirect the user to the home page of Éric Taillard from which one can download the following source files:

(i) a Pascal and a C++ implementation of the robust TS algorithm by Taillard [631] (see Section 8.2.5);

(ii) a C++ implementation of the FANT algorithm by Taillard [629] (see Section 8.2.8);

(iii) a C++ implementation of the SA algorithm by Connolly [194] (see Section 8.2.4).

An interactive Java applet for solving small QAPs has been developed by the Optimization and Technology Center at Northwestern University and is available at the Case Studies in the NEOS Guide.

The source file of a Fortran implementation of the Li and Pardalos [456] generator for QAP instances with known optimal solution is available at the home page of the Computational Optimization and Applications Software Forum. Other instances with known optimal solutions, generated with the algorithms developed by Gintaras Palubeckis [522, 523], are available at his home page.

8.4 Easy and hard special cases

For $\mathcal{NP}$-hard problems the question arises under which assumptions on the structure and on the data a problem becomes solvable in polynomial time. In this section we survey such results for QAPs in Koopmans–Beckmann form. We shall not, however, deal with special cases of the TSP and other combinatorial optimization problems which can be formulated as QAPs such as those reviewed in Section 7.1. For TSPs we refer the reader to the survey by Burkard, Deĭneko, van Dal, van der Veen, and Woeginger [144]. Some additional special cases of the other problems can be found in the survey by Burkard, Çela, Demidenko, Metelski, and Woeginger [139] (on which this section is based) as well as in the monograph by Çela [176].

In what follows we consider quadratic assignment problems $QAP(A, B)$ in Koopmans–Beckmann form. Since $QAP(A, B)$ is equivalent to $QAP(A_\psi, B_\varphi)$ with permuted matrices $A_\psi = (a_{\psi(i)\psi(k)})$ and $B_\varphi = (b_{\varphi(j)\varphi(l)})$, all of the following results also hold for permuted coefficient matrices. In Lemma 7.24 we already showed the following first result.

Proposition 8.1. *If matrix A is symmetric and matrix B is skew symmetric, then all solutions of $QAP(A, B)$ have the same value 0.*

8.4.1 Sum and product matrices

Another simple case occurs when the matrices A and B are diagonal matrices (i.e., the entries outside the main diagonal are all zero): in this case we just have to compute a minimum scalar product by ordering the diagonal entries according to Proposition 5.8. A similar situation occurs if matrix A is a sum matrix.

We call A a *sum matrix* if there are two vectors $u = (u_1, u_2, \ldots, u_n)$ and $v = (v_1, v_2, \ldots, v_n)$ such that

$$a_{ik} = u_i + v_k \quad (i, k = 1, 2, \ldots, n).$$

Proposition 8.2. *A Koopmans–Beckmann problem $QAP(A, B)$ in which A is a sum matrix can be solved via a linear assignment problem.*

Proof. The objective function value $z(A, B; \varphi)$ of $QAP(A, B)$ corresponding to permutation φ can be written as

$$z(A, B; \varphi) = \sum_{i=1}^{n}\sum_{k=1}^{n}(u_i + v_k)b_{\varphi(i)\varphi(k)} = \sum_{i=1}^{n} u_i \sum_{k=1}^{n} b_{\varphi(i)\varphi(k)} + \sum_{k=1}^{n} v_k \sum_{i=1}^{n} b_{\varphi(i)\varphi(k)}.$$

Thus $QAP(A, B)$ is equivalent to a linear assignment problem with cost coefficients c_{ij} defined by

$$c_{ij} = u_i \sum_{l=1}^{n} b_{jl} + v_i \sum_{l=1}^{n} b_{lj} \quad (i, j = 1, 2, \ldots, n). \quad \square \tag{8.6}$$

Example 8.3. Let us consider the input matrices below, where u and v are shown on the left and on the top of the sum matrix A, while $\sum_{l=1}^{n} b_{jl}$ and $\sum_{l=1}^{n} b_{lj}$ are shown on the left and on the top of matrix B:

$$\begin{array}{c} \\ 2 \\ 0 \\ 1 \end{array}\begin{array}{c} \begin{array}{ccc} 3 & 0 & 2 \end{array} \\ \begin{pmatrix} 5 & 2 & 4 \\ 3 & 0 & 2 \\ 4 & 1 & 3 \end{pmatrix} \\ A \end{array} \qquad \begin{array}{c} \\ 102 \\ 2 \\ 8 \end{array}\begin{array}{c} \begin{array}{ccc} 102 & 6 & 4 \end{array} \\ \begin{pmatrix} 99 & 2 & 1 \\ 1 & 1 & 0 \\ 2 & 3 & 3 \end{pmatrix} \\ B \end{array}.$$

The linear assignment problem matrix is obtained as

$$C = \begin{pmatrix} 204 & 4 & 16 \\ 0 & 0 & 0 \\ 102 & 2 & 8 \end{pmatrix} + \begin{pmatrix} 306 & 18 & 12 \\ 0 & 0 & 0 \\ 204 & 12 & 8 \end{pmatrix} = \begin{pmatrix} 510 & 22 & 28 \\ 0 & 0 & 0 \\ 306 & 14 & 16 \end{pmatrix}.$$

The optimal solution, of value 38, is given by the permutation $\varphi = (2, 1, 3)$. ∎

The result of Proposition 8.2 can be sharpened if one of the two matrices is symmetric or skew-symmetric.

Proposition 8.4. *If matrix A is a sum matrix and if either A or B is symmetric or skew-symmetric, then the Koopmans–Beckmann problem $QAP(A, B)$ can be solved in $O(n^2)$ time.*

Proof. We consider the case where A is a sum matrix and B is skew-symmetric. Let us define

$$\beta_j = \sum_{l=1}^{n} b_{jl} \quad (j = 1, 2, \ldots, n).$$

Since B is skew symmetric, using (8.6) we get

$$c_{ij} = (u_i - v_i)\beta_j \quad (i, j = 1, 2, \ldots, n). \tag{8.7}$$

According to Proposition 5.8, a linear assignment problem with such coefficients can be solved by reordering the vectors $(u_i - v_i)$ and β_j, which can be done in $O(n \log n)$ time. This time requirement is majorized by the time $O(n^2)$ needed to compute the coefficients β_j $(j = 1, 2, \ldots, n)$.

The other three cases (B symmetric, A symmetric, A skew symmetric) can be proved in a similar way. □

Example 8.5. We consider the same sum matrix A of Example 8.3 and a symmetric matrix B (for which we show the β values on the left):

$$\begin{array}{c} \\ 2 \\ 0 \\ 1 \end{array} \begin{array}{c} \begin{array}{ccc} 3 & 0 & 2 \end{array} \\ \begin{pmatrix} 5 & 2 & 4 \\ 3 & 0 & 2 \\ 4 & 1 & 3 \end{pmatrix} \\ A \end{array} \qquad \begin{array}{c} 103 \\ 3 \\ 4 \end{array} \begin{array}{c} \begin{pmatrix} 99 & 2 & 1 \\ 2 & 1 & 0 \\ 1 & 0 & 3 \end{pmatrix} \\ B \end{array}.$$

Since in this case B is not skew symmetric but symmetric, equation (8.7) becomes

$$c_{ij} = (u_i + v_i)\beta_j \quad (i, j = 1, 2, \ldots, n). \tag{8.8}$$

Hence, we get the linear assignment problem matrix

$$C = \begin{pmatrix} 515 & 15 & 20 \\ 0 & 0 & 0 \\ 309 & 9 & 12 \end{pmatrix}$$

(whose optimal solution is clearly given by the permutation $\varphi = (2, 1, 3)$). By sorting $(u_i + v_i) = (5, 0, 3)$ increasingly and $\beta = (103, 3, 4)$ decreasingly, we get the optimal objective function value $(0, 3, 5) \cdot (103, 4, 3) = 12 + 15$. ■

A matrix A is called a *product matrix* if there are two vectors $u = (u_1, u_2, \ldots, u_n)$ and $v = (v_1, v_2, \ldots, v_n)$ such that

$$a_{ik} = u_i v_k \quad (i, k = 1, 2, \ldots, n).$$

Matrix A is called a *negative product matrix* if

$$a_{ik} = -u_i v_k \quad (i, k = 1, 2, \ldots, n).$$

If $u = v$ holds, we say that A is a *symmetric product matrix*. In particular, if $u = v$ with $u_i = (-1)^i$ we get a so-called *chessboard matrix*. A negative chessboard matrix is defined accordingly. For example, for $n = 4$ we get the following chessboard matrix A and negative chessboard matrix A':

$$A = \begin{pmatrix} 1 & -1 & 1 & -1 \\ -1 & 1 & -1 & 1 \\ 1 & -1 & 1 & -1 \\ -1 & 1 & -1 & 1 \end{pmatrix}; \quad A' = \begin{pmatrix} -1 & 1 & -1 & 1 \\ 1 & -1 & 1 & -1 \\ -1 & 1 & -1 & 1 \\ 1 & -1 & 1 & -1 \end{pmatrix}. \tag{8.9}$$

It can be shown that a QAP with a chessboard matrix B is equivalent to a graph partitioning problem where the vertices of a graph are partitioned into two sets V_1 and V_2 with $|V_1| - |V_2| \leq 1$ (see Section 7.1.4 and Example 7.2).

The following result was shown by Burkard, Çela, Demidenko, Metelski, and Woeginger [139].

Proposition 8.6. *A quadratic assignment problem QAP(A, B) where A is a chessboard matrix and B is either a chessboard or a negative chessboard matrix can be solved in polynomial time.*

Proof. If B is a negative chessboard matrix, the identical permutation yields $-n^2$ as the objective function value, which is obviously the smallest value possible.

For a chessboard matrix B we get

$$\begin{aligned} z(A, B; \varphi) &= \sum_{i=1}^{n} \sum_{k=1}^{n} (-1)^{i+k} (-1)^{\varphi(i)+\varphi(k)} = \left(\sum_{i=1}^{n} (-1)^{i+\varphi(i)} \right)^2 \\ &= (\,|\{i \,:\, i + \varphi(i) \text{ even}\}| - |\{i \,:\, i + \varphi(i) \text{ odd}\}|\,)^2 \\ &= (n - 2 \cdot |\{i \,:\, i + \varphi(i) \text{ odd}\}|\,)^2 = (n - 4m_\varphi)^2, \end{aligned}$$

where m_φ is the number of even indices i with odd $\varphi(i)$. For any m with $0 \leq m \leq \lfloor n/2 \rfloor$ there exists a permutation φ with $m_\varphi = m$. Thus we obtain the optimal solution value:

$$\min_{\varphi} z(A, B; \varphi) = \min_{0 \leq m \leq \lfloor n/2 \rfloor} (n - 4m)^2 = \begin{cases} 1 & \text{for } n \text{ odd}, \\ 4 & \text{for } n = 2 \pmod 4, \\ 0 & \text{for } n = 0 \pmod 4. \end{cases}$$

An optimal solution can be obtained by constructing a permutation φ such that $(n - 4m_\varphi)^2$ is minimized. □

Example 8.7. For $n = 2$, permutation $\varphi = (1, 2)$ gives $m_\varphi = 0$, and permutation $\varphi = (2, 1)$ gives $m_\varphi = 1$. In both cases the solution value is $(2 - 4m_\varphi)^2 = 4$.

For $n = 3$, permutations $\varphi = (1, 2, 3)$ and $\varphi = (3, 2, 1)$ give $m_\varphi = 0$ and hence a solution value $(3 - 4m_\varphi)^2 = 9$. All other permutations φ give $m_\varphi = 1$; hence, the optimal solution value is given by $(3 - 4m_\varphi)^2 = 1$.

For $n = 4$ (see (8.9)) an optimal permutation is $\varphi = (1, 2, 4, 3)$, which yields $m_\varphi = 1$; hence, $(4 - 4m_\varphi)^2 = 0$. ■

On the other hand, the following $\mathcal{NP}$-hardness result is shown in Burkard, Çela, Rote, and Woeginger [143].

Proposition 8.8. *The QAP(A, B) where A is a symmetric product matrix and B is a chessboard matrix is $\mathcal{NP}$-hard.*

Proof. We show this result by reduction from the $\mathcal{NP}$-complete problem PARTITION.

PARTITION: Given $2n$ positive numbers $u_1, u_2, \ldots, u_{2n}$, is there a subset $I \subset \{1, 2, \ldots, 2n\}$ such that

$$\sum_{i \in I} u_i = \sum_{i \notin I} u_i? \tag{8.10}$$

Without loss of generality we can assume $u_1 \leq u_2 \leq \cdots \leq u_{2n}$. We define the $(2n \times 2n)$ matrix $A = (a_{ik})$ by $a_{ik} = u_i u_k$ $(i, k = 1, 2, \ldots, 2n)$. Then the objective function value $z(A, B; \varphi)$ can be written as

$$z(A, B; \varphi) = \sum_{i=1}^{2n} \sum_{k=1}^{2n} (-1)^{\varphi(i)+\varphi(k)} u_i u_k = \left(\sum_{i=1}^{2n} (-1)^{\varphi(i)} u_i \right)^2 \geq 0.$$

Therefore $z(A, B; \varphi) = 0$ holds if and only if

$$\sum_{i=1}^{2n} (-1)^{\varphi(i)} u_i = 0,$$

i.e., if and only if (8.10) holds and the answer to PARTITION is "yes." $\square$

This also implies that a QAP(A, B) with two symmetric product matrices A and B is $\mathcal{NP}$-hard in general. However, it is shown in Burkard, Çela, Demidenko, Metelski, and Woeginger [139] that, if all components of the generating vectors u and v have the same sign, then the problem is polynomially solvable.

Proposition 8.9. *Let QAP(A, B) be a QAP where A and B are symmetric product matrices generated by vectors u and v whose components have all the same sign. Then QAP(A, B) can be solved in $O(n \log n)$ time.*

Proof. We have

$$z(A, B, \varphi) = \sum_{i=1}^{n} \sum_{k=1}^{n} u_i u_k v_{\varphi(i)} v_{\varphi(k)} = \left(\sum_{i=1}^{n} u_i v_{\varphi(i)} \right)^2.$$

Proposition 5.8 states that the minimum is attained by ordering vector u increasingly and vector v decreasingly. $\square$

The same idea as in the proof of Proposition 8.9 shows (see [139]) the following.

Proposition 8.10. *A quadratic assignment problem QAP(A, B) where A is a symmetric product matrix and B is a negative symmetric product matrix can be solved in $O(n \log n)$ time.*

Proof. Since

$$z(A, B; \varphi) = -\sum_{i=1}^{n}\sum_{k=1}^{n} u_i u_k v_{\varphi(i)} v_{\varphi(k)} = -\left(\sum_{i=1}^{n} u_i v_{\varphi(i)}\right)^2,$$

$QAP(A, B)$ is equivalent to

$$\max_{\varphi}\left(\sum_{i=1}^{n} u_i v_{\varphi(i)}\right)^2. \tag{8.11}$$

By Proposition 5.8 the minimum and the maximum of $\sum_{i=1}^{n} u_i v_{\varphi(i)}$ can be computed. By squaring the maximum of (8.11) can be found. □

A matrix A is called *small* if there are two vectors $u = (u_1, u_2, \ldots, u_n)$ and $v = (v_1, v_2, \ldots, v_n)$ such that

$$a_{ik} = \min(u_i, v_k) \quad (i, k = 1, 2, \ldots, n).$$

Analogously, a matrix A is called *large* if

$$a_{ik} = \max(u_i, v_k) \quad (i, k = 1, 2, \ldots, n).$$

We obtain symmetric small (large) matrices by taking $u = v$. Burkard, Çela, Demidenko, Metelski, and Woeginger [139] have shown the following.

Proposition 8.11. *A quadratic assignment problem QAP(A, B) where A is a symmetric large matrix and B is a chessboard or a negative chessboard matrix can be solved in $O(n^2)$ time.*

Proof. Without loss of generality we can assume that the components of the vector u which generates matrix A are ordered so that $u_1 \le u_2 \le \cdots \le u_n$. This yields $a_{ik} = u_k$ $(i, k = 1, 2, \ldots, n; i \le k)$. Moreover, replacing B by $B^* = \frac{1}{2}(B + \mathbf{1})$, where $\mathbf{1}$ is the matrix with constant entries 1, leads to an equivalent $QAP(A, B^*)$, since $z(A, B; \varphi) = 2z(A, B^*; \varphi) - z(A, \mathbf{1}; \varphi)$ and $z(A, \mathbf{1}; \varphi)$ is a constant (equal to the sum of all entries in A).

Now let us assume that B is a chessboard matrix. This implies

$$b^*_{ij} = \begin{cases} 1 & \text{if } i + j \text{ is even,} \\ 0 & \text{if } i + j \text{ is odd.} \end{cases} \tag{8.12}$$

Let $E(\varphi) = \{i \; : \; \varphi(i) \text{ even}, \; 2 \le i \le n\}$. Using (8.12) the objective function value of $QAP(A, B^*)$ can be written as

$$z(A, B^*; \varphi) = \sum_{i,k \in E(\varphi)} \max(u_i, u_k) + \sum_{i,k \notin E(\varphi)} \max(u_i, u_k). \tag{8.13}$$

Equation (8.13) shows that the objective function value depends only on set E: any two permutations φ and ψ with $E(\varphi) = E(\psi)$ yield the same objective function value!

If set E is specified, the objective function value $z(A, B^*; E)$ of all permutations φ with $E(\varphi) = E$ can be written as

$$z(A, B^*; E) = 2\sum_{i=2}^{n} u_i m_i(E), \tag{8.14}$$

where

$$m_i(E) = \begin{cases} |\{k \in E \,:\, k < i\}| & \text{if } i \in E, \\ |\{k \notin E \,:\, k < i\}| & \text{if } i \notin E. \end{cases}$$

Equation (8.14) shows that we minimize $z(A, B; \varphi)$ when we solve

$$\min_E \sum_{i=2}^{n} u_i m_i(E). \tag{8.15}$$

The latter problem can be solved by dynamic programming. We define, for $r \geq s$,

$$z(r, s) = \min_{|E_r|=s} \sum_{i=2}^{r} u_i m_i(E_r),$$

where $E_r = E \cap \{1, 2, \dots, r\}$. Obviously,

$$z(n, \lfloor n/2 \rfloor) = \min_E \sum_{i=2}^{n} u_i m_i(E).$$

The start values for the dynamic programming procedure are $z(2, s) = u_2$ for $s = 0, 1, 2$ and

$$z(r, 0) = z(r-1, 0) + (r-1)u_r \quad \text{for } 3 \leq r \leq n.$$

This follows from the fact that, for $E = \emptyset$ $(s = 0)$, we have $m_i(E) = i - 1$. For infeasible pairs (r, s) we set $z(r, s) = \infty$. Let us suppose that values $z(r-1, 0), z(r-1, 1), \dots, z(r-1, s)$ are already known. For computing $z(r, s)$ we have to consider two possibilities: either index r is joined to set E, in which case we get

$$z(r, s) = z(r-1, s-1) + (s-1)u_r,$$

or index r is not joined to E, in which case we get

$$z(r, s) = z(r-1, s) + (r-s-1)u_r.$$

This leads to the recursion

$$z(r, s) = \min(z(r-1, s-1) + (s-1)u_r,\; z(r-1, s) + (r-s-1)u_r). \tag{8.16}$$

Whenever the minimum is taken for the first term, index r is joined to set E. Thus $z(n, \lfloor n/2 \rfloor)$ can be computed in $O(n^2)$ time. By backtracing, the corresponding set E can be computed. Any permutation which maps the indices in E to the even indices of $\{1, 2, \dots, n\}$ is an optimal solution.

The case of a negative chessboard matrix is treated in an analogous way. □

Example 8.12. Let us consider a problem with $n = 5$, a symmetric large matrix A produced by vectors $u = v = (0, 1, 2, 3, 4)$ and a chessboard matrix B. We want to find a permutation φ which minimizes $\langle A, B^*_\varphi \rangle$. By applying the dynamic programming approach above we get the table where the subscript tells which term determines the minimum.

$r \backslash s$	0	1	2
2	1	1	1
3	5	$\min(1+0, 1+2) = 1_1$	$\min(1+2, 1+0) = 1_2$
4	14	$\min(5+0, 1+6) = 5_1$	$\min(1+3, 1+3) = 4_1$
5	30	$\min(14+0, 5+12) = 14_1$	$\min(5+4, 4+8) = 9_1$

By backtracing we find $E = \{4, 5\}$. Thus an optimal permutation is $\varphi = (1, 3, 5, 2, 4)$. The corresponding matrix B^*_φ has the form

$$B^*_\varphi = \begin{pmatrix} 1 & 1 & 1 & 0 & 0 \\ 1 & 1 & 1 & 0 & 0 \\ 1 & 1 & 1 & 0 & 0 \\ 0 & 0 & 0 & 1 & 1 \\ 0 & 0 & 0 & 1 & 1 \end{pmatrix}. \quad \blacksquare$$

Since $QAP(A, B)$ is equivalent to $QAP(-A, -B)$, and a negative small matrix is a large matrix, we immediately get the following.

Corollary 8.13. *A quadratic assignment problem $QAP(A, B)$ with a symmetric small matrix A and a chessboard or negative chessboard matrix B can be solved in $O(n^2)$ time.*

The results described in this section are all related to Monge matrices, which were introduced in Section 5.2. Recall that a matrix A is a Monge matrix if

$$a_{ij} + a_{kl} \le a_{il} + a_{kj} \quad (i, k, j, l = 1, 2, \ldots, n; i < k; j < l),$$

while it is an inverse Monge matrix if

$$a_{ij} + a_{kl} \ge a_{il} + a_{kj} \quad (i, k, j, l = 1, 2, \ldots, n; i < k; j < l).$$

In particular, sum matrices are both Monge and inverse Monge matrices; product matrices $A = (u_i v_k)$ are Monge matrices provided the nonnegative numbers u_k are increasing and the nonnegative numbers v_k are decreasing. Moreover, large matrices $A = (\max(u_i, v_k))$ are Monge matrices where the numbers u_i and v_k are ordered increasingly. Further, if A is a Monge matrix, then $-A$ is an inverse Monge matrix.

The complexity status of Koopmans–Beckmann problems involving Monge matrices can be seen in Table 8.1, taken from [139], in which "**?**" stands for "open" and the terms referring to the matrix properties are self-explanatory.

When reading the first three rows of the table, keep in mind that every negative chessboard matrix is a permuted symmetric negative product matrix, which in turn is a permuted Monge matrix. Because of these inclusion relations, an entry "$\mathcal{NP}$-hard" turns all other entries which lie above or to the left of it also to "$\mathcal{NP}$-hard." Moreover, two "negatives" cancel each other. Therefore, all the $\mathcal{NP}$-hard results follow from Proposition 8.8, while

Table 8.1. *Complexity status of Koopmans–Beckmann problems involving Monge and inverse Monge matrices.*

	MONGE	NEGPRODSYM	NEGCHESS
MONGE	$\mathcal{NP}$-hard	$\mathcal{NP}$-hard	$\mathcal{NP}$-hard
NEGPRODSYM	$\mathcal{NP}$-hard	$\mathcal{NP}$-hard	$\mathcal{NP}$-hard
NEGCHESS	$\mathcal{NP}$-hard	$\mathcal{NP}$-hard	$\mathcal{P}$
INVMONGE	?	?	$\mathcal{P}$(even n), ?(odd n)
PRODSYM	?	$\mathcal{P}$	$\mathcal{P}$
CHESS	$\mathcal{P}$(even n), ?(odd n)	$\mathcal{P}$	$\mathcal{P}$

the only polynomiality result follows from Proposition 8.6. When reading the next three rows of the table, remember that every chessboard matrix is a permuted symmetric product matrix which in turn is a permuted inverse Monge matrix. Because of the inclusion relations mentioned before, an entry $\mathcal{P}$ turns all other entries which lie below or to the right of it to $\mathcal{P}$ also. Therefore, all polynomiality results follow from Proposition 8.10. The "$\mathcal{P}$ for even n" entries follow from the following result (see [139]).

Proposition 8.14. *Let A and B be $(n \times n)$ matrices with even $n = 2m$. If A is a Monge matrix and B is a chessboard matrix, then*

$$\varphi = (1, m + 1, 2, m + 2, \dots, n) \tag{8.17}$$

is an optimal solution.

Hence, a fixed permutation φ is optimal for any Monge matrix A with even size n. This is not the case for Monge matrices with odd size n as even simple examples with $n = 3$ show. Proposition 8.14 was implicitly proved by Pferschy, Rudolf, and Woeginger [545] in a paper on the maximum balanced bisection problem involving Monge matrices. An optimum bisection does not depend on the instance of the Monge matrix, as it always consists of the first half of rows (columns) versus the second half of rows (columns). This result on the bisection problem was solved using exchange arguments and directly implies Proposition 8.14. Another proof of Proposition 8.14 uses the extremal rays of the cone of nonnegative Monge matrices of size n. It can be shown that for any matrix R describing an extremal ray of this cone the permutation given by (8.17) is optimal. For details see, e.g., Çela [176].

8.4.2 Diagonally structured matrices

A matrix $A = (a_{ik})$ with constant entries on its diagonals is called a *Toeplitz matrix*. More formally, a Toeplitz matrix is defined by the $2n - 1$ constants $\gamma_{-(n-1)}, \gamma_{-(n-2)}, \dots, \gamma_0, \dots, \gamma_{n-1}$ with

$$a_{ik} = \gamma_{i-k} \quad (i, k = 1, 2, \dots, n).$$

In the case $\gamma_{-r} = \gamma_r$ we have a symmetric Toeplitz matrix. If $\gamma_{k-n} = \gamma_k$ for $k = 1, 2, \ldots, n-1$, the matrix is called a *circulant matrix*. We say that a matrix $A = (a_{ik})$ is *monotone* if

$$a_{ik} \le a_{i,k+1} \text{ and } a_{ik} \le a_{i+1,k} \text{ for all } i, k.$$

Example 8.15. For $n = 4$, A is a Toeplitz matrix defined by $\gamma = (7, 3, -1, 0, 1, 2, 4)$, B is a circulant matrix defined by $\gamma = (3, 2, 1, 0, 3, 2, 1)$, and C is a monotone matrix:

$$A = \begin{pmatrix} 0 & -1 & 3 & 7 \\ 1 & 0 & -1 & 3 \\ 2 & 1 & 0 & -1 \\ 4 & 2 & 1 & 0 \end{pmatrix}; \quad B = \begin{pmatrix} 0 & 1 & 2 & 3 \\ 3 & 0 & 1 & 2 \\ 2 & 3 & 0 & 1 \\ 1 & 2 & 3 & 0 \end{pmatrix}; \quad C = \begin{pmatrix} 1 & 2 & 3 & 4 \\ 2 & 3 & 4 & 4 \\ 2 & 4 & 6 & 7 \\ 3 & 5 & 6 & 8 \end{pmatrix}. \quad \blacksquare$$

Symmetric Toeplitz matrices occur, for example, when n rooms of equal size are to be assigned along a corridor in a hospital (see Picard and Queyranne [547]). If the distance between two neighboring rooms is just one unit, we get as distances $b_{jl} = |j - l|$ $(j, l = 1, 2, \ldots, n)$, which is a symmetric Toeplitz matrix. Other applications with such a distance matrix have been described by Sarker, Wilhelm, and Hogg [596] in connection with the design of a flowline in a manufactoring system and by Bhasker and Sahni [95] who arranged circuit elements along a straight line.

Another application involving a Toeplitz matrix is the so-called *turbine runner problem*, originally stated by Bolotnikov [106] and Stoyan, Sokolovskii, and Yakovlev [625]. Laporte and Mercure [446] observed that this problem can be formulated as a QAP (see also Mosevich [499] and Schlegel [598]). A turbine consists of a cylinder around which the blades are welded in regular spacings. Due to the manufactoring process, the blades have slightly different masses. In the turbine runner problem we want to arrange the blades on the cylinder such that the distance between the gravity center and the rotation axis is minimized. We can model this problem as follows. The masses of the n blades are positive numbers $0 < m_1 \le m_2 \le \cdots \le m_n$. The regular spacings on the cylinder are modeled by the vertices $v_1, v_2, \ldots, v_n$ of a regular polygon on the unit circle in the Euclidean plane. Thus the vertices have the coordinates

$$v_i = \left(\sin\frac{2i\pi}{n}, \cos\frac{2i\pi}{n}\right) \quad (i = 1, 2, \ldots, n).$$

We want to find a permutation φ which minimizes the Euclidean norm of the vector

$$v = \left(\sum_{i=1}^{n} m_{\varphi(i)} \sin\frac{2i\pi}{n}, \sum_{i=1}^{n} m_{\varphi(i)} \cos\frac{2i\pi}{n}\right).$$

A simple calculation reveals that this problem is equivalent to

$$\min_{\varphi} \sum_{i=1}^{n}\sum_{k=1}^{n} m_{\varphi(i)} m_{\varphi(k)} \cos\frac{2(i-k)\pi}{n}. \tag{8.18}$$

Problem (8.18) is a quadratic assignment problem *QAP*(A, B) with a symmetric Toeplitz matrix $A = (a_{ik}) = (\cos(2(i-k)\pi/n))$ and a product matrix $B = (b_{jl}) = (m_j m_l)$.

Moreover, since the masses m_j are ordered by nondecreasing values, matrix B is an inverse Monge matrix. In [143] the following has been shown, by reduction from the $\mathcal{NP}$-complete *even-odd partition problem*.

Theorem 8.16. *The turbine runner problem* (8.18) *is $\mathcal{NP}$-hard.*

If we consider the maximization version of the turbine runner problem, namely,

$$\max_{\varphi} \sum_{i=1}^{n} \sum_{k=1}^{n} m_{\varphi(i)} m_{\varphi(k)} \cos \frac{2(i-k)\pi}{n}, \tag{8.19}$$

we get the $QAP(-A, B)$ where

$$-A = -(a_{ik}) = \left(-\cos \frac{2(i-k)\pi}{n} \right) \tag{8.20}$$

is a symmetric Toeplitz matrix which fulfills the following additional property. An $(n \times n)$ Toeplitz matrix generated by $\gamma_{-(n-1)}, \dots, \gamma_0, \dots, \gamma_{n-1}$ is called *benevolent* if

$$\gamma_{-i} = \gamma_i \quad \text{for } i = 1, 2, \dots, n-1, \tag{8.21}$$

$$\gamma_i \le \gamma_{i+1} \quad \text{for } i = 1, 2, \dots, \lfloor n/2 \rfloor - 1, \tag{8.22}$$

$$\gamma_i \le \gamma_{n-i} \quad \text{for } i = 1, 2, \dots, \lceil n/2 \rceil - 1. \tag{8.23}$$

Example 8.17. For $n = 7$, A is a benevolent Toeplitz matrix generated by $\gamma = (2, 5, 8, 4, 3, 2, 6, 2, 3, 4, 8, 5, 2)$:

$$A = \begin{pmatrix} 6 & 2 & 3 & 4 & 8 & 5 & 2 \\ 2 & 6 & 2 & 3 & 4 & 8 & 5 \\ 3 & 2 & 6 & 2 & 3 & 4 & 8 \\ 4 & 3 & 2 & 6 & 2 & 3 & 4 \\ 8 & 4 & 3 & 2 & 6 & 2 & 3 \\ 5 & 8 & 4 & 3 & 2 & 6 & 2 \\ 2 & 5 & 8 & 4 & 3 & 2 & 6 \end{pmatrix}. \quad \blacksquare$$

Note that (8.21) implies that every benevolent Toeplitz matrix is symmetric. By a careful investigation of the extremal rays of the cone of inverse Monge matrices with positive entries and the cone of benevolent Toeplitz matrices, the following theorem has been proven in [143].

Theorem 8.18. *The quadratic assignment problem $QAP(A, B)$ where A is a benevolent Toeplitz matrix and B is a monotone inverse Monge matrix is solved by the cyclic permutation*

$$\varphi = \langle 1, 3, 5, \dots, n, \dots, 6, 4, 2 \rangle. \tag{8.24}$$

In particular, the permutation (8.24) solves the maximization version of the turbine runner problem 8.19, since the matrix $-A$ of (8.20) is benevolent.

The permutation (8.24) is well known in the theory of well-solvable TSPs, since Supnick [627] showed the following.

Proposition 8.19. *Every TSP with a symmetric Monge matrix as a distance matrix is solved by the tour*

$$1 \to 3 \to 5 \to \cdots \to n \to \cdots \to 6 \to 4 \to 2.$$

Theorem 8.18 directly provides proof of the above statement.

Theorem 8.18 also finds application in data arrangements in linear storage media. Consider n records $r_1, r_2, \ldots, r_n$ where record r_i is referenced repetitively with probability p_i. Without loss of generality the records can be numbered such that $p_1 \leq p_2 \leq \cdots \leq p_n$. The goal is to store the records in a linear array (e.g., on a tape) such that the expected distance between consecutively referenced records is a minimum. This leads to the QAP

$$\min_{\varphi} \sum_{i=1}^{n} \sum_{k=1}^{n} p_{\varphi(i)} p_{\varphi(k)} d_{ik}, \tag{8.25}$$

where d_{ik} is a distance function which only depends on $|i - k|$. Several authors (Timofeev and Litvinov [639], Burkov, Rubinstein, and Sokolov [160], Metelski [488], and Pratt [558]) proved (special cases of) the following.

Proposition 8.20. *If the distance function $d_{ik} = f(|i - k|)$ stems from a nondecreasing function f, then the data arrangement problem* (8.25) *is solved by the cyclic permutation* (8.24).

More generally, Hardy, Littlewood, and Pólya [363] showed the following in 1926.

Proposition 8.21. *Let A be a monotone product matrix and let B be a symmetric Toeplitz matrix generated by a function f that is nondecreasing on $\{0, 1, \ldots, n\}$. Then QAP(A, B) is solved by* (8.24).

In both propositions one of the matrices is a monotone inverse Monge matrix and the distance matrix is a benevolent Toeplitz matrix. Thus they are direct consequences of Theorem 8.18. Recently, Demidenko, Finke, and Gordon [224] derived two sets of conditions for matrices A and B which guarantee that the cyclic permutation (8.24) is an optimal solution of *QAP*(A, B). The new conditions rely on special orderings of the elements of A and B. In particular, A is not any more in general a Toeplitz matrix and B is not any more an inverse Monge matrix.

In [143] periodic continuations of benevolent Toeplitz matrices were also considered, which lead to further special cases (see also Çela [176]). Demidenko [222] generalized the results of QAPs with inverse Monge and Toeplitz matrices. Moreover, Demidenko and Dolgui [221, 223] recently proposed new conditions which guarantee that the QAP attains its optimum on a given permutation. Çela [176] proved the following.

Proposition 8.22. *Every permutation yields the same value in a QAP defined by a sum matrix and a circulant matrix.*

8.4.3 Ordered matrices

In this section we consider a $QAP(A, B)$ where the entries in the matrices A and B fulfill certain order requirements. A matrix $A = (a_{ik})$ is called *left-higher graded* if both its rows and columns are nondecreasing, i.e., $a_{ik} \leq a_{i,k+1}$ and $a_{ik} \leq a_{i+1,k}$ for all i, k. A matrix is called *right-lower graded* if both its rows and columns are nonincreasing, i.e., $a_{ik} \geq a_{i,k+1}$ and $a_{ik} \geq a_{i+1,k}$ for all i, k. The terms *left-lower graded* and *right-higher graded* are defined in an analogous way.

The following proposition generalizes an earlier result by Krushevski [436].

Proposition 8.23. *Let A be a left-higher graded matrix, and let B be a right-lower graded matrix. Then the identical permutation solves $QAP(A, B)$.*

Proof. Due to Proposition 5.8, the following inequalities hold for any pair of permutations φ and ψ:

$$\sum_{i=1}^{n}\sum_{k=1}^{n} a_{ik}b_{\varphi(i)\psi(k)} \geq \sum_{i=1}^{n}\sum_{k=1}^{n} a_{ik}b_{i\psi(k)} \geq \sum_{i=1}^{n}\sum_{k=1}^{n} a_{ik}b_{ik}. \tag{8.26}$$

Thus the identical permutation is an optimal solution. $\square$

Remark: The same proof shows that the identical permutation is also an optimal solution when A is left-lower graded and B is right-higher graded. The complexity of the apparently simple problem where matrix A is left-higher graded and matrix B is right-higher graded is not known. A generalization of Proposition 8.23 has been shown by Krushevski [437] (see also Burkard, Çela, Demidenko, Metelski, and Woeginger [140]).

Çela [176] studied QAPs with matrices having a small bandwidth. Among other results she showed the following proposition.

Proposition 8.24. *Let A be a left-higher graded inverse Monge matrix, and let B be a symmetric Toeplitz matrix generated by $\gamma_1 = \gamma_{-1} = 1$ and $\gamma_i = 0$ if $i \notin \{1, -1\}$. In this case $QAP(A, B)$ is solved by the* zig-zag permutation

$$\varphi(i) = \begin{cases} n-i & \text{if } i < n/2 \text{ and } i \text{ is odd;} \\ i & \text{if } i \leq n/2 \text{ and } i \text{ is even;} \\ i & \text{if } i > n/2 \text{ and } n-i \text{ is even;} \\ n-i & \text{if } i \geq n/2 \text{ and } n-i \text{ is odd.} \end{cases}$$

For a proof and further results in this area we refer the reader to [176].

8.4.4 Problems generated by a special underlying graph

Up until now we studied special Koopmans–Beckmann problems $QAP(A, B)$ where the matrices A and B fulfill certain order relations or algebraic conditions. In this section we view the matrices A and B as weighted adjacency matrices of undirected graphs G_1 and

G_2, and we require that the permutation $\varphi \in \mathcal{I}(G_1, G_2)$, i.e., is an isomorphism between G_1 and G_2 (see also Section 7.1.5). The study of such problems goes back to Christofides and Gerrard [187]. Their strongest result is the following.

Theorem 8.25. *Let T_1 and T_2 be two isomorphic trees with n vertices and arbitrary weights on the edges. Moreover, let A and B be the symmetric weighted adjacency matrices of T_1 and T_2, respectively. Then the QAP*

$$\min_{\varphi \in \mathcal{I}(G_1, G_2)} \sum_{i=1}^{n} \sum_{k=1}^{n} a_{ik} b_{\varphi(i)\varphi(k)} \tag{8.27}$$

can be solved by a polynomial-time dynamic programming algorithm.

On the other hand, it is shown in [187] that the cheapest embedding of a tree with n vertices in a weighted complete graph K_n is $\mathcal{NP}$-hard. Rendl [573] generalized Theorem 8.25 to a certain class of vertex series parallel digraphs. These graphs are defined recursively in terms of their minimal members, the so-called *minimal vertex series parallel digraphs (MVSP)*, namely:

(i) a digraph consisting of just a single vertex is an MVSP digraph;

(ii) given two vertex disjoint MVSP digraphs $G_i = (V_i, E_i)$ for $i = 1, 2$, the following parallel and series compositions lead again to an MVSP digraph:

parallel composition: $G_p = (V_1 \cup V_2, E_1 \cup E_2)$;
series composition: $G_s = (V_1 \cup V_2, E_1 \cup E_2 \cup (T_1 \times S_2))$, where T_1 is the set of terminal nodes in G_1 and S_2 is the set of sources in G_2.

A *vertex series parallel digraph* is a digraph whose transitive closure equals the transitive closure of an MVSP digraph. Rendl's main result is the following theorem.

Theorem 8.26. *Let A and B be the weighted adjacency matrices of two isomorphic MVSP digraphs G_1 and G_2. Then the QAP*

$$\min_{\varphi \in \mathcal{I}(G_1, G_2)} \sum_{i=1}^{n} \sum_{k=1}^{n} a_{ik} b_{\varphi(i)\varphi(k)} \tag{8.28}$$

is $\mathcal{NP}$-hard. However, if none of G_1 and G_2 contains the complete bipartite digraph $K_{2,2}$ as vertex induced subgraph, then (8.28) *can be solved in polynomial time.*

The proof of the second result uses the canonical decomposition tree of the isomorphic MVSP digraphs. The result of Theorem 8.26 can also be extended to edge series parallel digraphs. Similar results on other graph classes, for example, planar graphs, are not known.

Erdoğan and Tansel [258] identified the following special case of *QAP*(A, B). Let A be the adjacency matrix of a graph with *path structure* (i.e., every vertex of the underlying graph has degree 0, 1, or 2), and let B be the (weighted) adjacency matrix of a grid with a rows and b columns such that $ab = n$. In this case an optimal solution of *QAP*(A, B) can be found in polynomial time.

Chapter 9

Other types of quadratic assignment problems

9.1 Quadratic bottleneck assignment problem

In Section 7.1.6 we briefly introduced the *quadratic bottleneck assignment problem* (QBAP) in Koopmans–Beckmann form as

$$\min_{\varphi \in \mathcal{S}_n} \max_{1 \le i,k \le n} a_{ik} b_{\varphi(i)\varphi(k)}, \tag{9.1}$$

$\mathcal{S}_n$ being the set of all permutations of the integers $1, 2, \ldots, n$. An instance of this problem will be denoted as *QBAP*(A, B), while *QBAP*(A, B, C) will denote an instance of the problem in which a linear term is present, namely,

$$\min_{\varphi \in \mathcal{S}_n} \max \left(\max_{1 \le i,k \le n} a_{ik} b_{\varphi(i)\varphi(k)} \, , \, \max_{1 \le i \le n} c_{i\varphi(i)} \right). \tag{9.2}$$

The meaning of the input matrices is the same as for the QAP (see Section 7.1.1): a_{ik} is the flow from facility i to facility k, b_{jl} is the distance from location j to location l, and c_{ij} is the cost of placing facility i at location j. In addition, the QBAP, too, can be formulated in a more general version through a four-index cost array $D = (d_{ijkl})$:

$$\min_{\varphi \in \mathcal{S}_n} \max_{1 \le i,k \le n} d_{i\varphi(i)k\varphi(k)}. \tag{9.3}$$

The first occurrence of the QBAP is due to Steinberg [623] and arises as an application in backboard wiring when we want to minimize the maximum length of the involved wires. Another important application, the *bandwidth minimization problem*, was investigated by Kellerer and Wirsching [413]. We want to find a permutation of the rows and columns of a given symmetric matrix T such that the permuted matrix has a minimum bandwidth. Since the *bandwidth* of a symmetric matrix T is the maximum absolute difference $|i - j|$ for which $t_{ij} \neq 0$, it is easy to see that this problem can be modeled as a special QBAP with a 0-1 matrix A which has $a_{ij} = 1$ if and only if $t_{ij} \neq 0$. Defining the distance matrix $B = (b_{jl})$ with $b_{jl} = |j - l|$ and solving the corresponding *QBAP*(A, B) leads to an optimal solution of the bandwidth minimization problem.

QBAPs occur in many other applications (such as VLSI design and quadratic assignment under time aspects). Virtually any application of the QAP corresponds to a QBAP model, because it often makes sense to minimize the largest cost instead of the overall cost incurred by some decision.

From a practical point of view, QBAPs are often easier to solve than QAPs, although they are $\mathcal{NP}$-hard, too. Since the problem of checking whether an undirected graph contains a Hamiltonian cycle can be modeled as a *QBAP*(A, B) where A is the complementary adjacency matrix of an undirected graph and B is the matrix representing a cyclic permutation (see Section 7.1.2), QBAP is $\mathcal{NP}$-hard even in the case of (0, 1) matrices A and B.

We see in Section 6.2.2 that a natural approach for solving an LBAP is to use a threshold technique. The same is true for the quadratic case. For the Lawler's formulation, cost coefficients of the form d_{iijl} (with $j \neq l$) or d_{ijkk} (with $i \neq j$) can never occur, so we can have at most $n^4 - 2(n^3 - n^2)$ different values. For Koopmans–Beckmann problems the coefficients which can occur are of the form $a_{ii}b_{jj}$ or products of off-diagonal elements of the matrices A and B. In the first case we get n^2 products for the diagonal elements, while in the second case we have $n(n-1)$ off-diagonal elements in A which are multiplied by the same number of off-diagonal elements in B. In both cases $n^2(n-1)^2 + n^2$ different objective function values can occur.

An important question is, therefore, again the computation of strong lower bounds. We show in the next sections how the Gilmore–Lawler bounds turn over to QBAPs.

9.1.1 Gilmore–Lawler bound for the Koopmans–Beckmann form

Let us consider a *QBAP*(A, B, C). Throughout this section we assume that all cost coefficients a_{ik}, b_{jl}, and c_{ij} are nonnegative numbers. We shall make use of Proposition 6.15, which replaces the result on minimum scalar products. For any two nonnegative vectors a and b of the same dimension n, where the elements of a are ordered increasingly and the elements of b are ordered decreasingly, we define

$$[a, b]^- = \max_{1 \le k \le n} a_k b_k. \tag{9.4}$$

Due to Proposition 6.15 we have for any permutation φ

$$[a, b]^- \le \max_{1 \le k \le n} a_k b_{\varphi(k)}. \tag{9.5}$$

Since the diagonal elements of matrix A can only be mapped to the diagonal elements of matrix B, we can replace matrix $C = (c_{ij})$ by

$$c_{ij} = \max(c_{ij}, a_{ii} b_{jj}). \tag{9.6}$$

Moreover, we can delete the diagonal elements in matrices A and B. As in Section 7.5.1 we denote the $(n-1)$-dimensional vector obtained from the ith row of A by deleting the element a_{ii} by $\hat{a}_i$. For every row j of matrix B the vector $\hat{b}_j$ is defined in an analogous way. Let us suppose that a solution φ maps i to $j = \varphi(i)$. We can fix the indices i and j, and get a lower bound for

$$\max_{1 \le k \le n-1} \hat{a}_{ik} \hat{b}_{j\varphi(k)}$$

by ordering the elements of $\hat{a}_i$ increasingly and the elements of $\hat{b}_j$ decreasingly and forming

$$[\hat{a}_i, \hat{b}_j]^-. \tag{9.7}$$

So, in order to compute a lower bound on the optimum value of a QBAP instance, we first compute the n^2 values $[\hat{a}_i, \hat{b}_j]^-$ and define a new cost matrix $L = (l_{ij})$ of an LBAP by

$$l_{ij} = \max(c_{ij}, [\hat{a}_i, \hat{b}_j]^-) \quad (i, j = 1, 2, \ldots, n). \tag{9.8}$$

We obtain the bottleneck Gilmore–Lawler bound B-GLB for the Koopmans–Beckmann QBAP by solving an LBAP with cost matrix L.

Example 9.1. Let us consider a Koopmans–Beckmann bottleneck problem with the same data as in Example 7.12:

$$A = \begin{pmatrix} 7 & 2 & 1 & 7 \\ 0 & 0 & 4 & 2 \\ 3 & 0 & 5 & 6 \\ 2 & 3 & 4 & 0 \end{pmatrix}, \quad B = \begin{pmatrix} 1 & 7 & 3 & 1 \\ 2 & 1 & 0 & 6 \\ 0 & 4 & 2 & 6 \\ 0 & 0 & 3 & 7 \end{pmatrix}, \quad \text{and } C = 0.$$

Now we get

$$\begin{aligned} \hat{a}_1 &= (2, 1, 7), \quad \hat{b}_1 = (7, 3, 1); \\ \hat{a}_2 &= (0, 4, 2), \quad \hat{b}_2 = (2, 0, 6); \\ \hat{a}_3 &= (3, 0, 6), \quad \hat{b}_3 = (0, 4, 6); \\ \hat{a}_4 &= (2, 3, 4), \quad \hat{b}_4 = (0, 0, 3). \end{aligned}$$

Therefore, the matrix

$$([\hat{a}_i, \hat{b}_j]^-) = \begin{pmatrix} 7 & 6 & 8 & 3 \\ 6 & 4 & 8 & 0 \\ 9 & 6 & 12 & 0 \\ 14 & 12 & 12 & 6 \end{pmatrix}$$

together with matrices C and

$$(a_{ii}b_{jj}) = \begin{pmatrix} 7 & 7 & 14 & 49 \\ 0 & 0 & 0 & 0 \\ 5 & 5 & 10 & 35 \\ 0 & 0 & 0 & 0 \end{pmatrix}$$

yields

$$L = \begin{pmatrix} 7 & 7 & 14 & 49 \\ 6 & 4 & 8 & 0 \\ 9 & 6 & 12 & 35 \\ 14 & 12 & 12 & 6 \end{pmatrix}.$$

Since the smallest element in the third column is 8, the LBAP has at least the objective value 8, and, indeed, this is the optimal value of the LBAP, as the permutation $\varphi = (1, 3, 2, 4)$ shows. This permutation yields for the given data the objective function value 36 which is, therefore, an upper bound for the optimum value. In order to find an optimal solution a branch-and-bound algorithm might now be entered. ∎

Note that in the bottleneck case the B-GLB cannot be improved by reduction methods due to the absorption property of the maximum-operator.

9.1.2 Gilmore–Lawler bound for the general form

This section is based on Burkard [127]. In this section we consider bottleneck QAPs with a general cost array D. As in the sum case, we use the Kronecker product formulation (7.17)–(7.19) where the cost coefficients d_{ijkl} are arranged in an $n^2 \times n^2$ matrix D such that the entry d_{ijkl} lies in row $(i-1)n+k$ and column $(j-1)n+l$. Remember that we can write D as a matrix (D^{ij}) where every D^{ij} is the matrix (d_{ijkl}) with fixed indices i and j. The coefficients d_{ijil} with $j \neq l$ and the coefficients d_{ijkj} with $i \neq k$ can never occur in the objective function since φ is a one-to-one mapping. Therefore, we can again assume

$$d_{ijkl} = \infty \quad \text{for } (i = k \text{ and } j \neq l) \text{ or } (i \neq k \text{ and } j = l). \tag{9.9}$$

For every pair (i, j) we solve the LBAP with cost matrix D^{ij}. Let the optimal value of this problem be l_{ij}. Next, we collect the values l_{ij} in a cost matrix L and solve an LBAP with this cost matrix. The optimum value of this problem constitutes a lower bound for the quadratic bottleneck problem, since we just relaxed the property that the solution of the quadratic problem must be a Kronecker product of an $(n \times n)$ permutation matrix.

As in the sum case, a strengthening of the bound is possible by taking into account that $x_{ij}x_{kl} = x_{kl}x_{ij}$. Therefore, we can replace d_{ijkl} as well as d_{klij} by the maximum of these two entries and apply afterwards the bounding procedure described above.

Example 9.2. Let the cost coefficients of a general QBAP be given as in Example 7.20:

$$D = \left(\begin{array}{ccc|ccc|ccc}
10 & * & * & * & 2 & * & * & * & 3 \\
* & 5 & 3 & 7 & * & 4 & 4 & 1 & * \\
* & 1 & 2 & 3 & * & 2 & 8 & 9 & * \\
\hline
* & 1 & 2 & 1 & * & 5 & 4 & 2 & * \\
2 & * & * & * & 2 & * & * & * & 3 \\
* & 5 & 5 & 8 & * & 9 & 1 & 8 & * \\
\hline
* & 7 & 8 & 9 & * & 1 & 4 & 3 & * \\
* & 1 & 4 & 1 & * & 8 & 1 & 2 & * \\
9 & * & * & * & 9 & * & * & * & 0
\end{array}\right).$$

By applying the symmetrization step we obtain the matrix

$$\left(\begin{array}{ccc|ccc|ccc}
10 & * & * & * & 2 & * & * & * & 3 \\
* & 5 & 4 & 7 & * & 4 & 4 & 5 & * \\
* & 9 & 4 & 7 & * & 3 & 8 & 9 & * \\
\hline
* & 7 & 4 & 5 & * & 5 & 4 & 4 & * \\
2 & * & * & * & 2 & * & * & * & 3 \\
* & 5 & 5 & 8 & * & 9 & 4 & 8 & * \\
\hline
* & 7 & 8 & 9 & * & 9 & 4 & 3 & * \\
* & 8 & 4 & 5 & * & 8 & 5 & 9 & * \\
9 & * & * & * & 9 & * & * & * & 0
\end{array}\right).$$

Solving the n^2 LBAPs with the cost matrices D^{ij} yields the matrix

$$L = \begin{pmatrix} 10 & 7 & 8 \\ 5 & 8 & 4 \\ 9 & 9 & 5 \end{pmatrix}.$$

The minimum element in the first row of L is 7, and, indeed, there exists a (unique) optimal solution of the LBAP with cost matrix L, namely, the permutation $\varphi = (2, 1, 3)$. The value of the quadratic bottleneck objective function for this permutation is 7, too. (This is easy to check using the Kronecker product $X_\varphi \otimes X_\varphi$.) Therefore, the optimal value of the given QBAP is 7 and $\varphi = (2, 1, 3)$ is an optimal solution of the given problem. ∎

9.2 Asymptotic results

In Section 5.1.1 we discuss random LSAPs. If the cost coefficients c_{ij} of an LSAP are i.i.d. random variables generated according to an exponential distribution with mean 1, Aldous theorem, Theorem 5.2, says that the expected optimum value of the LSAP tends to $\pi^2/6$. On the other hand, the worst solution value tends to infinity, as already the greedy solution grows with n like $\log n$ (see Proposition 5.1). Random QAPs show a completely different behavior, namely, the ratio between the optimal solution and an arbitrary feasible solution tends to one almost surely, as the size of the problem tends to infinity.

Specifically, Burkard and Fincke [147, 146] showed, for the Koopmans–Beckmann QAP and the bottleneck QAP, that the ratio of the worst and the optimal feasible solution tends to 1 in probability. Frenk, van Houweninge, and Rinnooy Kan [281] strengthened this result for QAPs to almost sure convergence (see also Rhee [579, 580]). Burkard and Fincke [148] extended the above convergence in probability result to a class of combinatorial optimization problems characterized by a specific combinatorial condition. This result was strengthened to almost sure convergence by Szpankowski [628]. Following an idea of Bonomi and Lutton [107], Albrecher, Burkard, and Çela [23] recently showed that the asymptotic behavior of QAPs can be deduced by means of arguments from statistical mechanics. Finally, sharp convergence rates of the relative difference between best and worst solutions of QBAPs were recently obtained by Albrecher [22]. In the following we first describe the asymptotic analysis for generic combinatorial optimization problems and then deduce from these results the asymptotic behavior of QAPs. Finally, we report results on random quadratic bottleneck problems in Section 9.2.3.

9.2.1 Generic combinatorial optimization problem

A generic combinatorial optimization problem P may be defined as follows. Let a *ground set* E be given. A *feasible solution* S is a subset of the ground set E. The set of all feasible solutions is denoted by $\mathcal{F}$. With every (feasible) subset S of the ground set we associate a (real) cost $F(S)$. In particular, if we associate with any element $e \in E$ of the ground set E the cost $c(e)$, we may define the *sum objective function* as

$$F(S) = \sum_{e \in S} c(e) \tag{9.10}$$

and the *bottleneck objective function* as

$$F(S) = \max_{e \in S} c(e).$$

For fixed n we denote by $BV(n)$ and $WV(n)$, respectively, the best and worst objective function value of a problem of size n:

$$BV(n) = \min_{S \in \mathcal{F}} F(S),$$
$$WV(n) = \max_{S \in \mathcal{F}} F(S).$$

In the case of a QAP of size n we have the ground set

$$E_n = \{(i, j, k, l) : i, j, k, l = 1, 2, \dots, n; i = k \text{ if and only if } j = l\}.$$

A feasible solution of the QAP is given by

$$S_\varphi = \{(i, k, \varphi(i), \varphi(k)) : i, k = 1, 2, \dots, n\},$$

where φ is a permutation of $N = \{1, 2, \dots, n\}$ (i.e., $\varphi \in \mathcal{S}_n$; see Section 1.1). Finally, we have

$$\mathcal{F}_n = \{S_\varphi : \varphi \text{ is a permutation of } N\}$$

as the set of all feasible solutions.

On the other hand, a linear assignment problem of size n can be described as follows. The ground set $\bar{E}_n$ is given by $\bar{E}_n = \{(i, j) : i, j = 1, 2, \dots, n\}$, a feasible solution is represented by the set $\bar{S}_\varphi = \{(i, \varphi(i)) : i = 1, 2, \dots, n\}$ for some permutation φ of N, and the set of all feasible solutions is

$$\bar{\mathcal{F}}_n = \{\bar{S}_\varphi : \varphi \text{ is a permutation of } N\}.$$

Both problems, the linear and the quadratic assignment problems, have the property that for fixed n the size of a feasible solution is fixed, namely, $|S_\varphi| = n^2$ for any permutation φ in the case of a QAP, and $|\bar{S}_\varphi| = n$ in the case of a linear assignment problem. The number of feasible solutions, however, is $n!$ in both problems.

The different asymptotic behavior of linear and quadratic assignment problems is due to the condition

$$\exists \lambda > 0 : \quad \lim_{n \to \infty} (\lambda |S| - \log |\mathcal{F}|) = \infty. \tag{9.11}$$

It is clear that (9.11) is fulfilled for the QAP ($S = S_\varphi$) since we have $\lim_{n\to\infty}(\log(n!)/n^2) = 0$, but it is not fulfilled for the linear assignment problem ($S = \bar{S}_\varphi$) since we have $\lim_{n\to\infty} (\log(n!)/n) = \infty$.

We are now going to prove the following theorem by Burkard and Fincke [148].

Theorem 9.3. *Consider a sequence of generic combinatorial optimization problems with sum objectives* (9.10)*, where for fixed n all feasible solutions have the same cardinality* $|S_n|$*. For any n, let* $c(e)$ *be identically distributed random variables in* $[0, 1]$ *with expected value* μ *and variance* $\sigma^2 > 0$*, where the* $c(e)$ *values are independent for every fixed solution. For a given* ε *let* δ *fulfill*

$$0 < \delta \leq \sigma^2 \text{ and } 0 < \frac{\mu + \delta}{\mu - \delta} \leq 1 + \varepsilon. \tag{9.12}$$

Moreover, let condition (9.11) *be fulfilled for*

$$\lambda = 2\left(\frac{\delta\sigma}{\delta + 2\sigma^2}\right)^2.$$

Then

$$\lim_{n\to\infty} \mathbb{P}\left\{\frac{WV(n)}{BV(n)} < 1 + \varepsilon\right\} \geq \lim_{n\to\infty} (1 - 2|\mathcal{F}_n| \exp(-\lambda|S_n|)) = 1. \tag{9.13}$$

For proving this theorem we make use of the following Lemma by Rényi [577] (Theorem 1, page 387).

Lemma 9.4. *Consider n independent random variables* $X_1, X_2, \ldots, X_n$ *such that* $|X_k - \mathbb{E}(X_k)| \leq 1$ *for* $k = 1, 2, \ldots, n$*. Let*

$$L = \sqrt{\sum_{k=1}^{n} \sigma^2(X_k)} \tag{9.14}$$

and let ν *be a positive number with* $\nu \leq L$*. Then*

$$\mathbb{P}\left\{\left|\sum_{k=1}^{n}(X_k - \mathbb{E}(X_k))\right| \geq \nu L\right\} \leq 2\exp\left(-\frac{\nu^2}{2(1 + \nu/2L)^2}\right). \tag{9.15}$$

We now prove Theorem 9.3.

Proof. Let $S = S_n$, and define

$$\nu = \frac{\delta\sqrt{|S|}}{\sigma} \text{ and } \lambda = 2\left(\frac{\delta\sigma}{\delta + 2\sigma^2}\right)^2.$$

Note that in the case of random variables $c(e)$ with $e \in S$ we get

$$L = \sqrt{\sum_{k=1}^{n} \sigma^2(c(e))} = \sigma\sqrt{|S|}. \tag{9.16}$$

We now consider the following chain of inequalities:

$$
\begin{aligned}
&\mathbb{P}\left\{\exists S \in \mathcal{F}_n : \left|\sum_{e\in S}(c(e)-\mu)\right| \geq \delta|S|\right\}\\
&\leq \sum_{S\in\mathcal{F}_n} \mathbb{P}\left\{\left|\sum_{e\in S}(c(e)-\mu)\right| \geq \delta|S|\right\}\\
&\leq |\mathcal{F}_n|\mathbb{P}\left\{\left|\sum_{e\in S}(c(e)-\mu)\right| \geq \frac{\delta\sqrt{|S|}}{\sigma}\sigma\sqrt{|S|}\right\}\\
&\leq 2|\mathcal{F}_n|\exp\left(-\left(\frac{\delta\sqrt{|S|}}{\sigma}\right)^2 \frac{1}{2\left(1+\frac{\delta\sqrt{|S|}}{\sigma}\frac{1}{2\sqrt{|S|}\sigma}\right)^2}\right) \qquad \text{(by Lemma 9.4)}\\
&= 2|\mathcal{F}_n|\exp(-\lambda|S|) \quad \text{(by definition of } \lambda).
\end{aligned}
$$

Applying condition (9.11) yields

$$
\lim_{n\to\infty} \mathbb{P}\left\{\exists S \in \mathcal{F}_n : \left|\sum_{e\in S}(c(e)-\mu)\right| \geq \delta|S|\right\} = 0
$$

or, equivalently,

$$
\lim_{n\to\infty} \mathbb{P}\left\{\forall S \in \mathcal{F}_n : \left|\sum_{e\in S}(c(e)-\mu)\right| < \delta|S|\right\} = 1. \tag{9.17}
$$

If all $S \in \mathcal{F}_n$ fulfill $\left|\sum_{e\in S}(c(e)-\mu)\right| < \delta|S|$, then we have

$$
WV(n) < (\mu+\delta)|S| \quad \text{and} \quad BV(n) > (\mu-\delta)|S|.
$$

Thus by (9.12) we get

$$
\frac{WV(n)}{BV(n)} < \frac{(\mu+\delta)}{(\mu-\delta)} \leq 1+\varepsilon.
$$

It immediately follows from (9.17) that

$$
\lim_{n\to\infty} \mathbb{P}\left\{\frac{WV(n)}{BV(n)} < 1+\varepsilon\right\} = 1. \quad \square \tag{9.18}
$$

In 1995, Szpankowski [628] applied a different analysis to random generic combinatorial optimization problems and showed the following stronger result. Remember that $f(n) = o(g(n))$ means that $\lim_{n\to\infty}(f(n)/g(n)) = 0$.

Theorem 9.5. *Let the cost coefficients $c(e)$ of a generic combinatorial optimization problem be i.i.d. random variables with finite mean μ, finite variance, and finite third moment. Moreover, let*

$$
|\mathcal{F}_n| \leq |\mathcal{F}_{n+1}| \tag{9.19}
$$

and

$$\log |\mathcal{F}_n| = o(|S|). \tag{9.20}$$

Then

$$|S|\mu - o(|S|) = BV(n) \leq WV(n) = |S|\mu + o(|S|) \text{ almost surely.} \tag{9.21}$$

Spankowski noted that only convergence in probability can be shown if the growth condition (9.19) is dropped. Condition (9.20) is a little stronger than (9.11).

A related result under slightly different probabilistic assumptions is shown in Albrecher, Burkard, and Çela [23]. In particular it is assumed that the costs are random variables drawn from a finite interval $[0, M]$ and that the size of the feasible solutions, $|S_n|$, grows monotonically with n as $\log n = o(|S_n|)$. In this case the Chernoff–Hoëffding bound (see Chernoff [184], Hoëffding [371]) yields

$$\mathbb{P}\left(\sup_{S \in \mathcal{F}_n} \left|\frac{F(S)}{|S|}\right| > \varepsilon\right) \leq |\mathcal{F}_n| \exp\left(-\frac{2\varepsilon^2|S|}{M^2}\right). \tag{9.22}$$

Due to the growth condition $\log n = o(|S_n|)$, the right-hand side of (9.22) is summable for all $\varepsilon > 0$ which yields, by using the Borel–Cantelli lemma (see, e.g., Billingsley [97]),

$$\mathbb{P}\left(\lim_{n\to\infty} \frac{BV(n)}{|S_n|} = \mu\right) = 1. \tag{9.23}$$

It is interesting to note that the last result can also be derived using methods from statistical mechanics. Bonomi and Lutton [107] applied the framework of statistical mechanics to the QAP. For a generic combinatorial optimization problem one can proceed as follows. The feasible solutions of a combinatorial optimization problem correspond to the states of a physical system, the objective function to the energy of the corresponding state. In statistical mechanics, the partition function $Q(T)$ of a system at temperature T is defined by

$$Q(T) = \sum_j \exp\left(-\frac{E_j}{k_B T}\right), \tag{9.24}$$

where k_B denotes the Boltzmann constant and we sum over all possible states. The thermal equilibrium of a thermodynamic system is characterized by the *Boltzmann distribution* where the probability for the system of being in state i with energy E_i at temperature T is given by

$$\frac{1}{Q(T)} \exp\left(-\frac{E_i}{k_B T}\right). \tag{9.25}$$

This formalism can be turned over to a combinatorial optimization problem. The probability

$$\mathbb{P}(S) = \frac{\exp(-F(S)\tau)}{Q(\tau)} \tag{9.26}$$

is assigned to each feasible solution $S \in \mathcal{F}$, where τ mimics $1/T$. In analogy to (9.24) a partition function is defined by

$$Q(\tau) = \sum_{s \in \mathcal{F}} \exp(-F(S)\tau). \tag{9.27}$$

If $\mathbb{E}(F(S), \tau)$ denotes the expected value of the objective function $F(S)$ in the above probabilistic model for fixed τ, it can easily be shown that

$$\mathbb{E}(F(S), \tau) = -\frac{d}{d\tau}(\log Q(\tau)). \tag{9.28}$$

In Albrecher, Burkard, and Çela [23] a careful investigation of $\log Q(\tau)/|S|$ finally yields (9.23).

9.2.2 Quadratic assignment problem

When we assume that the cost coefficients a_{ik} and b_{jl} of a Koopmans–Beckmann problem are i.i.d. in $[0, 1]$ with positive variance, we immediately get that the $c(e)$ values for $e \in S$ are independently distributed. (Note that this is not the case for the products $a_{ik}b_{jl}$.) Moreover, since $|S| = n^2$ and $|\mathcal{F}_n| = n!$, condition (9.11) is fulfilled for every $\lambda > 0$. Thus Proposition 9.3 implies the following.

Proposition 9.6. *Let the cost coefficients a_{ik} and b_{jl} of a Koopmans–Beckmann problem be i.i.d. random variables in $[0, 1]$ with positive variance. Then, for all $\varepsilon > 0$,*

$$\lim_{n\to\infty} \mathbb{P}\left(\frac{WV(n)}{BV(n)} < 1 + \varepsilon\right) = 1. \tag{9.29}$$

An analogous result holds for the general QAP, namely,

$$\lim_{n\to\infty} \mathbb{P}\left(\frac{\max\limits_{\varphi\in\mathcal{F}_n} \sum_{i=1}^{n}\sum_{k=1}^{n} d_{i\varphi(i)k\varphi(k)}}{\min\limits_{\varphi\in\mathcal{F}_n} \sum_{i=1}^{n}\sum_{k=1}^{n} d_{i\varphi(i)k\varphi(k)}} < 1 + \varepsilon\right) = 1. \tag{9.30}$$

Frenk, van Houweninge, and Rinnooy Kan [281] improved these results by showing that the convergence holds almost surely. In particular they showed the following.

Proposition 9.7. *Let the cost coefficients a_{ik} and b_{jl} be mutually i.i.d. random variables in $(0, \infty)$ with $\mathbb{E}(a_{ik}b_{ik}) > 0$ and finite $\mathbb{E}(\exp(-\lambda a_{ik}b_{ik}))$ in a neighborhood of 0. Then there exists a constant K such that*

$$\limsup_{n\to\infty} \sqrt{\frac{n}{\log n}} \left| \frac{\sum_{i=1}^{n}\sum_{k=1}^{n} a_{ik}b_{\varphi(i)\varphi(k)}}{n^2\mathbb{E}(a_{ik}b_{ik})} - 1 \right| \le K \quad \textit{almost everywhere} \tag{9.31}$$

for all permutations φ.

The almost sure convergence can immediately be deduced from Szpankowski's theorem 9.5. For improved convergence rates see Rhee [580].

Special attention found the case of planar QAPs, where a_{ik} is the distance between independent and uniformly distributed random vectors in the unit square (see Burkard and

Fincke [147], Frenk, van Houweninge, and Rinnooy Kan [281], and Rhee [579]). The strongest result is due to Rhee, who showed, for planar QAPs, that

$$\limsup_{n\to\infty} \frac{\sqrt{n}}{\log n} \left| \frac{\sum_{i=1}^{n}\sum_{k=1}^{n} a_{ik}b_{\varphi(i)\varphi(k)}}{n^2\mathbb{E}(a_{ik})\mathbb{E}(b_{jl})} - 1 \right| \leq K \quad \text{almost everywhere.} \tag{9.32}$$

These results are remarkable as they show that the objective function values of a random QAP lie closely together, i.e., they form a very flat landscape. This explains why it is so difficult to find the true optimum, whereas it is rather simple to find good suboptimal solutions. Indeed, using Proposition 9.6, Dyer, Frieze, and McDiarmid [244] showed the following.

Theorem 9.8. *Consider a branch-and-bound algorithm for solving a Koopmans–Beckmann QAP whose cost coefficients a_{ik} and b_{jl} are i.i.d. random variables in* $[0, 1]$ *with positive variance, and assume that the bounds are computed by using the Frieze–Yadegar linearization* (7.45)–(7.50). *Then the number of branched nodes is at least $n^{(1-o(1))n/4}$ with probability 1 as the size n of the QAP tends to infinity.*

9.2.3 Quadratic bottleneck assignment problem

The asymptotic analysis of generic combinatorial optimization problems becomes even easier in the case of a bottleneck objective; see Burkard and Fincke [148] for a convergence in probability result, as well as Szpankowski [628] who showed that Theorem 9.5 holds for the case

$$BV(n) = \min_{S\in\mathcal{F}} \max_{e\in S} c(e),$$
$$WV(n) = \max_{S\in\mathcal{F}} \max_{e\in S} c(e).$$

Recently, Albrecher [22] sharpened the convergence rates of QBAP given by Burkard and Fincke [146] proving the following.

Proposition 9.9. *Let the cost coefficients of a QBAP be i.i.d. random variables in* $[0, 1]$. *Then*

$$\frac{WV(n) - BV(n)}{BV(n)} \leq \sqrt{\frac{2\log n}{n}} \left(1 - \frac{1}{2\log n} - \frac{1}{8(\log n)^2}\right) \quad \textit{almost surely.}$$

9.3 Cubic and quartic assignment problem

In 1963 Lawler [447] suggested the consideration of cubic, quartic, ..., n-adic assignment problems. But it was only 30 years later that such models found a closer interest due to an application in VLSI synthesis, where programmable logic arrays have to be implemented. The encoding of states such that the actual implementation by flip-flops is of minimum size leads in the case of data flip-flops to a QAP, but in case of toggle flip-flops it leads to quartic assignment problems. For details of this application see Burkard, Çela, and Klinz [141], who studied biquadratic (= quartic) assignment problems, derived lower bounds, and investigated the asymptotic probabilistic behavior of such problems. Burkard and Çela [137] developed metaheuristics for quartic assignment problems and compared their computational performance. Recently, cubic and quartic assignment problems found a role in sharpening bounds for the QAP, as in Adams, Guignard, Hahn, and Hightower [4] and Hahn [356]. Winter and Zimmermann [665] used a cubic assignment problem for finding the minimum shunting of trams in a storage yard. (Note that the objective function (2.1.1) in [665] contains some typos in the indices, but is actually the objective function of a cubic AP.)

The *cubic assignment problem* (Cubic AP) can be stated as follows. Let two arrays $A = (a_{ikp})$ and $B = (b_{jlq})$ $(i, j, k, l, p, q = 1, 2, \dots, n)$ be given. The problem is then

$$\min_{\varphi \in \mathcal{S}_n} \sum_{i=1}^{n} \sum_{k=1}^{n} \sum_{p=1}^{n} a_{ikp} b_{\varphi(i)\varphi(k)\varphi(p)}, \tag{9.33}$$

where $\mathcal{S}_n$ denotes the set of all permutations of the integers $1, 2, \dots, n$. This problem is used in Section 7.3.5 to obtain improved bounds for the QAP by a level-2 reformulation.

Similarly, let two arrays $A = (a_{ikpr})$ and $B = (b_{jlqs})$ $(i, j, k, l, p, q, r, s = 1, 2, \dots, n)$ be given. The *quartic assignment problem* (Quartic AP) can then be stated as

$$\min_{\varphi \in \mathcal{S}_n} \sum_{i=1}^{n} \sum_{k=1}^{n} \sum_{p=1}^{n} \sum_{r=1}^{n} a_{ikpr} b_{\varphi(i)\varphi(k)\varphi(p)\varphi(r)}. \tag{9.34}$$

Both problems can be written as integer linear programs with objective function

$$\min \sum_{i=1}^{n} \sum_{j=1}^{n} \sum_{k=1}^{n} \sum_{l=1}^{n} \sum_{p=1}^{n} \sum_{q=1}^{n} a_{ikp} b_{jlq} x_{ij} x_{kl} x_{pq}$$

or

$$\min \sum_{i=1}^{n} \sum_{j=1}^{n} \sum_{k=1}^{n} \sum_{l=1}^{n} \sum_{p=1}^{n} \sum_{q=1}^{n} \sum_{r=1}^{n} \sum_{s=1}^{n} a_{ikpr} b_{jlqs} x_{ij} x_{kl} x_{pq} x_{rs},$$

respectively, and constraints

$$\begin{aligned}
&\sum_{i=1}^{n} x_{ij} = 1 \quad (j = 1, 2, \dots, n),\\
&\sum_{j=1}^{n} x_{ij} = 1 \quad (i = 1, 2, \dots, n),\\
&x_{ij} \in \{0, 1\} \quad (i, j = 1, 2, \dots, n).
\end{aligned}$$

Since both the cubic and the quartic AP contain the QAP as a special case, they are $\mathcal{NP}$-hard. Linearizations similar to those discussed in Section 7.3 can also be applied to quartic APs by setting

$$y_{ijkl} = x_{ij}x_{kl} \text{ and } z_{ijklpqrs} = y_{ijkl}y_{pqrs}$$

(see, e.g., the equivalents of the Lawler linearization and of the Kaufman–Broeckx linearization in [141]). Burkard, Çela, and Klinz [141] also suggested two reduction bounds in analogy to the reduction bounds of Section 7.5.2. The idea is to shift as much information as possible from the quartic AP to simpler problems like the QAP and the linear assignment problem. Using instances with known objective function value (constructed along the same line as, in the case of QAPs, by Palubeckis [524] and Li and Pardalos [456]), computational tests were performed, showing that these reduction bounds are rather weak.

Since cubic and quartic APs are very hard to solve exactly, there is a need for efficient suboptimal algorithms. Several QAP metaheuristics were adapted by Burkard and Çela [137] to the quartic AP, in particular deterministic pairwise exchange methods, three versions of simulated annealing, and a tabu search combined with simulated annealing. Computational tests on instances with known optimal solutions showed that a specialized implementation of simulated annealing yields the best performance. For $n = 14$, for example, an optimal solution was found in 98.4% of all cases (in about 86 CPU seconds on a DECstation 5000/240).

The relatively good performance of metaheuristics suggests that cubic and quartic APs show a probabilistic asymptotic behavior similar to that of the QAP. Indeed they fulfill condition (9.11), which is essential to show that the ratio between worst and best solution value tends to 1 as the problem size increases. Burkard, Çela, and Klinz [141] showed a result analogous to Proposition 9.6 for the quartic AP. The result remains true even if some, but not too many, coefficients a_{ikpr} and b_{jlqs} are zero. In this case it is required that the best value $BV(n)$, given by (9.34), grows faster than $n \log n$, i.e., that

$$\lim_{n\to\infty} \frac{BV(n)}{n \log n} = \infty. \tag{9.35}$$

Then we get, for all $\varepsilon > 0$, that the best value and the worst value $WV(n)$ (given by (9.34) with "min" replaced by "max") satisfy

$$\lim_{n\to\infty} \mathbb{P}\left\{\frac{WV(n)}{BV(n)} < 1 + \varepsilon\right\} = 1. \tag{9.36}$$

9.4 Quadratic semi-assignment problem

The *quadratic semi-assignment problem* (semi-QAP), originally introduced by Greenberg [340], has the same objective function as the QAP, whereas the solutions are not permutations, but functions mapping the set of integers $N = \{1, 2, \ldots, n\}$ to the set $M = \{1, 2, \ldots, m\}$, with $n > m$. In the facility-location context (see Section 7.1.1) this means that there are n facilities but only m locations and there is no limit to the number of facilities we can assign to the same location. Similarly to what we have done for the QAP, we define semi-$QAP(A, B, C)$ by means of an $n \times n$ flow matrix $A = (a_{ik})$, an $m \times m$ distance matrix $B = (b_{jl})$, and an $n \times m$ cost matrix $C = (c_{ij})$. Using binary variables x_{ij} taking value 1 if facility i is assigned to location j, and 0 otherwise, we can model the semi-QAP in Koopmans–Beckmann form as

$$\min \quad \sum_{i=1}^{n}\sum_{j=1}^{m}\sum_{k=1}^{n}\sum_{l=1}^{m} a_{ik}b_{jl}x_{ij}x_{kl} + \sum_{i=1}^{n}\sum_{j=1}^{m} c_{ij}x_{ij} \tag{9.37}$$

$$\text{s.t.} \quad \sum_{j=1}^{m} x_{ij} = 1 \qquad (i = 1, 2, \ldots, n), \tag{9.38}$$

$$x_{ij} \in \{0, 1\} \qquad (i = 1, 2, \ldots, n;\ j = 1, 2, \ldots, m). \tag{9.39}$$

In Lawler's form (see again Section 7.1.1) of the semi-QAP the objective function (9.37) is replaced by

$$\min \quad \sum_{i=1}^{n}\sum_{j=1}^{m}\sum_{k=1}^{n}\sum_{l=1}^{m} d_{ijkl}x_{ij}x_{kl}. \tag{9.40}$$

An instance of semi-$QAP(A, B, C)$ can be transformed into an equivalent instance of $QAP(\tilde{A}, \tilde{B}, \tilde{C})$ by introducing a true facility and $m-1$ dummy facilities for each facility of the semi-QAP and n identical locations for each location of the semi-QAP. In the $nm \times nm$ matrix $\tilde{A} = (\tilde{A}^{ik})$ the unique nonzero element of each $m \times m$ submatrix $\tilde{A}^{ik}$ is $\tilde{A}^{ik}_{11} = a_{ik}$. This implies that there can be a positive flow only between two true facilities. Matrix $\tilde{B} = (\tilde{B}^{jl})$ is obtained by filling the elements of each $n \times n$ submatrix $\tilde{B}^{jl}$ with the value b_{jl}, whereas for matrix $\tilde{C} = (\tilde{C}^{ij})$, each $m \times n$ submatrix $\tilde{C}^{ij}$ has the first row filled with the value c_{ij} and zeroes in the remaining elements. Observe that a facility i of semi-$QAP(A, B, C)$ maps into the $((i-1)m+1)$th facility of $QAP(\tilde{A}, \tilde{B}, \tilde{C})$ (the true facility), while the jth location of semi-$QAP(A, B, C)$ is associated with locations $(j-1)n+1, (j-1)n+2, \ldots, jn$ of $QAP(\tilde{A}, \tilde{B}, \tilde{C})$. Given a solution φ of $QAP(\tilde{A}, \tilde{B}, \tilde{C})$, the corresponding solution X of semi-$QAP(A, B, C)$ is thus

$$x_{ij} = \begin{cases} 1 & \text{if } j = \lceil \varphi((i-1)m+1)/n \rceil, \\ 0 & \text{otherwise.} \end{cases}$$

Example 9.10. Let us consider a Koopmans–Beckmann semi-$QAP(A, B, C)$ with $n = 4$, $m = 2$, and input matrices

$$A = \begin{pmatrix} 7 & 2 & 1 & 7 \\ 0 & 0 & 4 & 2 \\ 3 & 0 & 5 & 6 \\ 2 & 3 & 4 & 0 \end{pmatrix}, \quad B = \begin{pmatrix} 0 & 2 \\ 1 & 5 \end{pmatrix}, \quad \text{and } C = \begin{pmatrix} 0 & 0 \\ 0 & 0 \\ 0 & 0 \\ 0 & 0 \end{pmatrix}.$$

By transformation, we obtain an instance of $QAP(\tilde{A}, \tilde{B}, \tilde{C})$ with

$$\tilde{A} = \left(\begin{array}{cc|cc|cc|cc} 7 & 0 & 2 & 0 & 1 & 0 & 7 & 0 \\ 0 & 0 & 0 & 0 & 0 & 0 & 0 & 0 \\ \hline 0 & 0 & 0 & 0 & 4 & 0 & 2 & 0 \\ 0 & 0 & 0 & 0 & 0 & 0 & 0 & 0 \\ \hline 3 & 0 & 0 & 0 & 5 & 0 & 6 & 0 \\ 0 & 0 & 0 & 0 & 0 & 0 & 0 & 0 \\ \hline 2 & 0 & 3 & 0 & 4 & 0 & 0 & 0 \\ 0 & 0 & 0 & 0 & 0 & 0 & 0 & 0 \end{array}\right), \quad \tilde{B} = \left(\begin{array}{cccc|cccc} 0 & 0 & 0 & 0 & 2 & 2 & 2 & 2 \\ 0 & 0 & 0 & 0 & 2 & 2 & 2 & 2 \\ 0 & 0 & 0 & 0 & 2 & 2 & 2 & 2 \\ 0 & 0 & 0 & 0 & 2 & 2 & 2 & 2 \\ \hline 1 & 1 & 1 & 1 & 5 & 5 & 5 & 5 \\ 1 & 1 & 1 & 1 & 5 & 5 & 5 & 5 \\ 1 & 1 & 1 & 1 & 5 & 5 & 5 & 5 \\ 1 & 1 & 1 & 1 & 5 & 5 & 5 & 5 \end{array}\right),$$

and $\tilde{C} = 0$. Consider, e.g., the feasible solution of $QAP(\tilde{A}, \tilde{B}, \tilde{C})$ given by the identical permutation $\varphi(i) = i$ for all i, of value

$$z = 2 + 14 + 8 + 4 + 3 + 25 + 30 + 2 + 3 + 20 = 111.$$

The corresponding solution of semi-$QAP(A, B, C)$ is $x_{11} = x_{21} = x_{32} = x_{42} = 1$, $x_{12} = x_{22} = x_{31} = x_{41} = 0$. ■

The semi-QAP is $\mathcal{NP}$-hard, as an instance of $QAP(A, B, C)$ can be transformed into an equivalent instance of semi-QAP by adding a large number to the elements of the main diagonal of B, so as to avoid that two facilities are assigned to the same location in the optimal solution. (Different transformations can be obtained from $\mathcal{NP}$-hard clustering problems studied by Gonzalez [335].) The structure of the semi-QAP polytope and that of the symmetric semi-QAP polytope have been studied, respectively, by Saito, Fujie, Matsui, and Matuura [595] and Saito [594], who extended the results on the structure of the QAP polytope (see Section 7.4).

Billionet and Elloumi [98] studied admissible transformations for the semi-QAP (see Section 7.6). They showed that the best reduction is obtained by applying to the Lawler's formulation (9.37)–(9.39) the Adams–Johnson linearization of Section 7.3.4, removing the integrality constraints, and adopting for the entries of the reduced matrix the reduced costs of the solution to the associated continuous problem.

9.4.1 Applications

Several real-world problems can be modeled as semi-QAPs.

Supply support for space bases

The first application of the semi-QAP is probably the one described by Freeman, Gogerty, Graves, and Brooks [279]. A space base needs to receive from Earth a number of supply modules for day-to-day living. Each module i has an associated delivery time window and a weight w_i. The problem is to find the optimal assignment of the supply modules to a set of cargo trips, while satisfying the schedule requirements and minimizing a transportation cost function. Let us define the *interference* between two modules i and k as $I_{ik} = w_i w_k$ if the two time windows overlap, and $I_{ik} = \infty$ otherwise. By minimizing the total interference between all pairs of modules assigned to the same trip, we reach two goals: (a) we ensure that all modules are delivered in time; (b) we minimize the total cost needed to transport the modules. We can model the problem as a semi-$QAP(A, B)$ where N is the set of modules and M is the set of trips. The elements of the flow matrix $A = (a_{ij})$ are defined as

$$a_{ik} = \begin{cases} I_{ik} & \text{if } i \neq k, \\ 0 & \text{otherwise,} \end{cases}$$

while the distance matrix B is the unit matrix.

Schedule synchronization in transit networks

This problem arises in the design of optimal transit networks (see, e.g., Daduna and Voss [209] and Malucelli [477]). We are given n lines, each of which transports people between two transit points. The lines may be operated by different means of transport like metro, bus, airplane, etc. For each pair of lines (i, k), we know the estimated amount of passengers p_{ik} that want to travel first on line i and then on line k. The synchronization problem requires a definition of the departure time of each line i, within a given time window $[s_i, e_i]$, so that the average waiting time of the passengers at the transit points is minimized. We can model this problem as a semi-QAP by defining N as the set of lines and M as the set of all possible line departure times (i.e., $M = \{\min_{i\in N}\{s_i\}, \min_{i\in N}\{s_i\} + 1, \dots, \max_{i\in N}\{e_i\}\}$). Let r_i be the traveling time for line i. Using the Lawler's form we define the elements of matrix $D = (d_{ijkl})$, with j and l being, respectively, the starting time of line i and the starting time of line k:

$$d_{ijkl} = \begin{cases} \infty & \text{if } j + r_i > l, \\ p_{ik}(l - j - r_i) & \text{otherwise.} \end{cases} \tag{9.41}$$

In this way the cost d_{ijkl} is infinite if a departure time j for line i does not allow us to reach the starting point of line k before its departure time l. If instead the two departure times determine a feasible connection, we have as cost the waiting time $(l - j - r_i)$ at the transit point multiplied by the number of passengers changing from line i to line k.

Task scheduling on distributed computing systems

A modular program consists of n tasks that must be executed on a computing system with m processors. The program can be described through an acyclic *task digraph* $G = (N; \hat{A})$ where N is the task set and $\hat{A}$ models the task precedences: an arc $(i, k) \in \hat{A}$ implies that task k requires as input the information elaborated by task i. The amount of information exchanged between i and k is f_{ik}, while t_{jl} denotes the time to transfer one unit of information from processor j to processor l. We want to find a feasible assignment of the tasks to the processors that minimizes a function of the total time required for the execution of the program.

First, observe that the orientation of the arcs is fundamental to describe the program flow, but it does not matter for the assignment problem. Indeed, once we have partitioned the tasks among the processors, we can easily reorder and schedule them so as to satisfy the precedences induced by the arcs. Therefore, many authors assume that the graph is not oriented. Magirou and Milis [472] considered a system with processors of different speeds, and defined as e_{ij} the execution time of task i on processor j. Moreover, they assumed that, before each transfer of information between two processors j and l, it is necessary to spend time θ_{jl} to set up an appropriate communication channel. Using binary variables x_{ij} taking the value 1 if and only if task i is assigned to processor j, this problem can be modeled as a semi-QAP in Lawler's form with

$$d_{ijkl} = f_{ik}t_{jl} + \theta_{jl} + e_{ij}.$$

Malucelli and Pretolani [478] considered the same application without setup costs. In this case we can model the problem as a Koopmans–Beckmann semi-$QAP(A, B, C)$ with $a_{ik} = f_{ik}$ $(i, k = 1, 2, \dots, n)$, $b_{jl} = t_{jl}$ $(j, l = 1, 2, \dots, m)$, and $c_{ij} = e_{ij}$ $(i = 1, 2, \dots, n;\ j = 1, 2, \dots, m)$.

Billionnet, Costa, and Sutter [99] described a problem in which the transfer rate between two processors does not depend on the processor themselves but on the two tasks involved in the transfer. So we can define the time $\hat{t}_{ik}$ required to transfer all the information of task i to task k, with $\hat{t}_{ik} = 0$ if i and k are executed on the same processor. The problem is then modeled as a Koopmans–Beckmann semi-$QAP(A, B, C)$ with $a_{ik} = \hat{t}_{ik}$ $(i, k = 1, 2, \dots, n)$, $b_{jl} = 1$ (resp., $b_{jl} = 0$) if $j \neq l$ (resp., $j = l$) $(j, l = 1, 2, \dots, m)$, and $c_{ij} = e_{ij}$ $(i = 1, 2, \dots, n;\ j = 1, 2, \dots, m)$.

Minimizing the mean flow time in parallel processors scheduling

A classical problem in scheduling theory is to assign n tasks to m unrelated parallel processors so that no two tasks are executed at the same time on the same processor and a given objective function is minimized. Let w_i and p_{ij} denote, respectively, the weight of task i and its processing time when executed on processor j. Moreover, for a given schedule, let C_i denote the completion time of task i. Using the three field notation introduced by Graham, Lawler, Lenstra, and Rinnooy Kan [337], we denote by $R||\sum w_i C_i$ the problem of scheduling the tasks by minimizing the weighted sum of the completion times, which is equivalent to minimize the mean flow time for identical weights.

Skutella [612] modeled this problem as a special case of semi-QAP. In order to solve the problem we have to (a) partition the tasks among the processors; and (b) determine the sequence of the tasks on each processor. Once the partition is known, the sequencing problem on a single processor can be optimally solved through the famous Smith's rule [614]: schedule i before k whenever $w_i/p_i > w_k/p_k$ (breaking ties arbitrarily). Let us denote by $\mathcal{A}(i, j)$ the set of tasks that must precede task i on processor j, according to

Smith's rule. The following integer program models problem $R||\sum w_i C_i$:

$$\min \quad \sum_{i=1}^{n} w_i C_i \tag{9.42}$$

$$\text{s.t.} \quad C_i = \sum_{j=1}^{m} x_{ij} \left(p_{ij} + \sum_{k \in \mathcal{A}(i,j)} p_{kj} x_{kj} \right) \quad (i = 1, 2, \ldots, n), \tag{9.43}$$

$$\sum_{j=1}^{m} x_{ij} = 1 \qquad (i = 1, 2, \ldots, n), \tag{9.44}$$

$$x_{ij} \in \{0, 1\} \qquad (i = 1, 2, \ldots, n;\ j = 1, 2, \ldots, m), \tag{9.45}$$

where $x_{ij} = 1$ if and only if task i is assigned to processor j. Substituting C_i in (9.42) by means of equation (9.43) we obtain the quadratic objective function

$$\min \sum_{i=1}^{n} \sum_{j=1}^{m} \left(w_i p_{ij} x_{ij} + w_i x_{ij} \sum_{k \in \mathcal{A}(i,j)} p_{kj} x_{kj} \right). \tag{9.46}$$

Using the notation introduced in Section 7.2.3 we can reformulate the problem as a quadratic convex program. Let vec(D) denote the vector formed by the columns of D and let $x =$ vec(X). Equation (9.46) can then be written as

$$\min q^T x + \frac{1}{2} x^T D x. \tag{9.47}$$

Vector $q^T \in \mathbb{R}^{nm}$ is given by $q = vec(Q)$, with $Q = (q_{ij}) = (w_i p_{ij})$, while $D = (d_{ijkl})$ is a symmetric $nm \times nm$ matrix whose element in row $(i-1)m + j$ and column $(k-1)m + l$ $(i, k = 1, 2, \ldots, n;\ j, l = 1, 2, \ldots, m)$ is given by

$$d_{ijkl} = \begin{cases} 0 & \text{if } i = k \text{ or } j \neq l, \\ w_k p_{ij} & \text{if } i \in \mathcal{A}(k, j) \text{ and } j = l, \\ w_i p_{kj} & \text{if } k \in \mathcal{A}(i, j) \text{ and } j = l. \end{cases}$$

The only nonzero values of D correspond to entries with $j = l$. Thus we can permute the rows and columns of D so that the $O(n^2)$ positive elements associated with a given value j are stored in an $n \times n$ submatrix D^{jj}. As a consequence the new matrix D becomes a block diagonal matrix (see Figure 9.1(a)) with blocks $D^{11}, D^{22}, \ldots, D^{mm}$ on the main diagonal (and zeros elsewhere). Moreover, if we reorder the rows and columns of each submatrix D^{jj} so that the kth row (resp., column) precedes the ith row (resp., column) if and only if $k \in \mathcal{A}(i, j)$, then D^{jj} assumes the form depicted in Figure 9.1(b).

Parallel processors scheduling

Consider a transportation network $\mathcal{N} = (V, E)$ where the vertex set V is partitioned into a set H of m *hub vertices* (*hubs* for short) and a set S of n *nonhub vertices* (usually called

$$
D = \begin{pmatrix} D^{11} & 0 & \dots & 0 \\ 0 & D^{22} & & \vdots \\ \vdots & & \ddots & 0 \\ 0 & \dots & 0 & D^{mm} \end{pmatrix} \qquad D^{jj} = \begin{pmatrix} 0 & w_2 p_{1j} & w_3 p_{1j} & \dots & w_n p_{1j} \\ w_2 p_{1j} & 0 & w_3 p_{2j} & \dots & w_n p_{2j} \\ w_3 p_{1j} & w_3 p_{2j} & 0 & & w_n p_{3j} \\ \vdots & \vdots & & \ddots & \vdots \\ w_n p_{1j} & w_n p_{2j} & w_n p_{3j} & \dots & 0 \end{pmatrix}
$$

(a) (b)

Figure 9.1. *Form of the permuted cost matrix for* $R||\sum w_i C_i$.

spokes), while $E = V \times V$ is the set of all possible links. For each pair of vertices $(i, j) \in E \setminus (H \times H)$, let f_{ij} be the nonnegative amount of flow required from i to j. Moreover, for any pair of vertices $(i, j) \in E \setminus (S \times S)$, let t_{ij} be the nonnegative unit transportation cost from i to j. In this problem the transportation of flow between two vertices is allowed only via hub vertices. We are asked to assign each spoke to a hub in such a way that the total transportation cost of the required flow is minimized. The problem has application in the airline industry, telecommunications, and postal and parcel delivery.

Iwasa, Saito, and Matsui [387] modeled this problem as a binary quadratic problem, with variables x_{ij} $(i = 1, 2, \dots, n;\ j = 1, 2, \dots, m)$ taking the value 1 if spoke i is assigned to hub j, and the value 0 otherwise:

$$
\begin{aligned}
\min \quad & \sum_{i\in S}\sum_{k\in S} f_{ik}\left(\sum_{j\in H} t_{ij}x_{ij} + \sum_{j\in H}\sum_{l\in H} t_{jl}x_{ij}x_{kl} + \sum_{l\in H} t_{lk}x_{kl}\right) \\
& + \sum_{i\in S}\sum_{h\in H} f_{ih}\sum_{j\in H}(t_{ij}+t_{jh})x_{ij} + \sum_{h\in H}\sum_{k\in S} f_{hk}\sum_{l\in H}(t_{hl}+t_{lk})x_{kl} \qquad (9.48)\\
\text{s.t.} \quad & \sum_{j\in H} x_{ij} = 1 \quad (i = 1, 2, \dots, n), \qquad (9.49)\\
& x_{ij} \in \{0, 1\} \quad (i = 1, 2, \dots, n;\ j = 1, 2, \dots, m). \qquad (9.50)
\end{aligned}
$$

The first term in (9.48) models the spoke-to-spoke quadratic transportation costs passing through the two associated hubs (or through a single hub, if it is common to the two spokes). Similarly, the second and third term model the spoke-to-hub and hub-to-spoke transportation costs, respectively.

Problem (9.48)–(9.49) can be transformed into a semi-QAP in Lawler's form as follows. By executing the multiplications in the first term of (9.48), we obtain an objective function with four linear terms and a quadratic term. Since in any optimal solution (9.49) holds, we can rewrite the first linear term as

$$
\sum_{i\in S}\sum_{k\in S} f_{ik}\sum_{j\in H} t_{ij}x_{ij} = \sum_{i\in S}\sum_{k\in S} f_{ik}\sum_{j\in H} t_{ij}x_{ij}\sum_{l\in H} x_{kl} = \sum_{i\in S}\sum_{k\in S}\sum_{j\in H}\sum_{l\in H} f_{ik}t_{ij}x_{ij}x_{kl}
$$

and the third linear term as

$$
\sum_{i\in S}\sum_{h\in H} f_{ih}\sum_{j\in H}(t_{ij}+t_{jh})x_{ij} = \frac{1}{n}\sum_{i\in S}\sum_{k\in S}\sum_{j\in H}\sum_{l\in H}\sum_{h\in H} f_{ih}(t_{ij}+t_{jh})x_{ij}x_{kl}.
$$

By rewriting in the same way the second and fourth linear terms, we finally obtain the cost matrix $D = (d_{ijkl})$ $(i, k = 1, 2, \ldots, n,\ j, l = 1, 2, \ldots, m)$ as

$$d_{ijkl} = f_{ik}(t_{ij} + t_{jl} + t_{lk}) + \frac{1}{n}\sum_{h\in H} f_{ih}(t_{ij} + t_{jh}) + \frac{1}{n}\sum_{h\in H} f_{hk}(t_{hl} + t_{lk}).$$

9.4.2 Solution methods

Few special cases of semi-QAP are known to be solvable in polynomial time. These problems are quite specific and not very useful by themselves, but can be used to obtain lower bounds for the general case.

Easy cases

Bokhari [103] studied the problem of scheduling tasks on a distributed computing system where the task digraph G is a tree. He solved the problem by expanding the tree to obtain a new digraph where each vertex of G is substituted by a *layer* of m copies (one per processor), and each arc (i, k) is substituted by m^2 arcs joining all the copies of i to all the copies of k. An arc, say, the one joining the jth copy of i to the lth copy of k, is thus associated with a quadruple (i, j, k, l) and is given a cost depending on these four elements. One can see that any feasible solution of the corresponding semi-QAP is associated with a tree that uses exactly one vertex from each layer. The optimal tree is computed in $O(nm^2)$ time by backtracing from the leaves to the root (i.e., visiting each arc exactly once).

Due to the generality of the costs associated with the arcs, Bokhari's algorithm is not restricted to solving only the task assignment problem above, but also a more general class of semi-QAPs that can be described as follows. Let us call a *flow graph* the graph whose adjacency matrix is obtained by setting the nonzero entries of the $n \times n$ matrix $\widetilde{D} = (\widetilde{d}_{ik}) = (\sum_{j=1}^{m}\sum_{l=1}^{m} d_{ijkl})$ to 1. (If the semi-QAP is given in Koopmans–Beckmann form, the adjacency matrix is obtained by setting to 1 the nonzero entries of the flow matrix A.) It easily follows that Bokhari's algorithm solves any semi-QAP whose flow graph is a tree.

Chhajed and Lowe [185] considered a semi-QAP with a flow graph having an edge series-parallel structure, as defined by Richey [582]. (See also Section 8.4.4 for QAPs generated by special series-parallel digraphs.) A graph is called *series-parallel* if it can be reduced to a single vertex by means of the following operations:

(i) *tail reduction*: remove a leaf and its incident edge;

(ii) *series reduction*: given a vertex i with degree 2, remove i and its incident edges, say, $[i, h]$ and $[i, k]$, and add edge $[h, k]$;

(iii) *parallel reduction*: given a set of parallel edges remove all of them but one.

(Note that parallel edges may appear in the graph due to the series reductions.) Chhajed and Lowe designed an $O(nm^3)$ time algorithm to solve this special case of semi-QAP. Malucelli and Pretolani [478] independently proposed the same reduction for the task scheduling case and, using the expansion of the graph in an implicit way, introduced by Bokhari [103], developed an equivalent $O(nm^3)$ algorithm to solve the problem.

Malucelli [477] considered the problem of scheduling synchronization in transit networks (see Section 9.4.1) and proved that it can be solved in polynomial time through network flow techniques. He introduced a bipartite graph $G = (U, V; E)$ where U contains one vertex for each line $i \in N$ and V contains one vertex for each line departure time $j \in M$. The edge set E has an edge $[i, j]$ for each line i and each feasible starting time $j \in [s_i, e_i]$. Malucelli observed that G is a convex graph (remember that G is convex, see Section 3.5, if $[i, j] \in E$ and $[i, l] \in E$, with $j < l$, implies that $[i, h] \in E$ for all $j < h < l$). The problem requires the assignment of each vertex of U to a vertex of V by minimizing Lawler's objective function (9.40) with costs given by (9.41). Using binary variables x_{ij} to describe the possible assignment of line i to starting time j and variables $\pi_i = \sum_{j=s_i}^{e_i} j x_{ij}$ to denote the starting time of line i, this semi-QAP can be reformulated as

$$\min \quad \sum_{i=1}^{n} \sum_{k=1}^{n} p_{ik}(\pi_k - \pi_i - r_i) \tag{9.51}$$

$$\text{s.t.} \quad \pi_k - \pi_i \geq r_i \qquad (i, k = 1, 2, \ldots, n;\ p_{ik} > 0) \tag{9.52}$$

$$s_i \leq \pi_i \leq e_i \qquad (i = 1, 2, \ldots, n), \tag{9.53}$$

$$\pi_i \geq 0, \text{ integer} \qquad (i = 1, 2, \ldots, n). \tag{9.54}$$

It is not difficult to see that the constraint matrix of (9.52)–(9.54) is totally unimodular, so the continuous relaxation of the problem has integer solutions. The dual of this relaxation is a network flow problem.

Lower bounds and exact algorithms

The first lower bound for the semi-QAP was proposed by Magirou and Milis [472] in the context of a task scheduling problem. We have seen that when the (not oriented) graph associated with the task digraph is a tree, one can solve the problem with the polynomial-time algorithm by Bokhari [103]. For a general graph, Magirou and Milis proposed reducing the graph to a tree, by removing a proper set of edges, so that the value of the optimal solution of the reduced instance is a lower bound for the original problem. There are several ways of selecting the set of edges that reduce the graph to a tree. The authors did not try to find the best set (i.e., the one that produces the maximum value of the bound). They simply used, as edge weighting, the amount of information to be exchanged between the tasks and removed the edges that leave the maximum spanning tree of the graph. Using this bound they implemented a simple branch-and-bound algorithm that was tested on small size instances with 6 processors and 15 tasks.

Malucelli and Pretolani [478] extended the above approach by considering a partition of the graph into trees and computing as the lower bound the sum of the optimal solution values of the resulting instances. They also proposed another bound obtained by partitioning the graph into reducible graphs and solving each resulting instance through their $O(nm^3)$ algorithm (see the above discussion on easy cases). Finally, a third bound was obtained by applying the tail, series, and parallel reductions described above, plus the following additional reduction:

(iv) *L-I reduction*: substitute a pair of edges $([i,k],[i,h])$ with a single edge $[h,k]$.

This operation does not guarantee that the graph remains connected, but it allows the reduction of any graph to a single vertex. A proper labeling technique was defined to extend the $O(nm^3)$ algorithm so that the L-I reductions are taken into account and the final value is a valid lower bound for the original semi-QAP. Computational experiments have shown that the resulting bound has a performance comparable with that of the two partition bounds, which, in turn, dominate the Magirou and Milis [472] bound.

Billionnet, Costa, and Sutter [99] studied a particular task scheduling problem where the cost for transferring information between two tasks assigned to different processors is independent of the two processors involved. Using a result by Rhys [581], they showed that in this case the problem obtained by relaxing constraints (9.38) in a Lagrangean fashion can be solved in polynomial time through network flow techniques. They embedded this bound into a lowest-first branch-and-bound algorithm and solved instances with up to 10 processors and 101 tasks.

Skutella [612] proposed a lower bound based on convex programming for the semi-QAP formulation of the $R||\sum w_i C_i$ scheduling problem discussed in Section 9.4.1. He considered the formulation given by objective function (9.47) with constraints (9.44) and (9.45) and relaxed the integrality constraints to obtain the continuous quadratic programming model

$$(CQP) \quad \min \quad q^T x + \frac{1}{2} x^T D x \tag{9.55}$$

$$\text{s.t.} \quad \sum_{j=1}^{m} x_{ij} = 1 \qquad (i = 1, 2, \ldots, n), \tag{9.56}$$

$$x_{ij} \geq 0 \qquad (i = 1, 2, \ldots, n; \; j = 1, 2, \ldots, m). \tag{9.57}$$

Kozlov, Tarasov, and Hačijan [429] and Chung and Murty [190] have shown that a quadratic program of the form $\min\{q^T x + \frac{1}{2} x^T D x : \tilde{A} x = \tilde{b}, x \geq 0\}$ can be solved in polynomial time when the objective function is convex. It is well known that $q^T x + \frac{1}{2} x^T D x$ is convex if and only if D is positive semidefinite (see Section 7.8). Unfortunately, this is not the case of the $R||\sum w_i C_i$ problem since, for example, an instance with two tasks and all weights and processing times equal to one has

$$D^{jj} = \begin{pmatrix} 0 & 1 \\ 1 & 0 \end{pmatrix} \qquad (j = 1, 2, \ldots, m),$$

which has a negative determinant. (It is known that all principal minors are nonnegative in a positive semidefinite matrix.)

Skutella proposed to increase the value of the main diagonal of D until the new matrix becomes positive semidefinite. In particular he considered the function

$$\min (1-\gamma) q^T x + \frac{1}{2} x^T (D + 2\gamma \operatorname{diag}(q)) x. \tag{9.58}$$

(Remember that $\operatorname{diag}(q)$ is a square matrix whose elements are all zero, but those on the main diagonal which have the values $q_1, q_2, \ldots, q_{mn}$.) He observed that, since $q \geq 0$, the linear term $q^T x$ is greater than or equal to the value of $x^T \operatorname{diag}(q)x$ for arbitrary $x \in [0, 1]^{mn}$ (with the equality holding for $x \in \{0, 1\}^{mn}$). Therefore, (9.58) underestimates (9.55), and the optimal solution of CQP with objective function (9.58) instead of (9.55) is a lower bound on the optimal solution value of $R||\sum w_i C_i$ for any value of γ Skutella [612] also showed that (9.55) is positive semidefinite if and only if $\gamma \geq 1/2$. Since (9.58) is nonincreasing in γ for each fixed x, the best polynomially computable lower bound is obtained for $\gamma = 1/2$. We observe that one can apply the above reasoning to any semi-QAP with nonnegative linear costs q, but the fact that $D + 2\gamma \operatorname{diag}(q)$ is positive semidefinite for $\gamma \geq 1/2$ depends on the particular structure of D. Therefore, in order to use this bounding technique for a generic semi-QAP instance, we need to identify the correct minimum γ value, which can turn out to be not an easy task.

Skutella [612] used the above bound to derive approximation algorithms with performance guarantee for $R||\sum w_i C_i$. Moreover, he extended the model, the bound, and the approximation algorithms to the generalization of the problem in which the tasks have release dates, usually denoted as $R|r_i|\sum w_i C_i$.

We finally mention that metaheuristic methods to solve the semi-QAP were also developed, mainly in the context of scheduling problems in transit networks: the simulated annealing algorithms by Domschke [236] and Voss [651] and the tabu search algorithm by Domschke, Forst, and Voss [237].

A survey on heuristics for the semi-QAP and other nonlinear assignment problems can be found in Voss [652].

Chapter 10

Multi-index assignment problems

10.1 Introduction

Multi-index assignment problems were introduced by Pierskalla [549] in 1968 as a natural extension of linear assignment problems. For a long time only 3-index assignment problems have been considered, while in recent years problems with more than 3 indices have been investigated, mainly in the context of multi-target tracking and data association problems (see, e.g., Poore [552, 553] and Poore, Rijavec, Liggins, and Vannicola [556]).

In the case of 3-index assignment problems two models have been investigated: the axial 3-index assignment problem and the planar 3-index assignment problem. (These names have been introduced by Schell [597] in 1955.) In the next section we describe the axial 3-index assignment problem, which in many respects resembles the classical assignment problem, but turns out to be $\mathcal{NP}$-hard. Therefore, we describe lower bound computations, polyhedral results, efficiently solvable special cases, and asymptotic results. The planar 3-index assignment problem is treated in Section 10.3. It has not been as thoroughly investigated as the axial 3-index assignment problem and is much harder to solve. In the last section we outline results on general multi-index assignment problems.

For surveys on multi-index assignment problems we refer the reader to Burkard and Çela [138] and Spieksma [618].

10.2 Axial 3-index assignment problem

The *axial 3-index assignment problem* (axial 3AP) can be stated in the following way. Let n^3 cost coefficients c_{ijk} $(i, j, k = 1, 2, \ldots, n)$ be given. We ask for two permutations φ and ψ such that $\sum_{i=1}^{n} c_{i\varphi(i)\psi(i)}$ is a minimum, i.e.,

$$\min_{\varphi,\psi\in\mathcal{S}_n} \sum_{i=1}^{n} c_{i\varphi(i)\psi(i)}, \tag{10.1}$$

where $\mathcal{S}_n$ denotes the set of all permutations of the integers $\{1, 2, \ldots, n\}$. Since the two permutations which describe a feasible solution can be chosen arbitrarily, the axial 3AP has

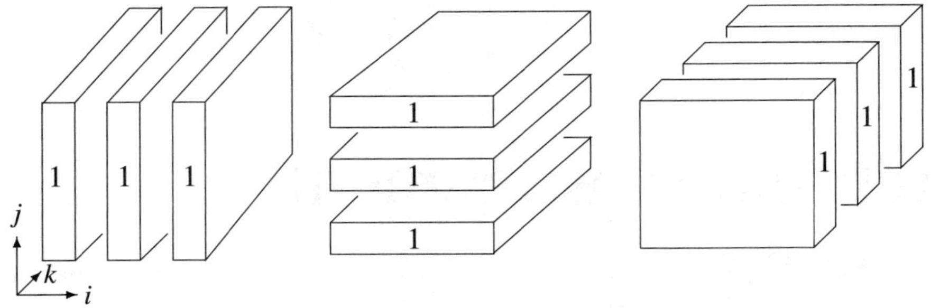

Figure 10.1. *Pictorial representation of the constraints of an axial 3AP.*

$(n!)^2$ feasible solutions. We can write this problem as an integer linear program:

$$\min \quad \sum_{i=1}^{n}\sum_{j=1}^{n}\sum_{k=1}^{n} c_{ijk}x_{ijk} \tag{10.2}$$

$$\text{s.t.} \quad \sum_{j=1}^{n}\sum_{k=1}^{n} x_{ijk} = 1 \qquad (i = 1, 2, \ldots, n), \tag{10.3}$$

$$\sum_{i=1}^{n}\sum_{k=1}^{n} x_{ijk} = 1 \qquad (j = 1, 2, \ldots, n), \tag{10.4}$$

$$\sum_{i=1}^{n}\sum_{j=1}^{n} x_{ijk} = 1 \qquad (k = 1, 2, \ldots, n), \tag{10.5}$$

$$x_{ijk} \in \{0, 1\} \qquad (i, j, k = 1, 2, \ldots, n). \tag{10.6}$$

Figure 10.1 gives a three-dimensional intuition of the constraints: a "1" on a face of the matrix means that exactly one 1 must be in that face.

As noted by Frieze [282], the axial 3AP can also be formulated as the following bilinear integer program with permutation matrices $Y = (y_{ij})$ and $Z = (z_{ik})$ (recall that $\mathbf{X}_n$ denotes the set of all $n \times n$ permutation matrices):

$$\begin{aligned} \min \quad & \sum_{i=1}^{n}\sum_{j=1}^{n}\sum_{k=1}^{n} c_{ijk}y_{ij}z_{ik} \\ \text{s.t.} \quad & Y \in \mathbf{X}_n, \\ & Z \in \mathbf{X}_n. \end{aligned}$$

Maximum weighted matchings in hypergraphs lead to another formulation of the axial 3AP. Consider the three-partite hypergraph $H = (V, E)$ whose $3n$ vertices correspond to the indices $i = 1, 2, \ldots, n$, $j = 1, 2, \ldots, n$, and $k = 1, 2, \ldots, n$ and whose edges are formed by triples (i, j, k). Let c_{ijk} be the cost of edge (i, j, k). A matching M in the hypergraph is a set of edges such that no two edges have a vertex in common. The axial 3AP asks for a maximum matching with minimum cost. This formulation is used in primal-dual bounding procedures which operate in an analogous way as the Hungarian method for the classical assignment problem.

Finally, formulation (10.2)–(10.6) suggests that we can consider the axial 3AP as a matroid intersection problem on the ground set $E = \{(i, j, k) : i, j, k = 1, 2, \dots, n\}$. The three blocks of constraints in (10.3)–(10.5) define a partiton matroid each. Thus the feasible solutions of the axial 3AP correspond to bases which lie in the intersection of the three matroids, and we ask for a basis which has minimum cost. This model is used for deriving lower bounds by subgradient methods related to a Lagrangean relaxation of the axial 3AP.

10.2.1 Applications

Applications of axial 3APs arise in quite a number of situations, for example, the investment of capital into different possible physical locations over some time horizon (Pierskalla [548]). Qi and Sun [567] mentioned the following application in a rolling mill: ingots are to be scheduled through soaking pits (temperature stabilizing baths) so as to minimize the idle time for the rolling mill. Crama, Oerlemans, and Spieksma [199] modeled the assembly of printed circuit boards as axial 3APs.

Minimizing the maximum cost instead of a sum of costs leads to bottleneck objective functions. Axial 3-index assignment problems with a bottleneck objective function have been considered by Malhotra, Bhatia, and Puri [476] as well as by Geetha and Vartak [307]. Time-cost trade-off problems in this context were studied by Geetha and Vartak [306].

10.2.2 Polyhedral aspects

R. Euler [260] started the investigation of the axial 3AP polytope, i.e., the convex hull of all feasible solutions of (10.2)–(10.6). He considered the role of odd cycles for a class of facets of this polytope. Independently, Balas and Saltzman [58] investigated in detail the polyhedral structure of the polytope. They showed that it has dimension $n^3 - 3n + 2$, and they described an $O(n^4)$ separation algorithm for facets induced by certain cliques. Further facets of the axial 3-index assignment polytope were described by Gwan and Qi [351], Balas and Qi [57], and Qi, Balas, and Gwan [566]. Qi and Sun [567] gave a survey on the facial structure of the axial 3-index polytope and described several classes of facets as well as separation algorithms for them.

A different kind of approach is used by the Byelorussian school in Minsk, which studies multi-index transportation polyhedra. If we relax constraints (10.6) to

$$x_{ijk} \geq 0 \qquad (i, j, k = 1, 2, \dots, n)$$

we get the relaxed axial 3AP polytope $M(1, 1, 1)$. In several papers this group studied integer points and non-integer vertices of such polyhedra. For example, Kravtsov, Kravtsov, and Lukshin [432] showed the following result with respect to axial 3APs. They call a vertex of $M(1, 1, 1)$ an *r-noninteger vertex* if it has exactly r fractional components. It is shown that $M(1, 1, 1)$ has r-noninteger vertices for each $r \in \{4, 6, 7, \dots, 3n - 2\}$ and only for these values (see also Kravtsov, Kravtsov, and Lukshin [433]).

10.2.3 Complexity and approximation

In contrast to the classical assignment problem the axial 3AP is $\mathcal{NP}$-hard (Karp [407]). Crama and Spieksma [200] even proved that no polynomial-time algorithm can achieve a constant performance ratio unless $\mathcal{P} = \mathcal{NP}$.

In the same paper Crama and Spieksma investigated the following graph theoretic version of the axial 3AP. They consider a complete 3-partite graph whose vertices correspond to the indices i, j, and k, respectively. Every edge $[i, j]$ (or $[i, k]$, or $[j, k]$, respectively) has a length $d(i, j)$ (or $d(i, k)$, or $d(j, k)$, respectively). Moreover, Crama and Spieksma assume that the edges fulfill the triangle inequality and consider two cost models. In the first cost model c_{ijk} is defined as

$$c_{ijk} = d(i, j) + d(i, k) + d(j, k).$$

In the second cost model, c_{ijk} is defined as the sum of the two shorter lengths in the triangle formed by the vertices i, j, and k. The authors proved that in both cases the corresponding axial 3AP is $\mathcal{NP}$-hard, but they designed approximation algorithms which yield a feasible solution whose value is not worse than $3/2$ of the optimal value in the first case, and $4/3$ of the optimal value in the second case. Computational experiments show a very good performance of these approximation algorithms in the case of randomly generated problems.

A similar model was investigated by Spieksma and Woeginger [617], who considered $3n$ points in the plane and defined the value of $d(i, j)$ as the Euclidean distance between point i and point j. They proved that if the cost coefficients of an axial 3AP are defined by $c_{ijk} = d(i, j) + d(i, k) + d(j, k)$, then the corresponding problem is $\mathcal{NP}$-hard. This also remains true if the sum of the areas of the single triangles is minimized instead of their perimeter.

Burkard, Rudolf, and Woeginger [157] investigated axial 3APs with decomposable cost coefficients, i.e., $c_{ijk} = a_i b_j d_k$ and a_i, b_j, and d_k are nonnegative real numbers. They showed that the minimization version of this problem is $\mathcal{NP}$-hard, whereas the maximization version is polynomially solvable (see Section 10.2.6).

The last statement shows that, in contrast to classical linear assignment problems, there are here essential differences between the minimization and maximization of a linear objective function. For the maximization case, the following approximability results are known. Since axial 3APs are special matroid intersection problems (see Section 10.3), a general theorem by Hausmann, Korte, and Jenkyns [365] states that a simple greedy algorithm yields a performance guarantee of $1/3$. Moreover, Arkin and Hassin [43] designed an $(1/2 - \varepsilon)$ approximation algorithm for maximizing a linear objective function subject to axial 3-index assignment constraints.

10.2.4 Lower bounds and exact solution methods

Hansen and Kaufman [361] proposed a primal-dual method for axial 3-index assignment problems which is similar to the Hungarian method for linear assignment problems. Initially as many zero elements as possible are generated in the cost array c_{ijk} by generalized row and column reductions. A three-partite hypergraph is then defined, whose vertices correspond to the indices $i = 1, 2, \ldots, n$, $j = 1, 2, \ldots, n$, and $k = 1, 2, \ldots, n$. The triple (i, j, k)

forms an edge in the hypergraph if $c_{ijk} = 0$, and the task is to find a maximum matching in this hypergraph. Unfortunately, this is already an $\mathcal{NP}$-hard problem. Instead of finding a maximum matching one can look for a minimum vertex cover in the hypergraph. Here we encounter two difficulties: on one hand the covering problem is $\mathcal{NP}$-hard, too, and on the other hand an analogue of König's theorem, Theorem 2.7, is not any more true in this setting: the minimum number of vertices in a vertex cover might be strictly larger than the cardinality of any matching in the hypergraph. Anyhow, if the covering number is less than n, one can either perform further reductions of the cost elements or branch.

As in the classical case, reductions of the cost coefficients can be found by means of admissible transformations, as shown below. If the minimum covering number equals n, then a maximum matching (or an approximation thereof as the problem is $\mathcal{NP}$-hard); if its cardinality is less than n one has again to branch, otherwise (i.e., if it is equal to n) an optimal solution to the axial 3-AP has been found.

The branching is mostly performed by fixing one variable x_{ijk} to 1 and to 0, respectively. Balas and Saltzman [59] introduced another branching strategy which exploits the structure of the problem and allows us to fix several variables at each branching node.

Burkard and Fröhlich [149] proved that admissible transformations for axial 3APs have the following form.

Proposition 10.1. *Let an axial 3AP with cost coefficients c_{ijk} $(i, j, k = 1, 2, \dots, n)$ be given. Consider three subsets I, J, K of $N = \{1, 2, \dots, n\}$ with $m = n - (|I| + |J| + |K|) \geq 1$, and let $\bar{I} = N \setminus I$, $\bar{J} = N \setminus J$, $\bar{K} = N \setminus K$. Let*

$$c = \min\{c_{ijk} : (i, j, k) \in \bar{I} \times \bar{J} \times \bar{K}\}$$

and

$$\bar{c}_{ijk} = \begin{cases} c_{ijk} - c & \text{if } (i, j, k) \in \bar{I} \times \bar{J} \times \bar{K}, \\ c_{ijk} + c & \text{if } (i, j, k) \in (\bar{I} \times J \times K) \cup (I \times \bar{J} \times K) \cup (I \times J \times \bar{K}), \\ c_{ijk} + 2c & \text{if } (i, j, k) \in I \times J \times K, \\ c_{ijk} & \text{otherwise.} \end{cases}$$

Then, for any feasible solution φ, ψ of the axial 3AP we have

$$\sum_{i=1}^{n} c_{i\varphi(i)\psi(i)} = \sum_{i=1}^{n} \bar{c}_{i\varphi(i)\psi(i)} + mc.$$

This result can be proved along the same lines as the corresponding Theorem 6.20 in Chapter 6. Row and column reductions, which were already considered by Vlach [646] in 1967 (see also Leue [453]), are special cases of the admissible transformations described above.

As in the case of the classical assignment problem, it is possible to consider different forms of objective functions like bottleneck objective functions, or more generally, algebraic objective functions as those in Section 6.3. In principle, the primal-dual approach outlined above can be used in these cases. It is also straightforward to generalize Proposition 10.1 to the case of bottleneck objective functions, or more generally, algebraic objective functions.

Other lower bounds for the axial 3AP can be computed through a Lagrangean relaxation approach. Let us dualize the two blocks of constraints (10.3) and (10.4) into the objective function via Lagrangean multipliers (see Burkard and Fröhlich [149]):

$$L(\lambda, \pi) = \min \left\{ \sum_{i=1}^{n} \sum_{j=1}^{n} \sum_{k=1}^{n} (c_{ijk} + \lambda_i + \pi_j) x_{ijk} - \sum_{i=1}^{n} \lambda_i - \sum_{j=1}^{n} \pi_j \right\} \tag{10.7}$$

$$\text{s.t.} \quad \sum_{i=1}^{n} \sum_{j=1}^{n} x_{ijk} = 1 \qquad (k = 1, 2, \ldots, n), \tag{10.8}$$

$$x_{ijk} \in \{0, 1\} \qquad (i, j, k = 1, 2, \ldots, n), \tag{10.9}$$

$$\lambda \in \mathbb{R}^n, \pi \in \mathbb{R}^n. \tag{10.10}$$

Since $L(\pi, \epsilon)$ is a concave function, we can use a subgradient method for finding its maximum. Burkard and Rudolf [156] report on satisfactory computational results obtained by an algorithm which uses the classical branching rule combined with a reduction step at every node of the branch-decision tree. The lower bound computation is done through a subgradient optimization procedure.

Another subgradient procedure for solving a Lagrangean relaxation of the axial 3AP, together with computational considerations, can be found in Frieze and Yadegar [286].

10.2.5 Heuristic algorithms

Pierskalla [548] was the first to propose a heuristic for solving axial 3APs. More recently, variations of exact methods have been applied as heuristics, and metaheuristics have been tailored for solving axial 3APs. Among the latter approaches we mention the work by Aiex, Resende, Pardalos, and Toraldo [15], who designed a greedy randomized adaptive search procedure with path relinking (see Sections 8.2.7 and 8.2.9) for solving axial 3APs. Their computational experiments showed very good results compared with previously proposed heuristics. Huang and Lim [379] proposed a hybrid genetic algorithm and reported on extensive computational experiments.

Bandelt, Crama, and Spieksma [67] and Burkard, Rudolf, and Woeginger [157] designed heuristic algorithms for the case of decomposable cost coefficients $c_{ijk} = a_i b_j c_k$. The former paper provides worst-case bounds for the algorithms, while the latter reports on computational experiments. Dell'Amico, Lodi, and Maffioli [215] provided lower bounds and heuristic algorithms for an axial 3AP with cost coefficients $c_{ijk} = a_i d_{jk}$, which is a special case of the general 3AP and a generalization of the 3AP with decomposable coefficients.

10.2.6 Special cases

Since axial 3APs are $\mathcal{NP}$-hard, the question about efficiently solvable special cases arises. One case, discussed by Burkard, Klinz, and Rudolf [152], is related to Monge arrays. A 3-dimensional array with cost coefficients c_{ijk} is called a *Monge array* if for every fixed index i (and analogously for fixed j or fixed k) the corresponding matrix fulfills the Monge property:

$$c_{ikl} + c_{irs} \le c_{iks} + c_{irl} \quad \text{for } 1 \le k < r \le n, 1 \le l < s \le n.$$

Proposition 10.2. *Let the cost array* (c_{ijk}) *of the axial 3AP* (10.1) *be a Monge array. Then the solution given by the entries* $\{(1,1,1),(2,2,2),\dots,(n,n,n)\}$ *is optimal.*

This proposition can be proved in the same way as the corresponding result for the classical assignment problem (see Proposition 5.7). The result remains true for axial 3-index bottleneck assignment problems if for every fixed index i (and analogously for fixed j or fixed k) the Monge property above is replaced by the bottleneck Monge property (see equation (6.10)):

$$\max(c_{ikl}, c_{irs}) \le \max(c_{iks}, c_{irl}) \quad \text{for } 1 \le k < r \le n,\ 1 \le l < s \le n,$$

as shown by Burkard, Klinz, and Rudolf [152]. Klinz and Woeginger [424] showed a related result. The identical permutations φ and ψ are optimal for the axial 3-index bottleneck assignment problem if the bottleneck Monge property is replaced by the following *wedge condition*:

$$c_{lll} < \min\{c_{ijk} \;:\; 1 \le i \le l; i \le j,k \le n; j+k \ne 2i\} \text{ for all } l = 1,2,\dots,n. \quad (10.11)$$

Gilbert and Hofstra [310], as well as Burkard, Rudolf, and Woeginger [157], investigated axial 3APs with decomposable cost coefficients $c_{ijk} = a_i b_j c_k$. Since the cost array $-a_i b_j c_k$ fulfills the Monge property provided the real, nonnegative numbers a_i, b_j, and c_k are sorted in nondecreasing order, the maximization version of the axial 3AP with decomposable cost coefficients can be solved in polynomial time. In [157], however, it is shown that the minimization version is in general $\mathcal{NP}$-hard. In the same paper several further polynomially solvable special cases of the minimization version are identified.

Barvinok, Johnson, Woeginger, and Woodroofe [72] considered $3n$ points in $\mathbb{R}^d$, with distance measured according to some fixed polyhedral norm. In this case the maximization version of the problem can be solved in polynomial time if the costs are defined by

$$c_{ijk} = d_{ij} + d_{jk} - d_{ik},$$

d_{xy} being the distance between point x and point y.

10.2.7 Asymptotic behavior

The probabilistic asymptotic behavior of the axial 3AP considerably differs from that of the classical assignment problem. In a series of papers, Grundel, Krokhmal, Oliveira, Pardalos, and Pasiliao [346, 344, 347, 435] studied the expected value of the optimal objective function. They proved that it converges to the lower bound of the distribution of the costs, hence, it shows behavior similar to that of the LBAP (see Section 6.2.7). In particular, let the cost coefficients c_{ijk} be independent and identically distributed random variables, uniformly distributed in [0, 1], and let

$$z_n = \min \sum_{i=1}^{n} c_{i\varphi(i)\psi(i)}.$$

Then

$$\lim_{n\to\infty} \mathbb{E}(z_n) = 0.$$

It is interesting to note that the authors used, as a basis of their proofs, the so-called *index-trees* which were introduced by Pierskalla [549] in one of the first papers on multi-index assignment problems. In a recent paper, Krokhmal, Grundel, and Pardalos [435] developed lower and upper bounds on the expected optimal objective function value.

10.3 Planar 3-index assignment problem

The *planar 3-index assignment problem* (planar 3AP) can be formulated as follows. We say that n permutations $\varphi_1, \varphi_2, \dots, \varphi_n$ are *mutually distinct* if $\varphi_r(i) \neq \varphi_s(i)$ for any $i = 1, 2, \dots, n$ and $r \neq s$. Given n^3 cost coefficients c_{ijk} $(i, j, k = 1, 2, \dots, n)$, the problem is to find n mutually distinct permutations such that

$$\sum_{k=1}^{n} \sum_{i=1}^{n} c_{i\varphi_k(i)k} \tag{10.12}$$

is a minimum. The corresponding integer linear program is

$$\min \quad \sum_{i=1}^{n} \sum_{j=1}^{n} \sum_{k=1}^{n} c_{ijk} x_{ijk} \tag{10.13}$$

$$\text{s.t.} \quad \sum_{k=1}^{n} x_{ijk} = 1 \qquad (i, j = 1, 2, \dots, n), \tag{10.14}$$

$$\sum_{i=1}^{n} x_{ijk} = 1 \qquad (j, k = 1, 2, \dots, n), \tag{10.15}$$

$$\sum_{j=1}^{n} x_{ijk} = 1 \qquad (i, k = 1, 2, \dots, n), \tag{10.16}$$

$$x_{ijk} \in \{0, 1\} \qquad (i, j, k = 1, 2, \dots, n). \tag{10.17}$$

Frieze [285] showed that the problem is $\mathcal{NP}$-hard. Planar 3APs are closely related to Latin squares. A *Latin square* is an $n \times n$ array with entries l_{ij} taking the values from 1 to n. Every row and every column of a Latin square contains exactly one entry of value k $(1 \leq k \leq n)$. For example, a Latin square of size 4 may have the form

4	2	1	3
1	4	3	2
2	3	4	1
3	1	2	4

.

Every feasible solution of a planar 3AP can be represented as a Latin square. Let $L = (l_{ij})$ be a Latin square of size n. Then, for $i, j = 1, 2, \dots, n$, l_{ij} is the (unique) index value k such that $x_{ijk} = 1$ in a feasible solution of the planar 3AP. Thus the Latin square above corresponds to the following solution of a planar 3AP with $n = 4$:

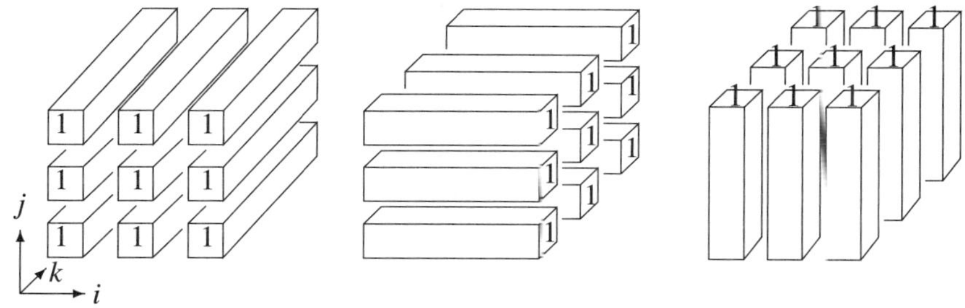

Figure 10.2. *Pictorial representation of the constraints of a planar 3AP.*

$$
\begin{aligned}
x_{114} = x_{122} = x_{131} = x_{143} = 1,\\
x_{211} = x_{224} = x_{233} = x_{242} = 1,\\
x_{312} = x_{323} = x_{334} = x_{341} = 1,\\
x_{413} = x_{421} = x_{432} = x_{444} = 1.
\end{aligned}
$$

This leads to the following geometric interpretation of the planar 3AP. Let us arrange the entries x_{ijk} in a cube. Then every plane in the cube, described either by i fixed, j fixed, or k fixed, respectively, must contain a (two-dimensional) assignment. Due to this interpretation, the number of feasible solutions of a planar 3AP of size n equals the number of Latin squares of order n, and hence increases very quickly. For example, the number of feasible solutions of a planar 3AP with $n = 9$ is $9! \cdot 8! \cdot 377,597,570,964,258,816 (\simeq 55 \cdot 10^{26})$ according to Bammel and Rothstein [66].

Figure 10.2 gives a three-dimensional intuition of the constraints: a "1" on a line of the matrix means that exactly one 1 must be in that line.

Constraints (10.14)–(10.16) show that the planar 3AP can be viewed as a matroid intersection problem on the ground set $E = \{(i, j, k) \ : \ i, j, k = 1, 2, \dots, n\}$. Every block of constraints defines a partiton matroid on E. For fixed indices i and j, let $P^{ij} = \{(i, j, k) \ : \ k = 1, 2, \dots, n\}$. Then $\mathcal{P}^{ij} = \{P^{ij} \ : \ i, j = 1, 2, \dots, n\}$ yields a partition of the ground set E. In a similar way, we get two other partitions of E, namely, $\mathcal{P}^{ik} = \{P^{ik} \ : \ i, k = 1, 2, \dots, n\}$ and $\mathcal{P}^{jk} = \{P^{jk} \ : \ j, k = 1, 2, \dots, n\}$. A subset $F \subseteq E$ is a basis of $(E, \mathcal{P}^{ij})$ if, for all $i, j = 1, 2, \dots, n$,

$$|F \cap P^{ij}| = 1.$$

In particular, $|F| = n^2$ holds. We get three partition matroids on the ground set E, namely, $(E, \mathcal{P}^{ij})$, $(E, \mathcal{P}^{ik})$, and $(E, \mathcal{P}^{jk})$. A common basis of these three matroids corresponds in a unique way to a feasible solution of a planar 3AP (Latin square) and vice versa. We ask for a basis in the intersection of the three partition matroids which has minimum cost. Similarly to the case of the axial 3AP, this model can be used for deriving lower bounds by subgradient methods related to a Lagrangean relaxation of the planar 3AP.

10.3.1 Applications and special cases

Planar 3APs play a crucial role in the context of timetabling problems. Consider an institute where n groups have to be taught by n lecturers in n different time slots. We assume that every lecturer has to teach each group just once. A feasible solution of this problem is an assignment of the lecturers to the groups for each fixed time and an assignment of the time slots to the lecturers for every fixed group. In short, a feasible solution can be represented as a planar 3AP (or as a Latin square).

In the case that the cost array has equal cost for each k, i.e., $c_{ijk} = c_{ij}$ for $k = 1, 2, \ldots, n$, the planar 3AP has a trivial solution: Any set of n mutually distinct permutations is a feasible solution yielding the constant objective function value

$$\sum_{i=1}^{n}\sum_{j=1}^{n} c_{ij}.$$

No other special case of the planar 3AP has yet been specified.

A slightly different model is of interest in the time slot assignment problems that are studied in Section 3.8.2. However, these problems have a modified objective function. Given an $n \times n$ matrix $C = (c_{ij})$, the problem is to find n mutually distinct permutations $\varphi_1, \varphi_2, \ldots, \varphi_n$ such that

$$\sum_{k=1}^{n} \max_{1 \le i \le n} c_{i\varphi_k(i)}$$

is a minimum. Rendl [571] showed that this problem is $\mathcal{NP}$-hard. Balas and Landweer [51] used planar 3APs for finding good approximate solutions for the time slot assignment problem (see Section 3.8.2).

10.3.2 Polyhedral aspects, lower bounds, and algorithms

A partial description of the polyhedral structure of the planar 3-index assignment polytope polytope can be found in Euler, Burkard, and Grommes [261]. In particular it is shown that the dimension of such a polytope is $(n-1)^3$. See also the related study on timetabling polyhedra by Euler and Verge [262]. New facets of the planar 3-index assignment polytope have been described in Appa, Magos, and Mourtos [39].

Since planar 3APs can be formulated as intersection problems of three partition matroids, admissible transformations similar to those of axial 3APs, discussed in Proposition 10.1, can be obtained (see Burkard [133]). In particular, a trivial lower bound for the optimum objective function value can be obtained by solving n linear assignment problems with fixed index k ($k = 1, 2, \ldots, n$). On the other hand, any family of n mutually distinct permutations φ_k ($k = 1, 2, \ldots, n$) yields an upper bound (10.12).

Not many algorithms are known for the planar 3AP. The first branch-and-bound algorithm dates back to Vlach [646], who computed lower bounds by applying (generalized) row and column reductions similar to those used in the case of axial 3APs. Another branch-and-bound procedure for the planar 3AP was described by Magos and Miliotis [474], who also reported computational results. Later, Magos [473] used similar ideas for implementing a tabu search algorithm for planar 3APs. A move in the neighborhood of some Latin square is completely determined by changing the contents of a certain cell (i, j). This affects, at least, between 3 and $2n-1$ other cells which have to be adapted accordingly. This neighborhood structure has two nice properties. First, the change in the objective function value after each move can be computed in linear time. Second, not all moves have to be evaluated at each iteration: all moves which put a certain element in a certain cell imply the same change in the objective function, independently of the solution to which they are applied. The numerical results of this algorithm show a good trade-off between computation time and solution quality for planar 3AP instances of size up to $n = 14$.

Kravtsov and Krachkovskii [431] designed a polynomial-time approximation algorithm for the planar 3AP and claimed it to be asymptotically optimal. Vozniul, Gimadi, and Fialtov [654] showed, however, that this claim is not correct. Gimadi and Korkishko [313] investigated a modification of the planar 3AP, the so-called *m-planar 3-index assignment problem* (m-planar 3AP): given an $n \times n \times m$ cost array $C = (c_{ijk})$, we ask for m mutually distinct permutations $\varphi_1, \varphi_2, \dots, \varphi_m$ such that

$$\sum_{i=1}^{n} \sum_{k=1}^{m} c_{i\varphi_k(i)k} \tag{10.18}$$

is a minimum. (The planar 3AP is obtained for $m = n$.) The m-planar 3AP can also be viewed as a special planar 3AP where all coefficients c_{ijk} for $k = m+1, m+2, \dots, n$ are 0. Gimadi and Korkishko [313] gave a simple polynomial-time approximation algorithm for the m-planar 3AP which is asymptotically optimal if m is $O(\ln n)$ and the cost coefficients are uniformly distributed. Glazkov [315] proved the asymptotical optimality of this algorithm for a special class of random instances.

10.4 General multi-index assignment problems

Driven by applications like data association problems and multitarget tracking (see, e.g., Poore [552, 554]), multi-index assignment problems attracted increasing interest in recent years.

A general multi-index assignment problem is specified by two parameters, k and s, where k is the number of indices and s describes the type of problem (such as axial problem, planar problem, or a generalization thereof). We can describe a (k, s) assignment problem in the following way. Let k index sets $I_1 = I_2 = \cdots = I_k = \{1, 2, \dots, n\}$ be given. To every k-tuple $(i_1, i_2, \dots, i_k)$, with $i_r \in I_r$ $(r = 1, 2, \dots, k)$, we associate a cost $c(i_1, i_2, \dots, i_k)$. Now we have to describe the constraints. For a fixed s $(1 \le s \le k-1)$, let $\mathcal{Q}_s$ be the class of all subsets of $K = \{1, 2, \dots, k\}$ with cardinality s, i.e., $\mathcal{Q}_s = \{Q : Q \subset K, |Q| = s\}$. A set $Q \in \mathcal{Q}_s$ determines "fixed" indices in the constraints. Therefore, we call the indices in $K \setminus Q$ "free" indices. For every fixed s-tuple $(i_{q_1}, i_{q_2}, \dots, i_{q_s}) \in I_{q_1} \times I_{q_2} \times \cdots \times I_{q_s}$ and for any $Q = \{q_1, q_2, \dots, q_s\} \in \mathcal{Q}_s$, we define $J_1^Q \times J_2^Q \times \cdots \times J_k^Q$ by

$$J_r^Q := \begin{cases} \{1, 2, \dots, n\} & \text{if } r \text{ is a free index,} \\ \{i_{q_l}\} & \text{if } r \text{ is the fixed index } q_l. \end{cases} \tag{10.19}$$

(Note that this definition is independent of the sequence $(q_1, q_2, \dots, q_s)$ but depends only on Q.) Then the corresponding constraint has the form

$$\sum_{r \in K} \sum_{i_r \in J_r^Q} x(i_1, i_2, \dots, i_k) = 1. \tag{10.20}$$

Thus a general (k, s) *assignment problem* can be stated as

$$\min \sum_{r\in K} \sum_{i_r\in I_r} c(i_1, i_2, \ldots, i_k)\, x(i_1, i_2, \ldots, i_k)$$

$$\text{s.t.} \sum_{r\in K} \sum_{i_r\in J_r^Q} x(i_1, i_2, \ldots, i_k) = 1 \quad \text{for all } (i_{q_1}, i_{q_2}, \ldots, i_{q_s}) \in I_{q_1} \times I_{q_2} \times \cdots \times I_{q_s} \text{ and for all } Q \in \mathcal{Q}_s;$$

$$x(i_1, i_2, \ldots, i_k) \in \{0, 1\} \quad \text{for all } (i_1, i_2, \ldots, i_k) \in I_1 \times I_2 \times \cdots \times I_k.$$

In particular, for $k \geq 3$ we call $(k, 1)$ assignment problems *axial k-index assignment problems* and $(k, 2)$ assignment problems *planar k-index assignment problems.* Axial k-index assignment problems can also be stated as follows: find $k-1$ permutations $\varphi_1, \varphi_2, \ldots, \varphi_{k-1}$ which minimize

$$\sum_{i=1}^{n} c_{i\varphi_1(i)\varphi_2(i)\ldots\varphi_{k-1}(i)}.$$

Several problems encountered in the previous sections and chapters can be classified according to this framework:

1. The classical assignment problem is the $(2, 1)$ assignment problem.
2. The axial 3-index assignment problem is the $(3, 1)$ assignment problem. We have three blocks of constraints. In the first block the last index is fixed (which corresponds to $Q = \{3\}$). In the other two blocks the second and the first index, respectively, are fixed, which corresponds to $Q = \{2\}$ and $Q = \{1\}$, respectively.
3. The planar 3-index assignment problem is the $(3, 2)$ assignment problem. We have again three blocks of constraints, but now every block of constraints contains n^2 equations. The blocks correspond to the sets $Q = \{2, 3\}$, $Q = \{1, 3\}$, and $Q = \{1, 2\}$.
4. The feasible solutions of the $(4, 2)$ assignment problem are pairs of orthogonal Latin squares. Due to a famous result of Euler, this problem has no feasible solution in the case $n = 6$ (see also Appa, Magos, and Mourtos [37]).

There also exist other ways to describe multi-index assignment problems. For example, they can be stated as special intersection problems of partition matroids, or they can be formulated as clique partition problems of complete k-partite graphs.

Appa, Magos, and Mourtos [40] studied (k, s) assignment polytopes and established the dimension of general axial and planar assignment problems. For $k \geq 2$ the dimension of $(k, 1)$ assignment problems (general axial assignment problems) is shown to be

$$\sum_{r=0}^{k-2} \binom{k}{r} \cdot (n-1)^{k-r}.$$

The dimension of $(k, 2)$ assignment problems (general planar assignment problems) is shown to be

$$\sum_{r=0}^{k-3} \binom{k}{r} \cdot (n-1)^{k-r},$$

provided that $k \geq 3$ and the polytope is nonempty. In the same paper the authors specify facets of axial k-index assignment problems induced by cliques. For further results on the facial structure of axial and planar k-index polytopes, see also the report by Magos, Mourtos, and Appa [475]. The polyhedral structure of the $(4, 2)$ assignment polytope was studied by Appa, Magos, and Mourtos [41, 38].

10.4.1 Applications

In surveillance systems there is the demand to identify and estimate targets in real time. In a series of papers, Poore (see, e.g., [552] and [523]) examined the assignment formulation for the data association problems arising in this context. Consider a radar or another surveillance system monitoring a certain region. The system creates measurements of the targets at discrete time units $t_1, t_2, \ldots, t_n$. The problem is to associate the measurements with the targets, i.e., to identify the tracks of the targets over time. The mathematical formulation of such problems leads to axial k-index assignment problems.

Another data association problem, stemming from high energy physics and leading to an axial 5-index assignment problem, has been described by Pusztaszeri, Rensing, and Liebling [565] (see also Pusztaszeri [564]). The authors tried to reconstruct the tracks of charged elementary particles generated by the large electron-positron collider at CERN in Geneva.

The *planar 4-index assignment problem* (Planar 4AP) occurs in the design of tournaments (see Appa, Magos, and Mourtos [36]). The feasible solutions of a planar 4AP are pairs of orthogonal Latin squares which can be used to schedule a tournament between two teams of n players each, where each player of the first team plays against any player of the second team, each player plays in each of the n rounds, and each player plays at each of the n locations just once during the tournament. (There are no games between players of the same team.) Other applications of the planar 4AP concern the conflict-free access to parallel memories and the design of error-correcting codes.

10.4.2 Algorithms and asymptotic behavior

As (k, s) assignment problems are $\mathcal{NP}$-hard for $k \geq 3$, only enumerative methods are known for their exact solution. Poore and Rijavec [555] designed a Lagrangean relaxation algorithm for axial multi-index assignment problems arising in the context of multitarget tracking. An improved version of Lagrangean relaxation approaches for this problem class has been discussed by Poore and Robertson [557].

For the planar 4APs, Appa, Magos, and Mourtos [36] designed a branch-and-cut algorithm which uses clique, odd-hole, and antiweb inequalities for the cuts. For each cut type, a polynomial-time separation algorithm is given. The computational results show the superiority of the branch-and-cut algorithm in comparison to branch-and-bound approaches.

Greedy randomized adaptive search heuristics for multidimensional assignment problems arising in multitarget tracking and data association were proposed by Murphey, Pardalos, and Pitsoulis [503] and Robertson [585]. A parallel version appeared in Murphey, Pardalos, and Pitsoulis [504].

A test problem generator for multidimensional assignment problems was published by Grundel and Pardalos [348].

Since the number of local minima greatly influences the performance of local search procedures, Grundel, Krokhmal, Oliveira, and Pardalos [345] undertook a study of such an issue for axial k-index assignment problems with random cost coefficients. The authors defined local minima with respect to different neighborhood structures and showed, among other things, the following.

Proposition 10.3. *In an axial k-index assignment problem with independent and identically distributed normal cost coefficients, the expected number M_2 of 2-exchange local minima is bounded as*

$$\frac{(n!)^{k-1}}{(k+1)^{n(n-1)/2}} \leq \mathbb{E}(M_2) \leq \frac{2(n!)^{k-1}}{n(n-1)k+2}.$$

A similar result applies to p-exchange neighborhoods.

An approximation algorithm for axial k-index assignment problems was designed by Gimadi and Kairan [314]. They showed that their sublinear approximation algorithm is asymptotically optimal provided the cost coefficients are i.i.d. according to a distribution which is minorized by the uniform distribution. Another asymptotical optimal approximation algorithm for axial k-index assignment problems was given by Kravtsov [434].

Some general approximation results apply in the case of axial k-index assignment problems when the objective function is *maximized*. Since such a problem can be formulated as a matroid intersection problem, the approach by Hausmann, Korte, and Jenkyns [365] yields an $1/k$ approximation algorithm, whereas the approach by Hurkens and Schrijver [383] yields a $(2/k - \varepsilon)$ approximation in the case of cost coefficients 0 and 1.

The only nontrivial special case known for (k, s) assignment problems concerns axial k-index assignment problems whose cost array fulfills the Monge property (see Burkard, Klinz, and Rudolf [152]). In this case the optimal solution is given by $\varphi_1(i) = \varphi_2(i) = \cdots = \varphi_{k-1}(i) = i$ for $i = 1, 2, \dots, n$. This can be shown by the same exchange arguments outlined in the proof of Proposition 5.7.

Krokhmal, Grundel, and Pardalos [435] investigated the probabilistic asymptotic behavior of axial k-index assignment problems. In particular, they studied the asymptotic behavior of the expected optimal value of problems whose cost coefficients are independent and identically distributed random variables with a given probability distribution. Let z_{kn} denote the optimal value of an axial k-index assignment problem of size n. They showed that, in the case of exponentially distributed cost coefficients, $\mathbb{E}(z_{kn})$ approaches zero when either k or n tends to infinity. Their approach also allows the derivation of asymptotic lower and upper bounds for z_{kn}. More generally it is shown that for a broad class of continuous distributions the expected optimal value tends to the left endpoint of the support set, unlike the case of the classical assignment problem, where it tends to $\pi^2/6$ (see Section 5.1.1). Axial multi-index assignment problems behave similarly to bottleneck assignment problems (see Section 6.2.7).

Bibliography

[1] E.H.L. Aarts and J.K. Lenstra. *Local Search in Combinatorial Optimization*. Wiley, Chichester, UK, 1997. (Cited on p. 259.)

[2] E.H.L. Aarts and J. Korst. *Simulated Annealing and Boltzmann Machines: A Stochastic Approach to Combinatorial Optimization and Neural Computing*. Wiley, Chichester, UK, 1989. (Cited on p. 259.)

[3] H. Achatz, P. Kleinschmidt, and K. Paparrizos. A dual forest algorithm for the assignment problem. In P. Gritzmann and B. Sturmfels, editors, *Applied Geometry and Discrete Mathematics*, volume 4 of *DIMACS Series*, pages 1–10. American Mathematical Society, Providence, RI, 1991. (Cited on pp. 119, 129.)

[4] W.P. Adams, M. Guignard, P.M. Hahn, and W.L. Hightower. A level-2 reformulation-linearization technique bound for the quadratic assignment problem. *European J. Oper. Res.*, 180:983–996, 2007. (Cited on pp. 222, 249, 251, 254, 292.)

[5] W.P. Adams and T.A. Johnson. Improved linear programming-based lower bounds for the quadratic assignment problem. In P.M. Pardalos and H. Wolkowicz, editors, *Quadratic Assignment and Related Problems*, volume 16 of *DIMACS Series*, pages 43–75. American Mathematical Society, Providence, RI, 1994. (Cited on pp. 221, 222, 223, 237.)

[6] W.P. Adams and H.D. Sherali. A tight linearization and an algorithm for zero-one quadratic programming problems. *Management Sci.*, 32:1274–1290, 1986. (Cited on p. 221.)

[7] W.P. Adams and H.D. Sherali. Linearization strategies for a class of zero-one mixed integer programming problems. *Oper. Res.*, 38:217–226, 1990. (Cited on p. 221.)

[8] V. Aggarwal, V.G. Tikekar, and L.-F. Hsu. Bottleneck assignment problems under categorization. *Computers & Oper. Res.*, 13:11–26, 1986. (Cited on p. 190.)

[9] R.K. Ahuja, O. Ergun, J.B. Orlin, and A.P. Punnen. A survey of very large-scale neighborhood search techniques. *Discr. Appl. Math.*, 123:75–102, 2002. (Cited on p. 265.)

[10] R.K. Ahuja, K.C. Jha, J.B. Orlin, and D. Sharma. Very large-scale neighborhood search for the quadratic assignment problem. *INFORMS J. Comput.*, 19:646–657, 2007. (Cited on p. 265.)

[11] R.K. Ahuja, T.L. Magnanti, and J.B. Orlin. *Network Flows*. Prentice-Hall, Englewood Cliffs, NJ, 1993. (Cited on pp. 79, 105, 123.)

[12] R.K. Ahuja and J.B. Orlin. The scaling network simplex algorithm. *Oper. Res.*, Suppl. 1:S5–S13, 1992. (Cited on p. 111.)

[13] R.K. Ahuja, J.B. Orlin, C. Stein, and R.E. Tarjan. Improved algorithms for bipartite network flow. *SIAM J. Comput.*, 23:906–933, 1994. (Cited on p. 124.)

[14] R.K. Ahuja, J.B. Orlin, and A. Tiwari. A greedy genetic algorithm for the quadratic assignment problem. *Computers & Oper. Res.*, 27(10):917–934, 2000. (Cited on p. 262.)

[15] R.M. Aiex, M.G.C. Resende, P.M. Pardalos, and G. Toraldo. GRASP with path relinking for three-index assignment. *INFORMS J. Comput.*, 17:224–247, 2005. (Cited on p. 310.)

[16] M. Akgül. A sequential dual simplex algorithm for the linear assignment problem. *Oper. Res. Lett.*, 7:155–158, 1988. (Cited on p. 118.)

[17] M. Akgül. Erratum. A sequential dual simplex algorithm for the linear assignment problem. *Oper. Res. Lett.*, 8:117, 1989. (Cited on p. 118.)

[18] M. Akgül. The linear assignment problem. In M. Akgül, H.W. Hamacher, and S. Tüfekçi, editors, *Combinatorial Optimization*, number 82 in NATO ASI Series F, pages 85–122. Springer-Verlag, Berlin, 1992. (Cited on pp. 79, 105, 126.)

[19] M. Akgül. A genuinely polynomial primal simplex algorithm for the assignment problem. *Discr. Appl. Math.*, 45:93–115, 1993. (Cited on pp. 78, 111, 128.)

[20] M. Akgül and O. Ekin. A criss-cross algorithm for the assignment problem. Working Paper IEOR 90–22, Bilkent University, 1990. (Cited on p. 126.)

[21] M. Akgül and O. Ekin. A dual feasible forest algorithm for the linear assignment problem. *RAIRO Rech. Opér.*, 25:403–411, 1991. (Cited on p. 119.)

[22] H. Albrecher. A note on the asymptotic behaviour of bottleneck problems. *Oper. Res. Lett.*, 33:183–186, 2005. (Cited on pp. 285, 291.)

[23] H. Albrecher, R.E. Burkard, and E. Çela. An asymptotical study of combinatorial optimization problems by means of statistical mechanics. *J. Computational and Appl. Math.*, 186:148–162, 2006. (Cited on pp. 285, 289, 290.)

[24] D.J. Aldous. Asymptotics in the random assignment problem. *Probab. Theory Related Fields*, 93:507–534, 1992. (Cited on pp. 145, 146, 146.)

[25] D.J. Aldous. The $\zeta(2)$ limit in the random assignment problem. *Random Structures Algorithms*, 18:381–418, 2001. (Cited on pp. 147, 148.)

[26] S.E. Alm and G.B. Sorkin. Exact expectations and distributions for the random assignment problem. *Combin. Probab. Comput.*, 11:217–248, 2002. (Cited on p. 147.)

[27] H. Alt, N. Blum, K. Mehlhorn, and M. Paul. Computing a maximum cardinality matching in a bipartite graph in time $O(n^{1.5}\sqrt{m/\log n})$. *Inform. Process. Lett.*, 37:237–240, 1991. (Cited on pp. 35, 47, 52, 127.)

[28] D. Andreou, K. Paparrizos, N. Samaras, and A. Sifaleras. Application of a new network-enabled solver for the assignment problem in computer-aided education. *J. Computer Sci.*, 1:19–23, 2005. (Cited on p. 129.)

[29] Y.P. Aneja and A.P. Punnen. Multiple bottleneck assignment problem. *European J. Oper. Res.*, 112:167–173, 1999. (Cited on p. 167.)

[30] E. Angel and V. Zissimopoulos. On the quality of local search for the quadratic assignment problem. *Discr. Appl. Math.*, 82:15–25, 1998. (Cited on p. 259.)

[31] E. Angel and V. Zissimopoulos. On the classification of NP-complete problems in terms of their correlation coefficient. *J. Computer Sci.*, 99:261–277, 2000. (Cited on p. 259.)

[32] E. Angel and V. Zissimopoulos. On the landscape ruggedness of the quadratic assignment problem. *Theoretical Computer Science*, 263:159–172, 2001. (Cited on p. 259.)

[33] E. Angel and V. Zissimopoulos. On the hardness of the quadratic assignment problem with metaheuristics. *J. of Heuristics*, 8:399–414, 2002. (Cited on p. 259.)

[34] K.M. Anstreicher and N.W. Brixius. A new bound for the quadratic assignment problem based on convex quadratic programming. *Math. Program.*, 89:341–357, 2001. (Cited on pp. 247, 251, 254.)

[35] K.M. Anstreicher, N.W. Brixius, J.P. Goux, and J. Linderoth. Solving large quadratic assignment problems on computational grids. *Math. Program.*, 91:563–588, 2002. (Cited on pp. 248, 249, 254.)

[36] G. Appa, D. Magos, and I. Mourtos. A branch & cut algorithm for the four-index assignment problem. *J. Oper. Res. Soc.*, 55:298–307, 2004. (Cited on p. 317.)

[37] G. Appa, D. Magos, and I. Mourtos. An LP-based proof for the non-existence of a pair of orthogonal Latin squares of order 6. *Oper. Res. Lett.*, 32:336–344, 2004. (Cited on p. 316.)

[38] G. Appa, D. Magos, and I. Mourtos. The wheels of the orthogonal Latin squares polytope: classification and valid inequalities. *J. Comb. Optim.*, 10:355–389, 2005. (Cited on p. 317.)

[39] G. Appa, D. Magos, and I. Mourtos. A new class of facets for the Latin square polytope. *Discr. Appl. Math.*, 154:900–911, 2006. (Cited on p. 314.)

[40] G. Appa, D. Magos, and I. Mourtos. On multi-index assignment polytopes. *Linear Algebra Appl.*, 416:224–241, 2006. (Cited on p. 316.)

[41] G. Appa, D. Magos, and I. Mourtos. On the orthogonal Latin squares polytope. *Discr. Math.*, 306:171–187, 2006. (Cited on p. 317.)

[42] J. Aráoz and J. Edmonds. A case of non-convergent dual changes in assignment problems. *Discr. Appl. Math.*, 11:95–102, 1985. (Cited on p. 83.)

[43] E.M. Arkin and R. Hassin. On local search for weighted packing problems. *Math. Oper. Res.*, 23:640–648, 1998. (Cited on p. 308.)

[44] E.M. Arkin, R. Hassin, and M. Sviridenko. Approximating the maximum quadratic assignment problem. *Inform. Process. Lett.*, 77:13–16, 2001. (Cited on p. 210.)

[45] R.D. Armstrong and Z. Jin. Solving linear bottleneck assignment problems via strong spanning trees. *Oper. Res. Lett.*, 12:179–180, 1992. (Cited on p. 186.)

[46] S. Arora, A.M. Frieze, and H. Kaplan. A new rounding procedure for the assignment problem with applications to dense graph arrangement problems. *Math. Program.*, 92:1–36, 2002. (Cited on p. 210.)

[47] A.A. Assad and W. Xu. On lower bounds for a class of quadratic 0-1 programs. *Oper. Res. Lett.*, 4:175–180, 1985. (Cited on pp. 237, 238.)

[48] M.J. Atallah and S.E. Hambrusch. On bipartite matchings of minimum density. *J. Algorithms*, 8:480–502, 1987. (Cited on p. 56.)

[49] D. Avis and L. Devroye. An analysis of a decomposition heuristic for the assignment problem. *Oper. Res. Lett.*, 3:279–283, 1985. (Cited on p. 149.)

[50] D. Avis and C.W. Lai. The probabilistic analysis of a heuristic for the assignment problem. *SIAM J. Comput.*, 17:732–741, 1988. (Cited on p. 149.)

[51] E. Balas and P.R. Landweer. Traffic assignment in commmunication satellites. *Oper. Res. Lett.*, 2:141–147, 1983. (Cited on pp. 197, 314.)

[52] E. Balas and J.B. Mazzola. Quadratic 0-1 programming by a new linearization. In *Joint ORSA/TIMS National Meeting*, Washington DC, 1980. (Cited on p. 249.)

[53] E. Balas and J.B. Mazzola. Nonlinear 0-1 programming: I. Linearization techniques. *Math. Program.*, 30:1–21, 1984. (Cited on pp. 219, 249.)

[54] E. Balas and J.B. Mazzola. Nonlinear 0-1 programming: II. Dominance relations and algorithms. *Math. Program.*, 30:22–45, 1984. (Cited on p. 249.)

[55] E. Balas, D.L. Miller, J.F. Pekny, and P. Toth. A parallel shortest augmenting path algorithm for the assignment problem. *J. ACM*, 38:985–1004, 1991. (Cited on pp. 140, 142, 142, 142.)

[56] E. Balas and W. Niehaus. Finding large cliques in arbitrary graphs by bypartite matching. In D. Johnson and M. Trick, editors, *Cliques, Coloring and Satisfability*, volume 26 of *DIMACS Series*, pages 29–52. American Mathematical Society, Providence, RI, 1996. (Cited on p. 262.)

[57] E. Balas and L. Qi. Linear-time separation algorithms for the three-index assignment polytope. *Discr. Appl. Math.*, 43:1–12, 1993. (Cited on p. 307.)

[58] E. Balas and M.J. Saltzman. Facets of the three-index assignment polytope. *Discr. Appl. Math.*, 23:201–229, 1989. (Cited on p. 307.)

[59] E. Balas and M.J. Saltzman. An algorithm for the three-index assignment problem. *Oper. Res.*, 39:150–161, 1991. (Cited on p. 309.)

[60] M.L. Balinski. On two special classes of transportation polytopes. *Math. Program. Study*, 1:43–58, 1974. (Cited on p. 106.)

[61] M.L. Balinski. The Hirsch conjecture for dual transportation polyhedra. *Math. Oper. Res.*, 9:629–633, 1984. (Cited on p. 117.)

[62] M.L. Balinski. Signature methods for the assignment problem. *Oper. Res.*, 33:527–536, 1985. (Cited on pp. 78, 114, 128.)

[63] M.L. Balinski. A competitive (dual) simplex method for the assignment problem. *Math. Program.*, 34:125–141, 1986. (Cited on pp. 118, 186.)

[64] M.L. Balinski and R.E. Gomory. A primal method for the assignment and transportation problems. *Management Sci.*, 10:578–593, 1964. (Cited on pp. 78, 104, 104, 105, 128.)

[65] M.L. Balinski and A. Russakoff. On the assignment polytope. *SIAM Rev.*, 16:516–525, 1974. (Cited on pp. 30, 31, 31, 33, 34.)

[66] S.E. Bammel and J. Rothstein. The number of 9×9 Latin squares. *Discr. Math.*, 11:93–95, 1975. (Cited on p. 313.)

[67] H.-J. Bandelt, Y. Crama, and F.C.R. Spieksma. Approximation algorithms for multi-dimensional assignment problems with decomposable cost coefficients. *Discr. Appl. Math.*, 49:25–50, 1994. (Cited on p. 310.)

[68] R.S. Barr, F. Glover, and D. Klingman. The alternating basis algorithm for assignment problems. *Math. Program.*, 13:1–13, 1977. (Cited on pp. 77, 87, 98, 106, 106, 106, 107, 107, 110, 114, 126, 128.)

[69] R.S. Barr, F. Glover, and D. Klingman. A new alternating basis algorithm for semi-assignment networks. In W. White, editor, *Computers and Mathematical Programming*, pages 223–232. National Bureau of Standards Special Publications, US Government Printing Office, Washington, DC, 1978. (Cited on p. 165.)

[70] R.S. Barr, F. Glover, and D. Klingman. Enhancements of spanning tree labelling procedures for network optimization. *INFOR*, 17:16–34, 1979. (Cited on p. 143.)

[71] R.S. Barr and B.L. Hickman. Parallel simplex for large pure network problems: Computational testing and sources of speedup. *Oper. Res.*, 42:65–80, 1994. (Cited on pp. 143, 144.)

[72] A.I. Barvinok, D.S. Johnson, G.J. Woeginger, and R. Woodroofe. The maximum traveling salesman problem under polyhedral norms. In R.E. Bixby, E.A. Boyd, and R.Z. Ríos-Mercado, editors, *Integer Programming and Combinatorial Optimization*, volume 1412 of *Lecture Notes in Comput. Sci.*, pages 195–201. Springer, Berlin-Heidelberg, 1998. (Cited on p. 311.)

[73] A.I. Barvinok. Combinatorial complexity of orbits in representations of the symmetric groups. *Advances of Soviet Math.*, 9:161–182, 1992. (Cited on pp. 223, 224, 224.)

[74] R. Battiti and G. Tecchiolli. The reactive tabu search. *ORSA J. Comput.*, 6:126–140, 1994. (Cited on p. 261.)

[75] R. Battiti and G. Tecchiolli. Simulated annealing and tabu search in the long run: A comparison on QAP tasks. *Computers and Mathematics with Applications*, 28:1–8, 1994. (Cited on p. 261.)

[76] J. Bautista, R. Companys, and A. Corominas. Modelling and solving the product rate variation problem. *TOP*, 5:221–239, 1997. (Cited on p. 189.)

[77] M.S. Bazaraa and O. Kirca. A branch-and-bound-based heuristic for solving the quadratic assignment problem. *Naval Res. Log. Quart.*, 30:287–304, 1983. (Cited on p. 252.)

[78] M.S. Bazaraa and H.D. Sherali. Benders' partitioning scheme applied to a new formulation of the quadratic assignment problem. *Naval Res. Log. Quart.*, 27:29–41, 1980. (Cited on pp. 217, 249.)

[79] M.S. Bazaraa and H.D. Sherali. On the use of exact and heuristic cutting plane methods for the quadratic assignment problem. *J. Oper. Res. Soc.*, 33:991–1003, 1982. (Cited on pp. 249, 250.)

[80] W.W. Bein and P.K. Pathak. A characterization of the Monge property and its connection to statistics. *Demonstratio Math.*, 29:451–457, 1996. (Cited on p. 151.)

[81] H. Bekker, E.P. Braad, and B. Goldengorin. Using bipartite and multidimensional matching to select the roots of a system of polynomial equations. In *Computational Science and Its Applications – ICCSA 2005*, volume 3483 of *Lecture Notes in Comput. Sci.*, pages 397–406. Springer, Berlin / Heidelberg, 2005. (Cited on p. 166.)

[82] M. Bellmore and J.F. Malone. Pathology of traveling salesman subtour elimination algorithms. *Oper. Res.*, 19:278–307, 1971. (Cited on p. 161.)

[83] J.F. Benders. Partitioning procedures for solving mixed variables programming problems. *Numerische Mathematik*, 4:238–252, 1962. (Cited on p. 250.)

[84] A. Bergamini. Computer codes and Java applets for assignment problems (in Italian). Master's thesis, Dipartimento di Elettronica, Informatica e Sistemistica, Università di Bologna, Italy, 2006. (Cited on p. 128.)

[85] C. Berge. Two theorems in graph theory. *Proc. Natl. Acad. Sci. USA.*, 43:842–844, 1957. (Cited on p. 36.)

[86] D.P. Bertsekas. A new algorithm for the assignment problem. *Math. Program.*, 21:152–171, 1981. (Cited on pp. 78, 103, 119, 121, 128, 130, 139.)

[87] D.P. Bertsekas. The auction algorithm: A distributed relaxation method for the assignment problem. In R. R. Meyer and S.A. Zenios, editors, *Parallel Optimization on Novel Computer Architectures*, volume 14 of *Ann. Oper. Res.*, pages 105–123. Baltzer, Basel, 1988. (Cited on pp. 125, 139.)

[88] D.P. Bertsekas. *Linear Network Optimization: Algorithms and Codes*. The MIT Press, Cambridge, MA, 1991. (Cited on pp. 79, 128, 129, 129, 129, 130, 130, 131, 165.)

[89] D.P. Bertsekas and D.A. Castañon. Parallel synchronous and asynchronous implementations of the auction algorithm. *Parallel Comput.*, 17:707–732, 1991. (Cited on pp. 139, 140, 140.)

[90] D.P. Bertsekas and D.A. Castañon. Parallel asynchronous Hungarian methods for the assignment problem. *ORSA J. Comput.*, 3:261–274, 1993. (Cited on pp. 123, 142.)

[91] D.P. Bertsekas, D.A. Castañon, and H. Tsaknakis. Reverse auction and the solution of inequality constrained assignment problems. *SIAM J. Optim.*, 3:268–297, 1993. (Cited on p. 123.)

[92] D.P. Bertsekas and J. Eckstein. Dual coordinate step methods for linear network flow problems. *Math. Program.*, 42:203–243, 1988. (Cited on pp. 78, 121, 121, 128, 140.)

[93] D.P. Bertsekas, S. Pallottino, and M.G. Scutellà. Polynomial auction algorithms for shortest paths. *Computational Opt. Appl.*, 2:99–125, 1995. (Cited on p. 160.)

[94] D.P. Bertsekas and J.N. Tsitsiklis. *Parallel and Distributed Computation: Numerical Methods*. Prentice-Hall, Englewood Cliffs, NJ, 1989. (Cited on pp. 138, 140.)

[95] J. Bhasker and S. Sahni. Optimal linear arrangement of circuit components. *J. VLSI and Computer Systems*, 2:87–109, 1987. (Cited on p. 276.)

[96] K.V.S. Bhat and B. Kinariwala. An algorithm for the $n \times n$ optimum assignment problem. *BIT*, 19:289–296, 1979. (Cited on pp. 98, 103, 128.)

[97] P. Billingsley. *Probability and Measure, 3rd Edition*. Wiley, New York, 1995. (Cited on p. 289.)

[98] A. Billionnet and S. Elloumi. Best reduction of the quadratic semi-assignment problem. *Discr. Appl. Math.*, 109:197–213, 2001. (Cited on p. 295.)

[99] A. Billionnet, M.C. Costa, and A. Sutter. An efficient algorithm for a task allocation problem. *J. ACM*, 39(3):502–518, 1992. (Cited on pp. 297, 302.)

[100] G. Birkhoff. Tres observaciones sobre el algebra lineal. *Revista Facultad de Ciencias Exactas, Puras y Aplicadas Universidad Nacional de Tucuman, Serie A (Matematicas y Fisica Teorica)*, 5:147–151, 1946. (Cited on pp. 25, 75.)

[101] A. Blanchard, S. Elloumi, A. Faye, and M. Wicker. Un algorithme de génération de coupes pour le problème de l'affectation quadratique. *INFOR*, 41:35–49, 2003. (Cited on p. 253.)

[102] R.G. Bland. New finite pivoting rules for the simplex method. *Math. Oper. Res.*, 2:103–107, 1977. (Cited on p. 106.)

[103] S.H. Bokhari. A shortest tree algorithm for optimal assignments across space and time in a distributed processor system. *IEEE Trans. Software Eng.*, 7:583–589, 1981. (Cited on pp. 300, 301, 301.)

[104] B. Bollobás. *Extremal Graph Theory*. Academic Press, London, UK, 1978. (Cited on p. 210.)

[105] B. Bollobás and A. Thomason. Random graphs of small order. In M. Karoński and A. Ruciński, editors, *Random Graphs '83*, volume 28 of *Ann. Discr. Math.*, pages 47–97. North-Holland, Amsterdam, 1983. (Cited on p. 60.)

[106] A.A. Bolotnikov. On the best balance of the disk with masses on its periphery (in Russian). *Problemi Mashinostroenia*, 6:68–74, 1978. (Cited on pp. 205, 276.)

[107] E. Bonomi and J. Lutton. The asymptotic behavior of quadratic sum assignment problems: A statistical mechanics approach. *European J. Oper. Res.*, 26:295–300, 1986. (Cited on pp. 285, 289.)

[108] K.S. Booth and G.S. Lueker. Testing for the consecutive ones property, interval graphs, and graph planarity using PQ-tree algorithms. *J. Comput. Syst. Sci.*, 13(3):335–379, 1976. (Cited on p. 53.)

[109] J. Bos. A quadratic assignment problem solved by simulated annealing. *J. Environmental Management*, 37:127–145, 1993. (Cited on p. 205.)

[110] R.C. Bose, S.S. Shrikhande, and E.T. Parker. Further results on the construction of mutually orthogonal Latin squares and the falsity of Euler's conjecture. *Canadian J. Math.*, 12:189–203, 1960. (Cited on p. 69.)

[111] F. Bourgeois and J.C. Lassalle. Algorithm 415: Algorithm for the assignment problem (rectangular matrices). *Commun. ACM*, 14:805–806, 1971. (Cited on p. 165.)

[112] F. Bourgeois and J.C. Lassalle. An extension of the Munkres algorithm for the assignment problem to rectangular matrices. *Commun. ACM*, 14:802–804, 1971. (Cited on p. 165.)

[113] M. Brady, K.K. Jung, H.T. Nguyen, R. Raghavan, and R. Subramonian. The assignment problem on parallel architectures. In D.S. Johnson and C.C. McGeoch, editors, *Network Flows and Matching: First DIMACS Implementation Challenge*, volume 12 of *DIMACS Series*, pages 469–517. American Mathematical Society, Providence, RI, 1993. (Cited on p. 138.)

[114] N.W. Brixius and K.M. Anstreicher. Solving quadratic assignment problems using convex quadratic programming relaxations. *Optimization Methods and Software*, 16:49–68, 2001. (Cited on pp. 251, 254.)

[115] W.L. Brogan. Algorithm for ranked assignments with applications to multiobject tracking. *J. Guidance*, 12:357–364, 1989. (Cited on pp. 158, 166, 190.)

[116] R.A. Brualdi and P.M. Gibson. The assignment polytope. *Math. Program.*, 11:97–101, 1976. (Cited on p. 34.)

[117] R.A. Brualdi and P.M. Gibson. Convex polyhedra of doubly stochastic matrices. IV. *Linear Algebra Appl.*, 15:153–172, 1976. (Cited on p. 34.)

[118] R.A. Brualdi and P.M. Gibson. Convex polyhedra of double stochastic matrices. II. Graph of Ω_n. *J. Combin. Theory Series B*, 22:175–198, 1977. (Cited on p. 34.)

[119] R.A. Brualdi and P.M. Gibson. Convex polyhedra of doubly stochastic matrices. I. Applications of the permanent function. *J. Combin. Theory Series A*, 22:194–230, 1977. (Cited on p. 34.)

[120] R.A. Brualdi and P.M. Gibson. Convex polyhedra of doubly stochastic matrices. III. Affine and combinatorial properties of Ω_n. *J. Combin. Theory Series A*, 22:338–351, 1977. (Cited on p. 34.)

[121] A. Brüngger, A. Marzetta, J. Clausen, and M. Perregaard. Joining forces in solving large-scale quadratic assignment problems in parallel. In *Proceedings of the 11th International Parallel Processing Symposium (IPPS)*, pages 418–427. IEEE Computer Society, Washington, DC, 1997. (Cited on pp. 249, 254.)

[122] A. Brüngger, A. Marzetta, J. Clausen, and M. Perregaard. Solving large-scale QAP problems in parallel with the Search Library ZRAM. *J. Parallel Distrib. Comput.*, 50:157–169, 1998. (Cited on pp. 249, 254.)

[123] E.S. Buffa, G.C. Armour, and T.E. Vollmann. Allocating facilities with CRAFT. *Harvard Bus. Rev.*, 42:136–158, 1964. (Cited on p. 258.)

[124] V.Y. Burdyuk and V.N. Trofimov. Generalization of the results of Gilmore and Gomory on the solution of the traveling salesman problem. *Engineering Cybernetics*, 14:12–18, 1976. (Cited on p. 151.)

[125] S. Burer and D. Vandenbussche. Solving lift-and-project relaxations of binary integer programs. *SIAM J. Optim.*, 16:726–750, 2006. (Cited on p. 247.)

[126] R.E. Burkard. Die Störungsmethode zur Lösung quadratischer Zuordnungsprobleme. *Oper. Res. Verf.*, 16:84–108, 1973. (Cited on pp. 229, 235.)

[127] R.E. Burkard. Quadratische Bottleneckprobleme. *Oper. Res. Verf.*, 18:26–41, 1974. (Cited on p. 284.)

[128] R.E. Burkard. Traveling salesman and assignment problems: A survey. In P.L. Hammer, E.L. Johnson, and B.H. Korte, editors, *Discrete Optimization I*, volume 4 of *Ann. Discr. Math.*, pages 193–215. North–Holland, Amsterdam, 1979. (Cited on p. 78.)

[129] R.E. Burkard. Time-slot assignment for TDMA systems. *Computing*, 35:99–112, 1985. (Cited on p. 70.)

[130] R.E. Burkard. Locations with spatial interactions: The quadratic assignment problem. In P.B. Mirchandani and R.L. Francis, editors, *Discrete Location Theory*. Wiley, Chichester, UK, 1990. (Cited on pp. 203, 252, 255.)

[131] R.E. Burkard. Shortest path algorithms. In M. Papageorgiou, editor, *Concise Encyclopedia of Traffic and Transportation Systems (CETTS)*, pages 461–468. Pergamon Press, Oxford, UK, 1991. (Cited on p. 179.)

[132] R.E. Burkard. Selected topics on assignment problems. *Discr. Appl. Math.*, 123:257–302, 2002. (Cited on p. 79.)

[133] R.E. Burkard. Admissible transformations and assignment problems. *Vietnam J. Math.*, 35:373–386, 2007. (Cited on p. 314.)

[134] R.E. Burkard and T. Bönniger. A heuristic for quadratic Boolean programs with applications to quadratic assignment problems. *European J. Oper. Res.*, 13:374–386, 1983. (Cited on pp. 219, 250, 256, 259.)

[135] R.E. Burkard and P. Butkovič. Finding all essential terms of a characteristic max-polynomial. *Discr. Appl. Math.*, 130:367–380, 2003. (Cited on p. 157.)

[136] R.E. Burkard and P. Butkovič. Max algebra and the linear assignment algorithm. *Math. Program. (B)*, 98:415–429, 2003. (Cited on pp. 153, 154, 155, 157.)

[137] R.E. Burkard and E. Çela. Heuristics for biquadratic assignment problems and their computational comparison. *European J. Oper. Res.*, 83:283–300, 1995. (Cited on pp. 292, 293.)

[138] R.E. Burkard and E. Çela. Linear assignment problems and extensions. In D. Z. Du and P.M. Pardalos, editors, *Handbook of Combinatorial Optimization, Supplement Volume A*, pages 75–149. Kluwer Academic Publishers, Dordrecht, The Netherlands, 1999. (Cited on pp. 79, 305.)

[139] R.E. Burkard, E. Çela, V.M. Demidenko, N.N. Metelski, and G.J. Woeginger. Perspectives of easy and hard cases of the quadratic assignment problem. Technical Report 104, Institute of Mathematics, Graz University of Technology, 1997. (Cited on pp. 267, 270, 271, 271, 272, 274, 275.)

[140] R.E. Burkard, E. Çela, V.M. Demidenko, N.N. Metelski, and G.J. Woeginger. A unified approach to simple special cases of extremal permutation. *Optimization*, 44:123–138, 1998. (Cited on p. 279.)

[141] R.E. Burkard, E. Çela, and B. Klinz. On the biquadratic assignment problem. In P.M. Pardalos and H. Wolkowicz, editors, *Quadratic Assignment and Related Problems*, volume 16 of *DIMACS Series*, pages 117–146. American Mathematical Society, Providence, RI, 1994. (Cited on pp. 292, 293, 293, 293.)

[142] R.E. Burkard, E. Çela, P.M. Pardalos, and L.S. Pitsoulis. The quadratic assignment problem. In D.Z. Du and P.M. Pardalos, editors, *Handbook of Combinatorial Optimization Vol. 3*, pages 241–337. Kluwer Academic Publishers, Boston, 1998. (Cited on p. 203.)

[143] R.E. Burkard, E. Çela, G. Rote, and G.J. Woeginger. The quadratic assignment problem with a monotone anti-Monge and a symmetric Toeplitz matrix: Easy and hard cases. *Math. Program. (B)*, 82:125–158, 1998. (Cited on pp. 205, 271, 277, 277, 278.)

[144] R.E. Burkard, V.G. Deĭneko, R. van Dal, J. van der Veen, and G.J. Woeginger. Well-solvable special cases of the traveling salesman problem: A survey. *SIAM Rev.*, 40:496–546, 1998. (Cited on p. 267.)

[145] R.E. Burkard and U. Derigs. *Assignment and Matching Problems: Solution Methods with FORTRAN Programs*. Springer-Verlag, Berlin-Heidelberg, 1980. (Cited on pp. 77, 78, 103, 128, 184, 266, 266.)

[146] R.E. Burkard and U. Fincke. On random quadratic bottleneck assignment problems. *Math. Program.*, 23:227–232, 1982. (Cited on pp. 285, 291.)

[147] R.E. Burkard and U. Fincke. The asymptotic probabilistic behaviour of the quadratic sum assignment problem. *Z. Oper. Res. (A)*, 27:73–81, 1983. (Cited on pp. 285, 291.)

[148] R.E. Burkard and U. Fincke. Probabilistic asymptotic properties of some combinatorial optimization problems. *Discr. Appl. Math.*, 12:21–29, 1985. (Cited on pp. 285, 286, 291.)

[149] R.E. Burkard and K. Fröhlich. Some remarks on 3-dimensioanl assignment problems. *Methods Oper. Res.*, 36:31–36, 1980. (Cited on pp. 309, 310.)

[150] R.E. Burkard, W. Hahn, and U. Zimmermann. An algebraic approach to assignment problems. *Math. Program.*, 12:318–327, 1977. (Cited on pp. 191, 192.)

[151] R.E. Burkard, S.E. Karisch, and F. Rendl. QAPLIB—A quadratic assignment problem library. *J. Glob. Optim.*, 10:391–403, 1997. On-line version at http://www.seas.upenn.edu/qaplib/. (Cited on pp. 203, 266.)

[152] R.E. Burkard, B. Klinz, and R. Rudolf. Perspectives of Monge properties in optimization. *Discr. Appl. Math.*, 70:95–161, 1996. (Cited on pp. 152, 187, 195, 310, 311, 318.)

[153] R.E. Burkard and J. Offermann. Entwurf von Schreibmaschinentastaturen mittels quadratischer Zuordnungsprobleme. *Z. Oper. Res. (B)*, 21:B121–B132, 1977. (Cited on p. 205.)

[154] R.E. Burkard and F. Rendl. A thermodynamically motivated simulation procedure for combinatorial optimization problems. *European J. Oper. Res.*, 17:169–174, 1984. (Cited on pp. 258, 259, 261, 267.)

[155] R.E. Burkard and F. Rendl. Lexicographic bottleneck problems. *Oper. Res. Lett.*, 10:303–307, 1991. (Cited on p. 198.)

[156] R.E. Burkard and R. Rudolf. Computational investigations on 3-dimensional axial assignment problems. *Belgian J. Operations Research*, 32:85–98, 1993. (Cited on p. 310.)

[157] R.E. Burkard, R. Rudolf, and G.J. Woeginger. Three-dimensional axial assignments problems with decomposable cost coefficients. *Discr. Appl. Math.*, 65:123–139, 1996. (Cited on pp. 308, 310, 311, 311.)

[158] R.E. Burkard and U. Zimmermann. Weakly admissible transformations for solving algebraic assignment and transportation problems. *Math. Program. Study*, 12:1–18, 1980. (Cited on p. 195.)

[159] R.E. Burkard and U. Zimmermann. Combinatorial optimization in linearly ordered semimodules: A survey. In B. Korte, editor, *Modern Applied Mathematics*, pages 392–436. North–Holland, Amsterdam, 1982. (Cited on p. 195.)

[160] V.N. Burkov, M.I. Rubinstein, and V.B. Sokolov. Some problems in optimal allocation of large-volume memories (in Russian). *Avtomatika i Telemekhanika*, 9:83–91, 1969. (Cited on p. 278.)

[161] P. Butkovič. Regularity of matrices in min-algebra and its time-complexity. *Discr. Appl. Math.*, 57:121–132, 1995. (Cited on pp. 155, 156.)

[162] P. Butkovič and R.A. Cuninghame-Green. On the linear assignment problem for special matrices. *IMA J. Management Mathematics*, 15:1–12, 2004. (Cited on p. 153.)

[163] L. Buš and P. Tvrdík. Look-back auction algorithm for the assignment problem and its distributed memory implementation. In *Proceedings of the 15th IASTED International Conference, Parallel and Distributed Computing and Systems (PDCS03), Marina del Rey*, pages 551–556, ACTA Press, Calgary, 2003. (Cited on p. 140.)

[164] G. Caron, P. Hansen, and B. Jaumard. The assignment problem with seniority and job priority constraints. *Oper. Res.*, 47:449–453, 1999. (Cited on p. 168.)

[165] G. Carpaneto, S. Martello, and P. Toth. Algorithms and codes for the assignment problem. In B. Simeone, P. Toth, G. Gallo, F. Maffioli, and S. Pallottino, editors, *Fortran Codes for Network Optimization*, volume 13 of *Ann. Oper. Res.*, pages 193–223. Baltzer, Basel, 1988. (Cited on pp. 77, 87, 99, 104, 127, 127, 129, 129, 132, 141.)

[166] G. Carpaneto and P. Toth. Algorithm 548: Solution of the assignment problem. *ACM Transactions on Mathematical Software*, 6:104–111, 1980. (Cited on pp. 83, 87, 87, 99.)

[167] G. Carpaneto and P. Toth. Algorithm for the solution of the bottleneck assignment problem. *Computing*, 27:179–187, 1981. (Cited on pp. 177, 184.)

[168] G. Carpaneto and P. Toth. Algorithm 50: Algorithm for the solution of the assignment problem for sparse matrices. *Computing*, 28:83–94, 1983. (Cited on pp. 99, 127, 127.)

[169] G. Carpaneto and P. Toth. Primal-dual algorithms for the assignment problem. *Discr. Appl. Math.*, 18:137–153, 1987. (Cited on pp. 99, 104, 128, 129, 129.)

[170] P. Carraresi and G. Gallo. A multi-level bottleneck assignment approach to the bus drivers' rostering problem. *European J. Oper. Res.*, 16:163–173, 1984. (Cited on p. 189.)

[171] P. Carraresi and F. Malucelli. A new lower bound for the quadratic assignment problem. *Oper. Res.*, 40:S22–S27, 1992. (Cited on pp. 237, 238.)

[172] P. Carraresi and C. Sodini. An efficient algorithm for the bipartite matching problem. *European J. Oper. Res.*, 23:86–93, 1986. (Cited on p. 98.)

[173] S.A. de Carvalho, Jr. and S. Rahmann. Microarray layout as a quadratic assignment problem. In D.H. Huson, O. Kohlbacher, A. Lupus, K. Nieselt, and A. Zell, editors, *Proceedings of the German Conference on Bioinformatics (GCB)*, volume P-38 of *Lecture Notes in Comput. Sci.*, pages 11–20. Springer, Berlin, Heidelberg, 2006. (Cited on p. 206.)

[174] D.A. Castañon. Reverse auction algorithms for assignment problems. In D.S. Johnson and C.C. McGeoch, editors, *Network Flows and Matching: First DIMACS Implementation Challenge*, volume 12 of *DIMACS Series*, pages 407–429. American Mathematical Society, Providence, RI, 1993. (Cited on pp. 123, 125.)

[175] A. Cayley. A theorem on trees. *Quart. J. Pure Appl. Math.*, 23:376–378, 1889. (Cited on p. 30.)

[176] E. Çela. *The Quadratic Assignment Problem: Theory and Algorithms.* Kluwer Academic Publishers, Dordrecht, The Netherlands, 1998. (Cited on pp. 203, 210, 267, 275, 278, 278, 279, 279.)

[177] V. Černý. Thermodynamical approach to the traveling salesman problem: An efficient simulation algorithm. *J. Optim. Theory Appl*, 45:41–51, 1985. (Cited on p. 259.)

[178] J. Chakrapani and J. Skorin-Kapov. Massively parallel tabu search for the quadratic assignment problem. *Ann. Oper. Res.*, 41:327–341, 1993. (Cited on p. 261.)

[179] J. Chakrapani and J. Skorin-Kapov. A constructive method to improve lower bounds for the quadratic assignment problem. In P.M. Pardalos and H. Wolkowicz, editors, *Quadratic Assignment and Related Problems*, volume 16 of *DIMACS Series*, pages 161–171. American Mathematical Society, Providence, RI, 1994. (Cited on pp. 228, 229.)

[180] S.F. Chang and S.T. McCormick. A fast implementation of a bipartite matching algorithm. Technical report, Faculty of Commerce and Business Administration, University of British Columbia, 1989. (Cited on p. 103.)

[181] C.R. Chegireddy and H.W. Hamacher. Algorithms for finding k-best perfect matchings. *Discr. Appl. Math.*, 18:155–165, 1987. (Cited on pp. 161, 163.)

[182] L. Cheng Cheng Sun. Algorithms for assignment problems (in Italian). Master's thesis, Dipartimento di Elettronica, Informatica e Sistemistica, Università di Bologna, Italy, 2004. (Cited on p. 128.)

[183] J. Cheriyan, T. Hagerup, and K. Mehlhorn. Can a maximum flow be computed in $O(nm)$ time? In *Automata, Languages and Programming (Coventry, 1990)*, volume 443 of *Lecture Notes in Comput. Sci.*, pages 235–248. Springer, New York, 1990. (Cited on p. 47.)

[184] H. Chernoff. A measure of asymptotic efficiency for tests of a hypothesis based on the sum of observations. *Annals of Mathematical Statistics*, 23:493–507, 1952. (Cited on p. 289.)

[185] D. Chhajed and T.J. Lowe. m-median and m-center problems with mutual communication: Solvable special cases. *Oper. Res.*, 40:S56–S66, 1992. (Cited on p. 300.)

[186] N. Christofides and E. Benavent. An exact algorithm for the quadratic assignment problem. *Oper. Res.*, 37:760–768, 1989. (Cited on p. 251.)

[187] N. Christofides and M. Gerrard. Special cases of the quadratic assignment problem. Management Science Research Report 361, Carnegie-Mellon University, Pittsburgh, PA, 1976. (Cited on p. 280.)

[188] N. Christofides and M. Gerrard. A graph theoretic analysis of bounds for the quadratic assignment problem. In P. Hansen, editor, *Studies on Graphs and Discrete Programming*, pages 61–68. North–Holland, Amsterdam, The Netherlands, 1981. (Cited on p. 227.)

[189] N. Christofides, A. Mingozzi, and P. Toth. Contributions to the quadratic assignment problem. *European J. Oper. Res.*, 4:243–247, 1980. (Cited on p. 229.)

[190] S.J. Chung and K.G. Murty. Polynomially bounded ellipsoid algorithms for convex quadratic programming. In O.L. Mangasarian, R.R. Meyer, and S.M. Robinson, editors, *Nonlinear Programming*, pages 439–485. Academic Press, Orlando, Fla., 1980. (Cited on p. 302.)

[191] J. Clausen, S.E. Karisch, M. Perregaard, and F. Rendl. On the applicability of lower bounds for solving rectilinear quadratic assignment problems in parallel. *Computational Opt. Appl.*, 10:127–147, 1998. (Cited on p. 238.)

[192] J. Clausen and M. Perregaard. Solving large quadratic assignment problems in parallel. *Computational Opt. Appl.*, 8:111–127, 1997. (Cited on pp. 249, 253.)

[193] A. Colorni, M. Dorigo, and V. Maniezzo. The ant system: Optimization by a colony of cooperating agents. *IEEE Trans. Syst. Man Cybernetics*, 26:29–41, 1996. (Cited on p. 263.)

[194] D.T. Connolly. An improved annealing scheme for the QAP. *European J. Oper. Res.*, 46:93–100, 1990. (Cited on pp. 259, 267.)

[195] K. Conrad. *Das Quadratische Zuweisungsproblem und zwei seiner Spezialfälle*. Mohr-Siebeck, Tübingen, Germany, 1971. (Cited on pp. 207, 229.)

[196] D. Coppersmith and G.B. Sorkin. Constructive bounds and exact expectations for the random assignment problem. *Random Structures Algorithms*, 15:113–144, 1999. (Cited on p. 146.)

[197] D. Coppersmith and S. Winograd. Matrix multiplication via arithmetic progressions. *J. Symbolic Computing*, 9:251–280, 1990. (Cited on pp. 57, 59, 174.)

[198] J.-F. Cordeau, P. Toth, and D. Vigo. A survey of optimization models for train routing and scheduling. *Transportation Sci.*, 32:380–404, 1998. (Cited on p. 66.)

[199] Y. Crama, A. Oerlemans, and F.C.R. Spieksma. *Production Planning in Automated Manufactoring*. Springer Verlag, Berlin, 1966. (Cited on p. 307.)

[200] Y. Crama and F.C.R. Spieksma. Approximation algorithms for three-dimensional assignment problems with triangle inequalities. *European J. Oper. Res.*, 60:273–279, 1992. (Cited on p. 308.)

[201] F. Della Croce, V.Th. Paschos, and A. Tsoukias. An improved general procedure for lexicographic bottleneck problems. *Oper. Res. Lett.*, 24:187–194, 1999. (Cited on p. 202.)

[202] V.-D Cung, T. Mautor, P. Michelon, and A. Tavares. A scatter search based approach for the quadratic assignment problem. In *Proceedings of IEEE International Conference on Evolutionary Computation*, pages 165–169. IEEE Computer Society, Washington, DC, 1997. (Cited on p. 265.)

[203] R.A. Cuninghame-Green. Process synchronisation in a steelworks—A problem of feasibility. In J. Banbury and J. Maitland, editors, *2nd Int. Conf. on Operational Research*, pages 323–328, English Universities Press Ltd., London, 1961. (Cited on p. 153.)

[204] R.A. Cuninghame-Green. *Minimax Algebra*, volume 166 of *Lecture Notes in Economics and Mathematical Systems*. Springer, Berlin, 1979. (Cited on pp. 154, 156, 157, 157.)

[205] W.H. Cunningham. A network simplex method. *Math. Program.*, 11:105–116, 1976. (Cited on pp. 77, 106, 106, 107, 107, 110, 114, 128.)

[206] W.H. Cunningham. Theoretical properties of the network simplex method. *Math. Oper. Res.*, 4:196–208, 1979. (Cited on pp. 110, 111, 111.)

[207] W.H. Cunningham. Private letter to M.L. Balinski, 1983. (Cited on p. 116.)

[208] W.H. Cunningham and A.B. Marsh, III. A primal algorithm for optimum matching. *Math. Program. Study*, 8:50–72, 1978. (Cited on pp. 78, 105, 128.)

[209] J.R. Daduna and S. Voss. Practical experiences in schedule synchronization. In J.R. Daduna, I. Branco, and J.M.P. Paixao, editors, *Compiter-Aided Transit Schedule*, Lecture Notes in Econ. Math. Syst., pages 41–55. Springer, Berlin, 1995. (Cited on p. 296.)

[210] O. Damberg, S. Storøy, and T. Sørevik. A data parallel augmenting path algorithm for the dense linear many-to-one assignment problem. *Computational Opt. Appl.*, 6:251–272, 1996. (Cited on p. 141.)

[211] G.B. Dantzig. *Linear Programming and Extensions*. Princeton University Press, Princeton, NJ, 1963. (Cited on pp. 104, 106, 106.)

[212] G.B. Dantzig, L.R. Ford, and D.R. Fulkerson. A primal-dual algorithm for linear programs. In H.W. Kuhn and A.W. Tucker, editors, *Linear Inequalities and Related Systems*, Annals of Mathematic Studies, pages 171–181. Princeton University Press, Princeton, NJ, 1956. (Cited on p. 79.)

[213] G.B. Dantzig and D.R. Fulkerson. On the max-flow min-cut theorem of networks. In H.W. Kuhn and A.W. Tucker, editors, *Linear Inequalities and Related Systems*, Annals of Mathematic Studies, pages 215–221. Princeton University Press, Princeton, NJ, 1956. (Cited on p. 18.)

[214] V.G. Deĭneko and V.L. Filonenko. On the reconstruction of specially structured matrices (in Russian). In *Aktualnyje Problemy EVM i programmirovanije*. DGU, Dnepropetrovsk, 1979. (Cited on p. 152.)

[215] M. Dell'Amico, A. Lodi, and F. Maffioli. Solution of the cumulative assignment problem with a well-structured tabu search method. *J. Heuristics*, 5:123–143, 1999. (Cited on p. 310.)

[216] M. Dell'Amico, A. Lodi, and S. Martello. Efficient algorithms and codes for the $k-$cardinality assignment problem. *Discr. Appl. Math.*, 110:25–40, 2001. (Cited on p. 164.)

[217] M. Dell'Amico and S. Martello. Open shop, satellite communication and a theorem by Egerváry (1931). *Oper. Res. Lett.*, 18:207–211, 1996. (Cited on pp. 25, 66, 67, 79.)

[218] M. Dell'Amico and S. Martello. The $k-$cardinality assignment problem. *Discr. Appl. Math.*, 76:103–121, 1997. (Cited on pp. 163, 164.)

[219] M. Dell'Amico and S. Martello. Linear assignment. In M. Dell'Amico, F. Maffioli, and S. Martello, editors, *Annotated Bibliographies in Combinatorial Optimization*, pages 355–371. John Wiley and Sons, Chichester, UK, 1997. (Cited on p. 79.)

[220] M. Dell'Amico and P. Toth. Algorithms and codes for dense assignment problems: The state of the art. *Discr. Appl. Math.*, 100:17–48, 2000. (Cited on pp. 79, 132, 132.)

[221] V.M. Demidenko and A. Dolgui. Quadratic assignment problem: easily solvable cases. In A. Dolgui, G. Morel, and S. Pereira, editors, *Information Control Problems in Manufactoring 2006: A Proceedings volume from the 12th IFAC International Symposium, St. Etienne, France, May 17–19, 2006*, volume 3, pages 441–446. Elsevier Science, Amsterdam and New York, 2007. (Cited on p. 278.)

[222] V.M. Demidenko. Generalizations of strong resolvability conditions of a quadratic assignment problem with anti-Monge and Toeplitz matrices (in Russian). *Doklady Natsionalnoj Akademii Nauk Belarusi*, 2:15–18, 2003. (Cited on p. 278.)

[223] V.M. Demidenko and A. Dolgui. Efficiently solvable cases of the quadratic assignment problem with generalized monotonic and incomplete anti-Monge matrices. *Cybernetics and Systems Analysis*, 43:112–125, 2007. (Cited on p. 278.)

[224] V.M. Demidenko, G. Finke, and V.S. Gordon. Well solvable cases of the quadratic assignment problem with monotone and bimonotone matrices. *J. Mathematical Modelling and Algorithms*, 5:167–187, 2006. (Cited on p. 278.)

[225] U. Derigs. Alternate strategies for solving bottleneck assignment problems—analysis and computational results. *Computing*, 33:95–106, 1984. (Cited on p. 184.)

[226] U. Derigs. The shortest augmenting path method for solving assignment problems—Motivation and computational experience. In C.L. Monma, editor, *Algorithms and Software for Optimization—Part I*, volume 4 of *Ann. Oper. Res.*, pages 57–102. Baltzer, Basel, 1985. (Cited on pp. 78, 95, 132, 141.)

[227] U. Derigs, O. Goecke, and R. Schrader. Monge sequences and a simple assignment algorithm. *Discr. Appl. Math.*, 15:241–248, 1986. (Cited on p. 152.)

[228] U. Derigs and A. Metz. An efficient labeling technique for solving sparse assignment problems. *Computing*, 36:301–311, 1986. (Cited on p. 98.)

[229] U. Derigs and A. Metz. An in-core/out-of-core method for solving large scale assignment problems. *Z. Oper. Res.*, 30:A181–A195, 1986. (Cited on p. 98.)

[230] U. Derigs and U. Zimmermann. An augmenting path method for solving linear bottleneck assignment problems. *Computing*, 19:285–295, 1978. (Cited on pp. 126, 178, 178, 184.)

[231] C. Derman and M. Klein. A note on the optimal depletion of inventory. *Management Sci.*, 5:210–213, 1959. (Cited on p. 167.)

[232] J.F. Desler and S.L. Hakimi. A graph-theoretic approach to a class of integer-programming problems. *Oper. Res.*, 17:1017–1033, 1969. (Cited on p. 77.)

[233] J.W. Dickey and J.W. Hopkins. Campus building arrangement using TOPAZ. *Transport. Res.*, 6:59–68, 1972. (Cited on p. 205.)

[234] E.W. Dijkstra. A note on two problems in connection with graphs. *Numerische Mathematik*, 1:269–271, 1959. (Cited on p. 80.)

[235] E.A. Dinic and M.A. Kronrod. An algorithm for the solution of the assignemt [sic] problem. *Sov. Math. Dokl.*, 10:1324–1326, 1969. (Cited on pp. 77, 78, 89, 89, 89, 90, 112, 120, 120, 128.)

[236] W. Domschke. Schedule synchronization for public transit networks. *OR Spektrum*, 11:17–24, 1989. (Cited on p. 303.)

[237] W. Domschke, P. Forst, and S. Voss. Tabu search techniques for the quadratic semi-assignment problem. In G. Fandel, T. Gulledge, and A. Jones, editors, *New Directions for Operations Research in Manufacturing*, pages 389–405. Springer-Verlag, 1992. (Cited on p. 303.)

[238] W.E. Donath. Algorithm and average-value bounds for assignment problems. *IBM J. Res. Develop.*, 13:380–386, 1969. (Cited on p. 146.)

[239] Z. Drezner. A new genetic algorithm for the quadratic assignment problem. *INFORMS J. Comput.*, 15:320–330, 2003. (Cited on p. 262.)

[240] Z. Drezner. Compounded genetic algorithms for the quadratic assignment problem. *Oper. Res. Lett.*, 33:475–480, 2005. (Cited on p. 262.)

[241] Z. Drezner, P.M. Hahn, and E.D. Taillard. Recent advances for the quadratic assignment problem with special emphasis on instances that are difficult for meta-heuristic methods. *Ann. Oper. Res.*, 139:65–94, 2005. (Cited on pp. 258, 259.)

[242] J.R. Driscoll, H.N. Gabow, R. Shrairman, and R.E. Tarjan. Relaxed heaps: An alternative to Fibonacci heaps with applications to parallel computation. *Commun. ACM*, 31:1343–1354, 1988. (Cited on p. 138.)

[243] R. Duncan. A survey of parallel computer architectures. *IEEE Computer*, 23:5–16, 1990. (Cited on p. 138.)

[244] M.E. Dyer, A.M. Frieze, and C.J.H. McDiarmid. On linear programs with random costs. *Math. Program.*, 35:3–16, 1986. (Cited on pp. 146, 291.)

[245] T.E. Easterfield. A combinatorial algorithm. *J. London Math. Soc.*, 21:219–226, 1946. (Cited on pp. 77, 128.)

[246] J. Edmonds. Maximum matching and a polyhedron with (0,1) vertices. *J. Res. Nat. Bur. Stand.*, 69B:125–130, 1965. (Cited on p. 105.)

[247] J. Edmonds. Paths, trees and flowers. *Canadian J. Math.*, 17:449–467, 1965. (Cited on pp. 13, 105.)

[248] J. Edmonds. Systems of distinct representatives and linear algebra. *J. Res. Nat. Bur. Stand.*, 718:241–245, 1967. (Cited on p. 57.)

[249] J. Edmonds and D.R. Fulkerson. Bottleneck extrema. *J. Combin. Theory*, 8:299–306, 1970. (Cited on p. 173.)

[250] J. Edmonds and R.M. Karp. Theoretical improvements in algorithmic efficiency for network flow problems. *J. ACM*, 19:248–264, 1972. (Cited on pp. 77, 87, 93, 94, 128.)

[251] C.S. Edwards. The derivation of a greedy approximator for the Koopmans-Beckmann quadratic assignment problem. In *Proceedings of the 77th Combinatorial Programming Conference (CP77)*, pages 55–86, 1977. (Cited on p. 205.)

[252] C.S. Edwards. A branch and bound algorithm for the Koopmans-Beckmann quadratic assignment problem. *Math. Program. Study*, 13:35–52, 1980. (Cited on pp. 212, 229.)

[253] E. Egerváry. Matrixok kombinatorius tulajdonságairol. *Matematikai és Fizikai Lapok*, 38:16–28, 1931. (English translation by H.W. Kuhn [439].). (Cited on pp. 25, 66, 79, 79.)

[254] H.A. Eiselt and G. Laporte. A combinatorial optimization problem arising in dartboard design. *J. Oper. Res. Soc.*, 42:113–118, 1991. (Cited on p. 206.)

[255] M. Engquist. A successive shortest path algorithm for the assignment problem. *INFOR*, 20:370–384, 1982. (Cited on p. 98.)

[256] R. Enkhbat and Ts. Javzandulam. A global search algorithm for the quadratic assignment problem. *J. Mongolian Math. Soc.*, 4:16–28, 2000. (Cited on p. 249.)

[257] P. Erdös and A. Rényi. On random matrices. *Magyar Tud. Akad. Mat. Kutató Int. Közl*, 8:455–461, 1964. (Cited on p. 150.)

[258] G. Erdoğan and B. Tansel. A note on a polynomial time solvable case of the quadratic assignment problem. *Discr. Optim.*, 6:382–384, 2006. (Cited on p. 280.)

[259] G. Erdoğan and B. Tansel. A branch-and-cut algorithm for quadratic assignment problems based on linearizations. *Computers & Oper. Res.*, 34:1085–1106, 2007. (Cited on p. 253.)

[260] R. Euler. Odd cycles and a class of facets of the axial 3-index assignment polytope. *Applicationes Mathematicae*, 19:375–386, 1987. (Cited on p. 307.)

[261] R. Euler, R.E. Burkard, and R. Grommes. On Latin squares and the facial structure of related polytopes. *Discr. Math.*, 62:155–181, 1986. (Cited on p. 314.)

[262] R. Euler and H. Le Verge. Time-tables, polyhedra and the greedy algorithm. *Discr. Appl. Math.*, 65:207–221, 1996. (Cited on p. 314.)

[263] T.A. Ewashko and R.C. Dudding. Application of Kuhn's Hungarian assignment algorithm to posting servicemen. *Oper. Res.*, 19:991, 1971. (Cited on p. 165.)

[264] U. Faigle. Some recent results in the analysis of greedy algorithms for assignment problems. *Oper. Res. Spekt.*, 15:181–188, 1994. (Cited on p. 79.)

[265] A. Faye and F. Roupin. A cutting planes algorithm based upon a semidefinite relaxation for the quadratic assignment problem. In G.S. Brodal and S. Leonardi, editors, *Algorithms – ESA 2005*, volume 3669 of *Lecture Notes in Comput. Sci.*, pages 850–861. Springer, Berlin, Heidelberg, 2005. (Cited on p. 247.)

[266] M. Fayyazi, D. Kaeli, and W. Meleis. Parallel maximum weight bipartite matching algorithms for scheduling in input-queued switches. In *Proceedings of the 18th International Parallel and Distributed Processing Symposium*, IPDPS'04. IEEE Computer Society, Washington, DC, 2004. (Cited on p. 139.)

[267] M. Fayyazi, D. Kaeli, and W. Meleis. An adjustable linear time parallel algorithm for maximum weight bipartite matching. Working paper, Northeastern University, Boston, 2005. (Cited on p. 139.)

[268] T. Feder and R. Motwani. Clique partitions, graph compression and speeding-up algorithms. In *Proc. 23rd Annual ACM Symp. Theory of Comput.*, pages 122–133, 1991. Also in *J. Comput. Syst. Sci.* 51, 261–272 (1995). (Cited on pp. 52, 127.)

[269] T.A. Feo and M.G.C. Resende. Greedy randomized adaptive search procedures. *J. Glob. Optim.*, 6:109–133, 1995. (Cited on p. 262.)

[270] G. Finke, R.E. Burkard, and F. Rendl. Quadratic assignment problems. In S. Martello, G. Laporte, M. Minoux, and C. Ribeiro, editors, *Surveys in Combinatorial Optimization*, volume 31 of *Ann. Discr. Math.*, pages 61–82. North–Holland, Amsterdam, 1987. (Cited on pp. 212, 238, 239.)

[271] M. Fischetti, A. Lodi, S. Martello, and P. Toth. A polyhedral approach to simplified crew scheduling and vehicle scheduling problems. *Management Sci.*, 47:833–850, 2001. (Cited on p. 66.)

[272] M. Fischetti, A. Lodi, and P. Toth. Exact methods for the asymmetric traveling salesman problem. In G. Gutin and A.P. Punnen, editors, *The Traveling Salesman Problem and Its Variations*, pages 169–205. Kluwer Academic Publishers, Dordrecht, The Netherlands, 2002. (Cited on p. 165.)

[273] C. Fleurent and J.A. Ferland. Genetic hybrids for the quadratic assignment problem. In P.M. Pardalos and H. Wolkowicz, editors, *Quadratic Assignment and Related Problems*, volume 16 of *DIMACS Series*, pages 173–187. American Mathematical Society, Providence, RI, 1994. (Cited on p. 262.)

[274] C. Fleurent and F. Glover. Improved constructive multistart strategies for the quadratic assignment problem using adaptive memory. *INFORMS J. Comput.*, 11:189–204, 1999. (Cited on p. 265.)

[275] M.J. Flynn. Some computer organizations and their effectiveness. *IEEE Trans. Comput.*, C-21:948–960, 1972. (Cited on p. 138.)

[276] L.R. Ford and D.R. Fulkerson. Maximal flow through a network. *Canadian J. Math.*, 8:399–404, 1956. (Cited on p. 19.)

[277] A. Frank. On Kuhn's Hungarian method—a tribute from Hungary. *Naval Res. Log. Quart.*, 52:2–5, 2004. (Cited on p. 77.)

[278] M.L. Fredman and R.E. Tarjan. Fibonacci heaps and their uses in improved network optimization algorithms. In *Proc. 25th Annual IEEE Symp. Found. Comput. Sci.*, pages 338–346, 1984. Also in *J. ACM* 34: 596–615 (1987). (Cited on pp. 98, 112, 126, 150.)

[279] R.L. Freeman, D.C. Gogerty, G.W. Graves, and R.B.S. Brooks. A mathematical model of supply for space operations. *Oper. Res.*, 14:1–15, 1966. (Cited on p. 295.)

[280] J.B.G. Frenk, M. van Houweninge, and A.H.G. Rinnooy Kan. Order statistics and the linear assignment problem. *Computing*, 39:165–174, 1987. (Cited on p. 148.)

[281] J.B.G. Frenk, M. van Houweninge, and A.H.G. Rinnooy Kan. Asymptotic properties of the quadratic assignment problem. *Math. Oper. Res.*, 10:100–116, 1985. (Cited on pp. 285, 290, 291.)

[282] A.M. Frieze. A bilinear programming formulation of the 3-dimensional assignment problem. *Math. Program.*, 7:376–379, 1974. (Cited on p. 306.)

[283] A.M. Frieze. An algorithm for algebraic assignment problems. *Discr. Appl. Math.*, 1:253–259, 1979. (Cited on p. 195.)

[284] A.M. Frieze and G.B. Sorkin. The probabilistic relationship between the assignment and asymmetric traveling salesman problems. In *Proc. 12th Annual ACM-SIAM Symp. Discr. Algorithms*, pages 652–660, ACM, New York, SIAM, Philadelphia, 2001. (Cited on p. 150.)

[285] A.M. Frieze. Complexity of a 3-dimensional assignment problem. *European J. Oper. Res.*, 13:161–164, 1983. (Cited on p. 312.)

[286] A.M. Frieze and J. Yadegar. An algorithm for solving 3-dimensional assignment problems with applications to scheduling a teaching practice. *J. Oper. Res. Soc.*, 32:989–995, 1981. (Cited on p. 310.)

[287] A.M. Frieze and J. Yadegar. On the quadratic assignment problem. *Discr. Appl. Math.*, 5:89–98, 1983. (Cited on pp. 221, 229, 229, 232, 232, 235.)

[288] A.M. Frieze, J. Yadegar, S. El-Horbaty, and D. Parkinson. Algorithms for assignment problems on an array processor. *Parallel Comput.*, 11:151–162, 1989. (Cited on pp. 258, 260.)

[289] U. Frisch and A. Sobolevskiĭ. Application of optimal transport theory to reconstruction of the early universe. *J. Math. Sci.*, 133:1539–1542, 2006. (Cited on p. 166.)

[290] F.G. Frobenius. Über zerlegbare Determinanten. *Sitzungsberichte der Königlich Preussischen Akademie der Wissenschaften zu Berlin*, XVIII:274–277, 1917. (Cited on p. 16.)

[291] K. Fukuda and T. Terlaky. Criss-cross methods: A fresh view on pivot algorithms. *Math. Program.*, 79:369–395, 1997. (Cited on p. 126.)

[292] D.R. Fulkerson, I. Glicksberg, and O. Gross. A production line assignment problem. Tech. Rep. RM-1102, The Rand Corporation, Santa Monica, CA, 1953. (Cited on p. 172.)

[293] H.N. Gabow and R.E. Tarjan. Algorithms for two bottleneck optimization problems. *J. Algorithms*, 9:411–417, 1988. (Cited on pp. 45, 184.)

[294] H.N. Gabow and R.E. Tarjan. Almost-optimum speed-ups of algorithms for bipartite matching and related problems. In *Proc. 20th Annual ACM Symp. Theory of Comput.*, pages 514–527, 1988. (Cited on p. 138.)

[295] H.N. Gabow. Scaling algorithms for network problems. *J. Comput. Syst. Sci.*, 31:148–168, 1985. (Cited on pp. 77, 87, 88, 88, 128.)

[296] H.N. Gabow and R.E. Tarjan. A linear-time algorithm for a special case of disjoint set union. *J. Comput. Syst. Sci.*, 30:209–221, 1985. (Cited on p. 54.)

[297] H.N. Gabow and R.E. Tarjan. Faster scaling algorithms for network problems. *SIAM J. Comput.*, 18:1013–1036, 1989. (Cited on pp. 77, 78, 87, 88, 123, 127, 128, 139, 139.)

[298] D. Gale and L.S. Shapley. College admissions and the stability of marriage. *American Math. Monthly*, 69:9–15, 1962. (Cited on p. 15.)

[299] L.M. Gambardella, E.D. Taillard, and M. Dorigo. Ant colonies for the QAP. *J. Oper. Res. Soc.*, 50:167–176, 1999. (Cited on p. 263.)

[300] M.R. Garey and D.S. Johnson. *Computers and Intractability: A Guide to the Theory of NP-Completeness*. W.H. Freeman and Company, New York, NY, 1979. (Cited on pp. 8, 207.)

[301] R. Garfinkel. An improved algorithm for the bottleneck assignment problem. *Oper. Res.*, 19:1747–1751, 1971. (Cited on p. 174.)

[302] E. Gassner and B. Klinz. A fast parametric assignment algorithm with applications in max-algebra. Tech. Rep. 336, Spezialforschungsbereich F003, Graz University of Technology, Graz, Austria, 2004. *Networks*, to appear. (Cited on p. 157.)

[303] S. Gaubert, P. Butkovič, and R.A. Cuninghame-Green. Minimal (max,+) realization of convex sequences. *SIAM J. Control Optim.*, 36:137–147, 1998. (Cited on p. 153.)

[304] J.W. Gavett and N.V. Plyter. The optimal assignment of facilities to locations by branch and bound. *Oper. Res.*, 14:210–232, 1966. (Cited on pp. 228, 252.)

[305] B. Gavish, P. Schweitzer, and E. Shlifer. The zero pivot phenomenon in transportation and assignment problems and its computational implications. *Math. Program.*, 12:226–240, 1977. (Cited on pp. 104, 106.)

[306] S. Geetha and M. Vartak. Time-cost trade-off in a three dimensional assignment problem. *European J. Oper. Res.*, 38:255–258, 1989. (Cited on p. 307.)

[307] S. Geetha and M. Vartak. The three-dimensional bottleneck assignment problem with capacity constraints. *European J. Oper. Res.*, 73:562–568, 1994. (Cited on p. 307.)

[308] A.M. Geoffrion. Lagrangean relaxation and its uses in integer programming. *Math. Program. Study*, 2:82–114, 1974. (Cited on pp. 232, 237.)

[309] A.M. Geoffrion and G.W. Graves. Scheduling parallel production lines with changeover costs: Practical application of a quadratic assignment/LP approach. *Oper. Res.*, 24:595–610, 1976. (Cited on p. 206.)

[310] K. Gilbert and R. Hofstra. Multidimensional assignment problems. *Decision Sciences*, 19:306–321, 1988. (Cited on p. 311.)

[311] P.C. Gilmore. Optimal and suboptimal algorithms for the quadratic assignment problem. *SIAM J. Appl. Math.*, 10:305–313, 1962. (Cited on pp. 225, 251, 255.)

[312] P.C. Gilmore, E.L. Lawler, and D.B. Shmoys. Well-solved special cases of the TSP. In E.L. Lawler, J.K. Lenstra, A.H.G. Rinnooy Kan, and D.B. Shmoys, editors, *The Traveling Salesman Problem*, pages 87–143. Wiley, Chichester, UK, 1985. (Cited on p. 151.)

[313] E.K. Gimadi and N. Korkishko. On some modifications of the three index planar assignment problem. In A.V.Eremeev, editor, *DOM'2004. Proceedings of the Second International Workshop on Discrete Optimization Problems in Production and Logistics, Omsk-Irkutsk, 2004*, pages 161–165, Omsk, Russia, 2004. Nasledie Dialog-Sibir Pbs. (Cited on p. 315.)

[314] E.K. Gimadi and N.M. Kairan. Multi-index assignment problem: An asymptotically optimal approach. In *Proceedings of the 8th IEEE International Conference on Emerging Technologies and Factory Automation*, volume 2, pages 707–709. Antibes-Juan les Pins, 2001. (Cited on p. 318.)

[315] Y. Glazkov. On asymptotically optimal algorithm for one modification of planar 3-dimensional assignment problem. In K.-H. Waldmann and U.M. Stocker, editors, *Operations Research Proceedings 2006*, pages 175–179, Springer, Berlin, Heidelberg, 2007. (Cited on p. 315.)

[316] F. Glover. Maximum matching in a convex bipartite graph. *Naval Res. Log. Quart.*, 14:313–316, 1967. (Cited on pp. 35, 53, 55.)

[317] F. Glover. Improved linear integer programming formulations of nonlinear integer problems. *Management Sci.*, 22:455–460, 1975. (Cited on p. 219.)

[318] F. Glover. Heuristics for integer programming using surrogate constraints. *Decision Science*, 8:156–166, 1977. (Cited on p. 264.)

[319] F. Glover. Tabu search—Part I. *ORSA J. Comput.*, 1:190–206, 1989. (Cited on p. 260.)

[320] F. Glover. Tabu search—Part II. *ORSA J. Comput.*, 2:4–32, 1990. (Cited on p. 260.)

[321] F. Glover. Scatter search and path relinking. In D. Corne, M. Dorigo, F. Glover, D. Dasgupta, P. Moscato, R. Poli, and K.V. Price, editors, *New ideas in optimization*, pages 297–316. McGraw-Hill Ltd., UK, Maidenhead, UK, 1999. (Cited on p. 262.)

[322] F. Glover, R. Glover, and D. Klingman. Threshold assignment algorithm. *Math. Program. Study*, 26:12–37, 1986. (Cited on p. 98.)

[323] F. Glover, D. Karney, and D. Klingman. Implementation and computational comparisons of primal, dual and primal-dual computer codes for minimum cost network flow problems. *Networks*, 4:191–212, 1974. (Cited on p. 104.)

[324] F. Glover, D. Karney, D. Klingman, and A. Napier. A computational study on start procedures, basis change criteria, and solution algorithms for transportation problems. *Management Sci.*, 20:793–813, 1974. (Cited on p. 104.)

[325] F. Glover and M. Laguna. *Tabu Search*. Kluwer, Dordrecht, The Netherlands, 1997. (Cited on p. 260.)

[326] F. Glover, E.D. Taillard, and D. de Werra. A user's guide to tabu search. *Ann. Oper. Res.*, 41:12–37, 1993. (Cited on p. 260.)

[327] M.X. Goemans and M. Kodialan. A lower bound on the expected value of an optimal assignment. *Math. Oper. Res.*, 18:267–274, 1993. (Cited on p. 146.)

[328] A.V. Goldberg and R. Kennedy. An efficient cost scaling algorithm for the assignment problem. *Math. Program.*, 71:153–177, 1995. (Cited on pp. 78, 123, 124, 124, 124, 124, 125, 125, 128, 128, 129, 131, 132, 132, 132.)

[329] A.V. Goldberg and R. Kennedy. Global price updates help. *SIAM J. Discr. Math.*, 10:551–572, 1997. (Cited on pp. 126, 150.)

[330] A.V. Goldberg, S.A. Plotkin, D.B. Shmoys, and E. Tardos. Using interior point methods for fast parallel algorithms for bipartite matching and related problems. *SIAM J. Comput.*, 21:140–150, 1992. (Cited on p. 139.)

[331] A.V. Goldberg, S.A. Plotkin, and P.M. Vaidya. Sublinear-time parallel algorithms for matching and related problems. In *Proc. 29th Annual IEEE Symp. Found. Comput. Sci.*, pages 174–184, IEEE Computer Society, Washington, DC, 1988. (Cited on pp. 124, 139, 139.)

[332] A.V. Goldberg, S.A. Plotkin, and P.M. Vaidya. Sublinear-time parallel algorithms for matching and related problems. *J. Algorithms*, 14:180–213, 1993. (Cited on pp. 124, 139, 139.)

[333] A.V. Goldberg and R.E. Tarjan. Finding minimum-cost circulation by successive approximation. *Math. Oper. Res.*, 15:430–466, 1990. (Cited on p. 124.)

[334] D. Goldfarb. Efficient dual simplex algorithms for the assignment problem. *Math. Program.*, 33:187–203, 1985. (Cited on pp. 78, 116, 118, 118, 118, 128.)

[335] T.F. Gonzalez. On the computational complexity of clustering and related problems. In *System Modeling and Optimization: Proceedings of the 10th IFIP Conference*, volume 38 in Lecture Notes in Control and Information Sciences, pages 174–182. Springer, Berlin, Heidelberg, 1982. (Cited on p. 295.)

[336] J.P. Goux, S. Kulkarni, J. Linderoth, and M. Yoder. An enabling framework for master-worker applications the computational grid. In *Proceedings of Ninth IEEE International Symposium on High Performance Distributed Computing Computation*, pages 43–50. IEEE Computer Society, Washington, DC, 2000. (Cited on p. 254.)

[337] R.L. Graham, E.L. Lawler, J.K. Lenstra, and A.H.G. Rinnooy Kan. Optimization and approximation in deterministic sequencing and scheduling: A survey. In P.L. Hammer, E.L. Johnson, and B.H. Korte, editors, *Discrete Optimization II*, volume 5 of *Ann. Discr. Math.*, pages 169–231. North–Holland, Amsterdam, 1979. (Cited on p. 297.)

[338] A. Grama, A. Gupta, G. Karypis, and V. Kumar. *Introduction to Parallel Computing*. Addison–Weslesy, Reading, MA, 2003. (Cited on p. 138.)

[339] G.W. Graves and A.B. Whinston. An algorithm for the quadratic assignment problem. *Management Sci.*, 17:453–471, 1970. (Cited on pp. 216, 217, 256.)

[340] H. Greenberg. A quadratic assignment problem without column constraints. *Naval Res. Log. Quart.*, 16:417–421, 1969. (Cited on p. 293.)

[341] M. Grönkvist. *The Tail Assignment Problem*. Ph.D. thesis, Department of Computing Science, Chalmers University of Technology, Gothenburg, Sweden, 2005. (Cited on p. 66.)

[342] O. Gross. The bottleneck assignment problem. Tech. Rep. P-1630, The Rand Corporation, Santa Monica, CA, 1959. (Cited on pp. 173, 178.)

[343] M. Grötschel, L. Lovász, and A. Schrijver. *Geometric Algorithms and Combinatorial Optimization*, volume 2 of *Algorithms and Combinatorics*. Springer, Berlin, 1988. (Cited on p. 223.)

[344] D.A. Grundel, P.A. Krokhmal, C.A.S. Oliveira, and P.M. Pardalos. On the average case behavior of the multidimensional assignment problem. *Pacific J. Optim.*, 1:39–57, 2005. (Cited on pp. 148, 311.)

[345] D.A. Grundel, P.A. Krokhmal, C.A.S. Oliveira, and P.M. Pardalos. On the number of local minima for the multidimensional assignment problem. *J. Comb. Optim.*, 13:1–18, 2007. (Cited on p. 318.)

[346] D.A. Grundel, C.A.S. Oliveira, and P.M. Pardalos. Asymptotic properties of random multidimensional assignment problems. *J. Optim. Theory Appl.*, 122:487–500, 2004. (Cited on p. 311.)

[347] D.A. Grundel, C.A.S. Oliveira, P.M. Pardalos, and E.L. Pasiliao. Asymptotic results for random multidimensional assignment problems. *Computational Optimization and Applications*, 31:275–293, 2005. (Cited on p. 311.)

[348] D.A. Grundel and P.M. Pardalos. Test problem generator for the multidimensional assignment problem. *Computational Opt. Appl.*, 30:133–146, 2005. (Cited on p. 318.)

[349] G. Grygiel. Algebraic $\sum_k$-assignment problems. *Control and Cybernetics*, 10:155–165, 1981. (Cited on p. 195.)

[350] D. Gusfield and R.W. Irving. *The Stable Marriage Problem, Structure and Algorithms*. MIT Press, Cambridge, MA, 1989. (Cited on p. 15.)

[351] G. Gwan and L. Qi. On facets of the three-index assignment polytope. *Australasian J. Combinatorics*, 6:67–87, 1992. (Cited on p. 307.)

[352] S.W. Hadley, F. Rendl, and H. Wolkowicz. Symmetrization of nonsymmetric quadratic assignment problems and the Hoffman-Wielandt inequality. *Linear Algebra Appl.*, 167:53–64, 1992. (Cited on p. 244.)

[353] S.W. Hadley, F. Rendl, and H. Wolkowicz. Bounds for the quadratic assignment problem using continuous optimization techniques. In *Proceedings of the* 1*st IPCO Conference*, pages 237–248, Waterloo, Ontario, Canada, 1990. (Cited on p. 243.)

[354] S.W. Hadley, F. Rendl, and H. Wolkowicz. A new lower bound via projection for the quadratic assignment problem. *Math. Oper. Res.*, 17:727–739, 1992. (Cited on p. 243.)

[355] P.M. Hahn. Corrigendum to: Tree elaboration strategies in branch-and-bound algorithms for solving the quadratic assignment problem. *Yugoslav J. Operations Research*, 12:271–279, 2002. (Cited on pp. 251, 254.)

[356] P.M. Hahn. Private communication, 2007. (Cited on pp. 223, 292.)

[357] P.M. Hahn, T.L. Grant, and N. Hall. A branch-and-bound algorithm for the quadratic assignment problem based on the Hungarian method. *European J. Oper. Res.*, 108:629–640, 1998. (Cited on pp. 251, 254.)

[358] P.M. Hahn and T.L. Grant. Lower bounds for the quadratic assignment problem based upon a dual formulation. *Oper. Res.*, 46:912–922, 1998. (Cited on pp. 235, 237, 251.)

[359] P.M. Hahn, W.L. Hightower, T.A. Johnson, M. Guignard-Spielberg, and C. Roucairol. Tree elaboration strategies in branch and bound algorithms for solving the quadratic assignment problem. *Yugoslav J. Operations Research*, 11:41–60, 2001. (Cited on pp. 251, 254.)

[360] P. Hall. On representatives of subsets. *J. London Math. Soc.*, 10:26–30, 1935. (Cited on p. 14.)

[361] P. Hansen and L. Kaufman. A primal-dual algorithm for the three-dimensional assignment problem. *Cahiers du CERO*, 15:327–336, 1973. (Cited on p. 308.)

[362] J. Hao and G. Kocur. An implementation of a shortest augmenting path algorithm for the assignment problems. In D.S. Johnson and C.C. McGeoch, editors, *Network Flows and Matching: First DIMACS Implementation Challenge*, volume 12 of *DIMACS Series*, pages 453–468. American Mathematical Society, Providence, RI, 1993. (Cited on p. 103.)

[363] G.H. Hardy, J.E. Littlewood, and G. Pólya. The maximum of a certain bilinear form. *Proceedings of the London Mathematical Society*, 25:265–282, 1926. (Cited on p. 278.)

[364] G.H. Hardy, J.E. Littlewood, and G. Pólya. *Inequalities*. Cambridge University Press, Cambridge, UK, 1952. (Cited on p. 153.)

[365] D. Hausmann, B. Korte, and T. Jenkyns. Worst case analysis of greedy type algorithms for independence systems. *Math. Program.*, 12:120–131, 1980. (Cited on pp. 308, 318.)

[366] D.R. Heffley. Assigning runners to a relay team. In S.P. Ladany and R.E. Machol, editors, *Optimal Strategies in Sports*, pages 169–171. North–Holland, Amsterdam, The Netherlands, 1977. (Cited on p. 206.)

[367] C.H. Heider. An N-step, 2-variable search algorithm for the component placement problem. *Naval Res. Log. Quart.*, 20:699–724, 1973. (Cited on p. 258.)

[368] I. Heller and C.B. Tompkins. An extension of a theorem of Dantzig. In H.W. Kuhn and A.W. Tucker, editors, *Linear Inequalities and Related Systems*, pages 247–252. Princeton University Press, Princeton, NJ, 1956. (Cited on pp. 70, 74.)

[369] F.L. Hitchcock. The distribution of a product from several sources to numerous localities. *J. Math. Phys.*, 20:224–230, 1941. (Cited on pp. 6, 104.)

[370] C.A.R. Hoare. Monitors: An operating system structuring concept. *Commun. ACM*, 17:549–557, 1974. (Cited on p. 143.)

[371] W. Hoëffding. Probability inequalities for sums of bounded random variables. *J. American Statistical Association*, 58:13–30, 1963. (Cited on p. 289.)

[372] A.J. Hoffman. On simple linear programming problems. In V. Klee, editor, *Convexity*, volume 7 of *Proceedings of Symposia in Pure Mathematics*, pages 317–327. American Mathematical Society, Providence, RI, 1963. (Cited on p. 150.)

[373] A.J. Hoffman and H.M. Markowitz. A note on shortest paths, assignment, and transportation problems. *Naval Res. Log. Quart.*, 10:375–379, 1963. (Cited on pp. 93, 139.)

[374] A. Holder. Navy personnel planning and the optimal partition. *Oper. Res.*, 53:77–89, 2005. (Cited on pp. 168, 169.)

[375] J.H. Holland. *Adaptation in Natural and Artificial Systems.* University of Michigan Press, Ann Arbor, MI, 1975. (Cited on p. 261.)

[376] J. Hopcroft and R.M. Karp. An $n^{5/2}$ algorithm for maximum matchings in bipartite graphs. *SIAM J. Comput.*, 2:225–231, 1973. (Cited on pp. 3, 35, 42, 44, 88, 127.)

[377] W.A. Horn. Minimizing average flow time with parallel machines. *Oper. Res.*, 21:846–847, 1973. (Cited on p. 166.)

[378] J. Houdayer, J.H. Boutet de Monvel, and O.C. Martin. Comparing mean field and Euclidean matching problems. *Eur. Phys. J. (B)*, 6:383–393, 1998. (Cited on p. 147.)

[379] G. Huang and A. Lim. A hybrid genetic algorithm for the three-index assignment problem. *European J. Oper. Res.*, 172:249–257, 2006. (Cited on p. 310.)

[380] L.J. Hubert. *Assignment methods in combinatorial data analysis.* Volume 73 in Statistics: Textbooks and Monograph Series. Marcel Dekker, New York, NY, 1987. (Cited on p. 206.)

[381] M.S. Hung. A polynomial simplex method for the assignment problem. *Oper. Res.*, 31:595–600, 1983. (Cited on p. 111.)

[382] M.S. Hung and W.O. Rom. Solving the assignment problem by relaxation. *Oper. Res.*, 28:969–982, 1980. (Cited on pp. 78, 93, 98, 112, 117, 128.)

[383] C.A.J. Hurkens and A. Schrijver. On the size of systems of sets every t of which have an SDR, with an application to the worst-case ratio of heuristics for packing problems. *SIAM J. Discrete Math.*, 2:68–72, 1989. (Cited on p. 318.)

[384] O.H. Ibarra and S. Moran. Deterministic and probabilistic algorithms for maximum bipartite matching via fast matrix multiplication. *Inform. Process. Lett.*, 13:12–15, 1981. (Cited on pp. 58, 127, 174.)

[385] O.H. Ibarra, S. Moran, and R. Hui. A generalization of the fast LUP matrix decomposition algorithm and applications. *J. Algorithms*, 3:45–56, 1982. (Cited on pp. 57, 58, 58.)

[386] M. Iri. A new method for solving transportation-network problems. *J. Oper. Res. Soc. Japan*, 3:27–87, 1960. (Cited on p. 77.)

[387] M. Iwasa, H. Saito, and T. Matsui. Approximation algorithms for the single allocation problem in hub-and-spoke networks and related metric labeling problems. Technical Report METR 2006-52, Dep. of Mathematical Informatics, University of Tokyo, Tokyo, Japan, 2006. (Cited on p. 299.)

[388] T. James, C. Rego, and F. Glover. Sequential and parallel path-relinking algorithms for the quadratic assignment problem. *IEEE Intelligent Systems*, 20:58–65, 2005. (Cited on p. 265.)

[389] T. James, C. Rego, and F. Glover. Multi-start tabu search and diversification strategies for the quadratic assignment problem. Technical Report, Virginia Tech, Blacksburg, VA, 2007. (Cited on p. 261.)

[390] D.S. Johnson and C.C. McGeoch (eds.). *Network Flows and Matching: First DIMACS Implementation Challenge*, volume 12 of *DIMACS Series*. American Mathematical Society, Providence, RI, 1993. (Cited on pp. 79, 132.)

[391] R. Jonker and A.T. Volgenant. Improving the Hungarian assignment algorithm. *Oper. Res. Lett.*, 5:171–175, 1986. (Cited on pp. 87, 130.)

[392] R. Jonker and A.T. Volgenant. A shortest augmenting path algorithm for dense and sparse linear assignment problems. *Computing*, 38:325–340, 1987. (Cited on pp. 77, 99, 101, 102, 104, 120, 125, 128, 129, 130, 130, 132, 141, 141, 165, 191.)

[393] R. Jonker and A.T. Volgenant. Teaching linear assignment by Mack's algorithm. In J.K. Lenstra, H. Tijms, and A.T. Volgenant, editors, *Twenty-Five Years of Operations Research in the Netherlands: Papers Dedicated to Gijs de Leve*, volume 70 of *CWI Tract*, pages 54–60. Centre for Mathematics and Computer Science, Amsterdam, 1989. (Cited on pp. 84, 141, 141.)

[394] R. Jonker and A.T. Volgenant. Linear assignment procedures. *European J. Oper. Res.*, 116:233–234, 1999. (Cited on p. 104.)

[395] M. Jünger. *Polyhedral Combinatorics and the Acyclic Subdigraph Problem*. Heldermann Verlag, Berlin, 1985. (Cited on p. 207.)

[396] M. Jünger and V. Kaibel. On the SQAP-polytope. *SIAM J. Optim.*, 11:444–463, 2000. (Cited on pp. 223, 224, 224, 225.)

[397] M. Jünger and V. Kaibel. Box-inequalities for quadratic assignment polytopes. *Math. Program.*, 91:175–197, 2001. (Cited on pp. 223, 225.)

[398] M. Jünger and V. Kaibel. The QAP-polytope and the star transformation. *Discr. Appl. Math.*, 111:283–306, 2001. (Cited on p. 223.)

[399] D. Jungnickel. *Graphs, Networks and Algorithms*. Springer-Verlag, Berlin Heidelberg, 1999. (Cited on p. 79.)

[400] V. Kaibel. *Polyhedral Combinatorics of the Quadratic Assignment Problem*. Ph.D. thesis, Universität zu Köln, Cologne, Germany, 1997. (Cited on pp. 223, 224, 253.)

[401] V. Kaibel. Polyhedral combinatorics of quadratic assignment problems with less objects than locations. In R.E. Bixby, E.A. Boyd, and R.Z. Ríos-Mercado, editors, *Integer Programming and Combinatorial Optimization*, volume 1412 of *Lecture Notes in Comput. Sci.*, pages 409–422. Springer, Berlin, Heidelberg, 1998. (Cited on p. 253.)

[402] É. Kamáromi. A finite primal method for the assignment problem. *Problems of Control and Inform. Theory*, 3:157–166, 1974. (Cited on p. 110.)

[403] M.-Y. Kao, T.W. Lam, W.-K. Sung, and H.-F. Ting. A decomposition theorem for maximum weight bipartite matchings. *SIAM J. Comput.*, 31:18–26, 2001. (Cited on pp. 78, 127, 127, 128.)

[404] S.E. Karisch, E. Çela, J. Clausen, and T. Espersen. A dual framework for lower bounds of the quadratic assignment problem based on linearization. *Computing*, 63:351–403, 1999. (Cited on p. 238.)

[405] S.E. Karisch and F. Rendl. Lower bounds for the quadratic assignment problem via triangle decompositions. *Math. Program.*, 71:137–151, 1995. (Cited on pp. 243, 266.)

[406] N.K. Karmarkar. A new polynomial time algorithm for linear programming. *Combinatorica*, 4:373–395, 1984. (Cited on p. 126.)

[407] R.M. Karp. Reducibility among combinatorial problems. In R.E. Miller and J.W. Thatcher, editors, *Complexity of Computer Computations*, pages 85–103. Plenum Press, New York, 1972. (Cited on pp. 9, 207, 308.)

[408] R.M. Karp. An algorithm to solve the $m \times n$ assignment problem in expected time $O(mn \log n)$. *Networks*, 10:143–152, 1980. (Cited on p. 149.)

[409] R.M. Karp. An upper bound on the expected cost of an optimal assignment. In D.S. Johnson, T. Nishizeki, A. Nozaki, and H.S. Wilf, editors, *Discrete Algorithms and Complexity (Kyoto 1986)*, volume 15 of *Perspectives in Computing*, pages 1–4. Academic Press, Boston, 1987. (Cited on p. 146.)

[410] R.M. Karp, A.H.G. Rinnooy Kan, and R.V. Vohra. Average case analysis of a heuristic for the assignment problem. *Math. Oper. Res.*, 19:513–522, 1994. (Cited on p. 149.)

[411] R.M. Karp, E. Upfal, and A. Wigderson. Constructing a perfect matching is in random $\mathcal{NC}$. *Combinatorica*, 6:35–48, 1986. (Cited on p. 139.)

[412] L. Kaufman and F. Broeckx. An algorithm for the quadratic assignment problem using Benders' decomposition. *European J. Oper. Res.*, 2:207–211, 1978. (Cited on pp. 218, 249.)

[413] H. Kellerer and G. Wirsching. Bottleneck quadratic assignment problems and the bandwidth problem. *Asia-Pacific J. Operational Research*, 15:169–177, 1998. (Cited on p. 281.)

[414] D. Kempka, J.L. Kennington, and H. Zaki. Performance characteristics of the Jacobi and Gauss-Seidel versions of the auction algorithm on the Alliant FX/8. *ORSA J. Comput.*, 3:92–106, 1991. (Cited on pp. 140, 141.)

[415] J.L. Kennington and Z. Wang. An empirical analysis of the dense assignment problem: Sequential and parallel implementations. *ORSA J. Comput.*, 3:299–306, 1991. (Cited on pp. 140, 141.)

[416] J.L. Kennington and Z. Wang. A shortest augmenting path algorithm for the semi-assignment problem. *Oper. Res.*, 40:178–187, 1992. (Cited on p. 165.)

[417] G. Kindervater, A.T. Volgenant, G. de Leve, and V. van Gijlswijk. On dual solutions of the linear assignment problem. *European J. Oper. Res.*, 19:76–81, 1985. (Cited on p. 103.)

[418] S. Kirkpatrick, C.D. Gelatt, and M.P. Vecchi. Optimization by simulated annealing. *Science*, 220:671–680, 1983. (Cited on p. 259.)

[419] V. Klee and P. Kleinschmidt. The d-step conjecture and its relatives. *Math. Oper. Res.*, 12:718–755, 1987. (Cited on p. 34.)

[420] V. Klee and D.W. Walkup. The d-step conjecture for polyhedra of dimension $d < 6$. *Acta Math.*, 117:53–78, 1967. (Cited on p. 33.)

[421] M. Klein. A primal method for minimal cost flows with applications to assignment and transportation problems. *Management Sci.*, 14:205–220, 1967. (Cited on pp. 78, 105.)

[422] P. Kleinschmidt, C.W. Lee, and H. Schannath. Transportation problems which can be solved by the use of Hirsch-paths for the dual problems. *Math. Program.*, 37:153–168, 1987. (Cited on p. 118.)

[423] B. Klinz, R. Rudolf, and G.J. Woeginger. On the recognition of permuted bottleneck Monge matrices. *Discr. Appl. Math.*, 63:43–74, 1995. (Cited on pp. 186, 187.)

[424] B. Klinz and G.J. Woeginger. A new efficiently solvable case of the three-dimensional axial bottleneck assignment problem. In *Combinatorics and Computer Science (Brest, 1995)*, volume 1120 of *Lecture Notes in Comput. Sci.*, pages 150–162. Springer, Berlin, 1996. (Cited on p. 311.)

[425] D. König. Über Graphen und ihre Anwendungen. *Mathematische Annalen*, 77:453–465, 1916. (Cited on p. 16.)

[426] D. König. *Theorie der Endlichen und Unendlichen Graphen*. Akademische Verlagsgesellschaft, Leipzig, 1936. (Cited on p. 79.)

[427] T.C. Koopmans and M.J. Beckmann. Assignment problems and the location of economic activities. *Econometrica*, 25:53–76, 1957. (Cited on p. 203.)

[428] B. Korte and J. Vygen. *Combinatorial Optimization: Theory and Algorithms*. Springer-Verlag, Berlin, Heidelberg, 2000. (Cited on p. 79.)

[429] M.K. Kozlov, S.P. Tarasov, and L.G. Hačijan. Polinomial solvability of convex quadratic programming. *Sov. Math. Dokl.*, 20:1108–1111, 1979. (Cited on p. 302.)

[430] J. Krarup and P.M. Pruzan. Computer-aided layout design. *Math. Program. Study*, 9:75–94, 1978. (Cited on p. 206.)

[431] M.K. Kravtsov and A. Krachkovskii. O polinimialnomalgoritmenahozhdeniia asimptoticheski optimalnogo resheniia trehindeksnoi planarnoiproblemi vibora. *Zhurnal vichislitelnoi matematiki i matematicheskoi fiziki*, 41:342–345, 2001. (Cited on p. 315.)

[432] M.K. Kravtsov, V.M. Kravtsov, and E.V. Lukshin. On the number of non-integer vertices of a polytope of the three-axial assignment problem (in Russian). *Proc. of the Natl. Academy of Sciences of Belarus, Ser. Phys. Math.*, 4:59–65, 2000. (Cited on p. 307.)

[433] M.K. Kravtsov, V.M. Kravtsov, and E.V. Lukshin. On non-integer vertices of the polytope of the three-index axial assignment problem. *Discrete Mathematics and Applications*, 11:303–325, 2001. (Cited on p. 307.)

[434] V.M. Kravtsov. Polynomial algorithms for finding the asymptotically optimum plan of the multiindex assignment problem. *Cybernetics and Systems Analysis*, 41:940–944, 2005. (Cited on p. 318.)

[435] P.A. Krokhmal, D.A. Grundel, and P.M. Pardalos. Asymptotic behavior of the expected optimal value of the multidimensional assignment problem. *Math. Program. (B)*, 109:525–551, 2007. (Cited on pp. 311, 312, 318.)

[436] A.V. Krushevski. The linear programming problem on a permutation group (in Russian). In *Proceedings of the Seminar on Methods of Mathematical Modeling and Theory of Electrical Circuits*, volume 3, pages 364–371. Institute of Cybernetics of the Academy of Sciences of Ukraine, 1964. (Cited on p. 279.)

[437] A.V. Krushevski. Extremal problems for linear forms on permutations and applications (in Russian). In *Proceedings of the Seminar on Methods of Mathematical Modeling and Theory of Electrical Circuits*, volume 4, pages 262–269. Institute of Cybernetics of the Academy of Sciences of Ukraine, 1965. (Cited on p. 279.)

[438] H.W. Kuhn. The Hungarian method for the assignment problem. *Naval Res. Log. Quart.*, 2:83–97, 1955. (Cited on pp. 77, 79, 128.)

[439] H.W. Kuhn. On combinatorial properties of matrices. Logistic Papers 11, 4, George Washington University, Washington, DC, 1955. (Cited on pp. 79, 337.)

[440] H.W. Kuhn. Variants of the Hungarian method for the assignment problem. *Naval Res. Log. Quart.*, 3:253–258, 1956. (Cited on pp. 77, 79.)

[441] H.W. Kuhn. On the origin of the Hungarian method. In J.K. Lenstra, A.H.G. Rinnooy Kan, and A. Schrijver, editors, *History of Mathematical Programming*, pages 77–81. North–Holland, Amsterdam, 1991. (Cited on p. 79.)

[442] H.W. Kuhn. The Hungarian method for the assignment problem. *Naval Res. Log.*, 52:7–21, 2005. (Cited on p. 79.)

[443] H.W. Kuhn. Statement for Naval Research Logistics: The Hungarian method for the assignment problem. *Naval Res. Log.*, 52:6, 2005. (Cited on p. 79.)

[444] J.M. Kurtzberg. On approximation methods for the assignment problem. *J. ACM*, 9:419–439, 1962. (Cited on p. 149.)

[445] A.M. Land. A problem of assignment with interrelated costs. *Oper. Res. Quart.*, 14:185–198, 1963. (Cited on pp. 204, 214, 217, 235, 252.)

[446] G. Laporte and H. Mercure. Balancing hydraulic turbine runners: A quadratic assignment problem. *European J. Oper. Res.*, 35:378–381, 1988. (Cited on pp. 205, 276.)

[447] E.L. Lawler. The quadratic assignment problem. *Management Sci.*, 9:586–599, 1963. (Cited on pp. 225, 251, 291.)

[448] E.L. Lawler. *Combinatorial Optimization: Networks and Matroids*. Holt, Rinehart and Winston, New York, 1976. (Cited on pp. 7, 77, 85, 127, 128.)

[449] A.J. Lazarus. Certain expected values in the random assignment problem. *Oper. Res. Lett.*, 14:207–214, 1993. (Cited on p. 146.)

[450] Y. Lee and J.B. Orlin. On very large scale assignment problems. In W.W. Hager, D.W. Hearn, and P.M. Pardalos, editors, *Large Scale Optimization. State of the Art*, pages 206–244. Kluwer Academic Publishers, Dordrecht, The Netherlands, 1994. (Cited on p. 99.)

[451] J. van Leeuwen, editor. *Handbook of Theoretical Computer Science, Volume A: Algorithms and Complexity*. Elsevier, Amsterdam, and MIT Press, Cambridge, MA, 1990. (Cited on pp. 138, 208.)

[452] W. Leontief. *Input-Output Economics*. Oxford University Press, New York, 1966. (Cited on p. 207.)

[453] O. Leue. Methoden zur Lösung dreidimensionaler Zuordnungsprobleme. *Angewandte Informatik*, 14:154–162, 1972. (Cited on p. 309.)

[454] J.L. Lewandowski, J.W.S. Liu, and C.L. Liu. SS/TDMA time slot assignment with restricted switching modes. *IEEE Trans. Commun.*, COM-31:149–154, 1983. (Cited on p. 69.)

[455] X. Li and S.A. Zenios. Data-level parallel solution of min-cost network flow problems using ε-relaxations. *European J. Oper. Res.*, 79:474–488, 1994. (Cited on p. 140.)

[456] Y. Li and P.M. Pardalos. Generating quadratic assignment test problems with known optimal permutations. *Computational Opt. Appl.*, 1:163–184, 1992. (Cited on pp. 267, 293.)

[457] Y. Li, P.M. Pardalos, K.G. Ramakrishnan, and M.G.C. Resende. Lower bounds for the quadratic assignment problem. *Ann. Oper. Res.*, 50:387–410, 1994. (Cited on pp. 227, 231, 231, 231, 251.)

[458] Y. Li, P.M. Pardalos, and M.G.C. Resende. A greedy randomized adaptive search procedure for the quadratic assignment problems. In P.M. Pardalos and H. Wolkowicz, editors, *Quadratic Assignment and Related Problems*, volume 16 of *DIMACS Series*, pages 237–261. American Mathematical Society, Providence, RI, 1994. (Cited on pp. 257, 262.)

[459] M.H. Lim, Y. Yuan, and S. Omatu. Extensive testing of a hybrid genetic algorithm for solving quadratic assignment problems. *Computational Opt. Appl.*, 23:47–64, 2002. (Cited on p. 262.)

[460] S. Linusson and J. Wästlund. A generalization of the random assignment problem. arXiv, http://front.math.ucdavis.edu/, 2000. (Cited on pp. 147, 148.)

[461] S. Linusson and J. Wästlund. A proof of Parisi's conjecture on the random assignment problem. *Probab. Theory Related Fields*, 128:419–440, 2004. (Cited on p. 147.)

[462] W. Lipski and F.P. Preparata. Efficient algorithms for finding maximum matchings in convex bipartite graphs and related problems. *Acta Inform.*, 15:329–346, 1981. (Cited on pp. 54, 55, 56.)

[463] M. Livny, J. Basney, R. Raman, and T. Tannenbaum. Mechanisms for high throughput computing. *SPEEDUP*, 11, 1997. (Cited on p. 254.)

[464] E.M. Loiola, N.M.M. de Abreu, P.O. Boaventura-Netto, P.M. Hahn, and T. Querido. A survey for the quadratic assignment problem. *European J. Oper. Res.*, 176:657–690, 2007. (Cited on p. 203.)

[465] V. Lotfi. A labeling algorithm to solve the assignment problem. *Computers & Oper. Res.*, 16:397–408, 1989. (Cited on pp. 83, 127.)

[466] L. Lovász. On determinants, matchings and random algorithms. In L. Budach, editor, *Fundamentals of Computation Theory FCT'79*, volume 2 of *Mathematical Research*, pages 565–574. Akademie Verlag, Berlin, 1979. (Cited on p. 58.)

[467] R.E. Machol. An application of the assignment problem. *Oper. Res.*, 9:585–586, 1961. (Cited on p. 165.)

[468] R.E. Machol. An application of the assignment problem. *Oper. Res.*, 18:745–746, 1970. (Cited on p. 165.)

[469] R.E. Machol and M. Wien. A hard assignment problem. *Oper. Res.*, 24:190–192, 1976. (Cited on p. 133.)

[470] R.E. Machol and M. Wien. Errata. *Oper. Res.*, 25:364, 1977. (Cited on p. 133.)

[471] C. Mack. The Bradford method for the assignment problem. *New J. Statist. and Oper. Res.*, 1:17–29, 1969. (Cited on p. 84.)

[472] V.F. Magirou and J.Z. Milis. An algorithm for the multiprocessor assignment problem. *Oper. Res. Lett.*, 8:351–356, 1989. (Cited on pp. 297, 301, 302.)

[473] D. Magos. Tabu search for the planar three-index assignment problem. *J. Glob. Optim.*, 8:35–48, 1996. (Cited on p. 314.)

[474] D. Magos and P. Miliotis. An algorithm for the planar three-index assignment problem. *European J. Oper. Res.*, 77:141–153, 1994. (Cited on p. 314.)

[475] D. Magos, I. Mourtos, and G. Appa. Polyhedral results for assignment problems. CDAM Research Report LSE-CDAM-2002–01, London School of Economics, 2002. (Cited on p. 317.)

[476] R. Malhotra, H. Bhatia, and M. Puri. The three dimensional bottleneck assignment problem and its variants. *Optimization*, 16:245–256, 1985. (Cited on p. 307.)

[477] F. Malucelli. A polynomially solvable class of quadratic semi-assignment problems. *European J. Oper. Res.*, 91:619–622, 1996. (Cited on pp. 57, 296, 301.)

[478] F. Malucelli and D. Pretolani. Lower bounds for the quadratic semi-assignment problem. *European J. Oper. Res.*, 83(955):365–375, 1995. (Cited on pp. 297, 300, 301.)

[479] V. Maniezzo and A. Colorni. The ant system applied to the quadratic assignment problem. *IEEE Trans. Knowl. Data Engin.*, 11:769–778, 1999. (Cited on p. 263.)

[480] S. Martello, W.R. Pulleyblank, P. Toth, and D. de Werra. Balanced optimization problems. *Oper. Res. Lett.*, 3:275–278, 1984. (Cited on p. 195.)

[481] S. Martello and P. Toth. Linear assignment problems. In S. Martello, G. Laporte, M. Minoux, and C. Ribeiro, editors, *Surveys in Combinatorial Optimization*, volume 31 of *Ann. Discr. Math.*, pages 259–282. North–Holland, Amsterdam, 1987. (Cited on p. 78.)

[482] A. Marzetta and A. Brüngger. A dynamic programming bound for the quadratic assignment problem. In *Computing and Combinatorics: 5th Annual International Conference, COCOON'99*, volume 1627 of *Lecture Notes in Comput. Sci.*, pages 339–348. Springer, Berlin, Heidelberg, 1999. (Cited on p. 254.)

[483] T. Mautor and C. Roucairol. A new exact algorithm for the solution of quadratic assignment problems. *Discr. Appl. Math.*, 55:281–293, 1994. (Cited on p. 252.)

[484] E.J. McCormick. *Human Factors Engineering*. McGraw–Hill, New York, 1970. (Cited on p. 205.)

[485] L.F. McGinnis. Implementation and testing of a primal-dual algorithm for the assignment problem. *Oper. Res.*, 31:277–291, 1983. (Cited on p. 87.)

[486] G.M. Megson and D.J. Evans. A systolic array solution for the assignment problem. *The Computer J.*, 33:562–569, 1990. (Cited on p. 139.)

[487] N.S. Mendelsohn and A.L. Dulmage. Some generalizations of the problem of distinct representatives. *Canadian J. Math.*, 10:230–241, 1958. (Cited on p. 22.)

[488] N.N. Metelski. On extremal values of quadratic forms on symmetric groups (in Russian). *Vesti Akad. Navuk BSSR*, 6:107–110, 1972. (Cited on p. 278.)

[489] M. Mézard and G. Parisi. Replicas and optimization. *J. Physique Lett.*, 46:771–778, 1985. (Cited on p. 146.)

[490] M. Mézard and G. Parisi. On the solution of the random link matching problems. *J. Physique Lett.*, 48:1451–1459, 1987. (Cited on p. 146.)

[491] D.L. Miller, J.F. Pekny, and G.L. Thompson. Solution of large dense transportation problems using a parallel primal algorithm. *Oper. Res. Lett.*, 9:319–324, 1990. (Cited on p. 143.)

[492] G.A. Miller. On a method due to Galois. *Quarterly J. Mathematics*, 41:382–384, 1910. (Cited on p. 16.)

[493] G. Miranda, H.P.L. Luna, G.R. Mateus, and R.P.M. Ferreira. A performance guarantee heuristic for electronic components placement problems including thermal effects. *Computers & Oper. Res.*, 32:2937–2957, 2005. (Cited on p. 250.)

[494] P.B. Mirchandani and T. Obata. Locational decisions with interactions between facilities: The quadratic assignment problem, a review. Working Paper Ps-79-1, Rensselaer Polytechnic Institute, Troy, New York, 1979. (Cited on p. 252.)

[495] L. Mirsky. The spread of a matrix. *Mathematika*, 3:127–130, 1956. (Cited on p. 240.)

[496] A. Misevicius. An improved hybrid optimization algorithm for the quadratic assignment problem. *Mathematical Modelling and Analysis*, 9:149–168, 2004. (Cited on p. 262.)

[497] N. Mladenović and P. Hansen. Variable neighborhood search. *Computers & Oper. Res.*, 24:1097–1100, 1997. (Cited on p. 266.)

[498] N. Moreno and A. Corominas. Solving the minmax product rate variation problem (PRVP) as a bottleneck assignment problem. *Computers & Oper. Res.*, 33:928–939, 2006. (Cited on p. 189.)

[499] J. Mosevich. Balancing hydraulic turbine runners—A discrete combinatorial optimization problem. *European J. Oper. Res.*, 26:202–204, 1986. (Cited on pp. 205, 276.)

[500] H. Müller-Merbach. *Optimale Reihenfolgen*. Springer-Verlag, Berlin, 1970. (Cited on p. 255.)

[501] K. Mulmuley, U.V. Vazirani, and V.V. Vazirani. Matching is as easy as matrix inversion. *Combinatorica*, 7:105–113, 1987. (Cited on pp. 59, 139.)

[502] J. Munkres. Algorithms for the assignment and transportation problems. *J. SIAM*, 5:32–38, 1957. (Cited on pp. 77, 128, 165.)

[503] R. Murphey, P.M. Pardalos, and L.S. Pitsoulis. A GRASP for the multitarget multisensor tracking problem. In P.M. Pardalos and D.-Z. Du, editors, *Network Design: Connectivity and Facilities Location*, volume 40 of *DIMACS Series*, pages 277–302. AMS, Providence RI, 1998. (Cited on p. 317.)

[504] R. Murphey, P.M. Pardalos, and L.S. Pitsoulis. A parallel GRASP for the data association multidimensional assignment problem. In *Parallel Processing of Discrete Problems*, volume 106 of *The IMA Volumes in Mathematics and Its Applications*, pages 159–180. Springer, New York, 1998. (Cited on p. 317.)

[505] K.A. Murthy, P.M. Pardalos, and Y. Li. A local search algorithm for the quadratic assignment problem. *Informatica*, 3:524–538, 1992. (Cited on pp. 211, 257.)

[506] K.G. Murty. An algorithm for ranking all the assignments in order of increasing cost. *Oper. Res.*, 16:682–687, 1968. (Cited on pp. 158, 163.)

[507] K.G. Murty. *Network Programming*. Prentice-Hall, Englewood Cliffs, NJ, 1992. (Cited on p. 79.)

[508] D. Naddef. The Hirsch conjecture is true for (0,1)-polytopes. *Math. Program.*, 45:109–110, 1989. (Cited on p. 33.)

[509] C. Nair, B. Prabhakar, and M. Sharma. Proofs of the Parisi and Coppersmith-Sorkin random assignment conjectures. *Random Structures Algorithms*, 27:413–444, 2005. (Cited on p. 147.)

[510] W.M. Nawijn and B. Dorhout. On the expected number of assignments in reduced matrices for the linear assignment problem. *Oper. Res. Lett.*, 8:329–335, 1989. (Cited on p. 99.)

[511] B. Neng. Zur Erstellung von optimalen Triebfahrzeugplänen. *Z. Oper. Res.*, 25:B159–B185, 1981. (Cited on pp. 66, 166.)

[512] C.E. Nugent, T.E. Vollmann, and J. Ruml. An experimental comparison of techniques for the assignment of facilities to locations. *Oper. Res.*, 16:150–173, 1968. (Cited on pp. 247, 249, 252, 253, 266.)

[513] B. Olin. *Asymptotic properties of random assignment problems*. Ph.D. thesis, Division of Optimization and Systems Theory, Department of Mathematics, Royal Institute of Technology, Stockholm, Sweden, 1992. (Cited on pp. 146, 148, 148.)

[514] J.B. Orlin. On the simplex algorithm for networks and generalized networks. *Math. Program. Study*, 24:166–178, 1985. (Cited on p. 111.)

[515] J.B. Orlin and R.K. Ahuja. New scaling algorithms for the assignment and minimum cycle mean problems. *Math. Program.*, 54:41–56, 1992. (Cited on pp. 78, 88, 123, 123, 124, 128.)

[516] J.B. Orlin and C. Stein. Parallel algorithms for the assignment and minimum-cost flow problems. *Oper. Res. Lett.*, 14:181–186, 1993. (Cited on p. 139.)

[517] C.N.K. Osiakwan and S.G. Akl. A perfect speedup parallel algorithm for the assignment problem on complete weighted bipartite graphs. In *Proc. of IEEE Parbase 90*, pages 293–301. IEEE Computer Society, Washington, DC, 1990. (Cited on p. 139.)

[518] M.W. Padberg. The boolean quadric polytope: Some characteristics, facets and relatives. *Math. Program. (B)*, 45:139–172, 1989. (Cited on p. 224.)

[519] M.W. Padberg and M.P. Rijal. *Location, Scheduling, Design and Integer Programming*. Kluwer Academic Publishers, Boston, MA, 1996. (Cited on pp. 223, 224, 224, 225, 225, 253.)

[520] M.W. Padberg and G. Rinaldi. Optimization of a 532-city symmetric traveling salesman problem by branch and cut. *Oper. Res. Lett.*, 6:1–7, 1987. (Cited on p. 253.)

[521] E.S. Page. A note on assignment problems. *The Computer J.*, 6:241–243, 1963. (Cited on p. 178.)

[522] G. Palubeckis. Generating hard test instances with known optimal solution for the rectilinear quadratic assignment problem. *J. Glob. Optim.*, 15:127–156, 1999. (Cited on p. 267.)

[523] G. Palubeckis. An algorithm for construction of test cases for the quadratic assignment problem. *Informatica*, 11:281–296, 2000. (Cited on pp. 267, 317.)

[524] G.S. Palubetskis. A generator of test quadratic assignment problems with known optimal solution. *U.S.S.R. Comput. Maths. Math. Phys.*, 28:97–98, 1988. Translated from *Zh. Vychisl. Mat. Fiz.*, 28:1740–1743, 1988. (Cited on pp. 229, 293.)

[525] C.H. Papadimitriou and P.C. Kanellakis. Flowshop scheduling with limited temporary storage. *J. ACM*, 27:533–549, 1980. (Cited on p. 132.)

[526] C.H. Papadimitriou and D. Wolfe. The complexity of facets resolved. *J. Comput. Syst. Sci.*, 37:2–13, 1988. (Cited on p. 211.)

[527] C. Papamanthou, K. Paparrizos, and N. Samaras. A parametric visualization software for the assignment problem. *Yugoslav J. Operations Research*, 15:1–12, 2005. (Cited on p. 128.)

[528] C. Papamanthou, K. Paparrizos, N. Samaras, and K. Stergiou. Worst case examples of an exterior point algorithm for the assignment problem. *Discr. Opt.*, 5:605–614, 2008. (Cited on p. 119.)

[529] K. Paparrizos. A non-dual signature method for the assignment problem and a generalization of the dual simplex method for the transportation problem. *RAIRO Rech. Opér.*, 22:269–289, 1988. (Cited on p. 119.)

[530] K. Paparrizos. An infeasible (exterior point) simplex algorithm for assignment problems. *Math. Program.*, 51:45–54, 1991. (Cited on pp. 119, 129.)

[531] K. Paparrizos. A relaxation column signature method for assignment problems. *European J. Oper. Res.*, 50:211–219, 1991. (Cited on p. 119.)

[532] P.M. Pardalos, L.S. Pitsoulis, and M.G.C. Resende. A parallel GRASP implementation for the quadratic assignment problem. In A. Ferreira and José D.P. Rolim, editors, *Parallel Algorithms for Irregular Problems: State of the Art*, pages 115–133. Kluwer Academic Publishers, Geneva, Switzerland, 1995. (Cited on p. 263.)

[533] P.M. Pardalos, L.S. Pitsoulis, and M.G.C. Resende. Fortran subroutines for approximate solution of sparse quadratic assignment problems using GRASP. *ACM Trans. Math. Software*, 23:196–208, 1997. (Cited on p. 263.)

[534] P.M. Pardalos, K.G. Ramakrishnan, M.G.C. Resende, and Y. Li. Implementation of a variance reduction-based lower bound in a branch-and-bound algorithm for the quadratic assignment problem. *SIAM J. Optim.*, 7:280–294, 1997. (Cited on p. 251.)

[535] P.M. Pardalos, F. Rendl, and H. Wolkowicz. The quadratic assignment problem: a survey and recent developments. In P.M. Pardalos and H. Wolkowicz, editors, *Quadratic Assignment and Related Problems*, volume 16 of *DIMACS Series*, pages 1–42. American Mathematical Society, Providence, RI, 1994. (Cited on pp. 203, 211.)

[536] P.M. Pardalos and H. Wolkowicz. Quadratic assignment and related problem. In P.M. Pardalos and H. Wolkowicz, editors, *Proceedings DIMACS Workshop on the Quadratic Assignment Problem*, volume 16 of *DIMACS Series*. American Mathematical Society, Providence, RI, 1994. (Cited on p. 203.)

[537] P.M. Pardalos and J. Xue. The maximum clique problem. *J. Glob. Optim.*, 4:301–328, 1994. (Cited on p. 209.)

[538] G. Parisi. A conjecture on random bipartite matching. Physics e-Print Archive, 1998. http://xxx.lanl.gov/ps/cond-mat/9801176. (Cited on p. 147.)

[539] M. Pascoal, M.E. Captivo, and J. Clímaco. A note on a new variant of Murty's ranking assignments algorithm. *4OR: A Quart. J. Oper. Res.*, 1:243–255, 2003. (Cited on pp. 160, 163.)

[540] C.R. Pedersen, L.R. Nielsen, and K.A. Andersen. A note on ranking assignments using reoptimization. Technical Report 2005/2, Department of Mathematical Sciences, University of Aarhus, 2005. (Cited on pp. 158, 160.)

[541] D.W. Pentico. Assignment problems: A golden anniversary survey. *European J. Oper. Res.*, 176:774–793, 2007. (Cited on p. 79.)

[542] J. Peters. The network simplex method on a multiprocessor. *Networks*, 20:845–859, 1990. (Cited on pp. 143, 144.)

[543] U. Pferschy. The random linear bottleneck assignment problem. *RAIRO Rech. Opér.*, 30:127–142, 1996. (Cited on pp. 60, 185, 185, 187.)

[544] U. Pferschy. Solution methods and computational investigations for the linear bottleneck assignment problem. *Computing*, 59:237–258, 1997. (Cited on pp. 184, 229.)

[545] U. Pferschy, R. Rudolf, and G.J. Woeginger. Monge matrices make maximization manageable. *Oper. Res. Lett.*, 16:245–254, 1994. (Cited on p. 275.)

[546] C.A. Phillips and S.A. Zenios. Experiences with large scale network optimization on the connection machine. In R. Sharada, B.L. Golden, E. Wasil, W. Stuart, and O. Balci, editors, *Impact of Recent Computer Advances on Operations Research*, pages 169–180. North–Holland, Amsterdam, 1989. (Cited on p. 140.)

[547] J.-C. Picard and M. Queyranne. On the one-dimensional space allocation problem. *Oper. Res.*, 29:371–391, 1981. (Cited on p. 276.)

[548] W.P. Pierskalla. The tri-substitution method for the three-dimensional assignment problem. *Canadian Operational Research Society J.*, 5:71–81, 1967. (Cited on pp. 307, 310.)

[549] W.P. Pierskalla. The multidimensional assignment problem. *Oper. Res.*, 16:422–431, 1968. (Cited on pp. 305, 312.)

[550] J. Pitman. Coalescent random forests. *J. Combin. Theory*, A85:165–193, 1999. (Cited on p. 30.)

[551] M.A. Pollatschek, N. Gershoni, and Y.T. Radday. Optimization of the typewriter keyboard by simulation. *Angewandte Informatik*, 17:438–439, 1976. (Cited on p. 205.)

[552] A.B. Poore. Multidimensional assignment formulation of data association problems arising from multitarget and multisensor tracking. *Computational Opt. Appl.*, 3:27–57, 1994. (Cited on pp. 305, 315, 317.)

[553] A.B. Poore. Multidimensional assignment and multitarget tracking. In *Partitioning Data Sets*, volume 19 of *DIMACS Series*, pages 169–196. American Mathematical Society, Providence, RI, 1995. (Cited on p. 305.)

[554] A.B. Poore. Multidimensional assignment problems arising in multitarget and multisensor tracking. In P.M. Pardalos and L.S. Pitsoulis, editors, *Nonlinear Assignment Problems. Algorithms and Applications*, pages 13–38. Kluwer Academic Publishers, Dordrecht, The Netherlands, 2000. (Cited on p. 315.)

[555] A.B. Poore and N. Rijavec. A Lagrangian relaxation algorithm for multidimensional assignment problems arising from multitarget tracking. *SIAM J. Optim.*, 3:545–563, 1993. (Cited on p. 317.)

[556] A.B. Poore, N. Rijavec, M. Liggins, and V. Vannicola. Data association problems posed as multidimensional assignment problems: Problem formulation. In O. E. Drummond, editor, *Signal and Data Processing of Small Targets*, pages 552–561. SPIE, Bellingham, WA, 1993. (Cited on p. 305.)

[557] A.B. Poore and A.J. Robertson. A new Lagrangian relaxation based algorithm for a class of multidimensional assignment problems. *Computational Opt. Appl.*, 8:129–150, 1997. (Cited on p. 317.)

[558] V.R. Pratt. An $n \log n$ algorithm to distribute n records optimally in a sequential access file. In R.E. Miller and J.W. Thatcher, editors, *Complexity of Computer Computations*, pages 111–118. Plenum Press, New York, 1972. (Cited on p. 278.)

[559] F.P. Preparata and W. Lipski. Three layers are enough. In *Proceedings of the 23rd Annual IEEE Symposium on Foundations of Computer Science*, pages 350–357. IEEE Computer Society, Washington, DC, 1982. (Cited on p. 55.)

[560] A. Przybylski, X. Gandibleux, and M. Ehrgott. The biobjective assignment problem. Technical Report 05.07, LINA, Université de Nantes, 2005. (Cited on p. 158.)

[561] A.P. Punnen. On bottleneck assignment problems under categorization. *Computers & Oper. Res.*, 31:151–154, 2004. (Cited on p. 190.)

[562] A.P. Punnen and Y.P. Aneja. Categorized assignment scheduling: A tabu search approach. *J. Oper. Res. Soc.*, 44:673–679, 1993. (Cited on p. 167.)

[563] A.P. Punnen and K.P.K. Nair. Improved complexity bound for the maximum cardinality bottleneck bipartite matching problem. *Discr. Appl. Math.*, 55:91–93, 1994. (Cited on p. 184.)

[564] J. Pusztaszeri. The nonlinear assignment problem in experimental high energy physics. In P.M. Pardalos and L.S. Pitsoulis, editors, *Nonlinear Assignment Problems. Algorithms and Applications*, pages 55–89. Kluwer Academic Publishers, Dordrecht, The Netherlands, 2000. (Cited on p. 317.)

[565] J. Pusztaszeri, P.E. Rensing, and T.M. Liebling. Tracking elementary particles near their primary vertex: A combinatorial approach. *J. Glob. Optim.*, 16:422–431, 1995. (Cited on p. 317.)

[566] L. Qi, E. Balas, and G. Gwan. A new facet class and a polyhedral method for the three-index assignment problem. In D.Z. Du and J. Sun, editors, *Advances in Optimization and Approximaton*, pages 256–274. Kluwer Academic, Dordrecht, The Netherlands, 1994. (Cited on p. 307.)

[567] L. Qi and D. Sun. Polyhedral methods for solving three index assignment problems. In P.M. Pardalos and L.S. Pitsoulis, editors, *Nonlinear Assignment Problem: Algorithms and Application, Combinatorial Optimization Series*, pages 91–107. Kluwer Academic Publishers, Dordrecht, The Netherlands, 2000. (Cited on p. 307.)

[568] M. Queyranne. Performance ratio of polynomial heuristics for triangle inequality quadratic assignment problems. *Oper. Res. Lett.*, 4:231–234, 1986. (Cited on p. 210.)

[569] K.G. Ramakrishnan, N.K. Karmarkar, and A.P. Kamath. An approximate dual projective algorithm for solving assignment problems. In D.S. Johnson and C.C. McGeoch, editors, *Network Flows and Matching: First DIMACS Implementation Challenge*, volume 12 of *DIMACS Series*, pages 431–452. American Mathematical Society, Providence, RI, 1993. (Cited on pp. 125, 126, 127.)

[570] G. Reinelt. *The Linear Ordering Problem: Algorithms and Applications*, volume 8 of *Research and Exposition in Mathematics*. Heldermann Verlag, Berlin, 1985. (Cited on p. 207.)

[571] F. Rendl. On the complexity of decomposing matrices arising in satellite communication. *Oper. Res. Lett.*, 4:5–8, 1985. (Cited on pp. 69, 314.)

[572] F. Rendl. Ranking scalar products to improve bounds for the quadratic assignment problem. *European J. Oper. Res.*, 20:363–372, 1985. (Cited on p. 241.)

[573] F. Rendl. Quadratic assignment problems on series-parallel digraphs. *Z. Oper. Res.*, 30:161–173, 1986. (Cited on p. 280.)

[574] F. Rendl. The quadratic assignment problem. In Z. Drezner and H.W. Hamacher, editors, *Facility Location*, pages 439–457. Springer, Berlin, 2002. (Cited on pp. 203, 242.)

[575] F. Rendl and R. Sotirov. Bounds for the quadratic assignment problem using the bundle method. *Math. Program. (B)*, 109:505–524, 2007. (Cited on p. 247.)

[576] F. Rendl and H. Wolkowicz. Applications of parametric programming and eigenvalue maximization to the quadratic assignment problem. *Math. Program.*, 53:63–78, 1992. (Cited on p. 242.)

[577] A. Rényi. *Probability Theory*. North–Holland, Amsterdam, 1970. (Cited on p. 287.)

[578] M.G.C. Resende, K.G. Ramakrishnan, and Z. Drezner. Computing lower bounds for the quadratic assignment problem with an interior point algorithm for linear programming. *Oper. Res.*, 43:781–791, 1995. (Cited on p. 238.)

[579] W.T. Rhee. A note on asymptotic properties of the quadratic assignment problem. *Oper. Res. Lett.*, 7:197–200, 1988. (Cited on pp. 285, 291.)

[580] W.T. Rhee. Stochastic analysis of the quadratic assignment problem. *Math. Oper. Res.*, 16:223–239, 1991. (Cited on pp. 285, 290.)

[581] J. Rhys. A selection problem of shared fixed costs and networks. *Management Sci.*, 17:197–204, 1970. (Cited on p. 302.)

[582] M.B. Richey. Optimal location of a path or tree on a network with cycles. *Networks*, 20:391–407, 1990. (Cited on p. 300.)

[583] M.B. Richey and A.P. Punnen. Minimum perfect bipartite matchings and spanning trees under categorization. *Discr. Appl. Math.*, 39:147–153, 1992. (Cited on p. 167.)

[584] M.P. Rijal. *Scheduling, Design and Assignment Problems with Quadratic Costs*. Ph.D. thesis, New York University, New York, NY, 1995. (Cited on p. 223.)

[585] A.J. Robertson. A set of greedy randomized adaptive local search procedure (GRASP) implementations for the multidimensional assignment problem. *Computational Opt. Appl.*, 19:145–164, 2001. (Cited on p. 317.)

[586] N. Robertson, P.D. Seymour, and R. Thomas. Permanents, Pfaffian orientations, and even directed circuits. *Ann. Math.*, 150:929–975, 1999. (Cited on p. 156.)

[587] J. Robinson. On the Hamiltonian game (a traveling salesman problem). Technical Report RM-303, The RAND Corporation, Santa Monica, CA, 1949. (Cited on p. 78.)

[588] J.M. Rodríguez, F.C. MacPhee, D.J. Bonham, and V.C. Bhavsar. Solving the quadratic assignment and dynamic plant layout problems using a new hybrid meta-heuristic approach. *International J. High Performance Computing and Networking*, 4:286–294, 2006. (Cited on p. 262.)

[589] E. Roohy-Laleh. *Improvements to the theoretical efficiency of the network simplex method.* PhD thesis, Carleton University, Ottawa, Canada, 1980. (Cited on pp. 77, 111.)

[590] G. Rote and F. Rendl. Minimizing the density of terminal assignments in layout design. *Oper. Res. Lett.*, 5:111–118, 1986. (Cited on p. 56.)

[591] C. Roucairol. A reduction method for quadratic assignment problems. *Oper. Res. Verf.*, 32:183–187, 1979. (Cited on p. 229.)

[592] C. Roucairol. A parallel branch and bound algorithm for the quadratic assignment problem. *Discr. Appl. Math.*, 18:211–225, 1987. (Cited on p. 252.)

[593] S. Sahni and T.F. Gonzalez. P-complete approximation problems. *J. ACM*, 23:555–565, 1976. (Cited on p. 210.)

[594] H. Saito. The symmetric quadratic semi-assignment polytope. *IEICE Trans. on Fund. of Electr., Comm. and Comp. Sci.*, 89:1227–1232, 2006. (Cited on p. 295.)

[595] H. Saito, T. Fujie, T. Matsui, and S. Matuura. The quadratic semi-assignment polytope. Mathematical Engineering Technical Reports METR 2004-32, Dep. of Mathematical Informatics, University of Tokyo, 2004. (Cited on p. 295.)

[596] B.R. Sarker, W.E. Wilhelm, and G.L. Hogg. One-dimensional machine location problems in a multi-product flowline with equidistant location. *European J. Oper. Res.*, 105:401–426, 1998. (Cited on p. 276.)

[597] E. Schell. Distribution of a product by several properties. In H. A. Antosiewicz, editor, *Proceedings of the Second Symposium in Linear Programming*, pages 615–642. National Bureau of Standards and U.S. Air Force, Washington D.C., 1955. (Cited on p. 305.)

[598] D. Schlegel. *Die Unwucht-optimale Verteilung von Turbinenschaufeln als quadratisches Zuordnungsproblem.* Ph.D. thesis, ETH Zürich, 1987. (Cited on pp. 205, 276.)

[599] A. Schrijver. *Combinatorial Optimization: Polyhedra and Efficiency.* Springer-Verlag, Berlin, Heidelberg, 2003. (Cited on pp. 13, 77, 79.)

[600] A. Schrijver. On the history of combinatorial optimization (till 1960). In K. Aardal, G.L. Nemhauser, and R. Weismantel, editors, *Discrete Optimization*, volume 12 of *Handbooks in Operations Research and Management Science*, pages 1–68. Elsevier, Amsterdam, 2005. (Cited on pp. 16, 77.)

[601] C. Schütt and J. Clausen. Parallel algorithms for the assignment problem—an experimental evaluation of three distributed algorithms. In P.M. Pardalos, M.G.C. Resende, and K.G. Ramakrishnan, editors, *DIMACS Workshop on Parallel Processing of Discrete Optimization Problems*, volume 22 of *DIMACS Series*, pages 337–351. American Mathematical Society, Providence, RI, 1995. (Cited on p. 140.)

[602] B.L. Schwartz. A computational analysis of the auction algorithm. *European J. Oper. Res.*, 74:161–169, 1994. (Cited on pp. 123, 166.)

[603] J.T. Schwartz, A. Steger, and A. Weissl. Fast algorithms for weighted bipartite matching. In S.E. Nikoletseas, editor, *WEA 2005*, volume 3503 of *Lecture Notes in Comput. Sci.*, pages 476–487. Springer-Verlag, Berlin, Heidelberg, 2005. (Cited on p. 150.)

[604] J.T. Schwartz. Fast probabilistic algorithms for verification of polynomial identities. *J. ACM*, 27:701–717, 1980. (Cited on p. 59.)

[605] U. Schwiegelshohn and L. Thiele. A systolic array for the assignment problem. *IEEE Trans. Comput.*, 37:1422–1425, 1988. (Cited on p. 139.)

[606] H.D. Sherali and W.P. Adams. A hyerarchy of relaxations between the continuous and convex hull representations for zero-one programming problems. *SIAM J. Discrete Math.*, 3:411–430, 1990. (Cited on pp. 222, 254.)

[607] H.D. Sherali and W.P. Adams. A hyerarchy of relaxations and convex hull characterizations for mixed-integer zero-one programming problems. *Discr. Appl. Math.*, 52:83–106, 1994. (Cited on pp. 222, 254.)

[608] H.D. Sherali and W.P. Adams. *A Reformulation-Linearization Technique for Solving Discrete and Continuous Nonconvex Problems*. Kluwer Academic Publishers, Dordrecht, The Netherlands, 1999. (Cited on pp. 222, 254.)

[609] R. Silver. Algorithm 27: Assignment. *Commun. ACM*, 3:603–604, 1960. (Cited on p. 77.)

[610] R. Silver. An algorithm for the assignment problem. *Commun. ACM*, 3:605–606, 1960. (Cited on p. 77.)

[611] J. Skorin-Kapov. Tabu search applied to the quadratic assignment problem. *ORSA J. Comput.*, 2:33–45, 1990. (Cited on pp. 260, 261.)

[612] M. Skutella. Convex quadratic and semidefinite programming relaxations in scheduling. *J. ACM*, 48:206–242, 2001. (Cited on pp. 297, 302, 303, 303.)

[613] D.D. Sleator and R.E. Tarjan. A data structure for dynamic trees. *J. Comput. Syst. Sci.*, 26:362–391, 1983. (Cited on p. 112.)

[614] W.E. Smith. Various optimizers for single-stage production. *Naval Res. Log. Quart.*, 3:59–66, 1956. (Cited on p. 297.)

[615] P.T. Sokkalingam and Y.P. Aneja. Lexicographic bottleneck combinatorial problems. *Oper. Res. Lett.*, 23:27–33, 1998. (Cited on p. 198.)

[616] R. Sotirov and H. Wolkowicz. The *Simple* method for the SDP relaxation of the QAP. Technical report, Department of Combinatorics and Optimization, University of Waterloo, Waterloo, Ontario, Canada, 2006. (Cited on p. 247.)

[617] F.C.R. Spieksma and G.J. Woeginger. Geometric three-dimensional assignment problems. *European J. Oper. Res.*, 91:611–618, 1996. (Cited on p. 308.)

[618] F.C.R. Spieksma. Multi-index assignment problems: Complexity, approximation, applications. In P.M. Pardalos and L.S Pitsoulis, editors, *Nonlinear Assignment Problem: Algorithms and Application, Combinatorial Optimization Series*, pages 1–12. Kluwer Academic Publishers, Dordrecht, The Netherlands, 2000. (Cited on p. 305.)

[619] V. Srinivasan and G.L. Thompson. An operator theory of parametric programming for the transportation problem. I. *Naval Res. Log. Quart.*, 19:205–225, 1972. (Cited on pp. 78, 105.)

[620] V. Srinivasan and G.L. Thompson. An operator theory of parametric programming for the transportation problem. II. *Naval Res. Log. Quart.*, 19:227–252, 1972. (Cited on pp. 78, 105.)

[621] V. Srinivasan and G.L. Thompson. Cost operation algorithms for the transportation problem. *Math. Program.*, 12:372–391, 1977. (Cited on pp. 104, 105, 105.)

[622] J.M. Steele. *Probability Theory and Combinatorial Optimization*, volume 69 of *CBMS-NSF Regional Conference Series in Applied Mathematics*. SIAM, Philadelphia, 1997. (Cited on p. 146.)

[623] L. Steinberg. The backboard wiring problem: A placement algorithm. *SIAM Rev.*, 3:37–50, 1961. (Cited on pp. 205, 281.)

[624] S. Storøy and T. Sørevik. Massively parallel augmenting path algorithms for the assignment problem. *Computing*, 59:1–16, 1997. (Cited on p. 141.)

[625] Y.G. Stoyan, V.Z. Sokolovskii, and S.V. Yakovlev. A method for balancing discretely distributed masses under rotation (in Russian). *Energomashinostroenia*, 2:4–5, 1982. (Cited on pp. 205, 276.)

[626] T. Stützle and H. Hoos. Max-min ant system. *Future Generation Comp. Sys.*, 16:889–914, 2000. (Cited on p. 264.)

[627] F. Supnick. Extreme Hamiltonian lines. *Ann. Math.*, 66:179–201, 1957. (Cited on p. 278.)

[628] W. Szpankowski. Combinatorial optimization problems for which almost every algorithm is asymptotically optimal. *Optimization*, 33:359–367, 1995. (Cited on pp. 285, 288, 291.)

[629] E.D. Taillard. FANT: Fast ant system. Technical Report 46-98, IDSIA, Lugano, Switzerland, 1998. (Cited on pp. 259, 263, 267.)

[630] E.D. Taillard and L.M. Gambardella. Adaptive memories for the quadratic assignment problem. Technical Report I-87-97, IDSIA, Lugano, Switzerland, 1999. (Cited on p. 266.)

[631] E.D. Taillard. Robust taboo search for the quadratic assignment problem. *Parallel Comput.*, 17:443–455, 1991. (Cited on pp. 260, 261, 261, 267.)

[632] E.-G. Talbi, Z. Hafidi, and J.-M Geib. A parallel adaptive tabu search approach. *Parallel Comput.*, 24:2003–2019, 1998. (Cited on p. 261.)

[633] R.E. Tarjan. *Data Structures and Network Algorithms*, volume 44 of *CBMS-NSF Regional Conference Series in Applied Mathematics*. SIAM, Philadelphia, 1983. (Cited on p. 98.)

[634] G. Tarry. Le problème des 36 officiers. *Comptes Rendus de l'Association Française pour l'Avancement des Sciences*, 1:122–123, 1900. (Cited on p. 69.)

[635] D.M. Tate and A.E. Smith. A genetic approach to the quadratic assignment problem. *Computers & Oper. Res.*, 22:73–83, 1995. (Cited on p. 262.)

[636] T. Terlaky. A convergent criss-cross method. *Optimization*, 16:683–690, 1985. (Cited on p. 126.)

[637] G.L. Thompson. A recursive method for solving assignment problems. In P. Hansen, editor, *Studies on Graphs and Discrete Programming*, Ann. Discr. Math., pages 319–343. North–Holland, Amsterdam, 1981. (Cited on p. 126.)

[638] R.L. Thorndike. The problem of classification of personnel. *Psychometrica*, 15:215–235, 1950. (Cited on p. 77.)

[639] B.B. Timofeev and V.A. Litvinov. On the extremal value of a quadratic form. *Kibernetika*, 4:56–61, 1969. (Cited on p. 278.)

[640] N. Tomizawa. On some techniques useful for solution of transportation network problems. *Networks*, 1:173–194, 1971. (Cited on pp. 77, 78, 93, 94, 94, 128, 128.)

[641] P. Toth and D. Vigo. Branch-and-bound algorithms for the capacitated VRP. In P. Toth and D. Vigo, editors, *The Vehicle Routing Problem*, SIAM Monographs on Discrete Mathematics and Applications, pages 29–51. SIAM, Philadelphia, 2002. (Cited on p. 165.)

[642] L.Y. Tseng and S.C. Liang. A hybrid metaheuristic for the quadratic for the quadratic assignment problem. *Computational Opt. Appl.*, 85–113, 2006. (Cited on p. 264.)

[643] W.T. Tutte. The factorization of linear graphs. *J. London Math. Soc.*, 22:107–111, 1947. (Cited on pp. 35, 57.)

[644] I.K. Ugi, J. Bauer, J. Brandt, J. Friedrich, J. Gasteiger, C. Jochum, and W. Schubert. Neue Anwendungsgebiete für Computer in der Chemie. *Angewandte Chemie*, 91:99–111, 1979. (Cited on p. 206.)

[645] L.G. Valiant. The complexity of computing the permanent. *Theoretical Computer Science*, 8:189–201, 1979. (Cited on p. 14.)

[646] M. Vlach. Branch-and-bound method for the three-index assignment problem. *Ekonomicko-Matematicky Obzor*, 3:181–191, 1967. (Cited on pp. 309, 314.)

[647] A.T. Volgenant. Linear and semi-assignment problems: A core oriented approach. *Computers & Oper. Res.*, 23:917–932, 1996. (Cited on pp. 99, 104, 128, 128, 129, 130, 165, 165.)

[648] A.T. Volgenant. A note on the assignment problem with seniority and job priority constraints. *European J. Oper. Res.*, 154:330–335, 2004. (Cited on p. 168.)

[649] A.T. Volgenant. Solving the $k-$cardinality assignment problem by transformation. *European J. Oper. Res.*, 157:322–331, 2004. (Cited on p. 164.)

[650] A.T. Volgenant. A note on parametric analysis in linear assignment. *Oper. Res.*, 56:519–522, 2008. (Cited on p. 169.)

[651] S. Voss. Network design formulations in schedule synchronization. In M. Desrochers and J.M. Rousseau, editors, *Computer Aided Transit Scheduling*, pages 137–152. Springer, Berlin, 1992. (Cited on p. 303.)

[652] S. Voss. Heuristics for nonlinear assignment problems. In M. Desrochers and J.M. Rousseau, editors, *Nonlinear Assignment Problems*, volume 386 of *Lecture Notes in Econ. Math. Syst.*, pages 137–152. Springer, Berlin, 2000. (Cited on p. 303.)

[653] D.F. Votaw and A. Orden. The personnel assignment problem. In *Symposium on Linear Inequalities and Programming*, SCOOP 10, pages 155–163. U.S. Air Force, 1952. (Cited on pp. 77, 165.)

[654] I. Vozniuk, E.K. Gimadi, and M. Fialtov. Asimptoticheski tochni algoritm dlia resheniia zadachi razmesheniia s ogranichennmiobiemami proizvodstva. *Diskretnii analiz i issledovanie operatsii, Ser. 2*, 8:3–16, 2001. (Cited on p. 315.)

[655] B.L. van der Waerden. Ein Satz über Klasseneinteilungen in endlichen Mengen. *Abhandlungen aus dem Mathematischen Seminar Hamburg*, 5:185–188, 1927. (Cited on p. 16.)

[656] D.W. Walkup. On the expected value of a random assignment problem. *SIAM J. Comput.*, 8:440–442, 1979. (Cited on pp. 146, 148, 149.)

[657] D.W. Walkup. Matching in random regular bipartite digraphs. *Discr. Math.*, 31:59–64, 1980. (Cited on pp. 62, 64.)

[658] J. Wästlund. A proof of a conjecture of Buck, Chan, and Robbins on the expected value of the minimum assignment. *Random Structures Algorithms*, 26:237–251, 2005. (Cited on p. 147.)

[659] J. Wästlund. Random matching problems on the complete graph. *Electron. Commun. Probability*, 13:258–265, 2008. (Cited on p. 147.)

[660] J.M. Wein and S.A. Zenios. On the massively parallel solution of the assignment problem. *J. Parallel Distrib. Comput.*, 13:228–236, 1991. (Cited on p. 140.)

[661] A. Weintraub and F. Barahona. A dual algorithm for the assignment problem. Departamento de Industrias 79/02/C, Universidad de Chile-Sede Occidente, Santiago, 1979. (Cited on p. 98.)

[662] D.J.A. Welsh. *Matroid Theory*. Academic Press, London, 1976. (Cited on pp. 4, 27.)

[663] H. Weyl. Almost periodic invariant vector sets in a metric vector space. *American J. Math.*, 71:178–205, 1949. (Cited on p. 16.)

[664] M.R. Wilhelm and T.L. Ward. Solving quadratic assignment problems by simulated annealing. *IEEE Trans.*, 19:107–119, 1987. (Cited on p. 259.)

[665] T. Winter and U. Zimmermann. Dispatch of trams in storage yards. *Ann. Oper. Res.*, 96:287–315, 2000. (Cited on pp. 206, 292, 292.)

[666] H. Zaki. A comparison of two algorithms for the assignment problem. *Computational Opt. Appl.*, 4:23–45, 1995. (Cited on pp. 140, 141.)

[667] Q. Zhao, S.E. Karisch, F. Rendl, and H. Wolkowicz. Semidefinite programming relaxations for the quadratic assignment problem. *J. Comb. Optim.*, 2:71–109, 1998. (Cited on p. 246.)

[668] S. Zionts. The criss-cross method for solving linear programming problems. *Management Sci.*, 15:426–445, 1969. (Cited on p. 126.)

[669] S. Zionts. Some empirical tests of the criss-cross method. *Management Sci.*, 19:406–410, 1972. (Cited on p. 126.)

Author Index

Index